Chloride Channels
and Carriers
in Nerve, Muscle,
and Glial Cells

Chloride Channels and Carriers in Nerve, Muscle, and Glial Cells

Edited by

F. J. Alvarez-Leefmans

*Centro de Investigación y de Estudios Avanzados del IPN
and Instituto Mexicano de Psiquiatriá
Mexico City, Mexico*

and

John M. Russell

*University of Texas Medical Branch
Galveston, Texas*

Plenum Press • New York and London

Library of Congress Cataloging-in-Publication Data

Chloride channels and carriers in nerve, muscle, and glial cells /
edited by F.J. Alvarez-Leefmans and John M. Russell.
      p.   cm.
  "Grew from a workshop held under the auspices of the Second World
Congress of Neuroscience (IBRO) in Budapest in August 1987"--Pref.
  Includes bibliographical references.
  ISBN 0-306-43426-1
  1. Chlorides--Physiological transport--Congresses. 2. Chloride
channels--Congresses. 3. Nerves--Physiology--Congresses.
4. Muscles--Physiology--Congresses. 5. Neuroglia--Physiology-
-Congresses.   I. Alvarez-Leefmans, F. J. (Francisco J.)
II. Russell, John M., 1942-   . III. International Brain Research
Organization. Congress (2nd : 1987 : Budapest, Hungary)
  [DNLM: 1. Chlorides--metabolism--congresses. 2. Ion Channels-
-metabolism--congresses. 3. Muscles--metabolism--congresses.
4. Neuroglia--metabolism--congresses. 5. Neurons--metabolism-
-congresses.   WL 102.5 C544 1987]
QP535.C5C525   1990
591.87'5--dc20
DNLM/DLC
for Library of Congress   90-6729
                                                               CIP

Contributors

C. Claire Aickin • The University Department of Pharmacology, Oxford OX1 3QT, United Kingdom

Norio Akaike • Department of Neurophysiology, Tohoku University School of Medicine, Sendai 980, Japan

Francisco J. Alvarez-Leefmans • Departamento de Farmacología y Toxicología, Centro de Investigación y de Estudios Avanzados del IPN, Mexico 07000, D.F., and Departamento de Neurobiología, Instituto Mexicano de Psiquiatría, Mexico 14370, D.F.

Jeffrey L. Barker • Laboratory of Neurophysiology, National Institute of Neurological Disorders and Stroke, National Institutes of Health, Bethesda, Maryland 20892

Andrew L. Blatz • Department of Physiology, University of Texas Southwestern Medical Center, Dallas, Texas 75235

Joachim Bormann • Max Planck Institute for Biophysical Chemistry, D-3400 Göttingen, Federal Republic of Germany; *present address:* Merz & Co., D-6000 Frankfurt 1, Federal Republic of Germany

Walter F. Boron • Department of Cellular and Molecular Physiology, Yale University, School of Medicine, New Haven, Connecticut 06510

Peter M. Cala • Department of Human Physiology, School of Medicine, University of California, Davis, California 95616

Dominique Chesnoy-Marchais • Laboratory of Neurobiology, École Normale Superieure, 75005 Paris, France

Fernando Giraldez • Departamento de Bisquimica, Biologia Molecular y Fisiología, Universidad de Valladolid, 47005 Valladolid, Spain.

N. L. Harrison • Laboratory of Neurophysiology, National Institute of Neurological Disorders and Stroke, National Institutes of Health, Bethesda, Maryland 20892; *present address:* Department of Anesthesia and Critical Care, University of Chicago, Chicago, Illinois

K. Kaila • Department of Zoology, Division of Physiology, University of Helsinki, SF-00100 Helsinki, Finland

Vladimir N. Kazachenko • Institute of Biological Physics, USSR Academy of Sciences, Pushchino, Moscow Region, 142292, USSR

H. Kettenmann • Department of Neurobiology, University of Heidelberg, D-6900 Heidelberg, Federal Republic of Germany

H. K. Kimelberg • Division of Neurosurgery, Departments of Biochemistry and Pharmacology/Toxicology, and Program in Neuroscience, Albany Medical College, Albany, New York 12208

Mark L. Mayer • Unit of Neurophysiology and Biophysics, Laboratory of Developmental Neurobiology, National Institute of Child Health and Human Development, National Institutes of Health, Bethesda, Maryland 20892

Christopher Miller • Howard Hughes Medical Institute, Graduate Department of Biochemistry, Brandeis University, Waltham, Massachusetts 02154

David G. Owen • Laboratory of Neurophysiology, National Institute of Neurological Disorders and Stroke, National Institutes of Health, Bethesda, Maryland 20892; *present address:* Electrophysiology Section, Wyeth Research UK Ltd, Taplow Near Slough, Bucks, United Kingdom

Edwin A. Richard • Howard Hughes Medical Institute, Graduate Department of Biochemistry, Brandeis University, Waltham, Massachusetts 02154

John M. Russell • Department of Physiology and Biophysics, University of Texas Medical Branch, Galveston, Texas 77550

J. Voipio • Department of Zoology, Division of Physiology, University of Helsinki, SF-00100 Helsinki, Finland

Preface

This is a book about how Cl^- crosses the cell membranes of nerve, muscle, and glial cells. Not so very many years ago, a pamphlet rather than book might have resulted from such an endeavor! One might ask why Cl^-, the most abundant biological anion, attracted so little attention from investigators. The main reason was that the prevailing paradigm for cellular ion homeostasis in the 1950s and 1960s assigned Cl^- a thermodynamically passive and unspecialized role. This view was particularly prominent among muscle and neuroscience investigators. In searching for reasons for such a negative (no pun intended) viewpoint, it seems to us that it stemmed from two key experimental observations. First, work on frog skeletal muscle showed that Cl^- was passively distributed between the cytoplasm and the extracellular fluid. Second, work on Cl^- transport in red blood cells confirmed that the Cl^- transmembrane distribution was thermodynamically passive and, in addition, showed that Cl^- crossed the membrane extremely rapidly. This latter finding [for a long time interpreted as being the result of a high passive chloride electrical permeability (P_{Cl})] made it quite likely that Cl^- would remain at thermodynamic equilibrium. These two observations were generalized and virtually all cells were thought to have a very high P_{Cl} and a thermodynamically passive Cl^- transmembrane distribution. These concepts can still be found in some physiology and neuroscience textbooks. To further distract workers of the time, cations claimed center stage during this period with the Na^+ action potential and the Na^+–K^+ pump. Thus, the belief that Cl^- played no major role in membrane physiology and the attention directed to the study of cations combined to cause research on Cl^- transport to be largely neglected during this period. The relative importance accorded Cl^- by membrane researchers of that era can be deduced from the following two facts: (1) ^{36}Cl was often used to measure the extracellular space of tissues and (2) perhaps the ultimate indignity accorded Cl^-, it was often omitted from the Goldman Hodgkin–Katz equation for resting membrane potential! Lest we seem unduly sensitive, this lack of interest was also noted by R. D. Keynes in his introductory remarks to a Royal Society Meeting in 1982. He recalled a meeting held in 1970 in which "sodium . . . was in the limelight with thick black arrows, while chloride was relegated to the indignity of being passively transported and was shuffled off with dashed lines in an obscure corner of the diagram" (Keynes, 1982).

However, evidence gathered over the last 15 to 20 years is leading to a dramatic change in this simplistic point of view. We now know that a wide variety of cells exhibit a nonequilibrium distribution of Cl^- across their surface membranes. Some cells actively extrude Cl^-, others actively accumulate it, but few cells ignore it. By virtue of being distributed out of electrochemical equilibrium, Cl^- can (and does) serve as a key player in a variety of channel-mediated electrical events, most of which were simply undreamed of 20 years ago.

The importance of Cl^- to the function of excitable cells and their companion cells, the glia, is becoming increasingly recognized. In fact, far from being considered unimportant, Cl^- is now postulated to be crucial in a wide variety of cellular activities such as intracellular pH regulation, cell volume regulation, synaptic signaling (in both the de- and hyperpolarizing directions), membrane potential stabilization, and K^+ scavenging. The fact that Cl^- can apparently perform such a variety of functions will undoubtedly make the study of this anion much more fragmented than the study of the cations has been and is one reason we felt a monograph pulling together work on a variety of preparations using a variety of techniques was timely and necessary.

The recent increase in interest in Cl^- is probably largely the result of two factors: first, an important addition to the paradigm of cellular ion homeostasis and second, several crucial technical advances. Beginning in the 1960s, there developed an appreciation that transmembrane ion gradients possess potential energy that can be tapped for various forms of biological work, including active ion transport. Thus, the concept of secondary active solute transport was born. This concept has been very important for the study of Cl^- because it has stimulated research into areas in which Cl^- transport mechanisms were discovered. The second reason includes several technical advances in the last 10 to 15 years that have been important to the increased interest in Cl^-. Perhaps the key technical development was that of Cl^--selective microelectrodes. With this tool, it has been possible to unambiguously demonstrate that Cl^- is often distributed across cell membranes out of thermodynamic equilibrium. The combination of this kind of finding with the concept of secondary active transport has been pivotal in the upsurge of interest in Cl^-. Three other technical advances have made it possible to study the small cells where many of the interesting Cl^--dependent processes reside, including neurons, glia, and smooth muscle cells. The first is the ability to isolate and culture a wide variety of cells not previously accessible to investigators of ion transport. A second important technical advance has been the development of patch-clamp technology. This technique has permitted sophisticated analyses to be performed on very small cells, which were not amenable to previous techniques. Another development that arose about the same time as the patch-clamp technique was the suction pipette, internal-"dialysis" technique, which permits the investigator to change and/or control the intracellular solute environment via the so-called "dialysis" method. The latter three developments have been particularly important in stimulating the renewed interest in the role of Cl^- in synaptic physiology. But as we shall see in this volume, they have also contributed importantly to the study of glial cell Cl^- physiology.

This book grew from a workshop held under the auspices of the Second World Congress of Neuroscience (IBRO) in Budapest in August 1987. The variety of exciting findings presented at that workshop suggested to us that progress in the study of Cl^- transport by neurons, muscle, and glia was about to enter its logarithmic growth phase. We became convinced that a monograph covering the present status of Cl^- transport work in excitable tissue and glia would be very timely. Thus, the contributors to this volume were encouraged to present their fields of expertise not just by cataloging their own findings but to put all the relevant findings into a perspective so that others wishing to enter the field might quickly see the obvious places where more work is needed. We should point out that already the field of Cl^- transport by nerve, muscle, and glia has become too large to be adequately covered in a single book of this sort. We hope, however, that this volume highlights current "hot spots" of interest in Cl^-.

The editors would like to acknowledge the word processing skills of Mary Oblich and Lynette Durant as well as the kind support from Fundación Mexicana para la Salud.

Francisco J. Alvarez-Leefmans
John M. Russell

Mexico City and Galveston

REFERENCE

Keynes, R. D., 1982, Introductory remarks, *Philos. Trans. R. Soc. London Ser. B.* **299:**367–368.

Contents

II. Different Types of Cl^- Channels

A. Transmitter-Activated Anion Channels

I

CHLORIDE CARRIERS

Methods for Measuring Chloride Transport across Nerve, Muscle, and Glial Cells

Francisco J. Alvarez-Leefmans, Fernando Giraldez, and John M. Russell

1. INTRODUCTION

1.1. Why Is It Important to Measure Intracellular Cl^- Activity in Excitable Cells?

Intracellular Cl^- together with HCO_3^- is the most abundant free anion in living cells. Measuring intracellular chloride activity (a_{Cl}^i) and studying the mechanisms involved in regulation of intracellular Cl^- is particularly important in excitable cells for four main reasons: (1) a_{Cl}^i is a quantity needed to determine E_{Cl}, the Cl^- equilibrium potential. (2) Several transport mechanisms responsible for intracellular pH regulation are tightly coupled to Cl^-. (3) Cl^- is also involved in transport mechanisms implicated in cell volume regulation. (4) Knowledge of intracellular Cl^- homeostasis is crucial for understanding synaptic inhibition.

Evidence is accumulating that in most cells, whether excitable or not, Cl^- is not passively distributed, i.e., the value of the transmembrane potential, E_m, is not equal to E_{Cl}, implying the existence of active transport mechanisms. In the last 10 years, many significant advances have been made in elucidating the nature, mode of operation, and physiological role of these transport mechanisms that are responsible for a_{Cl}^i regulation. It follows that accurate assessment of E_{Cl} and E_m is of obvious importance in determining whether Cl^- is or is not passively distributed.

In order to calculate E_{Cl}, we need to know the extracellular Cl^- activity (a_{Cl}^o) as well as a_{Cl}^i, both of which can be directly measured with ion-selective microelectrodes. This latter technique has the unique advantage of allowing simultaneous measurements

Francisco J. Alvarez-Leefmans • Departamento de Farmacología y Toxicología, Centro de Investigacíon y de Estudios Avanzados deo I. P. N., Mexico 07000, D. F., and Departamento de Neurobiología, Instituto Mexicano de Psiquitría, Mexico 14370, D. F.   *Fernando Giraldez* • Departamento de Bioquimica, Biología Molecular y Fisiología, Universidad de Valladolid, 47005 Valladolid, Spain. *John M. Russell* • Department of Physiology and Biophysics, University of Texas Medical Branch, Galveston, Texas 77550.

of a_{Cl}^i and E_m. Having values for a_{Cl}^o and a_{Cl}^i, one can readily calculate E_{Cl} using the Nernst equation:

$$E_{Cl} = \frac{RT}{F} \ln \frac{a_{Cl}^i}{a_{Cl}^o} \tag{1}$$

where R, T, and F have their usual meanings.

Cl^- transport mechanisms in excitable cells have been generically referred to as "Cl^- pumps," a term that is misleading as it suggests an ATPase mechanism. We prefer the term "secondary active transport." Secondary active transport mechanisms for Cl^- have been identified that can be either outwardly or inwardly directed causing E_{Cl} to be either more negative than or more positive than (respectively) E_m. Recent advances in the characterization of these transport mechanisms in glial and excitable cells are one of the main subjects of the present book.

1.2. Which Methods Are Available for Measuring Intracellular Cl⁻ Activity and Transmembrane Cl⁻ Movements?

The two most widely used methods of measuring membrane transport mediated Cl^- movements are tracers and ion-selective microelectrodes. Recently, a halide-sensitive fluorescent probe, 6-methoxy-N-(3-sulfopropyl) quinolium (SPQ), has been designed and successfully used to study Cl^- transport in nonexcitable cell preparations (Illsley and Verkman, 1987; Krapf *et al.*, 1988). In the sections that follow, we will consider the theoretical and practical background of each of these three methods as well as their advantages and disadvantages in studying cellular Cl^- regulation. Special emphasis will be put on ion-selective microelectrodes because of new and promising developments that make it the technique of choice for small cells in culture or in *in vitro* preparations. From the thermodynamic point of view, the important parameters when considering membrane transport processes are the intracellular free Cl^- concentration, $[Cl^-]_i$ and a_{Cl}^i; therefore, we will not consider methods that measure *total* Cl^- concentrations (i.e., free plus bound and/or sequestered internal Cl^-), such as silver ion titration methods (Cotlove *et al.*, 1958; Bomsztyk *et al.*, 1988), indirect flame photometric methods (e.g., Casteels and Kuriyama, 1965; Aickin and Brading, 1982), or electron probe X-ray microanalysis (Allakhverdov *et al.*, 1980; Galvan *et al.*, 1984; Acker *et al.*, 1985). However, a brief account on the method for measuring total, exchangeable Cl^- content of cells using isotopic ^{36}Cl can be found in Section 7.2.1.

2. WHY USE ION-SELECTIVE MICROELECTRODES FOR MEASURING INTRACELLULAR Cl⁻?

Ion-selective microelectrodes (ISMs) are the most widely used devices for measuring a_{Cl}^i. The method allows not only measurement of steady-state a_{Cl}^i but also changes in a_{Cl}^i. If judiciously used, this technique also allows measurement of Cl^- net fluxes across the plasma membrane of individual cells. The spectacular miniaturization of ISMs achieved during the last 10 years has allowed the technique to be used not only in large vertebrate or invertebrate cells, but also in cells as small as smooth muscle, vertebrate neurons and glia.

Advantages and disadvantages of using ISMs in general have been extensively considered by Thomas (1978) and Ammann (1986). Some papers on the subject, including new applications and developments, can be found in a recent symposium (Morris and Krnjević, 1987). The general advantages of ISMs will be summarized below emphasizing Cl^- activity measurements.

2.1. Advantages of Using the ISM Technique

2.1.1. ISMs Allow Direct Measurement of Intracellular Ion Activities

This is one of the most important advantages of ISMs since the activity, not the total concentration of an ion j, is the parameter needed for explaining and predicting chemical equilibria and rates of chemical reactions (Ross, 1969). Intracellular activities, together with binding and activation constants, are the variables that describe the molecular effectiveness of intracellular ions. Since the activity (or its logarithmic function, the chemical potential) measures the relative driving force on chemical equilibria, ". . . analyzing ionic regulation without knowing activities is like studying electrical events without knowing voltages" (Tsien, 1983).

Having a value for the intracellular activity of ion j, a_j^i, it is possible to estimate its free intracellular concentration, $[j]_i$. This is usually done simply by dividing the measured a_j^i by the intracellular activity coefficient of j, γ_j^i (see below). Although we have little direct information about intracellular ion activity coefficients, there are reasons to believe that for most intracellular ions, including Cl^-, they are much the same inside as outside (Ammann, 1986).

2.1.2. Adequate Speed of Response

ISMs have good time resolution for studying most ion transport mechanisms. For currently available Cl^- sensors, response times are on the order of a few seconds even for microelectrodes having very fine tips. Factors determining response time of ISMs have been reviewed by Ammann (1986).

2.1.3. Continuous Measurement of Intracellular Ion Activities

ISMs permit continuous recording of a_j^i without consuming the sample. Furthermore, it is possible to measure the intracellular activities of two or more ions simultaneously in the same cell (e.g., Thomas, 1977; Alvarez-Leefmans *et al.*, 1986).

2.1.4. Measurement of Intracellular Activity of Cytosolic Compartment

Even from the biophysicist's point of view, a cell is more than a simple bag made out of a semipermeable material, filled with an electrolyte. The intracellular space of a typical eukaryotic cell can be divided into two main compartments. The first one is formed by a set of membrane-bound structures within the cytoplasm, i.e., organelles such as the nucleus, lysosomes, and mitochondria. The other compartment of the cytoplasm includes everything other than membrane-bound organelles and is usually referred to as the cytosol (e.g., Alberts *et al.,* 1989) or ground cytoplasm (Lev and Armstrong, 1975). The cytosol of a typical vertebrate cell represents only about 55% of the total cell volume. It is a concentrated and highly organized polyelectrolyte whose complex properties, composition, and organization have been extensively reviewed (Edzes and Berendsen, 1975; Lev and Armstrong, 1975; Fulton, 1982; Clegg, 1984; Ammann, 1986). As a consequence of such complexity of the intracellular space, ions are not distributed uniformly within a cell and are present in more than one form. For example, they can be sequestered into organelles, bound to proteins, or form ion pairs. In any of these conditions, they are not free in the cytosolic aqueous phase which directly interacts with the plasma membrane and therefore do not contribute directly to the transmembrane electrochemical potential. Analytical measurements of ion content of cells yield total (free plus bound plus sequestered) concentrations and do not discriminate between bound and free forms of a particular ion. However, what is important from the point of view of plasma membrane transport processes is the ionic activity or the free concentration of a particular ion in the cytosolic compartment and not its total concentration.

To this complexity we also have to add the fact that not all intracellular water seems to be available as a solvent for the electrolytes. In fact, 10 to 20% of total intracellular water is claimed to be "structured water." This water fraction can be considered as "bound" and has a different structure from normal bulk water, i.e., it is *nonsolvent* and therefore is believed to be osmotically inactive for any solute (Dick, 1979; Fulton, 1982; Derbyshire, 1982; Clegg, 1982).

The tip size of even the smallest ISM usually prevents its penetrating intracellular organelles, with the exception perhaps of the nucleus (e.g., Palmer and Civan, 1977; Wuhrmann *et al.,* 1979). Therefore, ISMs most likely record ion activities in the *unstructured* cytosolic water compartment. There seems to be agreement that intracellular Cl^- behaves in this compartment as it does in free solution (e.g., Keynes, 1963; Donahue and Abercrombie, 1987; for reviews see Lev and Armstrong, 1975; Tsien, 1983). Many proteins are known to be capable of binding Cl^- and this binding is pH dependent. However, binding is quite small at pH 7 but it increases as the pH decreases (Carr, 1968). Therefore, it is expected that Cl^- binding to proteins at physiological intracellular pH must be negligible. In support of this view are recent measurements done in rat sympathetic neurons where a *total* intracellular Cl concentration of 32 mmole/liter cell water was estimated from data obtained by using energy-dispersive electron microprobe analysis (Galvan *et al.,* 1984). In a separate study performed in the same preparation, an apparent mean a^i_{Cl} of about 30 mM was measured with Cl^--selective microelectrodes (Ballanyi and Grafe, 1985). Correcting for an intracellular interference of about 5 mM (see Section 5.2.1d), the true a^i_{Cl} would be 25 mM and therefore an intracellular apparent activity coefficient γ^{app}_{Cli} of 0.78 can be estimated from the relation $\gamma^{app}_{Cli} = a^i_{Cl}/[Cl^-]_i$. This apparent activity coefficient is similar to the reported mean activity coefficient γ_{Cl} of 0.77 for a pure aqueous solution of KCl at 25°C at an ionic strength of 0.1 M (Parsons, 1959). The fact that γ^{app}_{Cli} is close to γ_{Cl} indicates that Cl^- is not compartmentalized in the intracellular space and it is likely to be as free in the cytosol as it is in an aqueous solution.

2.1.5. Simultaneous Measurement of Ion Activities, Membrane Potential, and Electrical Membrane Properties

This is one of the most striking and unique features of ISMs. Intracellular ion activities, membrane potential, and electrical membrane properties can be simultaneously monitored in resting conditions and during dynamic states induced by changes in extracellular ionic composition, including drugs (e.g., Reuss *et al.*, 1987).

2.1.6. The Possibility of Performing Measurements in Situ of Single Identified Cells in Excitable Tissues

This is a potential advantage of the technique that has not been fully exploited, particularly in nervous tissue, e.g., brain slices maintained *in vitro*. In principle, it would be possible to put a dye in the reference barrel of a double-barreled ISM to identify the impaled cell type. An example of this approach to locate the electrode tip in the tissue has been provided by Frömter *et al.* (1981) in epithelia. Characterization of suitable dyes for intracellular use that would not interfere with ISM readings awaits investigation.

2.1.7. ISMs Can Be Used to Measure Net Ion Fluxes

The way in which this can be done and the theory behind it is the subject of Section 6. An example of estimates of net Cl^- fluxes using ISMs in a vertebrate neuron can be found in Alvarez-Leefmans *et al.* (1988).

2.1.8. ISMs Can Be Used to Measure Cell Volume Changes

There are cells showing volume regulatory adjustments when exposed to anisotonic media (Gilles *et al.*, 1987). Chloride transport systems have been implicated in some forms of this cellular homeostatic function (see reviews by Larson and Spring, 1987; Hoffmann, 1987; Hoffmann and Simonsen, 1989). A technique for measuring changes in cell volume using ISMs has been devised by Reuss (1985). It constitutes a promising tool to study volume changes and how they are regulated in excitable cells (e.g., Serve *et al.*, 1988). The technique is based on the fact that K^+-sensitive microelectrodes containing the K^+ exchanger (CORNING 477317) are far more responsive to quaternary ammonium ions such as tetramethylammonium than to K^+ (Fig. 1A). Cells are loaded with tetramethylammonium by transient exposure to the polyene antibiotic nystatin (Reuss, 1985) or directly through an intracellular microelectrode (e.g., Serve *et al.*, 1988). Since tetramethylammonium is an impermeant cation, it does not leak out of the cells. Its intracellular activity can then be measured with K^+-sensitive microelectrodes containing CORNING 477317 as the ion-sensing element.

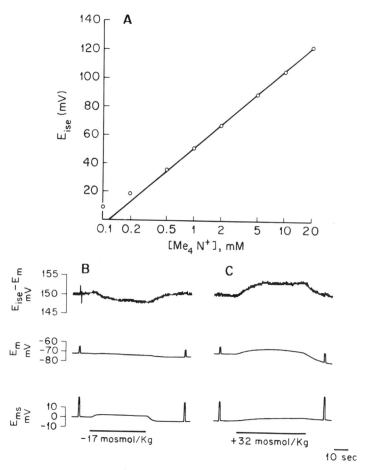

Figure 1. Changes in cell volume measured with an electrophysiological technique. (A) Electric potential of an ion-selective microelectrode (E_{ise})based on the lipophilic salt potassium tetrakis (*p*-chlorophenyl)borate, recorded in solutions containing tetramethylammonium (Me_4N^+) at the concentrations indicated on the abscissa. The sensor of the microelectrode was the exchanger ordinarily used to make K^+-sensitive, liquid-membrane microelectrodes. E_{ise} was recorded in a rapid-flow chamber with respect to a calomel half-cell connected to the solution by a KCl-saturated, flowing junction. The rate of exchange of the solution was high, to avoid changes in $[K^+]$. The calibration solutions consisted of 120 mM KCl plus the Me_4NCl concentration given on the abscissa. In the range 1–20 mM, the slope was 54 mV/log $[Me_4N^+]$. (B, C) Cell volume changes recorded intracellularly in epithelial cells of the gallbladder of *Necturus maculosus*. Cells had been loaded with Me_4N^+ by transient exposure of the apical (lumen-facing) surface to solutions containing these quaternary ammonium ions and nystatin. Cell volume changes were produced by rapid alterations in the osmolality of the mucosal solution. The records in B and C were obtained during impalements of the same cells with a conventional reference microelectrode (filled with 3 M KCl) and a "Me_4N^+-selective microelectrode." In each panel, the transepithelial potential (E_{ms}), the cell transmembrane potential ($E_m = E_{cell} - E_{serosa}$), and the difference between the potentials of the ion-selective and the reference microelectrode (E_{ise} - E_m) are depicted. Transepithelial current pulses were applied before the changes inosmolality for impalement validation. (B) Effect of reducing [NaCl] in the mucosal medium by 17 mosm/kg, equivalent to an 8.5% reduction of the control osmolality. Note the fall in $[Me_4N^+]_i$ (negative deflection in the differential, E_{ise} - E_m, trace), which indicates cell swelling. After about 20 sec, cell volume had increased to 109% of control. The initial rate of change of $[Me_4N^+]_i$ was $\simeq$ 0.7%/sec. (C) Effect of increasing [NaCl] in the mucosal medium by 32 mosm/kg, equivalent to a 16% increase in osmolality. Twenty seconds after the onset of the increase in $[Me_4N^+]_i$, cell volume had decreased to 85% of control. The initial rate of change of $[Me_4N^+]_i$ was $\simeq$ 1.8%/sec. (Modified from Reuss, 1985, and reproduced with permission.)

After loading the cells with tetramethylammonium, the intracellular content of this probe is constant, so that changes in its activity reflect changes in cell volume. The technique permits measurement of cell volume changes of 5% or less. Figure 1B and C show examples of how the technique has been used to measure changes in cell volume.

The principle on which this technique is based can in turn be considered a drawback for ISM measurements if microelectrodes are used uncritically. This is because cell volume changes might lead to spurious changes in physiological intracellular activities and therefore can also alter estimates of net fluxes and basal steady-state intracellular activities, including assessment of basal intracellular interference (e.g., Alvarez-Leefmans *et al.*, 1988; see also Spring and Ericson, 1982). The latter consideration is particularly important when studying changes in a_{Cl}^i when Cl^- is eliminated from the external bathing solution (see Section 6).

2.2. Disadvantages and Difficulties of ISM Techniques

Not all is Alice in Wonderland when considering ISMs. Like any technique, it has its drawbacks. However, when the technique is used critically, many of these drawbacks can be circumvented. In the following, we shall summarize disadvantages and difficulties of ISM techniques.

2.2.1. The Method Is Invasive

The cell under study needs to be penetrated with two microelectrodes assembled into either a fine-tipped double-barreled (Zeuthen, 1980; see also Ammann, 1986), a concentric type (Yamaguchi, 1986) or an eccentric type (Thomas, 1985), or two separate individual microelectrodes. This requires not only electrophysiological skill and training of some sort, but also a cell that allows such impalement without suffering irreversible damage. Several criteria for assessing cell viability and electrode performance have been established (reviewed by Ammann, 1986). These criteria must be fulfilled to reliably validate a measurement.

A reliable measurement of E_m is of utmost importance when studying intracellular Cl^- regulation. E_m is indeed a difficult parameter to measure accurately (Tasaki and Singer, 1968; Hironaka and Morimoto, 1979; Purves, 1981). As we have seen, the first clue for suspecting the presence of active Cl^- transport in a cell is finding a significant difference between E_m and E_{Cl}. Artifactual low values of E_m secondary to cell damage or liquid junction potentials, in the absence of changes in a_{Cl}^i, could lead to the erroneous conclusion that Cl^- is nonpassively distributed. Since E_m would be found to be more positive than E_{Cl}, an active Cl^- extrusion mechanism could be mistakenly postulated. Conversely, spuriously high readings of E_m, due to development of tip potentials inside cells (Purves, 1981), could lead the experimenter to think that E_{Cl} is more positive than E_m, and an inwardly directed active Cl^- uptake mechanism could be naively postulated.

It is important to bear in mind that once having a reliable measurement of E_m, a_{Cl}^i, and a_{Cl}^o, if E_m is significantly different from E_{Cl}, an active Cl^- transport mechanism must be suspected. However, there are cases in which $E_{Cl} \simeq E_m$, and due to a relatively high resting membrane permeability to Cl^-, the presence of a Cl^- transport system could be masked, as it is the case for skeletal muscle (Harris and Betz, 1987).

2.2.2. Lack of Perfect Selectivity for the Primary Ion

Nowadays, most ISMs use as the ion-sensing element a liquid membrane (see below). Unfortunately, not all currently available sensors for intracellular ionic activity measurements are perfectly selective for the ion whose activity they are designed to measure, i.e., the primary ion. Even though ISMs might lack perfect selectivity for the primary ion in mixed solutions, it is possible to reliably validate a measurement; if interfering ions are known, a quantitative correction for their contribution to the electrode response can be made. Examples of liquid sensors, mainly neutral carriers, suitable for intracellular ion activity measurements can be found in recent reviews (Ammann, 1986; Ammann *et al.*, 1987). It is important to emphasize that measurements done with sensors that suffer from interference from other naturally occurring or experimentally introduced cytosolic ions need to be critically assessed before they can be accepted as valid. An extreme case of inappropriate selectivity for intracellular measurements is that of the Mg^{2+} sensor based on the neutral ligand ETH 1117, which suffers from interference from K^+ and Na^+ at ionic activity levels at which these two cations are present in the cytosol. However, independent quantitative assessment of these interferences made possible the measurement of intracellular free Mg^{2+} concentrations in nerve and muscle cells under a variety of experimental conditions (Alvarez-Leefmans *et al.*, 1987).

Liquid ion exchangers for making Cl^--selective microelectrodes, as will be seen below in more detail, are adequate for reliable intracellular measurements although interference from intracellular HCO_3^- must be taken into account, particularly when a_{Cl}^i is expected to be low, say less than 10 mM. A novel, positively charged neutral carrier, suitable for measuring a_{Cl}^i, has been synthetized by W. Simon and his group at the ETH in Zurich. It shows more favorable selectivity coefficients than the commonly used CORNING 477913 liquid ion exchanger. In particular, it has better selectivity for Cl^- over HCO_3^-. Some of the properties of this novel sensor will be reviewed in Section 5.2.2. Here it is enough to say that if microelectrodes are used uncritically, imperfect selectivity of currently available Cl^- sensors may be a drawback when knowledge of an absolute steady-state value of a_{Cl}^i rather than a relative change is required. Fortunately, there are ways of assessing quantitatively possible intracellular anion interference on a Cl^--selective microelectrode response, as will be discussed in Section 5.2.1d. Other sources of intracellular interference on the Cl^--selective microelectrode response include cytosolic contamination by Cl^- itself or by other potentially interfering anions leaking from the reference microelectrode or coming from outside the cell, e.g., some anions used as Cl^- substitutes, or some drugs, including inhibitors of Cl^- transport systems such as furosemide and related compounds (see Section 5.2.1e).

2.2.3. They Do Not Measure Unidirectional Fluxes

Although it is possible to use Cl^--selective microelectrodes to measure *net fluxes*, they cannot be used to measure unidirectional fluxes as tracers do. How Cl^--selective microelectrodes can be used to estimate net fluxes is the subject of Section 6.

2.2.4. Point Sample Reading

Readings obtained with intracellular ISMs represent only the free ion level at the tip of the electrode. In the case of Cl^-, this does not seem to be a drawback since, unlike Ca^{2+}, it diffuses freely in the cytosol (see Section 2.1.4). Therefore, a point reading is likely to represent levels present in the bulk of the cytosol.

3. ION ACTIVITIES AND ION CONCENTRATIONS

We have just seen that one of the advantages of ISMs is that they sense ion activities and not concentrations. The activity of an ion species j in a solution is its thermodynamically *effective* concentration and not its *analytical* concentration. Thermodynamically *effective* concentration means that which makes mass action formulae quantitatively exact for explaining and predicting chemical equilibria. For example, in a 100 mM KCl solution at 25°C the analytical concentration of chloride will be 100 mM, but the activity will be only 77 mM. The inactive ions are those that due to ion–ion and solvent–ion interactions have a restricted freedom, i.e., are not freely mobile; they do not contribute to chemical potential change (see Bockris and Reddy, 1973). In general, the more concentrated the solution, the more ions are inactive. This is a well-known property of electrolytic solutions, which, although extensively studied, is not well understood by physical chemists. The classical treatment of the problem is the Debye–Hückel theory. The theory as well as its triumphs and limitations are well beyond the scope of this chapter but can be found in specialized books (e.g., Robinson and Stokes, 1965; Bockris and Reddy, 1973; see also Bates, 1973). What is relevant to consider here is that a "correcting factor" of some sort is required to go from free concentrations to activities and vice versa. This correction factor is known as the "activity coefficient."

An operational definition of the activity coefficient of species j that relates the activity a_j to the concentration [j] is given by

$$\gamma_j = a_j/[j] \tag{2}$$

where γ_j is the activity coefficient of ion j; it follows that

$$a_j = \gamma_j \, [j] \tag{2.1}$$

In thermodynamics it is customary to express [j] as molal concentration (m_j) and then Eq. (2) is also written as

$$\gamma_j = a_j/m_j \tag{2.2}$$

Equation (2) empirically defines the behavior of electrolytic solutions, but of course is not a theoretical expression of the activity coefficient γ_j. At a given temperature the activity coefficients of ions in aqueous solutions of low concentration vary chiefly with the interionic distances and with the number of charges borne by the ions. This is reflected by the fact that the activity of ion j is a complex function of the ionic strength I of the solution (see Fig. 2), which is defined by

$$I = \frac{1}{2} \sum_j m_j z_j^2 \tag{3}$$

where m_j is the molal concentration of ion j and z_j is its charge. The summation is made for all the ionic species present in the solution. Now, for very dilute solutions of a single salt (e.g., KCl) $\gamma_j \simeq 1$ and therefore the value of a_j approaches that of [j] [see Eq. (2.1) and Fig. 2]). In *some* electrolytes, γ_j falls from unity to lower values in a complex manner depending on I (Fig. 2). Certainly in biological fluids like the cytosol, in which $I = 0.1$ to 0.15, $\gamma_j < 1$.

Two kinds of activities (and activity coefficients) are considered in electrochemis-

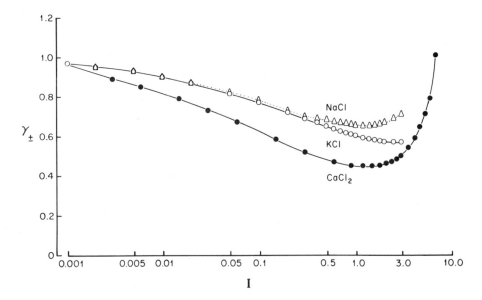

Figure 2. Mean activity coefficients ($\gamma\pm$) as a function of the ionic strength (I) for three different Cl^- salts of biological interest, at 25°C. The curves joining each symbol were drawn by eye. (Data from Parsons, 1959.)

try, namely mean activity (and mean activity coefficient) and single ion activities (and single ion activity coefficients). In Eqs. (2) to (2.2), γ_j is the *mean* activity coefficient and is often expressed as $\gamma\pm$; hence, Eq. (2.2) can be written as

$$\gamma\pm = a^\pm/m^\pm \tag{4}$$

Eq. (4) applies formally to ions, and one may write

$$\gamma+ = a^+/m^+ \qquad \text{for cations} \tag{5}$$

and

$$\gamma- = a^-/m^- \qquad \text{for anions} \tag{6}$$

where $\gamma+$ and $\gamma-$ are single ion activity coefficients, a^+ and a^- are single ion activities, and m^+ and m^- are their corresponding molalities. Single ion activities and activity coefficients cannot be thermodynamically defined and cannot be experimentally measured. The electroneutrality principle only permits a thermodynamic description of *mean activities*. On the other hand, mean activity coefficients have been unambiguously defined and are measurable. Precise values for mean activity coefficients of single-electrolyte solutions, also known in ISM parlance as "pure" solutions, are available (e.g., Parson, 1959). Although the potential difference at the boundary of an ion-selective membrane electrode is thermodynamically related to single ion activities, the potential E_{ise} of the complete electrochemical cell assembly (see Section 4.1) assess *mean* ion activities (see Ammann, 1986). Hence, the electromotive force of cell assemblies with liquid junctions depends on the logarithm of the *mean* ion activities, which hereinafter will be referred to simply as "ion activities."

The next problem we face is how to measure or calculate γ_j in biological fluids. As discussed by several authors (e.g., Lev and Armstrong, 1975; Tsien, 1983; Ammann, 1986), this is not straightforward for complex polyelectrolytes such as the cytosol. In fact, it is a problem that has not been solved. The discussion is relevant since activity measurements are not absolute but are always relative to some standard solution used to calibrate electrodes. The question arises as to which types of solutions are suitable for calibrating ISMs. What is usually done it to calibrate the ISM in solutions of known composition, measure the ISM potential (E_{ise}) outside the cell, impale the cell and read E_{ise} in the cytosol (subtracted from the transmembrane potential E_m), remove the ISM from the cell interior and go back to the calibration solutions. Then find by interpolation the value of the activity (or free ion concentration) of the sample solution, i.e., the cytosol. What is meant by solutions of known composition? This problem has been treated by Thomas (1978) and Tsien (1983). In brief, there are two groups of ISM users, namely those who express their results in terms of free ion concentrations and those who prefer using activities. Here we summarize the arguments put forward by the protagonists of each of these groups. Arguments in favor of expressing *free ion concentrations* are as follows.

1. The concentration of ions in the bathing and calibrating solutions are known. Therefore, to calculate intracellular free ion concentrations one only needs to assume that the activity coefficient for the ion in question is the same in the cytosol as it is in the extracellular fluid. This is a reasonable assumption considering that extracellular and intracellular ionic strengths must be equal or near equal. Besides, with this approach one does not need to know the actual value of the activity coefficients.
2. Free ion concentrations are used with other techniques such as indicator dyes.
3. Ion concentration changes are needed to estimate fluxes.
4. Biologists are more familiar with ion concentrations than with ion activities.

The most reasonable approach for expressing the readings of ISMs in terms of free ion concentrations is to calibrate the electrodes in solutions mimicking as closely as possible the composition of the cytosol. In this kind of solution, the free concentration of the primary ion is varied within the expected physiological range keeping the ionic strength constant. Some argue that what the electrode measures is activity and not free concentrations. However, as stated by Tsien (1983), ". . . free concentration is in fact a scale of activity in which the solvent is understood to include a specified constant ionic strength. Applying a Debye–Hückel activity coefficient does not make this value any more rigorous nor does it dispense with the practical need for calibration solutions that simulate the ionic background environment of the cytosol."

Arguments in favor of expressing *ion activities* are as follows.

1. The potential of a perfectly selective ISM assembly varies with the logarithm of the ion activity, not concentration.
2. Knowing the external activity coefficient, no assumption about the intracellular activity coefficient is needed to express measurements in terms of intracellular activity.
3. Ion activity is the physiologically and biochemically important parameter (see above).

For the above three reasons, most users of ISMs prefer to express their results in activities. There are two exceptions, Ca^{2+} and Mg^{2+}. For these two divalent cations, there is uncertainty as to the value of the extracellular activity coefficient so it is more convenient to express the results in terms of free ion concentrations (e.g., Alvarez-Leefmans *et al.,* 1981, 1986).

For Cl^-, activity scales are most commonly used. Calibration of Cl^--selective microelectrodes is usually done in "pure" solutions of known concentrations and their Cl^- activity is calculated using the values of mean Cl^- activity coefficients (γ_{Cl}) obtained from standard tables (e.g., Parsons, 1959). Measurements are then expressed in activities. The procedure requires knowledge of the γ_{Cl} of each calibration solution but not that of the cytosol (e.g., Bolton and Vaughan-Jones, 1977). Hence, no assumptions are made about the properties or composition of the cytosol. The procedure requires only knowledge of the mean activity coefficient of each of the single-electrolyte solutions used for calibration, usually KCl, and for this case both theory and experiment are reliable and well established.

It is not advisable to calibrate Cl^--selective microelectrodes with solutions of constant ionic strength or varying Cl^- free concentration. To make such solutions would require mimicking the cytosolic ionic composition, which is not feasible for Cl^- since little is known about the nature and free concentrations of intracellular anions other than Cl^-, with the possible exception of HCO_3^-.

Whatever convention the user prefers, it is crucial to specify the way in which electrodes are calibrated and the scale used. Regarding the nomenclature, it would be convenient to be in accord with The International Union of Pure and Applied Chemistry (IUPAC) recommendations (see Guibault *et al.*, 1976; IUPAC, 1979).

Finally, if once having a value for a_{Cl}^i, we want to calculate $[Cl^-]_i$, we need to know the mean activity coefficient of Cl^- in the cytosol (γ_{Cl}^i). There is simply no easy way around this problem. What is usually done is to give γ_{Cl}^i a value corresponding to a pure solution of similar or equal ionic strength to that we calculate for the cytosol. There is no formal justification for that, but it is all we can do, and it appears to give a fair approximation.

4. ION-SELECTIVE MEMBRANE MICROELECTRODES

Before attempting to explain what Cl^--selective membrane microelectrodes are and how they work, it is worth considering some general principles of membrane microelectrodes, some of their physicochemical properties such as permselectivity and ion-selectivity, how their selectivity is assessed and expressed in the current literature, and how all this is relevant to intracellular chloride ion activity measurements.

4.1. What Are Membrane Microelectrodes?

Any phase that separates two other phases to prevent mass movement between them but allows passage with various degrees of restriction of one of several species of the external phases may be defined as a membrane (Lakshminarayanaiah, 1976). Membranes that separate two electrolytes, and that are not equally permeable to all kinds of ions, are known as electrochemical membranes (Koryta, 1975). When a membrane of this kind is used as an electrode in an electrochemical cell, this constitutes a membrane electrode. It is with these that we are concerned here. The electrochemical membrane, which in fact is the active sensing element of an ion-selective electrode, is also referred to as "the sensor."

Miniaturization of ion-selective membrane electrodes led to the development of ISMs suitable for measuring intracellular ionic activities. In the early days of ISM technology, "microelectrodes" had tip diameters ranging from 15 to 100 μm, and could only be used on very large cells such as squid giant axons (Mauro, 1954; reviewed by Hinke, 1987) or crab giant muscle fibers (Caldwell, 1954, 1958). These early "microelectrodes" incorporated solid-state membranes. Today it is possible to fabricate submicrometer-tip microelectrodes selective for H^+, Na^+, K^+, Ca^{2+}, Mg^{2+}, and Cl^- among other ions (Ammann, 1986; Ammann *et al.*, 1987). The reduction in tip size was possible mainly due to the introduction of liquid membranes using liquid ion exchangers (Orme, 1969; Walker, 1971) or incorporating neutral carriers (Ammann, 1986).

The membrane or sensing element of an ion-selective electrode is held in a compact unit containing an electrolytic internal filling solution (Fig. 3A). The membrane can be either liquid or solid state. In the case of ISMs, the compact unit housing the sensor, whether liquid or solid, is a micropipette with tip diameter of 1 μm or less that is usually assembled into a fine-tipped double-barreled microelectrode (Fig. 3B). The behavior of an ion-selective electrode will be determined by the properties of the membrane and, in the case of microelectrodes, by other factors such as tip size, geometry, electrical properties of the glass, and shunt resistance, the latter being determined mainly by the effectiveness of the silanization procedure used in their fabrication (for theory and practice of silanization, see Munoz *et al.,* 1983).

When an ISM is incorporated in a complete electrochemical cell it will measure the potential difference across the electrochemical membrane that separates two electrolytic solutions. These solutions are the so-called internal "filling solution" of the micropipette and the sample or test solution. A complete electrochemical cell commonly used for measuring intracellular (or extracellular) ion activities that incorporates a membrane electrode is schematically shown in Fig. 3 and may be represented as follows:

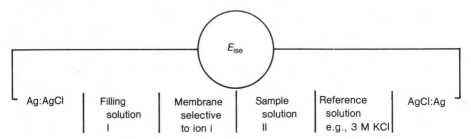

where the half cell Ag | AgCl | filling solution I | membrane selective to ion i, represents the membrane electrode. The reference electrode is the other half cell: Ag | AgCl | 3 M KCl. Assuming that the membrane is perfectly selective to ion i, the cell potential E_{ise} is given by the Nernst equation:

$$E_{ise} = E_o + \frac{RT}{zF} \ln \frac{a_i^{II}}{a_i^{I}} \tag{7}$$

where a_i^{I} and a_i^{II} are the thermodynamic activities in the filling solution (I) and in the test solution (II) respectively, of the ion i to which the membrane is ideally selective. R, T, z, and F have their usual meaning and E_o is a constant potential. Since the activity of the filling solution I is fixed, Eq. (7) becomes

$$E_{ise} = E_o + (RT/zF) \ln a_i^{II} \tag{8}$$

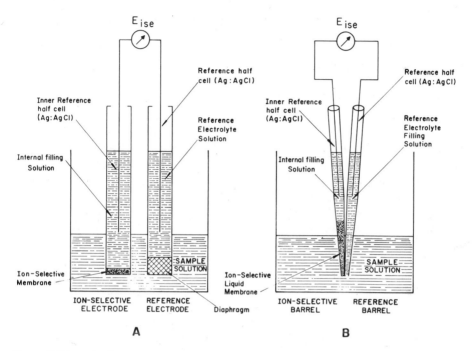

Figure 3. Schematic diagram of a macroelectrode (A) and a double-barreled microelectrode (B) liquid membrane cell assembly.

The origin of E_o will be explained below. Here it is enough to say that E_o can be set to zero and can be used as a reference potential. Its actual value corresponds to E_{ise} when $a_i^{II} = 1$. When E_o is set to 0 mV, Eq. (8) becomes

$$E_{ise} = (RT/zF) \ln a_i^{II} \qquad (9)$$

Equations (7)–(9) are valid only if within a given set of experimental conditions (e.g., physiological conditions) the electrode is perfectly selective for ion i, i.e., if the effects of other ions present in the calibration solutions on the electrode potential (E_{ise}) are negligible. This is true for a few electrodes, but for most cases the situation turns out to be a bit more complicated as seen in the following section.

4.2. Selectivity of ISMs

4.2.1. Permselectivity and Ion-Selectivity. The Nicolsky–Eisenman Equation

For a membrane to be useful as an ion-selective electrode, it has to be *permselective*. The exclusion of sample counterions from a membrane phase is a definition of permselectivity (Ammann, 1986). Permselective membranes, due to fixed charges within the membrane, distinguish among ions according to their charge but do not distinguish between ions of the same charge (Koryta, 1975). A membrane will be perfectly or ideally permselective if only the cationic or the anionic member of the electrolyte has finite mobility in the membrane phase. Ideal permselective behavior of electrochemical membranes is shown by a Nernstian electrode response in *pure* electrolytic solutions.

In addition to permselectivity, for a membrane to work as an ion-selective electrode it must have *ion-selectivity*. The ion selectivity of a membrane electrode is governed by both the mobility of the ions in the membrane and the physicochemical equilibrium conditions that exist at the membrane–solution interfaces (e.g., partition coefficients, ion-exchange equilibrium constants, or in the case of neutral carriers the topology of the carrier, the number and properties of binding sites, and so on). A detailed account of the factors responsible for ion-selectivity of electrochemical membranes is beyond the scope of the present review. The interested reader will find many references in the authoritative works of Eisenman (1967, 1969), Koryta and Štulík (1983), and Ammann (1986).

A membrane electrode will be ideally permselective and ion-selective when its response in potentiometric studies is Nernstian for the primary ion, i (i.e., the ion to be detected), even in the presence of any interfering ions. Only under certain restricted experimental conditions do some ion-selective electrodes show such an ideal behavior. In general, ideal behavior is seldom observed, particularly in mixed electrolyte solutions such as biological fluids. Therefore, it is important to consider the factors that determine the response of ion-selective electrodes in mixed electrolyte solutions. This is of particular relevance for the case of concentrated polyelectrolyte solutions such as those present in the cytosolic compartment of living cells.

The electrical potential response E_{ise} of an ion-selective electrode to its primary ion, i, in the presence of various interfering ions j, k, . . . , can be conveniently described by a semiempirical formalism known as the Nicolsky–Eisenman equation:

$$E_{ise} = E_o + S \cdot \log [a_i + K_{ij}^{POT} \cdot (a_j)^{z_i/z_j} + K_{ik}^{POT} \cdot (a_k)^{z_i/z_k} \cdots] \qquad (10)$$

where E_{ise} is the observed electrode potential; E_o is a constant potential difference (for a given temperature) comprising liquid junction potential and reference electrode potentials. More specifically, $E_o = E_I^o + E_R + E_D$ where E_I^o is a constant potential difference including the boundary potential difference between the internal filling solution and the electrochemical membrane; E_R is a constant potential that includes the potential difference between the metallic lead to the membrane electrode (Ag : AgCl wire) and the internal filling solution of this electrode, and of the potential difference between the reference electrolyte and the Ag : AgCl wire within the reference electrode. E_R is independent of changes in the composition of the sample solution. E_D is the liquid junction potential difference generated between the reference electrolyte and the sample solution. Note that E_D might change with the composition of the sample solution. S is the empirical slope of the ISM response. Its theoretical value is 2.303 (RT/z_iF). (It is also defined as $S = nRT/z_iF$ where n is a dimensionless empirical constant chosen so that nRT/z_iF is the *observed* slope of the line when E_{ise} is plotted as a function of ln a_i when $K_{ij}^{POT} \cdot a_j^{z_i/z_j} = 0$.) R, T, and F have their usual meaning. z_i is an integer with sign and magnitude that corresponds to the charge number of the primary ion i. a_i is the activity of the primary ion i in moles per liter (M) in the sample solution. a_j and a_k are the activities of the interfering ions, j and k, respectively. z_j and z_k are integers with sign and magnitude corresponding to the charge of interfering ions, j and k, respectively. K_{in}^{POT} is the potentiometric selectivity factor also referred to as the selectivity coefficient. It expresses the relative contribution of the primary ion i and an interfering ion j, k, . . . n to the electrode potential.

The Nicolsky–Eisenman equation can be expressed in a more general form as

$$E_{ise} = E_o + S \cdot \log \left[a_i + \sum_i K_{ij}^{POT} (a_j)^{z_i/z_j} \right] \tag{11}$$

Except for sign, this relation is the same for an anion-selective electrode (Srinivasan and Rechnitz, 1969):

$$E_{ise} = E_o - S \cdot \log \left[a_i + \sum_i K_{ij}^{POT} (a_j)^{z_i/z_j} \right] \tag{11.1}$$

where a_i is now the activity of the primary monovalent anion i and a_j is the activity of any other monovalent anion j to which the electrode responds.

4.2.2. The Meaning of Selectivity Coefficients

The selectivity coefficient K_{ij}^{POT} is a measure of the preference by the sensor for the interfering ion j relative to the ion i to be detected (i.e., the primary ion). In other words, a selectivity coefficient indicates the extent to which a "foreign" ion j interferes with the response of an electrode to its primary ion i. Therefore, the smaller the value of the selectivity coefficient K_{ij}^{POT}, the less is the interference from j ions experienced by an electrode sensitive to ion i. It follows that for an ideally selective membrane electrode or for sample solutions containing no other ions with the same sign of charge as the primary ion, all of the K_{ij}^{POT} values must be zero and the Nicolsky–Eisenman equation (10 and 11) reduces to the Nernst equation:

$$E_{ise} = E_o + S \cdot \log a_i \tag{12}$$

The selectivity of an electrode can be expressed as $1/K_{ij}^{POT}$, the reciprocal of the selectivity coefficient. For instance, under a given set of conditions, a selectivity coefficient $K_{Cl^--HCO_3^-}^{POT}$ of 0.05 was reported by Walker (1971) for the CORNING anion exchanger 477315. This indicates that the electrode has a selectivity for Cl^- over HCO_3^- of 20:1 or that it is 20 times more sensitive to Cl^- than to HCO_3^-.

Proper assessment of the selectivity of an ion-selective electrode is an absolutely necessary condition if one is to obtain meaningful, accurate, and reliable measurements of ionic activities inside cells. This becomes a fundamentally important consideration in deciding how to calibrate the electrodes, a point discussed in Section 3. It is important to very critically consider the selectivity coefficients being used when evaluating ion-selective electrode measurements in mixed electrolyte solutions. First, one has to bear in mind that selectivity coefficients are only empirical numbers having no strict physical interpretation. Second, the values of selectivity coefficients are not constant parameters characterizing electrode selectivity under all measuring conditions. In fact, the K_{ij}^{POT} values are dependent on: (1) the method used to measure them, e.g., separate or mixed solutions (see below), (2) the conditions under which the measurement is done (ionic strength, pH, ionic activities, and even the quality of the reagents used to prepare the solutions), and (3) the equation chosen to calculate them. All this implies that the value of a particular selectivity coefficient can be misleading without stating the method and the conditions used to determine it. An additional problem when considering the selectivity of an ISM is that the performance of the sensor can change with the microelectrode geometry, tip size, kind of glass, and silanization of the micropipette (Armstrong and García-Díaz, 1980; Tsien and Rink, 1980, 1981). This implies that in practice one should allow for certain variations in the selectivity of a given membrane microelectrode. Since electrode performance can also change with cell penetration, it is important to calibrate each ISM and if necessary, perform selectivity tests, before and after one or several impalements.

From the above discussion it should be obvious that there are no fixed values for selectivity coefficients but only rough guides of relative selectivity of ISMs. It is therefore highly advisable to calibrate each microelectrode in solutions resembling as closely as possible the background composition of the unknown solution. Since selectivity coefficients are so dependent on theoretical assumptions and experimental constraints, i.e., they are empirical, it becomes necessary to observe (and report) the effects of relevant interfering ions on electrode responses to solutions mimicking the unknown solution. This is possible in some instances (e.g., Ca^{2+}- or Mg^{2+}-selective microelectrodes) but unfortunately it is not feasible for Cl^--selective microelectrodes, since we do not know the nature and free concentrations of intracellular anions which might behave as interfering ions on electrode response with the possible exception of HCO_3^-. More detailed discussions of selectivity coefficients have been published (Moody and Thomas, 1971; Guibault *et al.*, 1976; Tsien and Rink, 1981; Ammann, 1986).

4.2.3. How Are Selectivity Coefficients Measured?

A detailed description of the methods used to determine selectivity coefficients is beyond the scope of this chapter. Complete accounts of this subject can be found in the literature (Srinivasan and Rechnitz, 1969; Moody and Thomas, 1971; Lakshminaraya-naiah, 1976; Armstrong and García-Díaz, 1980; Lee, 1981; Ammann, 1986).

In brief, there are two main approaches for experimental determination of selectivity factors: (1) those using separate solutions (Section 4.2.3a) and (2) those employing mixed solutions. The latter approach has at least three variations: the fixed interference method (Section 4.2.3b), the fixed primary ion method (Section 4.2.3c), and reciprocal dilutions keeping ionic strength (*I*) constant (Section 4.2.3d).

4.2.3a. The Separate Solution Method.   In this method, the potential of a cell comprising an ion-selective electrode and a reference electrode is measured with each of two separate solutions of pure salts, one containing only the primary ion, i, at the activity a_i, the other containing only the interfering ion, j, at the same activity, i.e., $a_i = a_j$. From the measured values of the electrode potential E_i in the solutions containing ion i and the electrode potential E_j in the solutions containing ion j, the value of the selectivity coefficient K_{ij}^{POT} can be calculated using the Nicolsky–Eisenman equation:

$$K_{ij}^{POT} = 10^{(E_j - E_i)z_i F/2.303 RT} \cdot \frac{a_i}{(a_j)^{z_i/z_j}} \tag{13}$$

For monovalent ions and when $a_i = a_j$, Eq. (13) becomes:

$$K_{ij}^{POT} = 10^{(E_j - E_i)F/2.303\,RT} \tag{14}$$

It has been customary to calculate K_{ij}^{POT} from values of E_i and E_j measured at a concentration of 0.1 M (see Srinivasan and Rechnitz, 1969). However, K_{ij}^{POT} can be obtained from measurements at any other concentration, as long as the two ions are present at the same activity in their respective solutions. Alternatively, if the activities of the solution of ion i and of ion j are so chosen that $E_i = E_j$, the selectivity coefficient can be calculated from the following expression:

$$K_{ij}^{POT} = a_i/a_j \tag{15}$$

The separate solution method has the merit of being simple. However, the selectivity coefficients obtained are not ideally representative for mixed sample solutions such as biological fluids. The method is illustrated in Fig. 4.

4.2.3b. The Fixed Interference Method. In this method the potential of the ion-selective electrode ensemble is measured in solutions containing a constant level of the interfering ion, j, and varying activity of the primary ion, a_i. The electrical potential values obtained are plotted against the activity of the primary ion, i. The value of a_i used to calculate K_{ij}^{POT}, which is denoted as a_x, is obtained from the intersection of the extrapolated parts of the linear portions of the electrode potential versus a_i function as illustrated in Fig. 5. The selectivity coefficient is then readily calculated using Eq. (16), where a_j is the constant background interference activity:

$$K_{ij}^{POT} = \frac{a_x}{(a_j)^{z_i/z_j}} \tag{16}$$

The fixed interference method is the one that most closely reproduces the experimental conditions under which the electrode is going to be used, particularly in biological fluids, which, as stated, are mixed electrolytes, and is the method assumed by IUPAC.

4.2.3c. The Fixed Primary Ion Method. This is the reverse of the fixed interference method. The concentration of the interfering ion j is varied at a constant

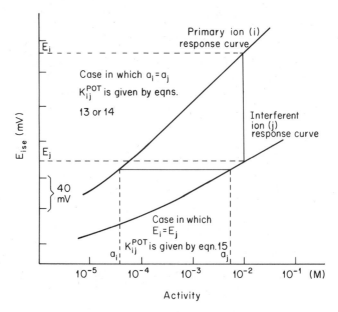

Figure 4. Illustration of the separate solution method to estimate selectivity coefficients. The case shown is for monovalent ions. E_{ise} is the potential of an electrochemical cell assembly comprising an ion-selective electrode and a reference electrode, measured in two separate solutions of pure salts of the activities indicated on the abscissa, as explained in the text.

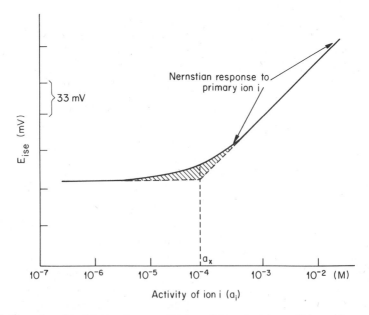

Figure 5. Illustration of the fixed interference method to estimate selectivity coefficients. The case shown is for monovalent ions. E_{ise} is the potential of an electrochemical cell assembly comprising an ion-selective electrode and a reference electrode, measured in solutions containing a constant level of the interfering ion, j, and varying activity of the primary ion, a_i, as explained in the text.

concentration of the primary ion i. This method is seldom used. For further details, see Ammann (1986).

4.2.3d. Mixed Solutions at Constant Ionic Strength: The Method of Reciprocal Dilutions. Many of the published selectivity values for liquid ion-exchanger microelectrodes have been evaluated by this method. The procedure has been described in detail by Brown *et al.* (1970) and Saunders and Brown (1977). These authors applied the method to estimate the selectivity coefficients for CORNING 477315 Cl$^-$ liquid exchanger microelectrodes. The solutions were made to maintain Cl$^-$ and one interfering anion (A$^-$) at constant ionic strength, e.g., [Cl$^-$] + [A$^-$] = 100 mM. A simple way of applying this method was published by Walker (1971), to evaluate also the selectivity coefficients of CORNING 477315 Cl$^-$ liquid exchanger microelectrodes. His method consisted of measuring the electrode potential in mixtures of constant ionic strength and finding the K_{ij}^{POT} (where i = Cl$^-$ and j = A$^-$) that made the data fit the Nicolsky–Eisenman formalism [Eq. (11)]. This method has been used for other liquid sensors and values are comparable with those obtained by other procedures (e.g., Giraldez, 1984; Alvarez-Leefmans *et al.*, 1986).

5. Cl$^-$-SELECTIVE MEMBRANE ELECTRODES THAT HAVE BEEN USED FOR INTRACELLULAR MEASUREMENTS

Two general types of Cl$^-$-selective microelectrodes have been used for intracellular measurements: those based on solid-state membranes and those containing liquid membranes. Except for one relatively recent report (McCaig and Leader, 1984), solid-state Cl$^-$-selective microelectrodes are no longer in use, but are responsible for many a_{Cl}^i values found in the literature at least until 1977. Those containing liquid membranes have been the best option since their introduction by Orme (1969) and Walker (1971).

5.1. Solid-State Cl$^-$-Selective Microelectrodes

Although solid-state Cl$^-$-selective microelectrodes are no longer used, it is important to explicitly consider their properties. They are responsible for a_{Cl}^i values reported 10–25 years ago and the uncritical use of the a_{Cl}^i values obtained by this methodology can lead to confusion in the field.

Perhaps the only advantage of solid-state over liquid ion-exchanger (LIX) microelectrodes is that the former are not sensitive to bicarbonate and other anions of biological importance such as propionate, acetate, and aspartate, at physiological or experimentally used concentrations (Saunders and Brown, 1977). However, as will be seen below, disadvantages of solid-state Cl$^-$-selective microelectrodes for measuring a_{Cl}^i far outweigh their advantages when compared with LIX microelectrodes.

5.1.1. Metallic Ag Microelectrodes

Based on the nature of the sensing element, solid-state microelectrodes fall into two main categories: those based on metallic Ag and those based on Ag : AgCl. The former type of sensor was introduced by Kerkut and Meech (1966). The sensitive material was a plug of metallic silver inside the open tip of a conventional glass microelectrode. The silver precipitate in the tip was achieved by chemical reduction of ammoniacal silver nitrate within the glass pipette. Details of their construction can be found in the original description of Kerkut and Meech (1966) and in the book of Thomas (1978). Since their introduction by Kerkut and Meech in 1966, no one else has reported a_{Cl}^i measurements, at least in excitable cells, using this kind of micro-electrode. Another Ag microelectrode design for measuring a_{Cl}^i was reported by Armstrong *et al.* (1977). It consisted of a protruding-tip microelectrode, which was made by sealing the tip of borosilicate micropipettes, coating them under vacuum with a 0.2- to 0.3-μm-thick layer of spectroscopic-grade silver and sealing them (except for the terminal 2–5 μm of the tip) inside a second borosilicate micropipette. They claim-ed that these silver-tipped microelectrodes gave more reliable measurements of a_{Cl}^i than those using Ag : AgCl microelectrodes. However, the values they reported for a_{Cl}^i in frog skeletal muscle were higher than those found in the more detailed study of Bolton and Vaughan-Jones (1977) using LIX microelectrodes. Besides technical difficulties in their fabrication, the theory of metallic Ag electrodes for Cl^- measurement is not well established and no information on yield, lifetime, or electrical resistance has been reported (see Thomas, 1978). This is in contrast to the situation for Ag : AgCl electrodes.

5.1.2. Ag : AgCl Microelectrodes

These electrodes are based on the well-known fact that a silver wire coated with AgCl behaves as a perfect Cl^--sensitive electrode when tested in pure Cl^- solutions. The theory of Ag : AgCl electrodes is well established (e.g., Janz and Ives, 1968; for further references see standard books of electrochemistry). Two main types of Ag : AgCl microelectrodes for intracellular measurement of Cl^- activities have been devised: recessed tip and protruding tip. Neild and Thomas (1973) designed a recessed-tip Ag : AgCl microelectrode basically consisting of an electrolytically sharpened and chlorided silver wire inserted and sealed into a borosilicate micropipette. The tips of the micropipette and the silver wire were within 10–50 μm of each other. This re-cessed-tip microelectrode has the disadvantages of a dead space around the silver wire, so that response to Cl^- activity changes is slow (1 min or more). Neild and Thomas (1974) used this type of microelectrode to measure a_{Cl}^i in snail neurons. They con-cluded that Cl^--selective electrodes with Ag : AgCl as the sensitive material are un-suitable for measuring a_{Cl}^i. Data were difficult to interpret because these micro-electrodes changed their properties when inside the cells. The main problem was that the electrode developed a significant electrical offset on the electrode potential which they claimed could not straightforwardly be attributed to interference from another ion (or ions). The results of Neild and Thomas (1974) represented the first proof that Ag : AgCl electrodes do not respond as expected in an intracellular environment, although previous workers had predicted problems in using them as probes for measur-ing a_{Cl}^i (Gesteland *et al.*, 1959; Tasaki and Singer, 1968; Janz and Ives, 1968). The arguments presented by all these authors can be summarized as follows: first, for a

AgCl electrode to function properly it must be surrounded by a saturated solution of AgCl, and so in solutions where there are no silver ions (Ag^+) the electrode will continuously lose AgCl. This might not be a serious problem for large electrodes, but miniaturized electrodes lose most of their AgCl very quickly (Janz and Ives, 1968). Second, the problem is aggravated when the system being investigated is a poly-electrolyte like the cytosol. In this case, the measurement of a_{Cl}^i is limited by the possible sorption of Ag^+ to the polyelectrolyte (Tasaki and Singer, 1968). For instance, sulfhydryl and amino groups of intracellular proteins might form highly stable complexes with Ag^+, which would make it impossible to achieve an equilibrium between the AgCl and its intracellular surroundings (Janz and Ives, 1968).

The theoretical factors considered above might explain why, after being in an intracellular environment, Ag : AgCl electrodes frequently do not return to their initial extracellular electrical potential value or do so very slowly, rendering meaningless the calibration before and after impalement (Neild and Thomas, 1974; Hinke and Gayton, 1971). In support of this idea is the observation that Cl^- microelectrodes that were left in the myoplasm of barnacle muscle fibers for an hour or more sometimes displayed an electrical drift in a calibrating solution for 1–5 min before stabilizing at the expected potential (Gayton and Hinke, 1968). However, Saunders and Brown (1977) argued that difficulties in recalibrating the electrodes after use, did not occur with any higher frequency with Ag : AgCl electrodes than with LIX electrodes. They attributed the irreversibility problem simply to plugging of the microelectrode tip. The reliability of results obtained with Ag : AgCl microelectrodes may depend on the cell type being investigated. Thus, it is difficult to assess the credibility of results reported in the literature obtained using Ag : AgCl microelectrodes. In view of the uncertainties involved in the technique, measurements reported so far with such microelectrodes must be interpreted with caution unless they are verified by more reliable techniques.

Another drawback of Ag : AgCl electrodes is that small amounts of I^- or Br^- can seriously interfere with their performance, causing them to behave erratically (Janz and Ives, 1968; Gayton and Hinke, 1968). It has been shown that I^- at concentrations as low as 0.1 mM is sufficient to irreversibly damage the electrodes (Saunders and Brown, 1977). Owen *et al.* (1975) using a Ag : AgBr microelectrode showed that the intracellular Br^- concentration in *Aplysia* neurons can be as high as 1 mM. They pointed out that Br^- in *Aplysia* giant cells would cause a_{Cl}^i to be overestimated. Similar considerations were made by Gayton and Hinke (1968) working with barnacle muscle fibers. This problem not only precludes the possibility of using either I^- or Br^- as a substitute for extracellular Cl^- but, more seriously, interference from these halides has to be kept in mind particularly when measuring a_{Cl}^i in marine animals or terrestrial invertebrates where Br^- or I^- might be present intracellularly at concentrations that would interfere with the electrode response (Hinke, 1969; Neild and Thomas, 1974).

From the report of Owen *et al.* (1975) in which measurements of a_{Cl}^i were made in *Aplysia* nerve cells with both LIX and Ag : AgCl electrodes, it was concluded that the former kind of electrode gave accurate readings of a_{Cl}^i and that other intracellular anions in these cells did not interfere with the LIX electrode response. In contrast, Ag : AgCl electrodes gave spuriously high readings of a_{Cl}^i, which were attributed to intracellular interference produced by Br^-. The average values of a_{Cl}^i were 38 and 152 mM for LIX and Ag : AgCl electrodes, respectively (see also Walker and Brown, 1977). Other studies in which Ag : AgCl electrodes were used to measure a_{Cl}^i in other excitable cells reached similar conclusions. For instance, Strickholm and Wallin (1965) measured a_{Cl}^i in the crayfish giant axon with an electrode consisting of a fine chlorided silver wire protruding from the end of a glass capillary. They found a_{Cl}^i to be 25 mM, but later on Wallin (1967) found by chemical analysis a value of 12.7 mM for the *total* intracellular Cl^- concentration, suggesting that the value obtained with the Ag : AgCl electrode was spuriously high. Three years later, Brown *et al.* (1970) measured a_{Cl}^i in the crayfish giant axon, using a liquid anion-exchanger microelectrode (Corning 477315). The value they found for a_{Cl}^i was 13.73 ± 0.4 mM, which was close to the value expected considering the measurements of total intracellular Cl^- concentration previously reported by Wallin (1967).

The general conclusion from the facts discussed above is that Cl^--selective microelectrodes that use either AgCl or Ag as the sensing element give erroneously high values of a_{Cl}^i and therefore it is important to reemphasize the necessity for extreme caution in the interpretation of results reported with these kinds of probes.

5.2. Liquid-Membrane Cl⁻-Selective Microelectrodes

The Cl^--selective microelectrodes that are currently used contain liquid membranes as the anion sensor element. They were introduced as early as 1969 by Frank Orme, at that time working at the University of California at Berkeley. Basically he showed how easily a liquid membrane microelectrode could be made by simply filling a 1-μm tip of a borosilicate glass micropipette with a liquid anion exchanger. However, the first intracellular Cl^- activity measurements reported with Cl^--LIX microelectrodes were those of the group of Arthur Brown and John Walker from the University of Utah in the early 1970s (Brown *et al.*, 1970; Cornwall *et al.*, 1970; Walker, 1971).

Besides the disadvantages of solid-state membrane microelectrodes, there are three main reasons for the popularity of liquid-membrane Cl^--selective microelectrodes. First, they are relatively easy to make. Second, the size of their tips can be minimized down to the submicrometer range allowing their use in very small cells. Third, in spite of their imperfect selectivity, if judiciously used, LIX microelectrodes give reliable measurements of a_{Cl}^i.

There are two basic types of liquid membranes available for measuring a_{Cl}^i with Cl^--selective microelectrodes: those containing an anion exchanger (e.g., CORNING 477913) and those containing anion-carrier molecules. At present, the most widely used liquid-state Cl^--selective microelectrodes are the ones containing an anion-exchanger membrane in which a lipophilic quaternary ammonium salt acts as the mobile exchanger site. Nowadays by far the most widely used of the anion exchangers is CORNING 477913. Although commonly used in many laboratories, LIX microelectrodes are not free from problems, which have to be critically evaluated before accepting a_{Cl}^i readings as reliable. Some of these problems as well as some of the electrochemical properties of liquid-membrane microelectrodes will be considered in the following sections.

Only recently have the first anion-carrier molecules with appropriate selectivities for a^i_{Cl} measurements been developed (see Section 5.2.2). Among them the most interesting are positively charged carrier molecules derived from vitamin B_{12}. An example of a^i_{Cl} measurements using one of these anion-carrier-based microelectrodes has been recently reported (Kondo *et al.*, 1989).

5.2.1. Liquid Ion-Exchanger Cl⁻-Selective Microelectrodes

5.2.1a. How Do Liquid Ion-Exchange Membranes Work? A liquid ion-exchange membrane is formed by dissolving a charged organic compound which acts as an ion exchanger, in a water-immiscible organic solvent. Experience has shown that a large number of hydrophobic ions act as ion-exchanger ions in liquid membranes (Lakshminarayanaiah, 1976; Koryta and Štulík, 1983). These liquid ion-exchange membranes can be considered in many ways as the liquid analogue of solid ion-exchange membranes and therefore they can be used as membrane electrodes (Sollner, 1968). Their principal difference results from the fact that the ion-exchanger sites (ionogenic groups), which are spatially fixed in the matrix of solid membranes, are free to move in the liquid systems (Eisenman, 1968).

The reactions that have been postulated to occur when the organic phase of LIX sensors contacts an aqueous electrolyte solution have been considered in detail by several authors (Eisenman, 1969; Koryta, 1981; Koryta and Štulík, 1983). The potentiometric response behavior of liquid anion-exchanger membrane electrodes in the presence of primary anions i^- and interfering anions j^- has been treated theoretically by Sandblom *et al.* (1967; Sandblom and Orme, 1972) and by others (Buck, 1975; Morf, 1981). A very didactic account of the affairs that take place at the tip of LIX microelectrodes has been given by Wright and McDougal (1972) and Tsien (1980). Here we will outline what LIX membranes are and how they work when introduced in glass micropipettes.

A property required of both the solvent and the ion-exchanger compound is a very low water solubility so that the exchanger solution can form a membranelike phase separating two aqueous electrolyte solutions (Sollner and Shean, 1964). The water-immiscible solvent makes it difficult for hydrophilic ions from the aqueous phase to enter the sensor. When no net current is permitted to flow across the LIX membrane, a condition that can be imposed by the experimenter by means of appropriate electronics, selectivity is achieved when one ionic species is much better than any of its rivals at entering the organic phase (Tsien, 1980). The selectivity of the sensor is determined not only by the properties of the solvent but also by the mobile sites of the exchanger compound. Finding the appropriate solvents and salts for anion- and cation-selective electrodes has been a matter of chance and design. For the specific case of anion-selective electrodes, it has been found that organic solvents with hydrogen-bonded protons, such as decanol, solvate anions better than cations, thereby favoring the entry of the former into the organic phase. The salt of the hydrophobic ion also influences the anion versus cation preference. For instance, if the salt consists of a highly hydrophobic cation and a rather hydrophilic anion, the marked propensity for the anions to leave the sensor (in the absence of net steady-state electrical current) will favor the entry of aqueous anions (and not cations) to the sensor in exchange. When a step change in the activity of the primary ion is imposed across the liquid exchange membrane (e.g., at the tip of a microelectrode filled with LIX), ions entering the organic phase are balanced by ions of the same sign leaving the phase until a new steady state is achieved.

5.2.1b. Liquid Anion-Exchange Membranes for Making Cl^--Selective Microelectrodes. Four liquid membranes based on anion exchangers have been used to make Cl^--selective microelectrodes for intracellular work. Although their exact composition has not been revealed, it is thought that they are made of chloride salts of very large hydrophobic cations, most probably quaternary ammonium salts, dissolved in 1-decanol. The manufacturers and code numbers of these four liquid anion exchangers are: Orion 92-17-02, CORNING 477315, CORNING 477913, and WPI IE-170. Orion 92-17-02 has been seldom used in microelectrodes (e.g., Spring and Kimura, 1978; Deisz and Lux, 1982). The same statement can be made about the WPI IE-170 (e.g., Ballanyi and Grafe, 1985; Harris and Betz, 1987). So far, most measurements of a_{Cl}^i in excitable cells have been made using CORNING 477315. Examples can be found in the literature for skeletal muscle (Bolton and Vaughan-Jones, 1977), cardiac muscle (Vaughan-Jones, 1979a,b, 1986; Baumgarten and Fozzard, 1981), smooth muscle (Aickin and Brading, 1985a), and nerve cells (Brown *et al.*, 1970; Russell and Brown, 1972a; Ascher *et al.*, 1976; Thomas, 1977; Deisz and Lux, 1982; Bührle and Sonnhof, 1983, 1985; Gardner and Moreton, 1985). CORNING 477315 LIX has been modified by increasing the concentration of organic salt fivefold (Baumgarten, 1981). Otherwise its composition is identical to that of CORNING 477315. The newer mixture, renumbered as CORNING 477913, has lower resistivity, gives lower-resistance microelectrodes (about fivefold less than that of microelectrodes made with CORNING 477315) with better slope, stability, and selectivity over some potentially interfering anions (see below). Unfortunately, for HCO_3^-, which is the most abundant intracellular free interfering anion, selectivity coefficients are the same for 477315 and 477913. Examples of use of this sensor to measure a_{Cl}^i in excitable cells are found in Baumgarten (1981), Moser (1985), and Alvarez-Leefmans *et al.* (1988).

5.2.1c. Selectivity Properties of Liquid Anion Exchangers Used for Making Cl^--Selective Microelectrodes to Measure a_{Cl}^i. Liquid membranes based on anion exchangers currently used for making "Cl^--selective" microelectrodes (e.g., CORNING 477913 or 477315) are not truly specific sensors toward Cl^-. As a matter of fact, they are sensitive to a wide variety of anions. This is why they should be more properly referred to as "liquid anion exchangers" rather than "Cl^--specific" or "Cl^--selective" exchangers. Regardless of the membrane composition, they select anions on a crude scale of hydrophobicity exhibiting the same selectivity sequence with a preference for lipophilic and a rejection of hydrophilic anions (Wegmann *et al.*, 1984). As a rule, the selectivity sequence of liquid anion-exchanger membranes follows the Hofmeister lyotropic series (Hofmeister, 1888):

$$R^- > ClO_4^- > SCN^- > I^- > NO_3^- > Br^- > Cl^- > HCO_3^- \sim$$
$$OAc^- \sim SO_4^{2-} \sim HPO_4^- \tag{17}$$

Theory and experimental facts indicate that liquid anion-exchanger membranes behave as dissociated anion exchangers, in which the complexation between the cationic sites and the counterions in the membrane phase is negligible (Wegmann *et al.*, 1984). The selectivity is completely described by the *distribution coefficients* of the various anions between the aqueous sample solution and the organic membrane phase (Morf *et al.*, 1985). The electromotive force of an electrode cell assembly (E_{ise}) for dissociated anion exchangers, where complexation between the cationic sites R^+ and the counterions is negligible, is given by the following expressions:

$$E_{ise} = Ex^o - (RT/F) \ln [ax' + K_{xy}^{POT} ay'] \tag{18}$$

$$K_{xy}^{POT} \simeq K_y/K_x \tag{19}$$

where Ex^o is a reference potential, R, T, and F have their usual meaning, ax' and ay' are the anion activities in the boundary zone of the sample solution contacting the membrane, and K_{xy}^{POT} is the potentiometric selectivity coefficient of the dissociated anion exchanger, being defined as the ratio of the distribution coefficients K_x and K_y of the anions. Since the logarithm of distribution coefficients between aqueous and organic phases is often linearly related to the free energies of hydration of the corresponding ions, ΔG_H^0, the selectivity coefficients K_{xy}^{POT} of liquid membrane electrodes based on dissociated anion exchangers can also be correlated with hydration energies. As we have seen, the observed selectivity sequence of liquid anion-exchanger membrane electrodes follows the Hofmeister lyotropic series (17). The sequence (17) evidently conforms to the order of decreasing ΔG_H^0 values for each anion (e.g., -210 kJ/mole for ClO_4^- and -330 kJ/mole for Cl^-). Table 1 shows selectivity coefficients for Cl^- against various interfering anions, estimated by different methods, for microelectrodes containing liquid anion exchangers.

Selectivity sequences different from the Hofmeister series have not been achieved by using different anion-exchanger sites (quaternary ammonium compounds) or by varying other membrane components such as solvents (e.g., decanol-1, octanol-1) and plasticizers such as polyvinyl chloride, the latter being used in solvent polymeric membranes (Wegmann *et al.*, 1984). In other words, regardless of the membrane composition, Cl^- sensors based on liquid ion exchangers are all bound to follow the same selectivity sequence. From these kinds of studies it can be concluded that novel anion-selective microelectrodes, with different selectivity sequences and better selectivities, will not emerge from liquid anion-exchanger compounds. However, as will be seen in the next section, contrasting behavior is expected for membrane electrodes based on anion-selective neutral or positively charged carriers, in which ion-selectivity highly depends on the *free energy of the interaction* of the ions with the ligand

Table 1. Selectivity Coefficients for Cl⁻ against Various Interfering Anions, Estimated by Different Methods, for Microelectrodes Containing Liquid Anion Exchangers

Interfering anion (j⁻)	Selectivity coefficient pot (K_{Cl-j-})	Liquid anion exchanger		Method[a]	References
Bicarbonate	0.14 ± 0.03	Corning	477315	SS	Saunders and Brown (1977)
(HCO_3^-)	~0.05	Corning	477315	SS	Baumgarten (1981)
	0.2	Corning	477315	SS	Bührle and Sonnhof (1983)
	0.05	Corning	477315	MSrd	Brown et al. (1970)
	0.12 ± 0.01	Corning	477315	MSrd	Saunders and Brown (1977)
	0.05	Corning	477315	MSrd	Walker (1971)
	0.12	Corning	477315	FI	Deisz and Lux (1982)
	0.13 ± 0.02	Corning	477315	FI	Saunders and Brown (1977)
	0.09	Corning	477315	NS	Spitzer and Walker (1980)
	0.2	Corning	477315	NS	Lux (1974)
	~0.04	Corning	477913	SS	Baumgarten (1981)
	0.11 ± 0.01	Corning	477913	SS	Ishibashi et al. (1988)
	0.13 ± 0.03	Corning	477913	FI	Greger et al. (1983)
	0.14	Corning	477913	NS	Greger and Schlatter (1984)
	0.09 ± 0.02	Corning	477913	SS	Wills (1985)
	0.21	Orion	921702	FI	Deisz and Lux (1982)
	0.1	Orion	921702	NS	Spring and Kimura (1978)
	0.15	Orion	921702	NS	Orme (1969)
	0.16	WPI	IE-170	NS	Harris and Betz (1987)
	0.05	WPI	IE-170	NS	World Precision Instruments (1988)
Acetate	0.30 ± 0.10	Corning	477315	SS	Baumgarten (1981)
	0.27 ± 0.32	Corning	477315	MSrd	Saunders and Brown (1977)
	0.27	Corning	477315	FI	Deisz and Lux (1982)
	0.22 ± 0.05	Corning	477913	SS	Baumgarten (1981)
	0.17 ± 0.01	Corning	477913	SS	Ishibashi et al. (1988)
	0.31 ± 0.14	Corning	477913	FI	Greger et al. (1983)
	0.24	Corning	477913	NS	Greger and Schlatter (1984)
	0.28	Orion	921702	FI	Deisz and Lux (1982)
	0.26	Orion	921702	NS	Orme (1969)
Propionate	0.47 ± 0.17	Corning	477315	SS	Baumgarten (1981)
	0.05	Corning	477315	MSrd	Brown et al. (1970)
	0.77–0.89	Corning	477315	MSrd	Saunders and Brown (1977)
	0.5	Corning	477315	MSrd	Walker (1971)
	0.59	Corning	477315	FI	Deisz and Lux (1982)
	0.40 ± 0.07	Corning	477913	SS	Baumgarten (1981)
	0.5	WPI	IE-170	NS	World Precision Instruments (1988)
Methanesulfonate	~0.19	Corning	477315	SS	Baumgarten (1981)
	0.22–0.29	Corning	477315	SS	Baumgarten and Fozzard (1981)
	0.21–0.31	Corning	477315	MSrd	Saunders and Brown (1977)
	~0.22	Corning	477913	SS	Baumgarten (1981)

Table 1. (Continued)

Interfering anion (j−)	Selectivity coefficient pot (K_{Cl-j-})	Liquid anion exchanger		Method[a]	References
Isethionate	0.2	Corning	477315	MSrd	Brown et al. (1970)
	0.13–0.35	Corning	477315	MSrd	Saunders and Brown (1977)
	0.2	Corning	477315	MSrd	Walker (1971)
	0.2	Corning	477315	MSrd	Brown (1976)
	0.13	Corning	477315	FI	Deisz and Lux (1982)
	0.14 ± 0.01	Corning	477913	SS	Ishibashi et al. (1988)
Sulfate (SO_4^{2-})	0.018–0.027	Corning	477315	SS	Baumgarten and Fozzard (1981)
	0.03	Corning	477315	MSrd	Brown et al. (1970)
	0.11	Corning	477315	FI	Deisz and Lux (1982)
	0.08 ± 0.003	Corning	477913	SS	Ishibashi et al. (1988)
	0.16 ± 0.06	Corning	477913	FI	Greger et al. (1983)
	0.09	Corning	477913	NS	Greger and Schlatter (1984)
	0.40	Orion	921702	FI	Deisz and Lux (1982)
	0.2	Orion	921702	NS	Spring and Kimura (1978)
Phosphates (HPO_4^{2-})	0.058 ± 0.005	Corning	477315	SS	Chao and Armstrong (1987)
	0.027 ± 0.014	Corning	477913	FI	Greger et al. (1983)
	0.06	Corning	477913	NS	Greger and Schlatter (1984)
	0.04	Orion	921702	NS	Spring and Kimura (1978)
	0.97	Orion	921702	NS	Orme (1969)
$H_2PO_4^-$	0.02 ± 0.01	Corning	477913	SS	Ishibashi et al. (1988)
	0.10 ± 0.05	Corning	477913	FI	Greger et al. (1983)
	0.03	Corning	477913	NS	Greger and Schlatter (1984)
	0.092	Orion	921702	NS	Orme (1969)
$H_2PO_4^-/HPO_4^{2-}$	0.04	Corning	477315	FI	Deisz and Lux (1982)
Thiocyanate	67	Corning	477315	FI	Deisz and Lux (1982)
	10	Corning	477913	NS	Cremaschi et al. (1987)
Gluconate	~0.03	Corning	477315	NS	Cassola et al. (1983)
	0.03 ± 0.002	Corning	477913	SS	Ishibashi et al. (1988)
	0.018 ± 0.001	Corning	477913	FI	Greger et al. (1983)
	0.015	Corning	477913	NS	Greger and Schlatter (1984)
Furosemide	157 ± 13	Corning	477315	FI	Chao and Armstrong (1987)
	306 ± 51	Corning	477913	FI	Greger et al. (1983)
Bumetanide	151 ± 18	Corning	477315	FI	Chao and Armstrong (1987)
SITS	1000	Corning	477315	FI	Chao and Armstrong (1987)
F−	0.07–0.09	Corning	477315	MSrd	Saunders and Brown (1977)
Br−	2.7–4.2	Corning	477315	MSrd	Saunders and Brown (1977)
I−	6.4–94	Corning	477315	MSrd	Saunders and Brown (1977)

(continued)

Table 1. (Continued)

Interfering anion (j⁻)	Selectivity coefficient pot (K_{Cl-j-})	Liquid anion exchanger	Method[a]	References
Aspartate	0.015–0.022	Corning 477315	MSrd	Saunders and Brown (1977)
	≤0.04	Corning 477315	FI	Deisz and Lux (1982)
Citrate	0.10	Corning 477315	FI	Deisz and Lux (1982)
Glutamate	0.01–0.089	Corning 477315	MSrd	Saunders and Brown (1977)
	≤0.04	Corning 477315	FI	Deisz and Lux (1982)
Tartrate	0.14	Corning 477315	FI	Deisz and Lux (1982)
Alanine	≤0.04	Corning 477315	FI	Deisz and Lux (1982)
Glycine	≤0.04	Corning 477315	FI	Deisz and Lux (1982)

[a]SS, separate solutions; MSrd, mixed solutions by reciprocal dilutions keeping the ionic strength constant ($I = 0.1$ M); FI, fixed interference; NS, method not specified by the authors.

(Wuthier *et al.,* 1984; Morf *et al.,* 1985; Ammann *et al.,* 1986). Hence, the potentiometric selectivity mainly reflects the anion binding affinity of the ionophores. This raises the possibility of tailoring carrier molecules with a given anion selectivity.

5.2.1d. Are There Cytosolic Anions That Might Interfere with Measurements of a_{Cl}^i?

In living cells, liquid anion exchangers can be used to reliably measure a_{Cl}^i because either there are few potentially interfering anions in sufficient free concentration in the cytosol as to cause significant "errors" in electrode readings or when present, the error introduced by these interferences can be quantified. This is particularly true for cells in which Cl^- is accumulated above its equilibrium distributions across the membrane. For instance, a_{Cl}^i values between 20 and 50 mM are found in some vertebrate neurons (Ballanyi and Grafe, 1985; Alvarez-Leefmans *et al.,* 1988), cultured astrocytes from rats (Kettenmann, 1987), vertebrate smooth muscle (Aickin and Brading, 1985b) and cardiac muscle (Vaughan-Jones, 1982). On the other hand, there is indirect evidence for many vertebrate central neurons having intracellular Cl^- activities of less than 10 mM, judging from their resting membrane potential, from the fact that they are hyperpolarized by GABA and glycine (two transmitters that are known to selectively open Cl^- channels), and from determinations of the reversal potential of the hyperpolarization and its underlying current (e.g., Thompson *et al.,* 1988; see also Chapter 4). All this suggests that in these neurons E_{Cl} is more negative than E_m. At relatively low values of a_{Cl}^i (i.e., ≤ 10 mM), interference from normally occurring intracellular anions on Cl^--LIX microelectrode responses might be thought to be significant. The most important of these intracellular interfering anions is HCO_3^-. Therefore, appropriate experiments and corrections have to be made to evaluate the degree of this interference. There are at least two possible ways of getting around this problem. First, it is possible in principle to lower the intracellular HCO_3^- concentration, $[HCO_3^-]_i$, by removing extracellular HCO_3^- and CO_2, and using, for example, HEPES-buffered solutions equilibrated with 100% O_2. When CO_2 is removed, the pH will increase as internal HCO_3^- takes up H^+ and leaves the cell as CO_2 (Thomas, 1976). Under these conditions, the internal and external levels of HCO_3^- should be negligibly low, although probably they will still be finite because of metabolic production of CO_2 by the cell (see Chapter 12). It is important to bear in mind that although this experimental maneuver indeed lowers $[HCO_3^-]_i$, it also blocks Cl^- transport systems such as Cl^-/HCO_3^- exchange, which is involved in pH_i and $[Cl^-]_i$ regulation, and of course it will not get rid of other potentially interfering anions whose nature and intracellular free concentration are still uncertain (e.g., Baumgarten and Fozzard, 1981). The maneuver can also be advantageous if one wishes to silence the Cl^-/HCO_3^- exchange system to study, say, a Na^+,K^+,Cl^- cotransport mechanism in isolation. A second strategy is to estimate $[HCO_3^-]_i$, and knowing $K_{Cl^-HCO_3^-}^{POT}$ for the Cl^--LIX microelectrode, plug these values in the Nicolsky–Eisenman formalism [Eq. (11)]. Reasonably accurate estimates of $[HCO_3^-]_i$ can be obtained if pH_i and CO_2 tension are known (e.g., Thomas, 1976; Bolton and Vaughan-Jones, 1977; see also Roos and Boron, 1981). Under physiological conditions, estimates of $[HCO_3^-]_i$ vary between 7 and 15 mM for *Helix* neurons and vertebrate tissues. Figure 6 shows that for

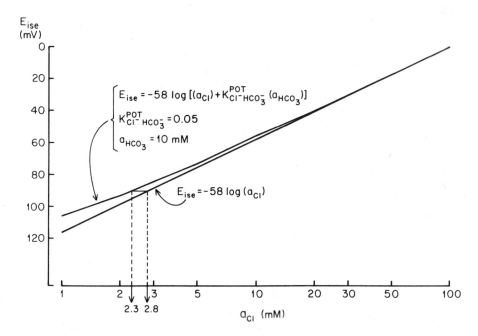

Figure 6. Illustration of the error that would be introduced in the measurement of a_{Cl}^i, if the interference produced by intracellular HCO_3^- on the Cl^- liquid—exchanger microelectrode is dimissed. The assumptions in the example shown are that the intracellular activity of $HCO_3^- = 10$ mM and $K_{Cl^-HCO_3^-}^{POT} = 0.05$. A jump of 90 mV in electrode potential (E_{ise}) on cell impalement (after E_m subtraction) would given an a_{Cl}^i reading of 2.8 mM, if the interference from intracellular HCO_3^- microelectrode response is ignored. However, if HCO_3^- interference is considered, the same jump in electrode potential would give an a_{Cl}^i reading of 2.3 mM. Therefore, a_{Cl}^i would have been overestimated by 0.5 mM, if HCO_3^- interference had been ignored. Obviously the possible error decreases as a_{Cl} increases.

a $K^{POT}_{Cl^--HCO_3^-}$ value of 0.05 as reported for CORNING Cl^--LIX sensors (Walker, 1971; Baumgarten, 1981; see Table 1) at $I = 0.1$ M, and assuming that the intracellular activities of HCO_3^- and Cl^- are 10 and 2 mM, respectively, the measured a^i_{Cl} would be overestimated by about 0.5 mM. As expected, the error decreases as a^i_{Cl} increases and vice versa. For $a^i_{Cl} \geq 10$ mM it is negligible. However, when external Cl^- is experimentally removed, a condition under which cells are likely to be depleted of Cl^-, the intracellular potential of the Cl^--selective microelectrode should reflect the contribution of intracellular interfering anions. Experimental evidence obtained from various types of cells shows that in Cl^--free media, the apparent a^i_{Cl} read in the final steady state by the microelectrode, $(a^i_{Cl})_{0Cl}$, never reaches zero. Furthermore, depending on cell type and experimental conditions (e.g., kind of anion used to substitute external Cl^-), the apparent $(a^i_{Cl})_{0Cl}$ usually ranges between 3 and 5 mM, although values as low as 0.8 mM and as high as more than 10 mM have been measured (see Table 2). This value might reflect intracellular interference due to HCO_3^- and/or other intracellular free anions. Equilibration of cells in 100% O_2 should reduce $[HCO_3^-]_i$ to negligible levels and therefore $(a^i_{Cl})_{0Cl}$ under these conditions should reflect the contribution of interfering anions other than HCO_3^- to the electrode potential. On the other hand, if P_{Cl} is relatively low (e.g., dorsal root ganglion neurons), in cells equilibrated with 100% O_2, the apparent $(a^i_{Cl})_{0Cl}$ might reflect Cl^- trapped inside the cell plus interference from other anions. For further discussion on this issue, see Spitzer and Walker (1980), Baumgarten and Fozzard (1981), and Alvarez-Leefmans et al. (1988, and Chapter 4, this volume). Another way of getting a value for $[HCO_3^-]_i$ is to measure it with HCO_3^--LIX microelectrodes (Khuri et al., 1974, 1976). Khuri and collaborators have used these electrodes to measure steady-state $[HCO_3^-]_i$ in skeletal muscle fibers of frog and rat and their estimates have corresponded to calculated values of pH_i (cf. Bolton and Vaughan-Jones, 1977; see Roos and Boron, 1981). However, measurements with HCO_3^--selective microelectrodes have to be viewed with caution because of a recent report by Kraig and Cooper (1987) showing that the HCO_3^- exchanger sold by World Precision Instruments (IE-310) actually predominantly sensed *carbonate* ions over bicarbonate. They suggest that this ion exchanger is presumably that originally described by Khuri et al. (1974). In fact, at the present time, IE-310 is sold by WPI as a carbonate sensor.

Other naturally occurring intracellular anions that might potentially interfere with Cl^--LIX microelectrode readings are propionate, acetate, and phosphate among others. Unfortunately, we know little about the free concentrations of anions other than Cl^- and HCO_3^- inside vertebrate cells. However, selectivity coefficients of Cl^--LIX for these anions (see Table 1) are relatively low and therefore little interference, if any, can be expected at a^i_{Cl} in the millimolar range. Interference from intracellular Br^- and I^- might be important in some invertebrate cells, particularly in marine species, since these halides are naturally present in seawater. This has been discussed in Section 5.1.2 for the case of solid-state Cl^--selective microelectrodes. For instance, in *Aplysia* nerve cells $[Br^-]_i$ is claimed to be about 1 mM (Owen et al., 1975). Since the $K^{POT}_{Cl^--Br^-}$ for Cl^--LIX (477315) microelectrodes is 3.3 (Saunders and Brown, 1977), the presence of Br^- in *Aplysia* giant neurons would, in principle, cause an overestimate of a^i_{Cl} in this particular cell type. Similar arguments hold for I^-, since the $K^{POT}_{Cl^--I^-}$ is 30 (Saunders and Brown, 1977). However, this putative interference from intracellular Br^- or I^- does not seem to preclude accurate measurements of a_{Cl}^i in *Aplysia* nerve cells. For instance, the L cells of the abdominal ganglion of *Aplysia* respond to acetylcholine (ACh) with a hyperpolarizing potential having two components, an early and a late one. The early component is due to selective opening of Cl^- channels. It was found that the reversal potential for ACh (E_{ACh}) corresponded to both the reversal potential of the inhibitory postsynaptic potential (E_{IPSP}) and E_{Cl}. E_{Cl} was calculated from direct measurements of a_{Cl}^i. Thus, it was concluded that measurements of a_{Cl}^i in these cells using Cl^--LIX (477315) microelectrodes are reliable and are not invalidated by some other anion to which the Cl^- ion exchanger may be sensitive (Brown and Kunze, 1974). Similar findings were reported for medial pleural ganglion neurons of *Aplysia* bathed with artificial seawater containing no bicarbonate (Ascher *et al.*, 1976). In this case, the mean value of E_{Cl} measured using Cl^--LIX (477315) microelectrodes was in agreement with the reversal potential of the "rapid" response to ACh, which was known to be mediated by Cl^- (Kehoe, 1972). In other words, estimates of E_{Cl} in the same cell, assessed with two different techniques, gave the same result, suggesting again the reliability of a_{Cl}^i measurements with Cl^--LIX microelectrodes. Therefore, the interference claimed for Br^- and/or I^- does not seem to affect a_{Cl}^i measurements even in this marine organism. Probably, intracellular levels of Br^- or I^- in marine species are not as high as thought by other authors. Fortunately, interference from these two halides cannot be considered as a problem in vertebrate cells.

5.2.1e. Intracellular Interferences with Cl^--Sensitive Microelectrodes Introduced by the Presence of Foreign Anions.

We have discussed possible sources of interference on Cl^--LIX microelectrode response produced by naturally occurring intracellular anions. We might call these "endogenous interferences." However, there are other sources of interference: those introduced by the experimenter. These we will call "exogenous interferences." They might arise from two sources: the filling solution of the intracellular reference microelectrode and foreign anions experimentally used in the external solution such as Cl^- substitutes and pharmacological substances like anion transport inhibitors. The latter interferences can reach the intracellular compartment either through leaks around the microelectrodes resulting from imperfect glass–membrane sealing around the micropipette tip or through cell membrane pathways. We will consider first the problem of interference arising from cytosolic contamination secondary to leakage of the electrolyte filling solution of the reference microelectrode.

When inside a cell, the potential measured by an ion-selective microelectrode will be the sum of the transmembrane potential (E_m) and the electromotive force due to the activity of the ion j to which the sensor is selective for (E_{jse}). Therefore, E_m must be subtracted from E_{jse} to obtain the differential signal $E_m - E_{jse}$ which is proportional to a_j. To measure E_m we need to penetrate the cell under study with a "reference" microelectrode in addition to the ISM. During an intracellular measurement the electrolyte of the reference microelectrode is in direct contact with the cytosol. Therefore, leakage of the electrolyte filling solution used in the reference microelectrode could

Table 2. Steady-State Intracellular Cl⁻ Activity Measured with Liquid Anion-Exchanger Microelectrodes in Various Cell Types Kept in Physiological Salines and after Removal of External Cl⁻

Preparation and species	$(a_{Cl}^i)_{app}$ (mM)	$(a_{Cl}^i)_{OCl}$ (mM)	Liquid anion exchanger	Reference microelectrode (filling solution and resistance MΩ)	External Cl⁻ substitute	Buffer	References
Neurons							
Pallial ganglion (snail, *Helix aspersa*)		4.0[a]	Corning 477315	0.6 M K₂SO₄	Sulfate		Thomas (1977)
		3.5	Corning 477315	0.6 M K₂SO₄	Sulfate		Thomas (1978)
Stretch receptor (crayfish)	12.7 ± 1.3	5.2	Corning 477315	85% 0.6 M K₂SO₄ +15% 1.5 M KCl (20–40 MΩ)	Isethionate + gluconate	H	Deisz and Lux (1982)
Stretch receptor (crayfish)		5.0	Corning 477913	85% 0.6 M K₂SO₄ +15% 1.5 M KCl (20–50 MΩ)	Isethionate and/or gluconate	H	Moser (1985)
Dorsal root ganglion (frog)	33.6 ± 0.9	10.7 ± 0.8	Corning 477913	3 M KCl or 4 M K-acetate (40–100 MΩ)	Gluconate	H	Alvarez-Leefmans *et al.* (1988)
Glial cells							
Olfactory cortex (guinea pig)	6.0 ± 1.5	3–4	WPI IE-170	0.5 M K₂SO₄	Gluconate + sulfate	B/P	Ballanyi *et al.* (1987)
Skeletal muscle cells							
Lumbrical muscle (rat)	7.8 ± 0.2	4.1 ± 0.2	WPI IE-170	0.5 M K₂SO₄ + 0.2 M KCl (20–40 MΩ)	Isethionate + sulfate	PI	Harris and Betz (1987)
		1.7 ± 1.0	Corning 477315	Organic sensor (RLIX) (15 MΩ when filled with 3 M KCl)	Isethionate + sulfate	PI	Aickin *et al.* (1989)
Smooth muscle cells							
Vas deferens (guinea pig)	41.2 ± 6.7	3.1 ± 0.7	Corning 477315	Organic sensor (RLIX) (18–20 MΩ when filled with 3 M KCl)	Gluconate or glucuronate + sulfate	B/P	Aickin and Brading (1982)
Cardiac muscle cells							
Ventricular papillary muscle (rabbit)	15.0 ± 4.4	4.8 ± 0.6	Corning 477315	3 M KCl (15–20 MΩ)	Methanesulfonate	H	Baumgarten and Fozzard (1981)

Purkinje fibers (sheep)	19	2.75	Corning 477315	3 M KCl (15 MΩ)	Glucuronate + gluconate + sulfate	H	Vaughan-Jones (1979a)
		3.37 ± 0.5		2.5–3 M KCl (8–20 MΩ) or 0.5 M K$_2$SO$_4$		H	
	14.06 ± 0.7			2.5–3 M KCl		H	
	12.83 ± 1.1			0.5 M K$_2$SO$_4$		H	
		4.64 ± 0.2		2.5–3 M KCl or 0.5 M K$_2$SO$_4$		B	
	19.75 ± 0.6			2.5–3 M KCl		B	
	17.7 ± 0.8			0.5 M K$_2$SO$_4$		B	
	19.0	4.7		2.5–3 M KCl		B	Vaughan-Jones (1979b)
Epithelial cells							
Renal proximal tubules							
Necturus maculosus	21.3 ± 2.9[d]	5.8 ± 0.8[c]	Orion	3 M KCl	Sulfate + phosphate + lactate	B/P → P[f]	Spring and Kimura (1978)
	24.5 ± 1.1	6.2 ± 0.5[c]					
Rat	18.0 ± 4.4	3.8 ± 0.2[c]	Corning 477913	2.7 M KCl[e] (30–80 MΩ)	Gluconate	B/P	Cassola et al. (1983)
Proximal straight tubule (rabbit kidney)	17.8 ± 0.5[b]	4.2 ± 0.4[c]	Corning 477913	0.5 M K$_2$SO$_4$ + 0.01 M KCl	Gluconate isethionate (or sulfate) + acetate	B	Ishibashi et al. (1988)
Descending colon epithelium (rabbit)	23 ± 2	0.8 ± 0.2[c]	Corning 477913	0.5 M KCl	Gluconate + methanesulfonate	B/P	Wills (1985)
Rectal gland (spiny dogfish, *Squalus acanthias*)	38 ± 4[b]	4–6[c]	Corning 477913	0.5 M KCl or 1.0 M Na$_2$SO$_4$ (100–200 MΩ) when filled with 1 M KCl)	Gluconate	B/P	Greger and Schlatter (1984)
Human airway epithelium	42.7 ± 2.0	8.0[c]	Corning 477913	3 M KCl or 0.5 M Na$_2$SO$_4$ (30–180 MΩ)	Gluconate	B/P	Willumsen et al. (1989)

Abbreviations: $(a_{Cl}^i)_{app}$ is the steady-state intracellular Cl$^-$ activity directly measured by the microelectrode (i.e., without correction for interference from intracellular anions on microelectrode response). Measurements were obtained when cells were bathed with solutions containing physiological concentrations of Cl$^-$ salts. $(a_{Cl}^i)_{DCl}$ is the steady-state apparent intracellular Cl$^-$ activity measured in the nominal absence of external Cl$^-$. It is assumed that this value represents the contribution of intracellular interfering anions to the microelectrode potential.

Buffers used in the bathing solutions: H, HEPES; PI, PIPES; B, bicarbonate; B/P bicarbonate–phosphate; P, phosphate. (a) Value expressed as ion intracellular concentration; (b) value corrected for intracellular interference assuming that $(a_{Cl}^i)_{DCl}$ represents that interference; (c) external Cl$^-$ replaced on both sides of the epithelium; (d) $(a_{Cl}^i)_{app}$ measured on recovery after cells were kept in a Cl$^-$–free solution containing sulfate; (e) V_m and a_{Cl}^i were not measured simultaneously; (f) $(a_{Cl}^i)_{DCl}$ was measured while cells were bathed with a P-buffered solution.

contaminate the cytosol upsetting intracellular ion activities and osmotic balance, particularly in relatively small cells. This leads to spurious readings of the ISM, which is also located in the cytosolic compartment. The leakage rate for a conical micropipette (in the absence of applied current) filled with an electrolytic solution will be determined by two processes: diffusion and bulk flow due to gradients of hydrostatic pressure. The theory underlying these two processes has been considered by Purves (1981). Here it is enough to say that the leakage rate will be determined mainly by the geometric parameters of the micropipette such as internal tip diameter and the angle of taper, and by the diffusion coefficient and concentration of the electrolyte. Leakage can be made relatively small by minimizing the geometric parameters and decreasing the concentration of the electrolyte. For instance, micropipettes having relatively large tips ($\sim 1 \mu m$) filled with 3 M KCl, the typical reference electrolyte, exhibit leakage rates greater than 100 fmole/sec, while submicrometer tip microelectrodes exhibit rates less than 10 fmole/sec (Ammann, 1986). By reducing the KCl concentration to 0.5 M, leakage rates are reduced to less than 2 fmole/sec (Fromm and Schultz, 1981; Ammann, 1986). Reference microelectrodes to measure E_m have been traditionally filled with 3 M KCl. The rationale here is that K^+ and Cl^- have similar diffusion coefficients and carry unit charge, and it is thus thought that KCl-filled microelectrodes have relatively small and constant junction potentials and if they do not develop tip potentials, they give reliable measurements of E_m. However, when measuring a_{Cl}^i the disadvantage of using concentrated KCl filling solutions in the reference microelectrode is leakage of Cl^- from the tip. This leakage can be large enough to alter a_{Cl}^i and consequently E_{Cl}, but this happens only when relatively *low-resistance*, large-tipped microelectrodes are used as shown in Fig. 7 (see Thomas, 1978; Thomas and Cohen, 1981). Clearly, under these conditions, artifactually high readings of a_{Cl}^i are obtained and reversal of polarity of inhibitory synaptic potentials generated by transmitters that selectively open Cl^- channels is frequently observed, as originally shown by Coombs *et al.* (1955). However, the problem is not so serious if relatively *high-resistance, submicrometer-tipped* reference microelectrodes are used, although such micropipettes often develop tip potentials and sudden changes in electrical resistance, which might ruin a laborious and precious experiment. Therefore, electrical properties of fine microelectrodes should be tested when possible, before, during, and after cell impalement. Those that develop tip potentials, excessive drift, or are plugged should be rejected, together with the experiment.

There are ways of checking for Cl^- leakage from a reference microelectrode filled with KCl. This can be better done if double-barreled microelectrodes are used, one barrel containing the Cl^--sensor and the other the KCl filling solution. If there is no significant Cl^- leakage from the reference barrel, on penetration of the cell under study, there must be an almost simultaneous step change in potential of both ion-selective and reference barrels, and the measured steady-state a_{Cl}^i should show no drift. There are abundant examples of reliable a_{Cl}^i measurements reported in the literature, using fine reference microelectrodes filled with KCl. We have successfully measured a_{Cl}^i in frog dorsal root ganglion neurons using double-barreled Cl^--selective microelectrodes. These cells have a mean diameter of about 60 μm and a volume-to-surface ratio of about 10^{-3} cm. The microelectrodes were beveled and the reference barrels

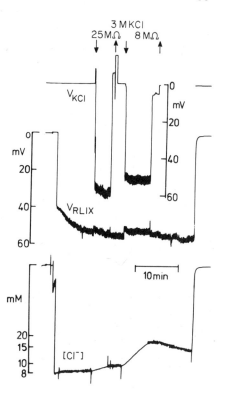

Figure 7. Experiment illustrating cytosolic contamination with Cl^- due to KCl leakage from a relatively low-resistance KCl-filled microelectrode. A snail neuron was first penetrated with a Cl^--selective microelectrode containing CORNING 477315 liquid anion exchanger. The signal recorded from this microelectrode is shown in the bottom trace already converted to $[Cl^-]$. Shortly after, the cell was impaled with a microelectrode filled with the organic reference liquid ion exchanger described by Thomas and Cohen (1981). The potential recorded from this microelectrode is shown in the middle trace (V_{RLIX}). Eight minutes later, a 25-$M\Omega$ KCl microelectrode was also inserted, left in for 4 min, withdrawn, broken to 8 $M\Omega$, and reinserted. The potential from this microelectrode is shown in the top trace (V_{KCl}). Arrows indicate impalement and withdrawal of this microelectrode. Note the increases in $[Cl^-]_i$ produced by the KCl microelectrode. (Modified from Thomas and Cohen, 1981, and reproduced with permission.)

had resistances between 40 and 100 $M\Omega$, and were filled with 3 M KCl. The steady-state a_{Cl}^i value was attained in less than 30 sec and remained stable for periods up to 1 hr. There was no evidence of cytosolic Cl^- contamination judging from several criteria: the step change in potential recorded by both barrels on crossing the membrane, the steadiness of the differential signal $E_m - E_{Clse}$, which was proportional to a_{Cl}^i, whose initial reading did not appreciably change with time after impalement (see Fig. 1, Chapter 4). Moreover, in damaged cells, $E_m - E_{Clse}$ changed indicating loss but not gain of the apparent a_{Cl}^i. Finally, the intracellular potential recorded through the ion-selective barrel E_{Clse} was independent of the composition of the filling solution of the reference barrel, which included 3 M KCl, or 4 M K-acetate, or the organic sensor of Thomas and Cohen (1981) (see below and Fig. 8).

From the foregoing discussion it follows that the ideal filling solution for reference microelectrodes when measuring a_{Cl}^i should not have any anions that might interfere with the response of the Cl^--selective microelectrode. In addition, the reference microelectrode must provide a reliable measurement of E_m. However, taking the appropriate precautions to avoid Cl^- leakage and tip potentials, 3 M KCl-filled microelectrodes continue to be a safe option for measuring E_m. To decrease Cl^- leakage, some workers use as filling solution KCl at less than 3 M. It is convenient to bear in mind that liquid junction potentials between the cytosol and the tip of the KCl microelectrode will change with dilution of the KCl, giving rise to spurious readings of E_m.

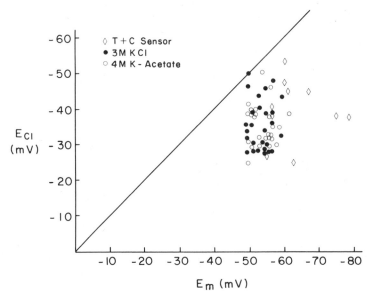

Figure 8. Relationship between E_{Cl} and E_m measured in dorsal root ganglion neurons with double-barreled Cl^--selective microelectrodes. Values of E_{Cl} calculated from steady-state measurements of a^i_{Cl} and a^o_{Cl} are plotted against the corresponding value of E_m for each neuron. The Cl^--selective barrel was filled with CORNING 477913 liquid anion exchanger. The reference barrel was filled with either of three different solutions: 3 M KCl (●), 4 M K^+-acetate (○), or Thomas and Cohen (T + C) organic reference liquid ion exchanger (◇). Note that the value of E_{Cl} is independent of the composition of the filling solution of the reference barrel. The solid line represents the expected relationship between E_{Cl} and E_m if Cl^- were distributed in electrochemical equilibrium across the membrane. Clearly, a^i_{Cl} in these cells departs from a passive distribution, E_{Cl} being more positive than E_m. (Data from Alvarez-Leefmans *et al.*, 1988, and Alvarez-Leefmans and Noguerón, unpublished observations.)

In a thorough study performed in skeletal muscle fibers, membrane potentials were measured with 0.03, 0.1, 0.3, 1.0, and 3 M KCl-filled microelectrodes; potentials were -62.4, -75.7, -83.0, -87.7, and -90.7 mV, respectively. This means that depending on the concentration of KCl, correction of the membrane potentials for liquid junction potentials in the electrode system is important (Hironaka and Morimoto, 1979).

Another way of circumventing the problem of Cl^- contamination of the cytosol due to reference microelectrode leakage is to use other filling solutions prepared with anions having selectivity coefficients such that they interfere less with Cl^--selective microelectrode response. The usual alternatives to KCl are 1 to 4 M K-acetate; K-sulfate; K-citrate (see Rosenberg, 1973; Thomas, 1978; Aickin and Brading, 1982) or more recently 1 M Na-formate containing 10 mM KCl (Oberleithner *et al.*, 1982; Chao and Armstrong, 1987; Reuss *et al.*, 1987). In snail neurons, Thomas (1978) reported the following errors in E_m measurements with microelectrodes filled with the following alternatives to 3 M KCl: K-sulfate (15 MΩ), 1–2 mV; K-sulfate (35 MΩ), 5–8 mV; K-citrate (40 MΩ), 1–9 mV. More recently, Thomas and Cohen (1981) introduced as an alternative to electrolytic filling solutions, an organic filling solution to be used in silanized glass microelectrodes. Essentially it is a liquid membrane microelectrode that senses E_m and therefore it can be used as a reference microelectrode (Fig. 9). The underlying assumption here is that in most animal cells and in most experimental situations, the total activity of monovalent cations intra- and extracellularly would be constant and essentially equal. Thus, on impaling of the cell a monovalent cation-sensitive microelectrode that did not distinguish between Na$^+$ and K$^+$ (i.e., $K^{POT}_{Na^+K^+}$ $\sim$ 1) would record the same membrane potential as a conventional KCl microelectrode (see Fig. 9). This can be achieved by using a 2% w/v solution of potassium tetrakis (*p*-chlorophenyl) borate in *n*-octanol, as filling solution for silanized microelectrodes. The micropipettes are silanized with tri-*n*-butyl chlorosilane at 200°C using the Tsien and Rink (1980) method or just by exposing the back end of the micropipettes to vapors of dimethyldichlorosilane as previously described (Alvarez-Leefmans *et al.*, 1986). These reference microelectrodes are safe to use in steady-state measurements of a^i_{Cl} as has been shown in snail neurons (Thomas and Cohen, 1981), in smooth muscle (Aickin and Brading, 1982, 1984, 1985a), in the epithelium of rabbit colon (Wills, 1985), and in dorsal root ganglion neurons (see Fig. 8). However, in experiments in which intra-cellular dynamic ion states are studied, they give reliable measurements as long as the sum ($a^i_{Na} + a^i_K$) is kept constant. We have used them to study dynamic a^i_{Cl} changes on removal of external Cl$^-$ in dorsal root ganglion neurons (Nogueron and Alvarez-Leefmans, unpublished observations). The results are indistinguishable from those obtained using KCl- or K-acetate-filled reference microelectrodes (cf. Aickin and Brading, 1982). In sum, the advantages of using the Thomas and Cohen liquid membrane reference microelectrodes are the absence of liquid junction and tip potentials, and the absence of cytosolic contamination due to the reference electrolytes. The disadvantages are that their accuracy in reporting E_m depends on the constancy of the sum ($a^i_{Na} + a^i_K$), and $K^{POT}_{Na^+K^+}$ must be equal to *one* under all experimental conditions. In addition, some leakage of 1-octanol should be expected, and the microelectrodes have relatively high resistances (10^{10}–10^{11} Ω), which makes them unsuitable for recording rapid E_m transients or passing current.

    Another possible "exogenous" source of interference on the response of Cl$^-$-selective microelectrodes based on liquid anion exchangers is that produced by foreign extracellular anions used under a variety of experimental situations (see below). Among these foreign anions are some substitutes of external Cl$^-$ and anionic Cl$^-$ transport inhibitors such as 5-sulfamoyl benzoic acid derivatives (e.g., furosemide and bumetanide) and the stilbene disulfonic acid derivative, 4-acetamido-4′-isothio-cyanostilbene 2,2′-disulfonic acid (SITS). These anions might have access to the cytosolic compartment either by crossing the cell membrane or by leaking around the microelectrodes.

    To study changes in a^i_{Cl} when the external Cl$^-$ is reduced or removed requires the replacement of Cl$^-$ in the bathing solution with another anion that ideally should fulfill the following criteria: (1) it should be impermeant through the cell membrane, (2) it should not directly alter intracellular pH, and (3) it should not be sensed by the Cl$^-$-selective microelectrode. In a study performed in dorsal root ganglion cells (Alvarez-Leefmans, Gamiño, and Giraldez, unpublished observations), it was found that many

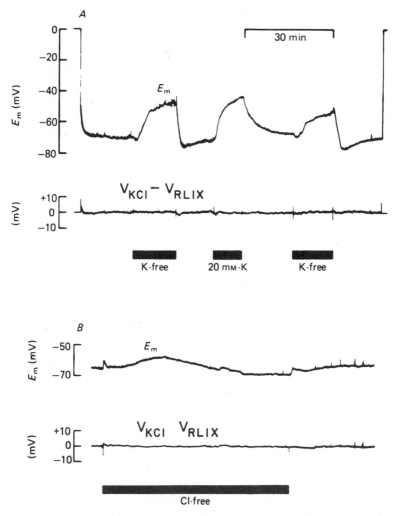

Figure 9. Pen recordings of two experiments in which the transmembrane potential (E_m) from smooth muscle cells of guinea pig vas deferens was measured by 3 M KCl and the organic reference liquid ion exchanger (RLIX) in a double-barreled microelectrode. E_m recorded by the KCl-filled barrel is shown in the top trace of each panel. The differential signal between KCl and RLIX barrels, $V_{KCl} - V_{RLIX}$, is shown in the lower trace of each panel. (A) An experiment showing impalement of a cell, the effect of removal and elevation of external K^+, and withdrawal of the double-barreled microelectrode. Changes in external $[K^+]$ were made with corresponding changes in external $[Na^+]$. (B) Part of an experiment showing the effect of complete removal and replacement of external Cl^-. Note that in A and B, the differential signal, $V_{KCl} - V_{RLIX}$ was unaltered through E_m changes produced by increasing and decreasing external $[K^+]$ and by removal and readdition of external Cl^-. This indicates that the RLIX microelectrode recorded the same transmembrane potential as the 3 M KCl-filled microelectrode under all experimental conditions depicted. (Modified from Aickin and Brading, 1982, and reproduced with permission.)

of the so-called "impermeant" anions, some of which are still used as Cl^- substitutes, actually were taken up by the cells. These anions were benzenesulfonate, methanesulfonate, and sulfate. In view of this, interference tests were performed on Cl^--selective microelectrode responses. The Cl^- sensor used in these microelectrodes was CORNING 477913. Among several Cl^- substitutes, gluconate produced the least interference on the Cl^--selective microelectrode (Fig. 10). The $K^{POT}_{Cl^-, \text{gluconate}}$ estimated from our data was ~ 0.01, which is close to the value of 0.018 reported by Greger *et al.* (1983) and that of 0.03 recently reported for the same Cl^- sensor by Ishibashi *et al.* (1988). The interference sequence derived from the data shown in Fig. 10 was benzenesulfonate $> Cl^- >$ methanesulfonate $>$ sulfate $>$ HEPES $\gg$ gluconate. This means that even if gluconate was able to cross the membrane, *which does not seem to*

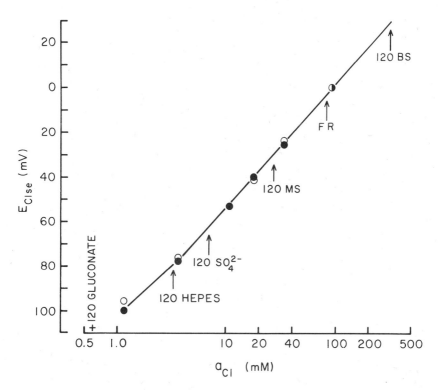

Figure 10. Interference tests of various anions on Cl^--selective microelectrode response. The Cl^--sensor was CORNING 477913 liquid anion exchanger. The bath reference electrode was a 1-mm polythene tube containing 3 M KCl gelled in agar. The solid line joining the circles is the response of the microelectrode to pure KCl calibration solutions having Cl^- activities as indicated on the abscissa. Open circles denote the response for ascending Cl^- activities and filled circles for descending Cl^- activities. The anions tested, excepting HEPES, are commonly used as Cl^- substitutes. Solutions having a constant concentration of 120 mM of each interfering anion were tested. The microelectrode potential (E_{ClSe}) measured in each anion test solution was referred to the calibration curve as indicated by arrows. 120 mM gluconate gave a reading equivalent to 0.5 mM Cl^- and benzenesulfonate (BS) produced the highest interference on electrode response. FR, standard "frog" Ringer; BS, benzenesulfonate; MS, methanesulfonate.

be the case (see MacKnight, 1985), it would only barely be sensed by the Cl^--selective microelectrode. On the other hand, in a systematic study on the effects of Cl^- substitutes on intracellular pH in crab muscle, Sharp and Thomas (1981) found that among many anion substitutes, gluconate, which has a pK_a' of 3.56, did not alter intracellular pH (Fig. 11). Glucuronate, which is also frequently used as a Cl^- substitute, had a small but variable effect on pH_i. In the same study, anions of weak acids ($pK_a' > 4.5$) such as acetate, butyrate and propionate produced large intracellular acidifications. Other anions of acids with an intermediate pK_a', such as lactate or formate, caused a relatively slow but substantial internal acidification (Fig. 11). Therefore, using the anions of weak acids to replace Cl^- should be avoided unless one also wishes to change pH_i.

More details on anion substitution experiments in excitable cells as well as the properties of several anions used to replace external Cl^- can be found in the literature (e.g., Woodbury and Miles, 1973; Kenyon and Gibbons, 1977; Saunders and Brown, 1977; Pollard *et al.*, 1977; Marranes and DeHemptinne, 1978; Bolton and Vaughan-Jones, 1977; Vaughan-Jones, 1979a,b). From the foregoing discussion, it can be concluded that the best choice for substituting external Cl^- is gluconate. However, when

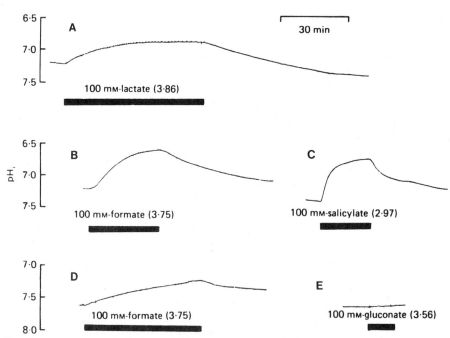

Figure 11. The effect on intracellular pH (pH_i) of replacing 20% (100 mM) of the external NaCl by the Na^+ salts of various anions of acids with an intermediate pK_a' range (2.97–4), as indicated by bars on each trace. pH_i was measured in crab muscle fibers using pH-selective miroelectrodes. The anions tested were lactate (A), formate (B, D), salicylate (C), and gluconate (E). The external pH was 7.5. Traces D and E were taken from the same experiment. (Modified from Sharp and Thomas, 1981, and reproduced with permission.)

using this anion it is important to remember that it complexes Ca^{2+} and Mg^{2+}. Therefore, the total concentration of these divalents in the bathing fluid has to be increased to compensate for complexation and have appropriate free divalent cation concentrations (Kenyon and Gibbons, 1977).

Experiments designed to identify the nature and properties of different Cl^- transport mechanisms often require using pharmacological inhibitors to isolate each transport process. Stilbene disulfonic acid derivatives such as SITS are used as inhibitors of the anion-exchanger transport system. Sulfamoyl benzoic acid derivatives such as furosemide and bumetanide are used as inhibitors of the Na^+,K^+,Cl^- cotransport mechanism. Because these inhibitors are anions in the physiological pH range and have relatively high lipid solubility, they might cross the cell membrane thereby interfering with the response to Cl^- of Cl^--selective microelectrodes containing either CORNING 477315 or 477913 liquid anion exchangers. For instance, Cl^--selective microelectrodes containing CORNING 477315 are $\sim$ 1000 times more sensitive toward SITS than they are toward Cl^- (Chao and Armstrong, 1987). Heinz and Grassl (1981) showed that 4,4′-diisothiocyanostilbene-2,2′-disulfonic acid (DIDS) also interferes with the measurement of a_{Cl} in solution by a similar electrode. In red blood cells, SITS and related compounds have been characterized as "nonpenetrating probes" (Grinstein et al., 1979). If this can be generalized to all cells, SITS interference on a^i_{Cl} readings might not be a problem if the drug is applied in the bath well after impalement, once the membrane resealing process around the microelectrode has taken place (see below).

Furosemide and bumetanide are also sensed by Cl^--selective microelectrodes based on CORNING anion exchangers. Heinz and Grassl (1981) found that the electrode was 16 times more sensitive to furosemide than to Cl^-. Later, Greger et al. (1983) determined a selectivity coefficient for Cl^- against furosemide, $K^{POT}_{Cl^-}$, of 305 for fine-tipped Cl^--selective microelectrodes. More recently, Chao and Armstrong (1987) found a $K^{POT}_{Cl^--Fu}$ of 157 for similar microelectrodes. In the same study, the selectivity coefficient for Cl^- against bumetanide, $K^{POT}_{Cl^--Bu}$, was 151. Fortunately, although Cl^--selective microelectrodes are more sensitive to bumetanide and furosemide than to Cl^-, the concentrations of these inhibitors used in transport studies are in general sufficiently low (up to 35×10^{-5} M for bumetanide and up to 0.5×10^{-3} M for furosemide) so that any interference with a^i_{Cl} measurements due to an intrinsic permeability of the cell membrane toward these compounds can be assumed to be negligible (Chao and Armstrong, 1987), although in Cl^--depleted cells interference can be detected (see Alvarez-Leefmans et al., 1988).

Leak of foreign anions (or cations) around the microelectrode due to incomplete sealing of the membrane with the microelectrode glass has been considered by several authors (reviewed by Ammann, 1986). Cytosolic contamination due to such leakage on microelectrode measurements of intracellular ion activities is prominent when using low-resistance microelectrodes, particularly when the electrochemical gradient for the interfering anion (or cation) is inwardly directed (e.g., Taylor and Thomas, 1984). However, under certain experimental conditions, outflow of ions from the cytosol might also alter the intracellular measurements. Cell membrane-to-glass sealing can be improved by silanizing the glass walls of the microelectrodes (Aickin and Brading,

1982; Munoz *et al.*, 1983), by increasing the concentration of external Ca^{2+} (e.g., Pelzer *et al.*, 1984), and obviously by minimizing the geometric parameters of the microelectrode such as tip diameter and the angle of taper. It is worth mentioning here that if the external $[Ca^{2+}]$ is increased aiming to improve membrane-to-glass sealing and bumetanide is used in the high-Ca^{2+} medium to block the Na^+,K^+,Cl^- cotransport, the inhibitor will bind Ca^{2+}, forming low-solubility Ca-bumetanide salts and therefore the effective concentration of bumetanide will be drastically reduced (Hoffmann *et al.*, 1986).

Table 1 summarizes the selectivity coefficients for Cl^- against several interfering anions, estimated for liquid anion-exchanger microelectrodes, using various methods.

5.2.2. Liquid Membrane Cl⁻-Selective Microelectrodes Based on Anion-Selective Carriers

We have seen that anion-selective liquid membrane microelectrodes based on ion exchangers such as quaternary ammonium salts all exhibit the same anion selectivity sequence with a preference for lipophilic and a rejection of hydrophilic anions. In this case, the free energy of transfer of the anions from the aqueous sample phase to the electrode membrane phase—and therefore the electrode membrane selectivity—is determined by the free energy of hydration of the anions. For these reasons, changes in selectivity sequences and improvement of selectivity for Cl^- over other anions is not expected to come from developing new liquid anion exchangers or from modifying the existing ones. This is in contrast to liquid membranes containing neutral or positively charged *carriers* for anions. These types of membranes show a wide variety of ion-selectivities that no longer correlate with the free energy of hydration of anions. In this case the potentiometric selectivity coefficient mainly reflects the anion-binding affinity of the ionophores. This raises the possibility of designing and developing carrier molecules with a given selectivity. The electrode response of this type of membrane is also described by Eq. (18) but the selectivity coefficient K_{xy}^{POT} is not defined by Eq. (19) but by the following expression (Morf *et al.*, 1985):

$$K_{xy}^{POT} \simeq \left[\frac{\Sigma_n \beta_y S_n \ C^n S + 1}{\Sigma_n \beta_x S_n \ C^n S + 1} \right] \cdot \frac{k_y}{k_x} \qquad (20)$$

where S represents the anion-selective carrier (ionophore). Note that the complex stability constants $\beta_y S_n$ and $\beta_x S_n$ and the concentration, CS, of uncomplexed carriers are the dominant selectivity-determining parameters, in addition to the pure ion-exchanger equilibrium constant K_y/K_x already found in Eq. (19).

So far there are two types of carrier molecules developed for anion-selective electrodes: neutral anion-carriers (Wuthier *et al.*, 1984) and charged anion-carriers (Schulthess *et al.*, 1984, 1985; Ammann *et al.*, 1986). Wuthier *et al.* (1984) found that trioctyltin chloride behaves as a neutral carrier for anions. Liquid membranes containing this carrier, a plasticizer, and polyvinyl chloride as membrane matrix show selectivity patterns that differ from those shown by anion-exchanger membranes (Fig. 12). Unfortunately, these neutral anion-carrier membranes are more sensitive to HCO_3^-

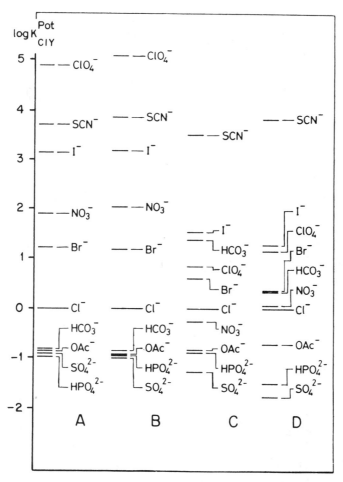

Figure 12. Logarithm of selectivity coefficients, log $K_{Cl\,y}^{POT}$, for ion-selective macroelectrodes based on a classical liquid anion exchanger (columns A and B) and on a tin organic compound (columns C and D) as determined by the separate solution method. The interfering anions (y) are indicated on each column. The composition of each membrane was as follows: (A) 6 wt% methyltri-*n*-dodecylammonium chloride (MTD-DACl), 65 wt% di-*n*-butyl phthalate (DBP), 29 wt% poly(vinyl chloride) (PVC); (B) 3 wt% MTDDACl, 49 wt% (*R,R*)-2,3-dimethoxysuccinic acid bis(1-butylpentyl) ester (DMSNE), 48 wt% PVC; (C) 3 wt% tri-*n*-octyltin chloride (TOTCl), 49 wt% DMSNE, 48 wt% PVC; (D) 20 wt% TOTCl, 40 wt% DMSNE, 40 wt% PVC. Membranes were mounted in Phillips electrode bodies IS 560 (N.V. Phillips, Gloeilampenfabrieken, Eindhoven, The Netherlands) and conditioned overnight in about 2 ml of a 0.01 M NaCl solution, which corresponded to the internal filling solution of these electrodes. (Modified from Wuthier *et al.*, 1984, and reproduced with permission.)

than to Cl^- and their selectivity toward other anions of biological importance (e.g., acetate) was not improved. Therefore, although they have the merit of being among the first anion carrier membranes to be designed, they are not suitable for intracellular measurements of Cl^-.

Schulthess *et al.* (1984) found that certain lipophilic derivatives of vitamin B_{12}

behave as positively charged carriers for anions. For instance, incorporation of the lipophilic cobalt (III)-cobyrinate octadecyl-cobester and its ionic aqua-cyano perchlorate derivative gives wonderful electrodes for NO_2^- and SCN^-, but their selectivity coefficients are still inappropriate for measuring a_{Cl}^i. More recently, Ammann et al. (1986) synthesized and characterized a set of metalloporphyrins that also act as positively charged carriers for anions, inducing anion selectivities in membranes, which clearly deviate from the sequence followed by classical anion exchangers (i.e., the Hofmeister series). It is thought that the relevant contribution of the metalloporphyrins to the anion selectivity of the membrane electrode is the tendency of the metal center to coordinate the anion. These compounds are lipophilic cobalt (III) and manganese (III) complexes of 5,10,15,20-tetrakis[4-hexyloxycarbonyl)phenyl] porphyrin. The most interesting of these compounds for intracellular Cl^- activity measurements is chloro {5,10,15,20-tetrakis[4-(hexyloxycarbonyl)phenyl] porphyrinato} manganese (III) whose chemical structure and properties in microelectrodes are shown in Fig. 13. Comparison of this compound, which will be referred to as Mn(III), with CORNING anion exchanger 477913 shows an improved selectivity for Cl^- over HCO_3^-, acetate, F^-, So_4^{2-}, and HPO_4^{2-} (Fig. 13). A Cl^- sensor containing a tetraphenyl porphin manganese (III) chloride derivative has been successfully used in double-barreled microelectrodes to measure a_{Cl}^i in mammalian proximal tubular cells (Kondo et al., 1989). The intercellular anion interference recorded with these microelectrodes when external Cl^- is replaced with gluconate, is significantly lower than that recorded with LIX microelectrodes. The selectivity properties of microelectrodes based on Mn (III) suggest that they are superior for a_{Cl}^i measurements than LIX microelectrodes.

6. MEASURING NET FLUXES WITH Cl⁻-SELECTIVE MICROELECTRODES

As with other methods for measuring ionic concentrations, it is possible, in principle, to calculate net transmembrane fluxes from ISM measurements of the change in the intracellular concentration of a given ion. ISMs were first used to measure changes in intracellular ion concentration by Russell and Brown (1972a,b) working with *Aplysia* neurons and by Thomas (1972) working with *Helix* neurons. Defining the net absolute flux across the cell membrane as the rate of change with time, in the *amount m* of the ion j in the cytoplasm, dm_j^i/dt (mole/sec), and knowing that the relation between the amount of j in the cell and its concentration $[j]_i$ (mole/cm^3) can be obtained from the cell volume v (cm^3):

$$m_j^i = [j]_i \, v \qquad (21)$$

we can get an equation for the net flux of ion j by simply differentiating Eq. (21):

$$dm_j^i/dt = d[j]_i/dt \cdot v + (dv/dt) \, [j]_i \qquad (22)$$

If we now want to express fluxes in a normalized way, i.e., per unit of membrane surface area, and to have in the equation our direct measurement of activity rather than

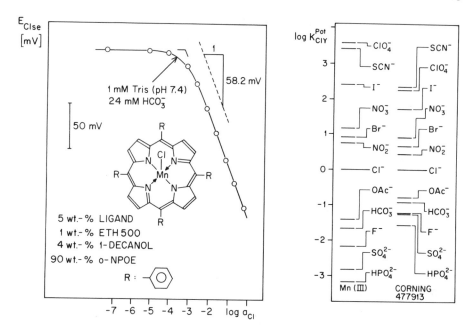

Figure 13. (Left) Chemical composition and properties in microelectrodes of an anion-selective membrane based on the positively charged carrier manganese III. The membrane composition was as follows: 5 wt% of the ligand chloro {5,10,15,20-tetrakis [4-(hexyloxycarbonyl)phenyl] porphirinato} manganese (III), whose structure is illustrated; 1 wt% tetradodecylammonium tetrakis (*p*-chlorophenyl) borate (ETH 500); 4 wt% 1-decanol; and 90 wt% *o*-nitrophenyl-*n*-octyl ether (*o*-NPOE). The microelectrode response (E_{Clse}) was determined in aqueous solutions containing 1 mM Tris buffer (pH 7.4), 24 mM HCO_3^-, and varying activities of CL^- as indicated on the abscissa. (Right) Logarithm of selectivity coefficients, log $K_{Cl^--y}^{POT}$, for microelectrodes based on Mn(III) compared with those containing the liquid anion exchanger CORNING 477913. Separate solution method ($I = 0.1$ M). (Figure and data kindly provided by Professor W. Simon.)

concentration, we can simply divide Eq. (22) by the surface area of the membrane and use the relation $a_j^i = \gamma_j^i [j]_i$ to obtain

$$J_j = 1/\gamma_j^i [(da_j^i/dt) \, h + (dh/dt) \, a_j^i] \qquad (23)$$

where J_j is the net flux of ion j per unit of membrane surface area, h is the volume-to-surface ratio of the cell (cm), and γ_j^i is the activity coefficient of ion j in the intracellular solution. There are several points worth noting now about the assumptions made to derive this expression and also about its use in experiments with ISMs.

Equation (22) assumes that changes in m_j^i are solely due to transmembrane fluxes of j and that no buffering, sequestration, or consumption/production of j occurs. What we are actually measuring with ISMs is the appearance and disappearance of j in the cytosol; therefore, any exchange between intracellular compartments or chemical reactions in which ion j participates can look just like net flux across the plasma membrane. The questions arise: which is the state of Cl^- with respect to these processes and how can we estimate the weight that they have in our calculations of fluxes? It is hard to exclude the occurrence of such processes in a cell, although some of them are less worrisome than others. Chemical buffering of Cl^-, for instance, does not seem to be very strong in animal cells. This can be judged from the comparison of measurements of Cl^- made on the same preparation using chemical and potentiometric methods, as discussed in Section 2.1.4. The emerging view is that binding or complexation of Cl^- within the cytosol is negligible. Another potential problem could arise if physically restricted intracellular compartments contain Cl^- at levels different from that in the cytosol. There is currently no clear example of this and therefore, as a first approximation, it seems safe to consider nerve cells as a single intracellular compartment, as has been done in the case of vertebrate peripheral ganglion cells (Alvarez-Leefmans *et al.*, 1988). Finally, the question of the production and/or consumption of Cl^- seems to be clearly of less importance since we know of no biochemical reactions in which Cl^- participates as a reactant.

Another point related to the above discussion is that, strictly speaking, what is relevant to the calculation of fluxes is the change in *concentration,* not the change in activity [Eq. (22)]. Since ISMs measure ionic activity, this might, at first glance, appear to be a problem. However, if one makes the not-unreasonable assumption that the intracellular activity coefficient γ_j^i is constant under most experimental conditions, meaningful flux data can still be obtained using Eq. (23).

We now face the problem of a possible change of cell volume, which appears in Eq. (23) as h and dh/dt. These terms simply reflect the fact that to know the net amount of j that has crossed the membrane, we need to know the cell volume. However, the equation also explicitly takes into account the possibility that the cell volume can change, i.e., there can be cell shrinkage or swelling. A value for h can be obtained by microscopic measurements of cell dimensions and by using assumptions or actual knowledge about cell geometry one can calculate cell volume and surface area. What is currently done is to model the cells as known geometric figures, spheres in the case of ganglion neurons, equivalent cylinders in the case of complex geometry nerve cells, cylinders for muscle fibers or axons (Jack *et al.*, 1975; Rall, 1977), and from the measured values of the radius the volume-to-surface ratio is worked out (e.g., Thomas, 1972; Vaughan-Jones, 1979a). The value of h for a sphere is given by $r/3$ and for a cylinder by $r/2$ and typical values for a squid axon and a dorsal root ganglion cell are 10^{-3} and 12.5×10^{-3} cm, respectively.

A more critical problem, however, is to evaluate how cell volume changes affect our measurements throughout the experiment. As changes in cell volume depend on the fraction of the total salt content of the cell that is lost or gained by the cell, if the cell is large and fluxes are small, then the *rate* of change in cell volume is small and, as a first approximation, one can ignore the right-hand side of Eqs. (22) and (23), and assume zero volume change. This can be safely done, for instance, in cardiac Purkinje muscle fibers where cells are large and changes in $[Cl^-]_i$ are slow (Vaughan-Jones, 1979a). However, for small cells or cells whose $[Cl^-]_i$ changes rapidly, one ideally needs a measure of dh/dt. If this is not possible, one at least must have an idea of maximal *limits* for the possible error introduced by changes in volume and to assess the *direction* of these changes (Alvarez-Leefmans *et al.*, 1988). A simple procedure to calculate dh/dt from measurements of ionic activities is the following (Giraldez, 1984). At any time t, for electroneutrality to hold, it is required that

$$\sum_{j}^{n} z_j F \, (dm_j^i/dt) = 0 \tag{24}$$

which simply states that no net charge is gained or lost by the cell. Combining Eqs. (22) and (24), we get

$$\left[v \sum_{j}^{n} z_j \, (d[j]_i/dt) + (dv/dt) \sum_{j}^{n} z_j \, [j]_i \right] = 0 \tag{25}$$

and rearranging:

$$(1/v) \, (dv/dt) = - \sum_{j}^{n} z_j (d[j]_i/dt) \, / \, \sum_{j}^{n} z_j [j]_i \tag{26}$$

which is a useful expression relating the fractional change in cell volume to the rate of change in cell concentrations and to the initial concentrations of the ions present in the cytosol. As a first approximation, it can be assumed that the n ions are Na^+, K^+, and Cl^-, all of which can be measured with ISMs. Even if HCO_3^- is involved, one can measure pH_i. The only problem now is doing the experiments. . . ! That is, to repeat the experiment we did with Cl^- but now measuring these other ions. If we succeed, we will have the required values to solve Eq. (26).

If one is lucky enough to obtain a Cl^- transient in the absence of changes in E_m, then Eq. (26) can be further simplified because the fluxes of Cl^- have to occur either in exchange for another permeant anion or they have to be accompanied by permeant cations or a combination of both. Therefore, the maximum possible change in cell volume will take place when all the Cl^- ions are accompanied by intracellular cations, say K^+ or Na^+. In this situation, for isotonicity to hold, the sum of $[K^+]_i + [Na^+]_i$ should remain constant and, therefore, $d[K^+]_i/dt + d[Na^+]_i/dt = 0$. In this case, Eq. (26) is reduced to

$$(1/v) \, (dv/dt) = (d[Cl^-]_i/dt) \, / \, [Na^+]_i + [K^+]_i - [Cl^-]_i \tag{27}$$

Here, taking average values for $[Na^+]_i$ and $[K^+]_i$ from the literature or even simpler, making $[K^+]_i + [Na^+]_i = [K^+]_o + [Na^+]_o$, we can directly estimate the maximal fractional change in cell volume from the experiment with Cl^- microelectrodes. Plugging this value into the flux equation we can determine the maximum possible effects of volume changes on our flux calculations (e.g., Alvarez-Leefmans *et al.*, 1988).

A very interesting way to tackle the problem of volume changes is the direct approach followed by Reuss (1985; see also Cotton *et al.*, 1989) and Serve *et al.* (1988). These authors have used a micropotentiometric method for measuring volume changes by incorporating an impermeant ion into the cell and monitoring the changes in its intracellular concentration. This method opens the very interesting possibility of studying cell-volume-regulation and maintenance with microelectrodes (see Section 2.1.8).

Describing the Transients

The study of Cl^- inside nerve, muscle, and glial cells has reached a state of maturity in which simply measuring steady-state a_{Cl}^i is of diminishing interest. We are now increasingly interested in how a_{Cl}^i is regulated or maintained and the mechanisms by which Cl^- crosses the membrane. To obtain this kind of information, we must perturb a_{Cl}^i and analyze the transients in a_{Cl}^i that result. This approach has been used with considerable success by Boron (1985) to analyze pH regulation in squid axon, by Vaughan-Jones (1979a,b) to study intracellular Cl^- regulation in cardiac muscle, and by Alvarez-Leefmans *et al.* (1988) to study the mechanism of Cl^- transport and accumulation in dorsal root ganglion neurons. This approach has several advantages (Zeuthen, 1985) over steady-state experiments in which secondary effects might take place and in which comparison of different situations that are not necessarily comparable is usually made. For example, say that one thinks that the mechanism that accumulates Cl^- in a cell is a Na^+-coupled transport system. One might then attempt to provide evidence for such a mechanism by measuring the steady-state value of a_{Cl}^i before and after the replacement of external Na^+ by an impermeant cation. The experimental results may show that intracellular Cl^- is indeed lower in the Na^+-free medium and one might happily, though naively, conclude that the entry of Cl^- is tightly coupled to Na^+. However, this is not necessarily true because a similar result could be obtained if the entry of Cl^- was coupled to HCO_3^- and, in parallel, a Na^+/H^+ exchanger was operating. In this case, the reduction of external Na^+ would produce a cell acidification, which, in turn, would reduce intracellular HCO_3^- and slow the operation of the anion exchanger with the consequent fall in internal Cl^-. In this latter case we have generated a new steady state in which intracellular pH is different and the link between Na^+ and Cl^- fluxes is a change in the *composition* of the cell and not a directly coupled membrane transport process. Therefore, even if we were able to measure net Cl^- fluxes before and after 0 Na^+ we would not be directly addressing the problem, because in the new steady state in 0 Na^+ the a_{Cl}^i will also be reduced regardless of the mechanism involved. The power of transient analysis experiments, on the contrary, resides in the fact that at $t = 0$, when it is possible to calculate initial rates of change in concentration or the instantaneous change in potential, everything is as it was in the steady state, so these transients help to characterize the original steady state. For instance, we can envision a parallel array of transport systems that in the steady-state keep the intracellular activity of Cl^- constant by the balance of all fluxes:

$$d[Cl^-]_i/dt = 0 = (1/h)\,(J_{Cl}^1 + J_{Cl}^2 + \ldots + J_{Cl}^n) \qquad (28)$$

Imagine that at $t = 0$ we are able to specifically block one transport system, say J_{Cl}^1, and look at $d[Cl^-]_i/dt$. If we are able to measure $d[Cl^-]_i/dt$ in an interval that is short enough so as to be sure that nothing else has changed in the cell, we can then take the remaining J_{Cl}^n as constants and hence $d[Cl^-]_i/dt$ becomes equal to J_{Cl}^1/h. Thus, we can look at a single transport system before the others have had time to realize that something has changed!

The critical problem now is to have good enough time resolution and a good empirical mathematical description of the transient so as to allow the measurement of the desired changes at $t = 0$. The following example, dealing with net Cl^- efflux from a cell, will be of some help. Say we perform a $0\ Cl^-$ experiment, substituting extracellular Cl^- with an impermeant anion while monitoring the fall in the intracellular Cl^- activity. We now want to calculate the initial rate of change in a^i_{Cl}, $(da^i_{Cl}/dt)_0$, which is proportional to the initial efflux of Cl^- and which we can use in the equations discussed above. The first possibility is to work it out directly from the trace of the recorder. Starting from the Nicolsky–Eisenman formalism [Eq. (11)] and differentiating it with respect to time, we get

$$(dE_{Clse}/dt) = d(S \log a^i_{Cl})/dt = (S/2.303\ a^i_{Cl})\ (da^i_{Cl}/dt) \tag{29}$$

where the selectivity coefficient $K^{POT}_{Cl^- y}$ for the interfering anion y has been taken as negligible.

Rearranging Eq. (29) and taking concentrations instead of activities:

$$d[Cl^-]_i/dt = (2.303/S)(dE_{Clse}/dt)\ [Cl^-]_i \tag{30}$$

that gives an easy way—"on-line"—of getting a measure of the initial $d[Cl^-]_i/dt$ by using a ruler and taking $dE_{Clse}/dt \simeq \Delta E_{Clse}/\Delta t$.

The other conventional way of obtaining initial rates is by plotting the instantaneous value of $[Cl^-]_i$ against time, and fitting a continuous function through these data (see Vaughan-Jones, 1979a,b, for examples). In general, intracellular space filling and emptying can be reasonably described by a single exponential of the form

$$([j]_i)_t = [([j]_i)_0 - ([j]_i)_\infty]\ e^{-kt} + ([j]_i)_\infty \tag{31}$$

which for the specific case of Cl^- can be written as follows:

$$([Cl^-]_i)_t = [([Cl^-]_i)_0 - ([Cl^-]_i)_\infty]\ e^{-kt} + ([Cl^-]_i)_\infty \tag{31.1}$$

where the subscripts 0 and ∞ refer to the initial and final steady-state values of $[Cl^-]_i$ and k is the rate constant ($k = 1/\tau$) that characterizes the process.

The foregoing merely describes what is seen and does not imply any specific mechanism of ion permeation. It is true, however, that if the net uptake or loss of Cl^- (or other ion considered) is dominated by a single mechanism that happens to follow first-order kinetics, then the transient *has to be* a single exponential. As an exercise, let us treat the case in which the filling of the intracellular space is due to electrodiffusion. Ignoring cell volume changes, we can write

$$(d[j]_i/dt)h = P_j V\ [([j]_o - [j]_i\ e^V)/(1 - e^V)] \tag{32}$$

which is simply the Goldman–Hodgkin–Katz equation where $V = (z_j F/RT)E_m$, subscripts o and i denote extracellular and intracellular, respectively, and P_j is the permeability coefficient of ion j. If E_m does not change at all or if it changes instantaneously and then remains more or less constant ($dE_m/dt \ll da_j^i/dt$), then Eq. (32) is of the general form

$$(d[j]_i/dt)h = \alpha[j]_i + \beta \tag{33}$$

where α and β are constants. Integrating Eq. (33) between $t = 0$ and $t = \infty$, for $([j]_i)_0$ at $t = 0$ and $([j]_i)_\infty$ at $t = \infty$, we obtain Eq. (31) where

$$k = (P_j V e^V)/[h(1 - e^V)] \tag{34}$$

and

$$[j]_i = ([j]_i)_0 e^{-V} \tag{35}$$

The same would hold if the dominating process were one following Michaelis-type kinetics with a relatively low affinity. In this case when $[j]_i < K_m$, the Michaelis–Menten equation is reduced to

$$J_j = (J^{max}/K_m) [j]_i \tag{36}$$

and we have a situation that is similar to the one just described for electrodiffusion. One can, in theory, separate as many rate constants as there are processes involved. However, this is not advisable since the combination of different parallel transport systems with complex kinetics *plus* the existence of volume changes can shape the transient in an unpredictable way. The safest option is to simply try to fit the transient data by a single exponential. If the fit is successful, then the resulting rate constant can be used to calculate the initial rate of change in concentration, $(d[j]_i/dt)_0$. It is worth remembering that we are using the exponential function as a mere description of the transient without implying a particular mechanism of ion transport.

From the exponential equation (31) the initial rate is obtained simply by differentiation and

$$(d[j]_i/dt)_0 = k [([j]_i)_0 - ([j]_i)_\infty] \tag{37}$$

What we do next is to use the whole transient of $[j]_i$ to get k and then use this value to calculate the initial rate.

Up to this point we have emphasized the problems involved in a quantitative approach to the experiments with Cl^--selective microelectrodes. It is worth remembering, however, that qualitative experiments are sometimes more interesting and do not require anything else except careful observation of the trace on the pen-recorder. For instance, to block the rise in Cl^- with loop diuretics or SITS may have a straightforward interpretation and certainly does not require any mathematical formalism. The problem arises when one wants to have numbers for fluxes and permeabilities. The efflux of Cl^- in dorsal root ganglion neurons exposed to Cl^--free solutions gives an example of how we can use a semiquantitative approach to get new information

(Alvarez-Leefmans *et al.*, 1988; see also Chapter 4). Looking at the decrease in a_{Cl}^i in Cl^--free solutions, the question arises as to whether the observed decrease in a_{Cl}^i occurs solely by electrodiffusion after stopping an inwardly directed Cl^- transport mechanism, or whether it implies the reverse operation of the transport mechanism. One possible answer is to test whether the observed fluxes can be accounted for by electrodiffusion. We can calculate "P_{Cl}" using the value of $d[Cl^-]_i/dt$ from the transient to work out the net efflux of Cl^-, and then the Goldman–Hodgkin–Katz equation. Using the "P_{Cl}" derived from the Goldman–Hodgkin–Katz equation, we can obtain a "g_{Cl}" by the following relation:

$$g_{Cl} = P_{Cl} \, (zF)^2/RT \, [Cl^-]_* \tag{38}$$

where $[Cl^-]_* = ([Cl^-]_o + [Cl^-]_i)/2$.

On the other hand, we can get a figure for the conductance of the cell membrane, g_m, from current clamp experiments by simply producing a current–voltage curve, and knowing the cell geometry, which, for instance, in the case of dorsal root ganglion cells can be approximated to a sphere. If we now compare "g_{Cl}" with the total membrane conductance g_m, in dorsal root ganglion neurons, "g_{Cl}" is larger than "g_m," which is nonsensical. The conclusion is that the high rate of net Cl^- efflux cannot be accounted for by electrodiffusion alone. This is a *reductio ad absurdum* argument that can be useful in situations like this in which we have a clear answer (see Chapter 4, Section 5.1.2a).

7. RADIOTRACER TECHNIQUES FOR STUDYING Cl^- FLUXES

7.1. ^{36}Cl

^{36}Cl is by far the most common radioisotope used to study Cl^- transport in nerve, muscle, and glial cells. Although a few studies have used ^{82}Br, we will limit our discussion to ^{36}Cl. ^{36}Cl decays by emitting a β (electron) particle to become the inert gas, argon. The energy of the emitted particle is relatively high (compared to ^{14}C and ^3H), 0.709 MeV, making it easily counted by liquid scintillation spectroscopy. It has a radioactive half-life of 3.07×10^5 years. This very long half-life means that the specific activity of ^{36}Cl (amount of radioactivity per mole of Cl^-) cannot be very high. Theoretically, the highest attainable specific activity is 1.17 Ci/mole; in practice one gets about 0.3 Ci/mole. This low specific activity represents a limitation in the use of the isotope from standpoints of cost and signal detection. Thus, in order to use ^{36}Cl to monitor Cl^- fluxes, one must have a preparation in which the flux is high and/or which has a large amount of membrane across which the flux is occurring.

7.2. Uses of ^{36}Cl

7.2.1. Estimate of Cellular Cl^- Content

One can measure the total, exchangeable Cl^- content by soaking a preparation in ^{36}Cl-containing medium until no further uptake of the radioisotope occurs. At that time all the Cl^- inside the cell will have equilibrated with the ^{36}Cl of the external fluid and will have the same specific activity [e.g., counts per minute (cpm)/mole]. One must subtract the cpm resulting from extracellular fluid contamination in order to obtain the amount of ^{36}Cl within the intracellular compartment. Estimates of (1) total tissue water and (2) extracellular water are made (i) by subtracting the wet weight of the tissue from the completely dry weight and (ii) by using a ^{14}C- or ^{3}H-labeled extracellular space marker such as inulin or mannitol. (It is of interest that in the older literature, the "^{36}Cl space" was often used as the extracellular fluid space. Since Cl^- is permeant and present inside all cells, this practice can no longer be condoned.) Thus, one can determine the moles of Cl^- in the intracellular fluid compartment and independently, the volume of that compartment. Together, this gives the exchangeable Cl^- concentration. Since this method depends upon several independent measurements, its precision and accuracy are rather poor. With the advent of Cl^--selective micro-electrodes, this use of ^{36}Cl has become rare.

7.2.2. Measurement of Fluxes

The most powerful use of ^{36}Cl is to measure Cl^- flux. The basic assumption is simple: The transport mechanism being studied cannot distinguish between ^{36}Cl and nonradioactive ^{35}Cl and hence transports ^{36}Cl just as fast as it does ^{35}Cl. At this time, there is no reason to suspect that this assumption is invalid.

7.2.2a. What Is Meant by "Flux"? A few years ago, the noted transport physiologist H. N. Christensen complained about the incorrect usage of the terms "influx" and "efflux" (Christensen, 1982). He correctly pointed out that fluxes are rates of *one-way* movements of substances and not *net* movements. This rather narrow definition is used in enzyme kinetics and the usefulness of enzyme kinetics to describe transport kinetics is well known. Thus, a flux is an amount of substance crossing a known area of membrane per unit time. In the case of most solutes (including Cl^-), flux is conveniently expressed in the following units: pmoles/cm^2 per sec, cm = centimeters of membrane surface area across which flux is occurring. Workers in our field often convert ionic current to flux or vice versa by making use of the Faraday constant (96,500 coulombs/mole) and the definition of an ampere (1 coulomb/sec). It is important to recognize that measurements of current are measurements of *net* movements and flux measurements (when performed correctly; see below) are measurements of unidirectional movements. The reason to emphasize this point is that sometimes workers do not clearly distinguish between the two kinds of measurements and such a failure can result in confusion. For instance, in comparing a current and a flux that one suspects of being mediated by the same mechanism, one might find that the flux exceeds the current by a large amount. This is because the current would be the algebraic sum of both unidirectional fluxes (influx and efflux). However, if the current exceeds the flux of the same direction, something is seriously wrong with one or more of the investigator's assumptions.

7.2.2b. Practical Problems of Making Flux Measurements. (1) How can we be certain we are measuring only a one-way movement? For example, if we put

^{36}Cl outside an intact (or nondialyzed) cell and measure the rate of ^{36}Cl influx, the only time ^{36}Cl back flux (efflux) is not occurring is right at the beginning, before any appreciable ^{36}Cl has entered the cell. Thus, unless we can immediately remove the ^{36}Cl as it enters (e.g., by intracellular dialysis), we must carefully measure the rate of entry (influx) during the very earliest portion of the uptake procedure. This is known as the "initial rate" measurement. Just how quickly we can make this measurement will be affected by a variety of practical matters (e.g., rapidity of sampling, amount of radioactivity that can be collected, the specific activity of the sample).

(2) Exchange Fluxes. Many transport mechanisms have modes that engage in self-exchange. This is a potential problem for the study of any transport mechanism that is reversible. Thus, all so-called secondary active transport processes may have self-exchange characteristics in which extracellular ^{36}Cl might simply exchange for intracellular ^{35}Cl. Although such self-exchange may be an inherent property of the transporter, it is almost never the property of interest from a functional point of view. Unfortunately, from a practical standpoint, such exchange fluxes can introduce errors into estimates of rate and of stoichiometry of coupled transport process. For example, estimates of Na$^+$,K$^+$,Cl$^-$ cotransporter stoichiometry could be seriously in error if a large portion of any one of the influxes measured were via a self-exchange mode (e.g., Cl$^-$/Cl$^-$, Na$^+$/Na$^+$, or K$^+$/K$^+$; Lauf *et al.,* 1987). Thus, if possible, the investigator must arrange conditions where such a possibility does not exist. That is, deplete the solution on the *trans* (or opposite) side of the membrane of the possible self-exchange ion (see Russell and Boron, this volume). A certain sign of such self-exchange fluxes is the phenomenon of *trans*-side acceleration of an isotopic flux. This is defined as the acceleration of *cis*-to-*trans* unidirectional movement of radiolabeled solute caused by the presence of the same, but nonradiolabeled solute at the *trans* face (Stein, 1986). Although this phenomenon is useful for distinguishing "carrier" transport mechanisms from channel mechanisms, it obviously can be a problem for correctly characterizing functional stoichiometry or even for determining whether the transporter is operating in a functional mode.

7.2.3. The Ability to Measure Unidirectional Fluxes Is the Most Powerful Theoretical Advantage of the Technique

Ease of performance is obviously an important practical advantage for using radiotracer methodology. But, from a standpoint of information content of the data acquired by the method, the importance resides in the fact that it is the only technique that can directly give unidirectional fluxes. By making careful initial rate measurements (or, if one's preparation permits it, by continuously internally dialyzing the cell) one can learn much about the kinetics of one-way movement of Cl$^-$ by a given transporter. This can be a very powerful advantage, particularly when there are multiple pathways capable of effecting net transport, which could obscure results obtained from techniques relying upon net flux measurements. Additionally, one can make measurements of Cl$^-$ fluxes under steady-state conditions, thus bypassing possible complications resulting from cell volume changes, changes of cell pH, and changes of ionic transmembrane gradients that might occur as a net measurement is being made.

8. *Cl⁻-SENSITIVE FLUORESCENT INDICATOR*

The fields of study of Ca^{2+} and H^+ transport have been stimulated enormously by the development and characterization of ion-sensitive fluorescent dyes. Their rapid response time, ease of use, and noninvasiveness have made possible the current explosion of information about these two ions.

Recently, Verkman and his colleagues introduced a Cl^--sensitive fluorescent compound, SPQ (6-methoxy-*N*-(3-sulfonatopropyl) quinolinium (Illsley and Verkman, 1987; Chen *et al.*, 1988; Chen and Verkman, 1988; Verkman *et al.*, 1988, 1989a). On the plus side, this agent possesses several very attractive properties in addition to those generally associated with ion-sensitive fluorescent dyes. It is apparently nontoxic (Krapf *et al.*, 1988; Chao *et al.*, 1989a,b). It does not act as a Cl^- ionophore or buffer (Chen *et al.*, 1988; Chen and Verkman, 1988; Krapf *et al.*, 1988). It is not sensitive to HCO_3^-, SO_4^{2-}, NO_3^-, HPO_4^{2-}, and physiological cations (1–100 mM) including H^+ (pH 5–9) (Illsley and Verkman, 1987; Verkman *et al.*, 1989a). It has a high water/lipid solubility ratio, and thus, is not partitioned into the membrane, but resides within the cytoplasmic compartment and it does not bind to intracellular membranes or proteins. SPQ has been used to quantitate Cl^- transport in membrane vesicles (Chen *et al.*, 1988; Chen and Verkman, 1988; Pearce and Verkman, 1989), in liposomes reconstituted with Cl^- transporters (Verkman *et al.*, 1989b), in the intact kidney proximal tubule (Krapf *et al.*, 1988), and in a variety of cultured cells (Chao *et al.*, 1989a,b). There are several limitations in the use of SPQ for the study of Cl^- transport by cells. First, the fluorophore leaks rapidly out of cells and the absolute signal depends on the dye concentration. The leakage rate appears to vary with cell type and SPQ-loading procedure (Krapf *et al.*, 1988; Chao *et al.*, 1989a,b). In rabbit proximal tubule cells the half-time of SPQ leakage at 38°C was 8.6 min. In cultured cells (fibroblasts, LC–PKl cells, and tracheal epithelia) reported leakage rates are $< 10\%$ in 60 min at 37°C. Another major limitation is the inability to make fluorescene ratio measurements. The lack of a second reference wavelength in the SPQ spectrum makes it difficult to determine accurately cell Cl^- activity if significant SPQ leakage occurs. Obviously this is a major drawback in relatively long-lasting experiments. Because the peak excitation of this dye occurs at about 350 nm, autofluorescence of cells becomes quantitatively important. In addition, such a peak excitation wavelength causes interference with Cl^- transport inhibitors such as disulfonic acid stilbenes. Fortunately, the dihydroderivatives (e.g., H_2-DIDS) do not suffer from this drawback. Finally, there is an as yet unexplained but significant fall in the sensitivity of the dye to changes of $[Cl^-]_i$ which occurs when the dye is inside cells. Thus, it is reported that the quenching constant (Stern–Volmer constant) falls from 118 M^{-1} in a cuvette to 12 M^{-1} inside rabbit proximal tubule cells (Krapf *et al.*, 1988) and 13 M^{-1} inside cultured fibroblasts, LLC-PKl cells (an epithelial cell line), and canine tracheal epithelial cells (Chao *et al.*, 1989a,b). This has been attributed to SPQ quenching by intracellular anions. The nature of these intracellular interfering anions remains to be elucidated. This means that intracellular sensitivity of this Cl^- indicator must be evaluated in every cell type. In other words, intracellular calibration of SPQ fluorescence is mandatory for each cell type. Detailed procedures for intracellular SPQ calibration have been described (Krapf *et al.*, 1988; Chao *et al.*, 1989a,b). Calibration of SPQ fluorescence against $[Cl^-]_i$ involves the use of a double ionophore technique. $[Cl^-]_i$ is set equal to $[Cl^-]_o$ using a solution which has a high $[K^+]$, the K^+/H^+ exchange ionophore nigericin, and the Cl^-/OH^- exchange ionophore tributyltin. The high K^+ solution in the presence of nigericin supposedly clamps pH_i at the external pH value, and depolarizes the cell. Tributyltin addition in pH-clamped cells is thought to result in equal intracellular and extracellular Cl^- activities (or concentrations). It is important to point out that the SPQ leakage rate significantly increases in the presence of ionophore. SPQ seems well suited for work on sealed membrane vesicles where the relatively short effective life of the agent is not a practical problem. In addition, although there is some loss of Cl^- sensitivity by the dye inside vesicles, it seems to be accounted for by the effect of the gluconate being used as the other anion (Chen *et al.*, 1988). In vesicles, Verkman and his colleagues have clearly demonstrated that their method can easily detect changes of $[Cl^-]$ as small as 2 mM (Chen *et al.*, 1988). The time resolution of the technique (in the msec range) is very suitable for net flux measurements in vesicles. Finally, the fact that an entire time course of flux can be obtained using as little as 10 μg of vesicle protein makes this fluorescent dye approach very attractive for use to study transport processes in biological membrane vesicles.

Improved Cl^- sensitive fluorescent indicators have been recently synthesized and characterized for the purpose of increasing Cl^- sensitivity, adding ester function for cell trapping, and red-shifting the fluorescence peak wavelengths (Verkman *et al.*, 1989a). The new compound, *N*-(6-methoxyquinolyl) acetoethyl ester (MQAE), is water soluble, has a higher Cl^- sensitivity than SPQ (Stern–Volmer constant 200 M^{-1}), with peak excitation and emission wavelengths of 355 and 460 nm. MQAE fluorescence is not altered by anions of physiological importance (e.g., HCO_3^-), by cations, or by pH. Unfortunately, MQAE leaks out of cells at a rate of $< 20\%$ in 60 min at 37°C. The introduction of Cl^--sensitive fluorescent indicators is certainly one of the most exciting recent developments in the field of study of Cl^- transport. Based on the influence of such dyes on the study of other ions, one can confidently predict that these agents (or their successors) will play an important role in the future study of Cl^- transport.

ACKNOWLEDGMENTS. We are very grateful to Mr. C. Rivera for preparation of the figures. The authors received travel grants from the following agencies: Fundación Mexicana para la Salud, Agency for International Development, and Spanish Ministerio de Educacion y Ciencia (Subprograma 4). F.J.A.L. was supported by CONACYT Grant PO28CCOX891556. F.G. was supported by DGICYT Grant PB 86/0326. J.M.R. was supported by NIH Grant NS-11946.

REFERENCES

Acker, H., Pietruschka, F., and Zierold, K., 1985, Comparative measurements of potassium and chloride with ion-sensitive microelectrodes and x-ray microanalysis in cultured skeletal muscle fibers, *In Vitro Cell Dev. Biol.* **21**:45–48.

Aickin, C. C., and Brading, A. F., 1982, Measurement of intracellular chloride in guinea-pig vas deferens by ion analysis, [36] chloride efflux and microelectrodes, *J. Physiol. (London)* **326**:139–154.

Aickin, C. C., and Brading, A. F., 1984, The role of chloride bicarbonate exchange in the regulation of intracellular chloride in guinea-pig vas deferens, *J. Physiol. (London)* **349:**587–606.

Aickin, C. C., and Brading, A. F., 1985a, The effects of bicarbonate and foreign anions on chloride transport in smooth muscle of the guinea-pig vas deferens, *J. Physiol. (London)* **366:**267–280.

Aickin, C. C., and Brading, A. F., 1985b, Advances in the understanding of transmembrane ionic gradients and permeabilities in smooth muscle obtained by using ion-selective micro-electrodes, *Experientia* **41:** 879–887.

Aickin, C. C., Betz, W. J., and Harris, G. L., 1989, Intracellular chloride and the mechanism for its accumulation in rat lumbrical muscle, *J. Physiol. (London)* **411:**437–455.

Alberts, B., Bray, D., Lewis, J., Raff, M., Roberts, K., and Watson, J. D., 1989, *Molecular Biology of the Cell,* 2nd ed., Garland, New York.

Allakhverdov, B. L., Burovina, I. V., Chmykhova, N. M., and Shapovalov, A. I., 1980, Electron probe x-ray microanalysis of intracellular sodium, potassium and chlorine contents in amphibian motoneurones, *Neuroscience* **5:**2023–2031.

Alvarez-Leefmans, F. J., Rink, T. J., and Tsien, R. Y., 1981, Free calcium ions in neurones of *Helix aspersa* measured with ion-selective micro-electrodes, *J. Physiol. (London)* **315:**531–548.

Alvarez-Leefmans, F. J., Gamiño, S. M., Giraldez, F., and Gonzalez-Serratos, H., 1986, Intracellular free magnesium in frog skeletal muscle fibres measured with ion-selective micro-electrodes, *J. Physiol. (London)* **378:**461–483.

Alvarez-Leefmans, F. J., Giraldez, F., and Gamiño, S. M., 1987, Intracellular free magnesium in excitable cells: Its measurement and its biological significance, *Can. J. Physiol. Pharmacol.* **65:**915–925.

Alvarez-Leefmans, F. J., Gamiño, S. M., Giraldez, F., and Noguerón, I., 1988, Intracellular chloride regulation in amphibian dorsal root ganglion neurones studies with ion-selective micro-electrodes, *J. Physiol. (London)* **406:**225–246.

Ammann, D., 1986, *Ion-Selective Microelectrodes,* Springer-Verlag, Berlin.

Ammann, D., Huser, M., Kräutler, B., Rusterholz, B., Schulthess, P., Lindemann, B., Halder, E., and Simon, W., 1986, Anion selectivity of metalloporphyrins in membranes, *Helv. Chim. Acta* **69:**849–854.

Ammann, D., Oesch, V., Bührer, T., and Simon, W., 1987, Design of ionophores for ion-selective micro-sensors, *Can. J. Physiol. Pharmacol.* **65:**879–884.

Armstrong, W. M., and Garcia-Diaz, J. F., 1980, Ion-selectivity microelectrodes: Theory and technique, *Fed. Proc. Fed. Am. Soc. Exp. Biol.* **39:**2851–2859.

Armstrong, W. M., Wojtkowski, W., and Bixennan, W. R., 1977, A new solid-state microelectrode for measuring intracellular chloride activities, *Biochim. Biophys. Acta* **465:**165–170.

Ascher, P., Kunze, D. L., and Neild, J. O., 1976, Chloride distribution in Aplysia neurones, *J. Physiol. (London)* **256:**441–464.

Ballanyi, K., and Grafe, P., 1985, An intracellular analysis of γ-aminobutyric-acid-associated ion movements in rat sympathetic neurones, *J. Physiol. (London)* **365:**41–58.

Ballanyi, K., Grafe, P., and Ten Bruggencate, G., 1987, Ion activities and potassium uptake mechanisms of glial cells in guinea-pig olfactory cortex slices, *J. Physiol. (London)* **382:**159–174.

Bates, R. G., 1973, *Determination of pH: Theory and Practice,* 2nd ed., Wiley, New York.

Baumgarten, C. M., 1981, An improved liquid ion exchanger for chloride ion-selective microelectrodes, *Am. J. Physiol.* **241:**C258–C263.

Baumgarten, C. M., and Fozzard, H. A., 1981, Intracellular chloride activity in mammalian ventricular muscle, *Am. J. Physiol.* **241:**C121–C129.

Bockris, J. O., and Reddy, A. K. N., 1973, *Modern Electrochemistry,* Volume 1, Plenum/Rosetta Edition, Plenum Press, New York.

Bolton, T. B., and Vaughan-Jones, R. D., 1977, Continuous direct measurement of intracellular chloride and pH in frog skeletal muscle, *J. Physiol. (London)* **270:**801–833.

Bomsztyk, K., Calalb, M. B., Smith, L., and Stanton, T. H., 1988, A microelectrometric titration method for measurement of total intracellular Cl⁻ concentration, *Am. J. Physiol.* **254:**C200–C205.

Boron, W. F., 1985, Intracellular pH-regulating mechanism of the squid axon. Relation between the external Na^+ and HCO_3^- dependences, *J. Gen. Physiol.* **85:**325–345.

Brown, A. M., and Kunze, D. L., 1974, Ion activities in identifiable Aplysia neurons, in: *Ion-Selective Microelectrodes* (H. J. Berman and N. C. Herbert, eds.), Plenum Press, New York, pp. 57–73.

Brown, A. M., Walker, J. L., Jr., and Sutton, R. B., 1970, Increased chloride conductance as the proximate cause of hydrogen ion effects in Aplysia neurons, *J. Gen. Physiol.* **56**:559–582.

Brown, H. M., 1976, Intracellular Na^+, K^+ and Cl^- activities in *Balanus* photoreceptors, *J. Gen. Physiol.* **68**:281–290.

Buck, R. P., 1975, Electroanalytical chemistry of membranes, *Crit. Rev. Anal. Chem.* **5**:323–419.

Bührle, C. P., and Sonnhof, U., 1983, Intracellular ion activities and equilibrium potentials in motoneurones and glial cells of the frog spinal cord, *Pfluegers Arch.* **396**:144–153.

Bührle, C. P., and Sonnhof, U., 1985, The ionic mechanism of postsynaptic inhibition in motoneurones of the frog spinal cord, *Neuroscience* **14**:581–592.

Caldwell, P. C., 1954, An investigation of the intracellular pH of crab muscle fibres by means of micro-glass and micro-tungsten electrodes, *J. Physiol. (London)* **126**:169–180.

Caldwell, P. C., 1958, Studies on the internal pH of large muscle and nerve fibres, *J. Physiol. (London)* **142**:22–62.

Carr, C. W., 1968, Applications of membrane electrodes, *Ann. N.Y. Acad. Sci.* **148**:180–190.

Cassola, A. C., Mollenhauer, M., and Frömter, E., 1983, The intracellular chloride activity of rat kidney proximal tubular cells, *Pfluegers Arch.* **399**:259–265.

Casteels, R., and Kuriyama, H., 1965, Membrane potential and ionic content in pregnant and non-pregnant rat myometrium, *J. Physiol. (London)* **177**:263–287.

Chao, A. C., and Armstrong, W. M., 1987, Cl^--selective microelectrodes: Sensitivity to anionic Cl^- transport inhibitors, *Am. J. Physiol.* **253**:C343–C347.

Chao, A. C., Dix, J. A., Sellers, M. C., and Verkman, A. S., 1989a, Fluorescence measurement of chloride transport in monolayer cultured cells: mechanisms of chloride transport in fibroblasts, *Biophys. J.* (in press).

Chao, A. C., Widdicombe, J. H., and Verkman, A. S., 1989b, Chloride conductive and cotransport mechanisms in cultures of canine tracheal epithelial cells measured by an entrapped fluorescent indicator, *J. Membrane Biol.* (in press).

Chen, P.-Y., and Verkman, A. S., 1988, Sodium-dependent chloride transport in basolateral membrane vesicles isolated from rabbit proximal tubules, *Biochemistry* **27**:655–660.

Chen, P.-Y., Illsley, N. P., and Verkman, A. S., 1988, Renal brush-border chloride transport mechanisms characterized using a fluorescent indicator, *Am. J. Physiol.* **254**:F114–F120.

Christensen, H. N., 1982, Efflux used as a fad word? *Trends Biosci.* **7**:134.

Clegg, J. S., 1982, Alternative views on the role of water in cell function, in: *Biophysics of Water* (F. Franks and S. F. Mathias, eds.), Wiley, New York, pp. 365–383.

Clegg, J. S., 1984, Properties and metabolism of the aqueous cytoplasm and its boundaries, *Am. J. Physiol.* **246**:R133–R151.

Coombs, J. S., Eccles, J. C., and Fatt, P., 1955, The specific ionic conductances and the ionic movements across the motoneuronal membrane that produce the inhibitory post-synaptic potential, *J. Physiol. (London)* **130**:326–373.

Cornwall, M. C., Peterson, D. F., Kunze, D. L., Walker, J. L., Jr., and Brown, A. M., 1970, Intracellular potassium and chloride activities measured with liquid ion exchanger microelectrodes, *Brain Res.* **23**:433–436.

Cotlove, E., Trantham, H. U., and Bowman, R. L., 1958, An instrument and method for automatic, rapid, accurate and sensitive titration of chloride in biological samples, *J. Lab. Clin. Med.* **51**:461–468.

Cotton, C. U., Weinstein, A. M., and Reuss, L., 1989, Osmotic water permeability of *Necturus* gallbladder epithelium, *J. Gen. Physiol.* **93**:649–679.

Cremaschi, D., Meyer, G., Botta, G., and Rossetti, C., 1987, The nature of the neutral Na^+-Cl^--coupled entry at the apical membrane of rabbit gallbladder epithelium: II. Na^+-Cl^--symport is independent of K^+, *J. Membr. Biol.* **95**:219–228.

Deisz, R. A., and Lux, H. D., 1982, The role of intracellular chloride in hyperpolarizing post-synaptic inhibition of crayfish stretch receptor neurones, *J. Physiol. (London)* **326**:123–138.

Derbyshire, W., 1982, Dynamics of water in cellular systems, in: *Biophysics of Water* (F. Franks and S. F. Mathias, eds.), Wiley, New York, pp. 249–253.

Dick, D. A. T., 1979, Structure and properties of water in the cell, in: *Mechanisms of Osmoregulation in Animals* (R. Gilles, ed.), Wiley, New York, pp. 3–45.

Donahue, B. S., and Abercrombie, R. T., 1987, Free diffusion coefficient of ionic calcium in cytoplasm, *Cell Calcium* **8**:437–448.

Edzes, H. T., and Berendsen, H. J. C., 1975, The physical state of diffusible ions in cells, *Annu. Rev. Biophys. Bioeng.* **68**:159–178.

Eisenman, G., 1967, *Glass Electrodes for Hydrogen and Other Cations: Principles and Practice*, Dekker, New York.

Eisenman, G., 1968, Similarities and differences between liquid and solid ion exchangers and their usefulness as ion specific electrodes, *Anal. Chem.* **40**:310–320.

Eisenman, G., 1969, Theory of membrane electrode potentials: An examination of the parameters determining the selectivity of solid and liquid ion exchangers and of neutral ion-sequestering molecules, in: *Ion-Selective Electrodes* (R. A. Durst, ed.), National Bureau of Standards Special Publication 314, pp. 1–56.

Fromm, M., and Schultz, S. G., 1981, Some properties of KCl-filled microelectrodes: Correlation of potassium "leakage" with tip resistance, *J. Membr. Biol.* **62**:239–244.

Frömter, E., Simon, M., and Gebler, B., 1981, A double-channel ion-selective microelectrode with the possibility of fluid ejection for localization of the electrode tip in the tissue, in: *Progress in Enzyme and Ion-Selective Electrodes* (D. W. Lübbers, H. Acker, R. P. Buck, G. Eisenman, M. Kessler, and W. Simon, eds.), Springer-Verlag, Berlin, pp. 35–44.

Fulton, A. B., 1982, How crowded is the cytoplasm? *Cell* **30**:345–347.

Galvan, M., Dörge, A., Beck, F., and Rick, R., 1984, Intracellular electrolyte concentrations in rat sympathetic neurones measured with an electron microprobe, *Pfluegers Arch.* **400**:274–279.

Gardner, D. R., and Moreton, R. B., 1985, Intracellular chloride in molluscan neurons, *Comp. Biochem. Physiol.* **80A**:461–467.

Gayton, D. C., and Hinke, J. A. M., 1968, The location of chloride in single striated muscle fibers of the giant barnacle, *Can. J. Physiol. Pharmacol.* **46**:213–219.

Gesteland, R. C., Howland, B., Lettvin, J. Y., and Pitts, W. H., 1959, Comments on microelectrodes, *Proc. Inst. Radio Electron. Eng. Aust.* **47**:1856–1861.

Gilles, R., Bolis, L., and Kleinzeller, A., eds., 1987, *Cell Volume Control: Fundamental and Comparative Aspects in Animal Cells, Curr. Top. Membr. Transp.* **30.**

Giraldez, F., 1984, Active sodium transport and fluid secretion in the gallbladder epithelium of Necturus, *J. Physiol. (London)* **348**:431–455.

Greger, R., and Schlatter, E., 1984, Mechanism of NaCl secretion in the rectal gland of spiny dogfish (Squalus acanthias), *Pfluegers Arch.* **402**:63–75.

Greger, R., Oberleithner, H., Schlatter, E., Cassola, A. C., and Weidtke, C., 1983, Chloride activity in cells of isolated perfused cortical thick ascending limbs of rabbit kidney, *Pfluegers Arch.* **399**:29–34.

Grinstein, S., McCulloch, L., and Rothstein, A., 1979, Transmembrane effects of irreversible inhibitors of anion transport in red blood cells: Evidence for mobile transport sites, *J. Gen. Physiol.* **73**:493–514.

Guibault, G. C., Durst, R. A. Frant, M. S., Freiser, H., Hansen, E. H., Light, T. S., Pungor, E., Rechnitz, G., Rice, N. M., Rohm, T. J., Simon, W., and Thomas, J. D. R., 1976, Recommendations for nomenclature of ion-selective electrodes, *Pure Appl. Chem.* **48**:127–132.

Harris, G. L., and Betz, W. J., 1987, Evidence for active chloride accumulation in normal and denervated rat lumbrical muscle, *J. Gen. Physiol.* **90**:127–144.

Heinz, E., and Grassl, S., 1981, Interference of furosemide and other anion transport inhibitors with liquid Cl^--exchanger electrodes, *Biophys. J.* **33**:222a.

Hinke, J. A. M., 1969, Glass microelectrodes in the study of binding and compartmentalization of intracellular ions, in: *Glass Microelectrodes* (M. Lavallee, O. F. Schanne, and N. C. Hebert, eds.), Wiley, New York, pp. 349–375.

Hinke, J. A. M., 1987, Thirty years of ion-selective microelectrodes: Disappointments and successes, *Can. J. Physiol. Pharmacol.* **65**:873–878.

Hinke, J. A. M., and Gayton, D. C., 1971, Transmembrane K$^+$ and Cl$^-$ activity gradients for the muscle fibre of the giant barnacle, *Can. J. Physiol. Pharmacol.* **49:**312–322.

Hironaka, T., and Morimoto, S., 1979, The resting membrane potential of frog sartorius muscle, *J. Physiol. (London)* **297:**1–8.

Hoffmann, E. K., 1987, Volume regulation in cultured cells, *Curr. Top. Membr. Transp.* **30:**125–180.

Hoffmann, E. K., and Simonsen, L. O., 1989, Membrane mechanisms in volume and pH regulation in vertebrate cells, *Physiol. Rev.* **69:**315–382.

Hoffmann, E. K., Schiødt, M., and Dunham, P., 1986, The number of chloride–cation cotransport sites on Ehrlich ascites cells measured with [^{3}H]bumetanide, *Am. J. Physiol.* **250:**C688–C693.

Hofmeister, F., 1888, Zur lehre von der wirkung der salze. Zweite mitteilung, *Arch. Exp. Pathol. Pharmakol.* **24:**247–260.

Illsley, N. P., and Verkman, A. S., 1987, Membrane chloride transport measured using a chloride-sensitive fluorescent probe, *Biochemistry* **26:**1215–1219.

Ishibashi, K., Sasaki, S., and Yoshiyama, N., 1988, Intracellular chloride activity of arbbit proximal straight tubule perfused in vitro, *Am. J. Physiol.* **255:**F49–F56.

IUPAC, 1979, Commission on analytical nomenclature (prepared for publication by G. G. Guibault). Recommendations for publishing manuscripts on ion-selective electrodes, *Ion-selective Electrode Rev.* 1, 139.

Jack, J. J. B., Noble, D., and Tsien, R. W., 1975, *Electric Current Flow in Excitable Cells.* Oxford University Press (Clarendon), London.

Janz, G. J., and Ives, D. J. G., 1968, Silver, silver chloride electrodes, *Ann. N.Y. Acad. Sci.* **148:**210–221.

Kehoe, J., 1972, Ionic mechanisms of a two component cholinergic inhibition in Aplysia neurones, *J. Physiol. (London)* **225:**85–114.

Kenyon, J., and Gibbons, W. R., 1977, Effects of low chloride solutions on action potentials of sheep cardiac Purkinje fibres, *J. Gen. Physiol.* **70:**635–660.

Kerkut, G. A., and Meech, R. W., 1966, The internal chloride concentration of H and D cells in the snail brain, *Comp. Biochem. Physiol.* **19:**819–832.

Kettenmann, H., 1987, K$^+$ and Cl$^-$ uptake by cultured oligodendrocytes, *Can. J. Physiol. Pharmacol.* **65:** 1033–1037.

Keynes, R. D., 1963, Chloride in the squid giant axon, *J. Physiol. (London)* **169:**690–705.

Khuri, R. N., Bogharian, K. K., and Agulian, S. K., 1974, Intracellular bicarbonate in single skeletal muscle fibers, *Pfluegers Arch.* **349:**285–299.

Khuri, R. N., Agulian, S. K., and Bogharian, K. K., 1976, Intracellular bicarbonate of skeletal muscle under different metabolic states, *Am. J. Physiol.* **230:**228–232.

Kondo, Y., Bührer, T., Seiler, K., Frömter, E. and Simon, W., 1989, A new double-barrelled, ionophore-based microelectrode for chloride ions, *Pfluegers Arch.* **414:**663–668.

Koryta, J., 1975, *Ion-Selective Electrodes,* Cambridge University Press, London.

Koryta, J., 1981, Theory of ion-selective electrodes, in: *Ion-Selective Microelectrodes and Their Use in Excitable Tissues* (E. Sykova, P. Hnik, and L. Vyklicky, eds.), Plenum Press, New York, pp. 3–11.

Koryta, J., and Štulík, K., 1983, *Ion-Selective Electrodes,* 2nd ed., Cambridge University Press, London.

Kraig, R. P., and Cooper, J. L., 1987, Bicarbonate and ammonia changes in brain during spreading depression, *Can. J. Physiol. Pharmacol.* **65:**1099–1104.

Krapf, R., Berry, C. A., and Verkman, A. S., 1988, Estimation of intracellular chloride activity in isolated perfused rabbit proximal convoluted tubules using a fluorescent indicator, *Biophys. J.* **53:**955–962.

Lakshminarayanaiah, N., 1976, *Membrane Electrodes,* Academic Press, New York.

Larson, M., and Spring, K., 1987, Volume regulation in epithelia, *Curr. Top. Membr. Transp.* **30:**105–123.

Lauf, P. K., McManus, T. J., Hass, M., Forbush, B., Duhn, J., Flatman, P. W., Saier, M. H., Jr., and Russell, J. M., 1987, Physiology and biophysics of chloride and cation cotransport across cell membranes, *Fed. Proc.* **46:**2377–2394.

Lee, C. O., 1981, Ionic activities in cardiac muscle cells and applications of ion-selective microelectrodes, *Am. J. Physiol.* **241:**H459–H478.

Lev, A. A., and Armstrong, W. M., 1975, Ionic activities in cells, *Curr. Top. Membr. Transp.* **6:**59–123.

Lux, H. D., 1974, Fast recording ion specific microelectrodes: Their use in pharmacological studies in the CNS, *Neuropharmacology* **13:**509–517.

McCaig, D., and Leader, J. P., 1984, Intracellular chloride activity in the extensor digitorum longus (EDL) muscle of the rat, *J. Membr. Biol.* **81:**9–17.

MacKnight, A. D. C., 1985, The role of anions in cellular volume regulation, *Pfluegers Arch.* **405**(Suppl. 1):S12–S16.

Marranes, R., and DeHemptinne, A., 1978, Conduction velocity of the action potential in isolated cardiac fibers; transient effects under influence of organic anions, *Arch. Int. Physiol.* **86:**1162–1163.

Mauro, A., 1954, Electrochemical potential difference of chloride ion in the giant squid axon–sea water system, *Fed. Proc. Fed. Am. Soc. Exp. Biol.* **13:**96.

Moody, G. J., and Thomas, J. D. R., 1971, *Selective Ion Sensitive Electrodes,* Merrow, Durham, England.

Morf, W. E., 1981, The principles of ion-selective electrodes and of membrane transport, in: *Studies in Analytical Chemistry,* Volume 2, Akadémiai Kiadó, Budapest, and Elsevier, Amsterdam.

Morf, W. E., Ruprecht, H., Oggenfuss, P., and Simon, W., 1985, Ion transport in asymmetric artificial membranes mediated by neutral carriers, in: *Ion Measurements in Physiology and Medicine* (M. Kessler, D. K. Harrison, and J. Höper, eds.), Springer-Verlag, Berlin, pp. 1–5.

Morris, M. E., and Krnjević, K., eds., 1987, *Ion-Selective Microelectrodes and Excitable Tissues, Can. J. Physiol. Pharmacol.* **65:**867–1110.

Moser, H., 1985, Intracellular pH regulation in the sensory neurone of the stretch receptor of the crayfish (Astacus fluviatilis), *J. Physiol. (London)* **362:**23–38.

Munoz, J.-L., Deyhimi, F., and Coles, J. A., 1983, Silanization of glass in the making of ion-sensitive microelectrodes, *J. Neurosci. Methods* **8:**231–247.

Neild, T. O., and Thomas, R. C., 1973, New design for a chloride-sensitive microelectrode, *J. Physiol. (London)* **231:**7P–8P.

Neild, T. O., and Thomas, R. C., 1974, Intracellular chloride activity and the effects of acetylcholine in snail neurones, *J. Physiol. (London)* **242:**453–470.

Oberleithner, H., Guggino, W., and Giebisch, G., 1982, Mechanism of distal tubular chloride transport in Amphiuma kidney, *Am. J. Physiol.* **242:**F331–F339.

Orme, F. N., 1969, Liquid ion-exchanger microelectrodes, in: *Glass Microelectrodes* (M. Lavallée, O. F. Schanne, and N. C. Hébert, eds.), Wiley, New York, pp. 376–395.

Owen, J. D., Brown, H. M., and Saunders, J. H., 1975, Chloride activity in the giant cell of Aplysia, *Biophys. J.* **15:**45a.

Palmer, L. G., and Civan, M. M., 1977, Distribution of Na^+, K^+ and Cl^- between nucleus and cytoplasm in Chironomus salivary gland cells, *J. Membr. Biol.* **33:**41–61.

Parsons, R., 1959, *Handbook of Electrochemical Constants,* Butterworths, London.

Pelzer, D., Trube, G., and Piper, H. M., 1984, Low resting potentials in single isolated heart cells due to membrane damage by the recording microelectrode, *Pfluegers Arch.* **400:**197–199.

Pollard, H. B., Creutz, C. E., Pazoles, C. J., and Hansen, J., 1977, Calcium binding properties of monovalent anions commonly used to substitute for chloride in physiological salt solutions, *Anal. Biochem.* **83:**311–314.

Purves, R. D., 1981, *Microelectrode Methods for Intracellular Recording and Iontophoresis,* Academic Press, New York.

Rall, W., 1977, Core conductor theory and cable properties of neurons, in: *Handbook of Physiology,* Volume 1, *The Nervous System,* Part I, (E. R. Kandel, ed.), American Physiological Society, Bethesda, Md., pp. 39–97.

Reuss, L., 1985, Changes in cell volume measured with an electrophysiologic technique, *Proc. Natl. Acad. Sci. USA* **82:**6014–6018.

Reuss, L., Constantin, J. L., and Bazile, J. E., 1987, Diphenylamine-2-carboxylate blocks Cl^--HCO_3^- exchange in Necturus gallbladder epithelium, *Am. J. Physiol.* **253:**C79–C89.

Robinson, R. A., and Stokes, R. H., 1965, *Electrolyte Solutions,* 2nd ed. rev., Butterworths, London.

Roos, A., and Boron, W. F., 1981, Intracellular pH, *Physiol. Rev.* **61:**296–434.

Rosenberg, M., 1973, A comparison of chloride and citrate filled microelectrodes for d-c recording, *J. Appl. Physiol.* **35:**166–168.

Ross, J. W., 1969, Solid-state and liquid membrane ion-selective electrodes, in: *Ion-Selective Electrodes* (R. A. Durst, ed.), National Bureau of Standards Special Publication 314, pp. 57–88.

Russell, J. M., and Brown, A. M., 1972a, Active transport of chloride by the giant neuron of the *Aplysia* abdominal ganglion, *J. Gen. Physiol.* **60**:499–518.

Russell, J. M., and Brown, A. M., 1972b, Active transport of potassium by the giant neuron of the *Aplysia* abdominal ganglion, *J. Gen. Physiol.* **60**:519–533.

Sandblom, J., and Orme, F., 1972, Liquid membranes as electrodes and biological models, in: *Membranes—A Series of Advances,* Volume 1 (G. Eisenman, ed.), Dekker, New York, pp. 125–177.

Sandblom, J., Eisenman, G., and Walker, J. L., Jr., 1967, Electrical phenomena associated with the transport of ions and ion pairs in liquid ion-exchange membranes. I. Zero current properties, *J. Physiol. Chem.* **71**:3862–3870.

Saunders, J. H., and Brown, H. M., 1977, Liquid and solid-state Cl⁻-sensitive microelectrodes, *J. Gen. Physiol.* **70**:507–530.

Schulthess, P., Ammann, D., Simon, W., Caderas, C. Stepánek, R., and Kräutler, B., 1984, A lipophilic derivative of vitamin B_{12} as a selective carrier for anions, *Helv. Chim. Acta* **67**:1026–1032.

Schulthess, P., Ammann, D., Kräutler, B., Caderas, C., Stepánek, R., and Simon, W., 1985, Nitrite selective liquid membrane electrode, *Anal. Chem.* **57**:1397–1401.

Serve, G., Endres, W., and Grafe, P., 1988, Continuous electrophysiological measurements of changes in cell volume of motoneurons in the isolated frog spinal cord, *Pfluegers Arch.* **411**:410–415.

Sharp, A. P., and Thomas, R. C., 1981, The effects of chloride substitution on intracellular pH in crab muscle, *J. Physiol. (London)* **312**:71–80.

Sollner, K., 1968, Membrane electrodes, *Ann. N.Y. Acad. Sci.* **148**:154–179.

Sollner, K., and Shean, G. M., 1964, Liquid ion-exchange membranes of extreme selectivity and high permeability for anions, *J. Am. Chem. Soc.* **86**:1901–1902.

Spitzer, K. W., and Walker, J. L., Jr., 1980, Intracellular chloride activity in quiescent cat papillary muscle, *Am. J. Physiol.* **238**:H487–H493.

Spring, K. R., and Ericson, A.-C., 1982, Epithelial cell volume modulation and regulation, *J. Membr. Biol.* **69**:169–176.

Spring, K. R., and Kimura, G., 1978, Chloride reabsorption by renal proximal tubules of Necturus, *J. Membr. Biol.* **38**:233–254.

Srinivasan, K., and Rechnitz, G. A., 1969, Selectivity studies on liquid membrane ion-selective electrodes, *Anal. Chem.* **40**:1203–1208.

Stein, W. D., 1986, *Transport and Diffusion across the Cell Membrane,* Academic Press, New York.

Strickholm, A., and Wallin, B. G., 1965, Intracellular chloride activity of crayfish giant axons, *Nature* **208**: 790–791.

Tasaki, I., and Singer, I., 1968, Some problems involved in electric measurements of biological systems, *Ann. N.Y. Acad. Sci.* **148**:36–53.

Taylor, P. S., and Thomas, R. C., 1984, The effect of leakage on micro-electrode measurements of intracellular sodium activity in crab muscle fibres, *J. Physiol. (London)* **352**:539–550.

Thomas, R. C., 1972, Intracellular sodium activity and the sodium pump in snail neurones, *J. Physiol. (London)* **220**:55–71.

Thomas, R. C., 1976, The effect of carbon dioxide on the intracellular pH and buffering power of snail neurones, *J. Physiol. (London)* **255**:715–735.

Thomas, R. C., 1977, The role of bicarbonate, chloride and sodium ions in the regulation of intracellular pH in snail neurons, *J. Physiol. (London)* **273**:317–338.

Thomas, R. C., 1978, *Ion-Sensitive Intracellular Microelectrodes,* Academic Press, New York.

Thomas, R. C., 1985, Eccentric double micropipette suitable both for pHi microelectrodes and for intracellular iontophoresis, *J. Physiol. (London)* **371**:24P.

Thomas, R. C., and Cohen, C. J., 1981, A liquid ion-exchanger alternative to KCl for filling intracellular reference microelectrodes, *Pfluegers Arch.* **390**:96–98.

Thompson, S. M., Deisz, R. A., and Prince, D. A., 1988, Relative contributions of passive equilibrium and active transport to the distribution of chloride in mammalian cortical neurons, *J. Neurophysiol.* **60**:105–124.

Tsien, R. Y., 1980, Liquid sensors for ion-selective microelectrodes, *Trends Neurosci.* **3**:219–221.

Tsien, R. Y., 1983, Intracellular measurements of ion activities, *Annu. Rev. Biophys. Bioeng.* **12**:91–116.

Tsien, R. Y., and Rink, T. J., 1980, Neutral carrier ion-selective microelectrodes for measurement of intracellular free calcium, *Biochim. Biophys. Acta* **599**:623–638.

Tsien, R. Y., and Rink, T. J., 1981, Ca^{2+}-selective electrodes: A novel PVC-gelled neutral carrier mixture compared with other currently available sensors, *J. Neurosci. Methods* **4**:73–86.

Vaughan-Jones, R. D., 1979a, Non-passive chloride distribution in mammalian heart muscle: Micro-electrode measurement of the intracellular chloride activity, *J. Physiol. (London)* **295**:83–109.

Vaughan-Jones, R. D., 1979b, Regulation of chloride in quiescent sheep-heart Purkinje fibres studied using intracellular chloride and pH-sensitive micro-electrodes, *J. Physiol. (London)* **295**:111–137.

Vaughan-Jones, R. D., 1982, Chloride activity and its control in skeletal and cardiac muscle, *Philos. Trans. R. Soc. London Ser. B* **299**:537–548.

Vaughan-Jones, R. D., 1986, An investigation of chloride–bicarbonate exchange in the sheep cardiac Purkinje fibre, *J. Physiol. (London)* **379**:377–406.

Verkman, A. S., Chen, P.-Y., Davis, B., Fong, P., Illsley, N. P., and Krapf, R., 1988, Development of chloride-sensitive fluorescent indicators, in: *Cellular and Molecular Basis of Cystic Fibrosis* (G. Mastella and P. M. Quinton, eds.), San Francisco Press, San Francisco, pp. 471–478.

Verkman, A. S., Sellers, M. C., Chao, A. C., Leung, T., and Ketcham, R., 1989a, Synthesis and characterization of improved chloride-sensitive fluorescent indicators for biological applications, *Anal. Biochem.* **178**:355–361.

Verkman, A. S., Takla, R., Sefton, B., Basbaum, C., and Widdicombe, J. H., 1989b, Fluorescence assay of chloride transport in liposomes reconstituted with chloride transporters, *Biochemistry* **28**:4240–4244.

Walker, J. L., Jr., 1971, Ion specific liquid ion exchanger microelectrodes, *Anal. Chem.* **43**(3):89A–93A.

Walker, J. L., Jr., and Brown, H. M., 1977, Intracellular ionic activity measurements in nerve and muscle, *Physiol. Rev.* **57**:729–778.

Wallin, B. G., 1967, Intracellular ion concentrations in single crayfish axons, *Acta Physiol. Scand.* **70**:419–430.

Wegmann, D., Weiss, H., Ammann, D., Morf, W. E., Pretsch, E., Sugahara, K., and Simon, W., 1984, Anion-selective liquid membrane electrodes based on lipophilic quaternary ammonium compounds, *Mikrochim. Acta* **3**:1–16.

Wills, N. K., 1985, Apical membrane potassium and chloride permeabilities in surface cells of rabbit descending colon epithelium, *J. Physiol. (London)* **358**:433–445.

Willumsen, N. J., Davis, C. W., and Boucher, R. C., 1989, Intracellular Cl^- activity and cellular Cl^- pathways in cultured human airway epithelium, *Am. J. Physiol.* **256**:C1033–C1044.

Woodbury, J. W., and Miles, P. R., 1973, Anion conductance of frog muscle membranes: One channel, two kinds of pH dependence, *J. Gen. Physiol.* **62**:324–353.

World Precision Instruments, 1988, Electrodes, accessories and supplies, Cat. #2.

Wright, F. S., and McDougal, W. S., 1972, Potassium-specific ion-exchanger microelectrodes to measure K^+ activity in the renal distal tubule, *Yale J. Biol. Med.* **45**:373–383.

Wuhrmann, P., Ineichen, H., Riesen-Willi, V., and Lezzi, M., 1979, Change in nuclear potassium electrochemical activity and puffing of potassium-sensitive salivary chromosome regions during Chironomus development, *Proc. Natl. Acad. Sci. USA* **76**:806–808.

Wuthier, U., Pham, H. V., Zünd, R., Welti, D., Funck, R. J. J., Bezegh, A., Ammann, D., Pretsch, E., and Simon, W., 1984, Tin organic compounds as neutral carriers for anion selective electrodes, *Anal. Chem.* **56**:535–538.

Yamaguchi, H., 1986, Recording of intracellular Ca^{2+} from smooth muscle cells by sub-micron tip, double-barrelled Ca^{2+}-selective microelectrodes, *Cell Calcium* **7**:203–219.

Zeuthen, T., 1980, How to make and use double-barreled ion-selective microelectrodes, *Curr. Top. Membr. Transp.* **13**:31–47.

Zeuthen, T., 1985, The advantages of transient experiments over steady-state experiments, in: *Ion Measurements in Physiology and Medicine* (M. Kessler, D. K. Harrison, and J. Höper, eds.), Springer-Verlag, Berlin, pp. 150–157.

2

Principles of Cell Volume Regulation
Ion Flux Pathways
and the Roles of Anions

Peter M. Cala

1. INTRODUCTION

The goal of this chapter is to discuss general principles of cell volume regulation (in response to osmotic swelling and shrinkage) with particular emphasis on the ion flux pathways. While other chapters in this volume deal with various aspects of neuronal, glial, or muscle cell function, the present chapter will draw upon data from studies performed on red blood cells. The emphasis upon blood cells reflects the relative ease with which volume regulation by cells in suspension can be studied and not the biological importance of volume regulation by such cells. While changes in the cell volume of circulating cells can result in changes in blood viscosity and altered rheologic properties, disruption of neural cell volume can result in altered cable properties and consequent effects upon neural integration.

1.1. Criteria for Cell Volume Regulation

Cell volume regulation as discussed in this chapter refers to the process whereby a cell is able to restore volume to normal subsequent to an osmotic swelling or shrinkage. Since the membranes of animal cells are unable to support significant differences in transmembrane hydrostatic pressure, volume regulation is a consequence of net solute flux and osmotically obliged H_2O flow. If a process is truly volume regulatory, as distinct from volume sensitive, the flux rate must be a graded function of the stimulus (volume perturbation) and must be of sufficient magnitude to restore normal volume. Implicit in the above are notions of solute abundance and energetic feasibility. That is, the solute serving as an osmotic effector must be available in sufficient quantity to effect volume changes of the magnitude necessary to restore normal volume. In addition, the thermodynamic forces must be of sufficient magnitude and in the appropriate direction to drive the volume regulatory process. These principles are reflected in the

Peter M. Cala • Department of Human Physiology, School of Medicine, University of California, Davis, California 95616.

volume regulatory strategies of vertebrate cells. That is, since, in general, free amino acid pools are small, vertebrate cells must rely upon inorganic ions, which are the major osmotic constituents, as osmotic effectors. Since most cells are of the high K^+, low Na^+ type, yet exist in an environment that is high $[Na^+]$, low $[K^+]$, the dissipative loss of K^+ and uptake of Na^+ are responsible for volume regulation following cell swelling and shrinkage, respectively. Thus, pathways responsible for volume regulation subsequent to cell swelling tend to dissipate energy stored in the K^+ gradient, while subsequent to osmotic shrinkage the energy stored in Na^+ and/or Cl^- gradients drives solute and H_2O uptake. In addition, the flux pathways responsible for volume regulation may be quiescent at normal volume, yet become activated in response to osmotic perturbation. Finally, the process of volume regulation should not disrupt normal cell function. This last point relates to the nature of the ion flux pathway employed in the regulation of cell volume.

1.2. Volume Regulatory Ion Flux Pathways

To date, cell volume regulation has been described for most cells studied. The most frequently observed ion flux pathways mediating volume regulation by swollen cells are K^+ and Cl^- conductance, K^+,Cl^- cotransport, or parallel K^+/H^+ and Cl^-/HCO_3^- exchange. The pathways most commonly implicated in volume regulation following osmotic shrinkage are Na^+,Cl^- cotransport, Na^+,K^+,Cl^- cotransport, or parallel Na^+/H^+ and Cl^-/HCO_3^- exchange. In all cases, net cation and anion fluxes are coupled. In the case of conductive pathways the coupling is functional as a result of changes in membrane voltage and serves to maintain macroscopic electroneutrality. In the case of volume regulation by parallel alkali metal/proton and Cl^-/HCO_3^- exchange pathways, the coupling is again functional but unlike the conductance case, coupling is a result of cation flux-dependent changes in pH (and therefore $[HCO_3^-]$) as opposed to membrane voltage. In contrast to the conductive or parallel exchange pathways, the coupling of cations and anions via the cotransport pathways is obligatory. That is, in the case of flux via conductive pathways, addition/removal of Cl^- through some external circuit (short circuit) can eliminate Cl^- flux through the membrane. So too in the case of parallel alkali metal/H^+ and Cl^-/HCO_3^- exchangers, pH stat or high buffer power can prevent net Cl^- flux through the anion pathway. In the case of the cotransport pathways, however, there is no means by which the net anion and cation fluxes can be dissociated since the ions are transported as a single carrier substrate complex.

In the sections to follow, I will attempt to develop a more detailed description of the characteristics of the various volume regulatory transport systems and try to outline principals that govern function.

2. VOLUME REGULATION SUBSEQUENT TO CELL SWELLING

2.1. Volume Decrease via Parallel K^+ and Cl^- Conductances

Human T lymphocytes (Grinstein *et al.*, 1984), platelets (Livne *et al.*, 1987), Ehrlich ascites tumor cells (Hoffmann *et al.*, 1984), and cultured intestinal epithelial cells (Hazama and Okada, 1987) regulate volume subsequent to swelling as a consequence of separate conductive K^+ and Cl^- pathways (Fig. 1). Cell volume regulation resulting from net conductive K^+ and Cl^- efflux is best described in studies performed on the human lymphocyte (Grinstein et al., 1982) and Ehrlich ascites tumor cell (Hoffmann et al., 1984). Both cell types have a high resting K^+ conductance (gK) and a relatively low Cl^- conductance (gCl). In response to osmotic swelling or exposure to Ca^{2+} and calcium ionophore (increased $[Ca^{2+}]_i$), the cells lose K^+, Cl^-, and H_2O. The primary event responsible for K, Cl, and therefore H_2O loss is an increase in the membrane Cl^- conductance which, in the steady state, is rate limiting. Thus, Cl^-, by serving as an electrical counterion for K^+, is permissive.

It is interesting that the volume-activated increase in gCl is highly time-, as opposed to volume-dependent, in both the lymphocyte and Ehrlich ascites tumor cell. Briefly, both human lymphocytes and Ehrlich ascites cells can be swollen in the presence of quinidine and net K^+ loss and cell volume regulation are prevented. If, however, the ionophore gramicidin is added (in order to create a pathway for K^+ loss) at different times following cell swelling, the net H_2O loss is inversely proportional to the time (after swelling) of gramicidin addition. Since the ability of gramicidin to mediate net cation and H_2O flux is not time-dependent but is dependent upon the net flux of an anion to serve as a counterion (assuming the cation gradient is favorable), it was concluded that the time dependence of gramicidin-induced H_2O flow is a reflection of the time dependence of the cell anion conductance. Thus, the authors concluded that swelling causes an increase in anion conductance which decreases with time even if cell volume remains expanded. Further support for the notion that the anion conductance is rate limiting to cation flux (see Grinstein et al., 1982; Hoffmann et al., 1984) is the fact that gramicidin addition to cells at control volume does not cause net particle and H_2O flux, as would be the case if resting anion conductance was high.

At present, the exact nature of the volume-sensitive K^+ and Cl^- conductances is poorly understood. There is currently some debate regarding channel type, i.e., single channel conductance, magnitude, Ca^{2+} versus voltage gating, and the relationship between the anion and cation channels. For example, the lymphocyte system responds

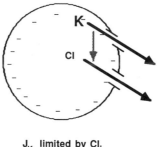

J_K limited by Cl_i

Figure 1. Cartoon depicting cell K^+ and Cl^- loss via conductive pathways. Cl^- flux is coupled to K^+ via the membrane potential. Net $K^+ : Cl^-$ stoichiometry = 1 and pH changes are not a consequence of net transport.

to cell volume and Ca^{2+}, yet $[Ca^{2+}]_i$ does not appear to increase in response to cell swelling. If, however, the cells are Ca^{2+}-depleted, net K^+, Cl^- loss is reduced (Grinstein *et al.*, 1984). Thus, while one set of observations argues for Ca^{2+}-dependent activation, the fact that Ca^{2+} does not appear to increase argues against Ca^{2+} involvement. One possible explanation for the seemingly contradictory results could be that the cell volume change causes the Ca^{2+} sensitivity of the mechanisms responsible for volume regulation to increase (see Cala *et al.*, 1986). If such were the case, then Ca^{2+} could play a role in volume-dependent activation in the absence of any increase in $[Ca^{2+}]_i$.

Common features of volume decrease due to ion conductance changes include (1) a transient change in the membrane potential, (2) a potentiation of K^+ and water loss by substitution of Cl^- with anions to which the membrane has higher conductance, such as NO_3^-, I^-, or SCN^-, (3) inhibition of swelling-activated K^+ conductance by quinine or cetiedyl, and (4) inhibition of swelling-activated Cl^- conductance by diphenylamine-2-carboxylate (DPC) and indacrinone (MK-196).

2.2. K^+,Cl^- Cotransport

A K^+,Cl^- cotransport mechanism has been implicated in the corrective loss of KCl and water from osmotically swollen red cells from humans (Kaji, 1986), sheep (Ellory and Dunham, 1980; Lauf, 1983), rats (Duhm and Gobel, 1984), dogs (Parker, 1983), and fish (Bourne and Cossins, 1984; Lauf, 1982), from Ehrlich ascites tumor cells (Thornhill and Laris, 1984; Kramhoft *et al.*, 1986), and from certain low-resistance epithelia (Corcia and Armstrong, 1983). While the pathway is volume sensitive in all of the above, it has yet to be established that it subserves a volume regulatory function (Fig. 2).

Strong evidence for electrically neutral K^+,Cl^- cotransport has been obtained in studies on duck red cells (Lytle and McManus, 1987). When swollen, these cells exhibit mutually interdependent net movements of K^+ and Cl^- in a fixed ratio of 1 : 1.

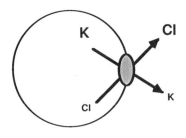

J_K limited by Cl_i

Figure 2. Cartoon depicting cell K^+ and Cl^- loss as a result of electroneutral K^+ plus Cl^- cotransport. As is the case for conductive loss, the stoichiometry of net $K^+ : Cl^-$ flux is unity and pH is unaffected by net transport.

As expected for a cotransport mechanism, the kinetic characteristics of K^+ activation critically depend on the prevailing concentration of Cl^-, and vice versa. The rapid K^+ extrusion evoked by swelling does not detectably alter the membrane potential as measured with a potentiometric dye (McManus *et al.*, 1985), even though calculations based on the observed K^+ flux and the measured anion conductance of these cells predict large, readily detectable changes in membrane voltage if the K^+ flux were conductive.

Features of K^+, Cl^- cotransport by duck red blood cells, which might serve as diagnostic criteria in other cell types, include: (1) sodium-independence, (2) a specific cation requirement (K^+ or Rb^+), (3) a specific anion requirement (Cl^- or Br^-), (4) a hyperbolic cation dependence with a low apparent affinity ($K_{0.5}$ for $[K^+]_o$ near 30 mM), (5) complete inhibition by relatively high concentrations (100–500 μM) of certain loop diuretics (furosemide > bumetanide) or stilbenedisulfonates (DIDS $\simeq$ SITS), with drug potency being critically dependent on $[K^+]_o$, and (6) acceleration by cell swelling.

K^+, Cl^- Cotransport and Conductance: Practical Considerations

As discussed in the Introduction, the regulation of cell volume is dependent upon net solute flux and therefore the availability of solute and sufficient force to drive net solute flux. With regard to both substrate availability and driving force, volume regulation via parallel K^+ and Cl^- conductance or K^+, Cl^- cotransport presents some interesting problems since, as a result of negatively charged intracellular protein and the constraint of macroscopic electroneutrality, cells contain less Cl^- than K^+. In excitable cells with large negative membrane potentials and relatively low $[Cl^-]_i$, the discrepancy between intracellular K^+ and Cl^- content is particularly pronounced. Thus, in the absence of some means of Cl^- recycling, the net K^+ and therefore H_2O loss is limited by cell Cl^- content. In addition to problems of substrate availability, there are also energetic considerations. For example during net K^+, Cl^- cotransport out of the cell, the net force driving KCl loss will approach zero as $[Cl^-]_i$ decreases and $[Cl^-]_o/[Cl^-]_i$ approaches $[K^+]_i/[K^+]_o$. If, however, the membrane is also capable of Cl^-/HCO_3^- exchange, the anion exchange pathway will serve as a mechanism for recycling Cl^-, thereby maintaining $[Cl^-]_i$ sufficiently high to support net transport (Fig. 3). That is, as Cl^-_i is lost with K^+, the ratio of $[Cl^-]_i/[Cl^-]_o$ will decrease relative to $[HCO_3^-]_i/[HCO_3^-]_o$. This will increase the force driving Cl^- into the cell and HCO_3^- out of the cell through the Cl^-/HCO_3^- exchanger. A diagnostic feature of KCl loss by either conductance (Fig. 3a) or cotransport (Fig. 3b) pathways in parallel with Cl^-/HCO_3^- exchange is that intracellular acidification (and ultimately net KCl loss) should be blocked by inhibitors of the Cl^-/HCO_3^- exchange such as the disulfonic stilbenes. Further, in the presence of Cl^-/HCO_3^- exchange inhibitors (absence of Cl^-/HCO_3^- exchange), the ratio of net Cl^- to K^+ efflux should approach its limiting value of unity, while in the absence of such inhibitors (presence of Cl^-/HCO_3^- exchange), the net Cl^-/K^+ stoichiometry should decrease.

Of additional concern is the effect of conductive-flux-mediated volume regulation on the electrical activity of excitable cells. The flux rates reported, for example in

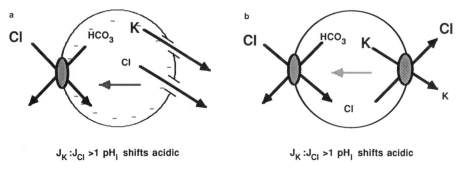

$J_K : J_{Cl} > 1$ pH$_i$ shifts acidic $J_K : J_{Cl} > 1$ pH$_i$ shifts acidic

Figure 3. Net KCl loss via conductive pathway (a) or electroneutral KCl cotransport (b) in parallel with Cl$^-$/HCO$_3^-$ exchange. The Cl$^-$/HCO$_3^-$ exchange is functionally coupled to the conductive and cotransport pathways as a result of changes in Cl$^-$. That is, as [Cl$^-$]$_i$ decreases, the force driving Cl$^-$ into the cell increases. Thus, the Cl$^-$/HCO$_c^-$ exchanger tends to buffer changes in [Cl$^-$]$_i$ and produce changes in pH$_i$.

studies of human lymphocytes or Ehrlich ascites cells are of such magnitude as to totally disrupt the normal electrical activity of excitable cells. If we accept the notion that cell volume regulation is necessary for the preservation of normal cell function, then it would seem unlikely that excitable cells should employ conductive transport pathways for volume regulation.

2.3. The K$^+$/H$^+$ Exchange Pathway

K$^+$/H$^+$ exchange-dependent cell volume regulation has been most extensively studied in the *Amphiuma* red blood cell (Cala, 1980, 1983a,b, 1985). The pathway mediates net K$^+$ efflux, the rate and magnitude of which are graded functions of the degree of cell swelling. Net K$^+$/H$^+$ exchange is able to effect a net decrease in osmotically active intracellular particles only if the H$^+$ is buffered (Fig. 4a), since to

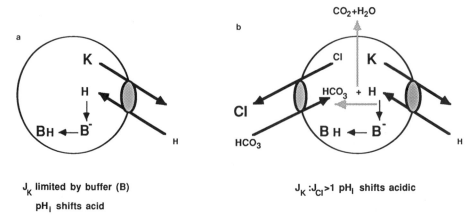

J_K limited by buffer (B) $J_K : J_{Cl} > 1$ pH$_i$ shifts acidic

pH$_i$ shifts acid

Figure 4. Cartoon depicting K$^+$/H$^+$ exchange-dependent volume regulation. Note that net solute and H$_2$O flux is dependent upon H$^+$ buffering by protein buffers (a) or Cl$^-$/HCO$_3^-$ exchange (b).

the extent that H is in free solution, it contributes to the osmolarity, although at any reasonable cell pH the concentration of H^+ is on the order of 0.1 μM. Consequently, net flux K^+/H^+ exchange in an unbuffered system will not mediate net particle flux and will not contribute to net H_2O flow. The *Amphiuma* red blood cell has a high protein or "fixed" buffer capacity, which is on the order of 200 mmoles H^+/unit change in pH_i per kg dry cell solid ($\sim$ 70 mmoles H^+/liter of cells or 100 mmoles H^+/liter cell H_2O). In addition to the protein buffers, the *Amphiuma* red blood cell has a robust Cl^-/HCO_3^- exchanger that functions as a dynamic H^+ buffering system (Fig. 4b). That is, the Cl^-/HCO_3^- exchanger responds to incipient decreases in pH_i and therefore decreased $[HCO_3^-]_i$ (resulting from net K^+/H^+ exchange) by adding HCO_3^- to the cell interior in exchange for Cl_i^-. Thus, H^+ and HCO_3^- are counterions in an electrical sense, serving to neutralize charge transferred by K^+ and Cl^-, respectively, yet the H^+ and HCO_3^- ions are without osmotic consequence since they combine and move back through the membrane as CO_2 and H_2O (see Cala, 1980, 1983a,b).

2.3.1. The Role of Cl^-

The efficiency of volume regulation via K^+/H^+ exchange is increased as a result of the parallel operation of the Cl^-/HCO_3^- exchanger. First, the anion exchange pathway prevents changes in pH_i which would tend to dissipate the force driving K^+/H^+ exchange, and second, when operating in parallel with K^+/H^+ exchange, the anion exchanger contributes to net particle (Cl^-) and therefore osmotically obliged H_2O loss. That is, when in parallel with Cl^-/HCO_3^- exchange, each cycle of K^+/H^+ exchange results in net transport of not one (K^+) but two (K^+ and Cl^-) particles and therefore the H_2O loss associated with each cycle of the K^+/H^+ exchanger is doubled. In the event that $[Cl^-]_i$ decreases to a point where net Cl^-/HCO_3^- exchange is no longer significant, or if the anion exchanger is inhibited with DIDS or SITS, the K^+/H^+ exchange pathway can continue to mediate volume flux to the extent that net H^+ flux is buffered by fixed buffers. This is in contrast to the K^+,Cl^- conductance and cotransport pathways where Cl^- depletion results in cessation of net K^+ flux.

2.3.2. Activation of K^+/H^+ Exchange

K^+/H^+ exchange is activated by cell volumes greater than control or by exposure of cells at normal volume to the Ca^{2+} ionophore A23187 (Cala, 1983b). If cells are swollen in A23187-containing media, the volume-induced net K^+ loss is increased, suggesting a role for $[Ca^{2+}]_i$ in the activation of the pathway. More suggestive of a role for $[Ca^{2+}]_i$ is the fact that exposure of cells in isotonic medium to subthreshold concentrations of A23187 followed by osmotic swelling in the absence of A23187 results in accelerated net K^+ loss as compared to that of cells swollen to the same degree but never exposed to A23187. Consistent with the notion that $[Ca^{2+}]_i$ is involved in activating the K^+/H^+ exchanger, experiments evaluating $[Ca^{2+}]_i$ using arsenazo III and digitonin-permeabilized cells revealed that osmotically swollen cells had lost Ca^{2+} buffer capacity and inferentially then that $[Ca^{2+}]_i$ increased in response to cell swelling (Cala *et al.*, 1986). That increased $[Ca^{2+}]_i$ is not solely responsible for activating volume-dependent K^+/H^+ exchange is illustrated by the observation that the apparent Ca^{2+} affinity of the K^+/H^+ exchanger activation/translocation apparatus is increased by cell swelling. Thus, in addition to increased $[Ca^{2+}]_i$ there are other events associated with cell swelling that are involved in the volume-dependent activation of *Amphiuma* red blood cell K^+/H^+ exchange.

2.3.3. Inhibitors of K^+/H^+ Exchange

It appears that some K^+/H^+ exchange inhibitors (e.g., calmodulin blockers, internal Ca^{2+} depletion, and disruptive anions like SCN^- or NO_3^-) interfere with biochemical events regulating volume-sensitive transport rather than the transport protein itself. Commonly used inhibitors of ion transport (ouabain, DIDS or SITS, and loop diuretics like bumetamide or furosemide) do not inhibit K^+/H^+ exchange and no specific inhibitor of K^+/H^+ exchange has been identified.

2.4. Force–Flow Analysis: Distinction between Various Modes of K^+ Efflux

As discussed above, volume regulation subsequent to cell swelling is most commonly the result of conductive K^+ and Cl^- loss, K^+,Cl^- cotransport, or K^+/H^+ exchange, the latter in parallel with Cl^-/HCO_3^- exchange. Attempts to distinguish between the various modes of K^+ flux based upon net flux measurements and simple kinetic criteria are complicated by the following: (1) the substrates are common to all pathways; (2) it is difficult to separate the effects of substrates on translocation from those on activation and deactivation; and (3) there are as yet no firmly established, discriminatory kinetic models around which studies can be designed. Consequently, this laboratory has found a thermodynamic or force–flow approach to pathway identification to be productive and relatively simple. The utility of the approach is based on the recognition that the ion flows are coupled to forces that are unique to each flux pathway. Therefore, if properly executed, force–flow studies can provide an unambiguous result. Briefly, the flux of K^+, regardless of the pathway, is always subject to the chemical force which reflects the energy stored in the K^+ concentration gradient. Thus, all modes of dissipative K^+ flux are, in part, driven by the chemical potential difference for K^+ ($\Delta\mu_{K^+} = RT \ln [K^+]_i/[K^+]_o$ where R and T are respectively the gas constant in joules/mole and temperature in degrees Kelvin) which is responsible for what might be thought of as diffusional drift. Because this term is common to all modes of K^+ flux, changes in K^+ flux corresponding to changes in $\Delta\mu_{K^+}$ alone are not discriminatory. However, each of the above modes of K^+ flux is subject to a second force term, which is responsible for *imposed* drift (as opposed to *diffusional* drift). Unlike $\Delta\mu_{K^+}$, this second term is unique to each pathway (Fig. 5). Using the more familiar example of K^+ conductance, the membrane voltage (E_m) is responsible for imposed drift. That is, the voltage can drive net K^+ flux in the absence of a K^+ concentration gradient or, in fact, against a K^+ concentration gradient. By analogy with the above, imposed drift of K^+ through the K^+,Cl^- cotransport pathway is due to the chemical potential difference for Cl^- ($\Delta\mu_{Cl^-}$), while that responsible for imposed drift of K^+ via K^+/H^+ exchange is the chemical potential difference for hydrogen ion. Thus, the relevant driving forces for

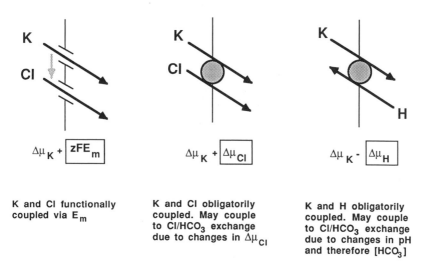

$$\Delta\mu_K + \boxed{zFE_m}$$ $$\Delta\mu_K + \boxed{\Delta\mu_{Cl}}$$ $$\Delta\mu_K - \boxed{\Delta\mu_H}$$

K and Cl functionally coupled via E_m

K and Cl obligatorily coupled. May couple to Cl/HCO$_3$ exchange due to changes in $\Delta\mu_{Cl}$

K and H obligatorily coupled. May couple to Cl/HCO$_3$ exchange due to changes in pH and therefore [HCO$_3$]

Figure 5. Cartoon depicting K^+ flux pathways involved in volume regulation in response to cell swelling and the relevant thermodynamic forces responsible for net transport by the various routes. The second term in each expression is that responsible for imposed drift and is unique to each pathway.

K^+ conductance, K^+,Cl^- cotransport, and K^+/H^+ exchange are respectively ($\Delta\mu_{K^+}$ + zFE_m), ($\Delta\mu_{K^+} + \Delta\mu_{Cl^-}$), and ($\Delta\mu_{K^+} - \Delta\mu_{H^+}$). Therefore, we reasoned that the least ambiguous means of identifying the K^+ flux pathway was to determine the relevant term responsible for imposed drift. Since red cell H^+ and Cl^- are distributed at electrochemical equilibrium, then by definition, $zFE_m = \Delta\mu_{H^+} = \Delta\mu_{Cl^-}$. Consequently, it was necessary to alter membrane voltage (using valinomycin) and displace H^+ from electrochemical equilibrium so that $zFE_m \neq \Delta\mu_{H^+}$ and, in analogy with a current–voltage plot (Fig. 6), evaluate the dependence of net K^+ flux upon ($\Delta\mu_{K^+}$ + zFE_m) and ($\Delta\mu_{K^+} - \Delta\mu_{H^+}$). Since K^+ flux is observed in media where all Cl^- is replaced by gluconate or p-aminohippurate (pAH), and therefore cell Cl^- is replaced by HCO_3^-, it was not necessary to evaluate correspondence of flux with ($\Delta\mu_{K^+} + \Delta\mu_{Cl^-}$). The power of the approach is derived from analysis of the correspondence of net flux and net force direction and not flux rate or magnitude. That is, one can imagine many scenarios in which K^+ flux or magnitude might respond to [H^+], [Cl^-], or voltage, yet in the absence of metabolic input reversal of the direction of K^+ flux will occur only if the term responsible for imposed drift is greater than and oppositely directed to $\Delta\mu_{K^+}$. Thus, we were able to establish that both the volume- and Ca^{2+}-induced net K^+ flux observed in *Amphiuma* red cells is via K^+/H^+ exchange (Cala, 1983a,b, 1985).

3. VOLUME REGULATION SUBSEQUENT TO CELL SHRINKAGE

In response to osmotic shrinkage, the most frequently implicated volume regulatory pathways are Na^+,Cl^- cotransport, Na^+,K^+,Cl^- cotransport, and Na^+H^+ exchange. In contrast to volume regulation following cell swelling, conductive trans-

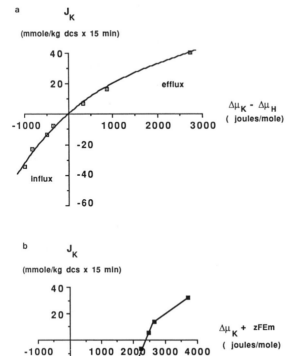

Figure 6. The relationship between net K^+ flux by osmotically swollen *Amphiuma* red blood cells and (a) the force driving K^+/H^+ exchange ($\triangle\mu_K - \triangle\mu_H$), or (b) that driving conductive K^+ flux ($\triangle\mu_K + zFE_m$). Note that in both a and b the flux decreases as the force decreases. Yet in b, flux reverses direction before the force driving conductive flux changes sign. In contrast, the magnitude and direction of net K^+ flux correlates with $\triangle\mu_K - \triangle\mu_H$. Thus, we conclude that net K^+ flux by swollen cells is due to K^+/H^+ exchange. (Note: the measurements were performed in the presence of valinomycin so that the membrane is hyperpolarized and $E_m \neq \triangle\mu_H$.)

port pathways do not appear to play a role in the volume regulation by shrunken cells. Further, while volume regulation by swollen cells can be limited by intracellular Cl^- availability, such is not the case for shrunken cells employing Na^+,Cl^- cotransport and Na^+,K^+,Cl^- cotransport, since net ion uptake is promoted by the availability of extracellular Cl^- as a substrate as well as the energy stored in the Cl^- concentration gradient. Volume regulation via Na^+,Cl^- cotransport, as described for the Ehrlich ascites tumor cell (Hoffmann, 1986), is highly energetically favorable since both the Na^+ and Cl^- chemical gradients are directed into the cell. In addition, because Na^+ and Cl^- are the primary osmotic constituents of plasma, substrate availability is not a problem. The pathway is sensitive to the loop diuretics furosemide and bumetanide and, as is the case for K^+,Cl^- cotransport, Na^+,Cl^- cotransport has an absolute anionic requirement for Cl^- or Br^-.

3.1. Na+,K+,Cl− Cotransport

Na^+,K^+,Cl^- cotransport (Fig. 7) has been identified in a variety of animal cells where it serves two related functions: (1) net transport of salt and water into the cell in an electrically neutral fashion, and/or (2) maintenance of internal $[Cl^-]$ above eleca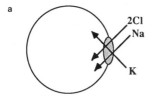

b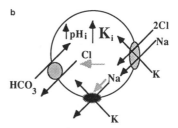

Increase volume and buffer K.

Increase volume buffer K_o and increase K_i and pH_i.

Figure 7. Na^+, K^+, Cl^- cotransport alone (a) and in parallel with Cl^-/HCO_3^- exchange (b). Note that when operating in parallel with Cl^-/HCO_3^- exchange, changes in cell and/or medium pH are a consequence of net cotransport.

trochemical equilibrium. Cells might exploit electroneutral Na^+, K^+, Cl^- entry to drive transepithelial salt transport (O'Grady *et al.*, 1987), to modulate cellular electrolyte and water content (Geck and Heinz, 1986), to energize acid extrusion via Cl^-/HCO_3^- exchange (Wieth and Brahm, 1985), or to buffer disturbances in $[K^+]_o$ (Haas and McManus, 1985).

The Na^+, K^+, Cl^- cotransport pathway is best described in the duck red blood cell where it is activated by cell shrinkage or manipulations that increase cell cAMP (propranolol norepinephrine exposure). While the pathway has been extensively studied in the duck red blood cell following activation by cell shrinkage, it is highly unlikely that it serves a volume regulatory function in this cell. That is, in the duck red blood cell, under conditions of normal transmembrane Na^+, K^+, and Cl^- distribution, the net force driving transport is zero. Therefore, activation of the pathway under these conditions leads to large increases in unidirectional flux, yet net flux remains zero (see Tosteson and Robertson, 1956; Kregenow, 1971). Thus, while the pathway exhibits volume-sensitivity necessary for a volume regulatory pathway, the driving forces are such as to preclude net ion and therefore H_2O flux. Typical studies of net transport in the shrinkage-activated duck red blood cell have been performed at $[K^+]_o$ between 15 and 20 mM (Kregenow, 1971; Schmidt and McManus, 1977). Clearly, an animal with plasma K^+ in the range of 15–20 mM has more pressing problems than cell volume control.

If we examine the expression for the force driving Na^+, K^+, Cl^- cotransport:

$$RT \ln [Na^+]_i/[Na^+]_o + RT \ln [K^+]_i/[K^+]_o + RT \ln [Cl^-]_i^2/[Cl^-]_o^2$$

it is immediately obvious that the force is exquisitely sensitive to changes in the $[Cl^-]$ gradient, since the force depends upon the square of the $[Cl^-]$ distribution ratio (see below). Thus, the Cl^- chemical gradient constitutes a major contributor to net driving force and will tend to promote salt and H_2O uptake by cells with low $[Cl^-]_i$ (i.e., excitable cells). Further, since $[Cl^-]_i$ is lower than $[Cl^-]_o$, the Cl^- ratio and therefore the driving force will be most sensitive to changes in $[Cl^-]_i$. Extending these same arguments to Na^+ and K^+, since $[Na^+]_i$ and $[K^+]_o$ are small relative to $[Na^+]_o$ and $[K^+]_i$, the force driving Na^+, K^+, Cl^- cotransport is going to be more sensitive to changes in $[Na^+]_i$ and $[K^+]_o$ than their trans-counterparts. Consequently, small changes in $[K^+]_o$ will substantially alter the magnitude of the force driving flux into the cell. Thus, while the Na^+,K^+,Cl^- cotransport pathway is not well suited to the task of volume regulation in the duck red blood cell (or other high $[Cl^-]_i$ cells), it is poised to serve as a mechanism for buffering plasma $[K^+]$.

Measurements of net ion movement indicate an electrically neutral coupling stoichiometry of $1 Na^+ : 1 K^+ : 2 Cl^-$ (Haas *et al.*, 1982; Geck *et al.*, 1980; Kracke *et al.*, 1988). An intriguing exception to this rule is found in the squid axon, where bidirectional $2 Na^+ + 1 K^+ + 3 Cl^-$ cotransport has been demonstrated based upon rigorous criteria (Altamirano and Russell, 1987). With the above-noted exception, the stoichiometry of net transport is $1 Na^+ : 1 K^+ : 2 Cl^-$, yet the ratio of unidirectional cation fluxes often exhibits a different and variable stoichiometry ranging from $5 K^+ : 1 Na^+$ to $1 K^+ : 3 Na^+$ (see Parker and Dunham, 1989). In avian and mammalian red cells, the discrepancy between net and unidirectional transport stoichiometry can be attributed to sectors of the cotransport cycle that mediate unproductive, one-for-one cation exchange, i.e., K^+/K^+ exchange in cells containing mostly K^+, and Na^+/Na^+ exchange in cells containing mostly Na^+ (Lytle and McManus, 1986; Duhm, 1987). Since at present such partial reactions have only been investigated in erythrocytes, it is not known if they are common features of all Na^+,K^+,Cl^- cotransport mechanisms.

The Na^+,K^+,Cl^- cotransporter can accept Rb^+ in lieu of K^+, Li^+ for Na^+ (albeit with low affinity), and Br^- for Cl^-. Partial replacement of Cl^- with permeant anions like nitrate, thiocyanate, and iodide retards Na^+,K^+,Cl^- cotransport and alters many other volume-sensitive transport systems (see Parker, 1984; Adorante and Cala, 1987). These anions, which are known to associate with proteins and alter their function, may directly compete with Cl^- for transport sites and/or they may interfere with biochemical events involved in transport regulation (discussed by Parker, 1988). Loop diuretics and stilbenedisulfonates, when used appropriately, can help define ion movements via cotransport. Bumetanide ($IC_{50} \sim 0.26 \mu M$), the current inhibitor of choice, blocks the Na^+,K^+,Cl^- cotransporter specifically at $5 \mu M$, although at concentrations in excess of $50 \mu M$ bumetanide has been observed to retard other ion transport pathways, including Cl^-/HCO_3^- exchange (Palfrey *et al.*, 1980) and K^+,Cl^- cotransport (Kaji, 1986). Furosemide, typically 10–100 times less potent than bumetanide, also lacks specificity at concentrations above $50 \mu M$. Optimal inhibition of cotransport by loop diuretics requires low concentrations ($\sim 20 mM$) of external Na_o^+, K_e^+, and Cl_e^- (Palfrey *et al.*, 1980). Stilbenedisulfonates like DIDS and SITS have little or no direct effect on Na^+,K^+,Cl^- cotransport (Palfrey *et al.*, 1980), but reversibly inhibit K^+,Cl^- cotransport in bird red cells provided that $[K^+]_o$ is elevated (Lytle and McManus, 1987), and irreversibly inhibit Cl^-/HCO_3^- exchange in a variety of cell types (Knauf, 1979).

3.2. The Na^+/H^+ Exchange Pathway

Probably the most commonly implicated and, by inference, the most widely distributed pathway responsible for regulating volume of shrunken cells is the Na^+

/H^+ exchange pathway (Fig. 8). The volume regulatory role of Na^+/H^+ exchange has been extensively studied in the *Amphiuma* (Cala, 1980, 1985) and dog erythrocyte (Parker, 1988) as well as the human lymphocyte (Grinstein and Rothstein, 1986). Since proton gradients are typically small and directed from cell to plasma while Na^+ gradients are steep from plasma to cell, net Na^+ uptake by Na^+/H^+ exchange is energetically favorable. The maintenance of a favorable driving force for net Na^+/H^+ exchange is, however, highly dependent upon both passive (chemical) and dynamic (ion transporters) H^+ buffers. That is, in a completely unbuffered system net H^+ flux on the Na^+/H^+ exchange would result in changes of pH_i that would dissipate the force driving Na^+/H^+ exchange prior to any significant H_2O flow. For example, assume a cell with $[Na^+]_i = 15$ mM and $pH_i = 7$ is suspended in media with a buffer capacity of zero at $pH_o = 7.4$ and $[Na^+]_o = 150$ mM. The net Na^+ (and) H^+ flux necessary to cause pH_o to drop to 6.0 and therefore dissipate the force driving Na^+/H^+ exchange is 9.6×10^{-7} mole/liter. Clearly, fluxes of the magnitude of 10^{-6} mole will not contribute to appreciable osmotically obliged H_2O flows in physiological solutions (osmolarity on the order of 300 mosm).

As described above, changes in pH can dissipate the energy stored in the combined gradients of Na^+ and H^+ and thus cause net Na^+ transport to cease. In addition to thermodynamic effects, the Na^+/H^+ exchange can be kinetically modulated by pH_i. Kinetic modulation by H^+ may be a reflection of the role played by Na^+/H^+ exchange in pH_i regulation. That is, in response to decreases in pH_i below some "set point" value, the Na^+/H^+ exchange of many cells is activated. Conversely, the Na^+/H^+ exchanger is deactivated as pH_i increases towards the "set point." Thus, failure of some cells to completely restore volume following cell shrinkage may reflect such pH_i-dependent inactivation of Na^+/H^+ exchange. It appears that such pH_i sensitivity is the reason that the Na^+/H^+ exchange of the human lymphocyte fails to completely restore cell volume since increases in pH_i lead to deactivation prior to volume recovery.

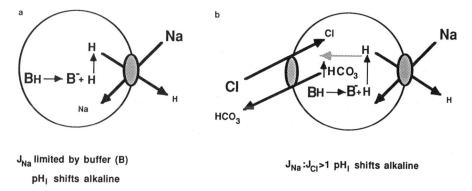

J_{Na} limited by buffer (B)
pH_I shifts alkaline

$J_{Na}:J_{Cl}>1$ pH_I shifts alkaline

Figure 8. Cartoon depicting volume and pH_i-regulating modes of the Na^+/H^+ exchanger. As was the case for K^+/H^+ exchange, net solute flux is dependent upon buffering H^+ flux resulting from net Na^+/H^+ exchanger. This is accomplished by fixed (a) or dynamic buffers (b).

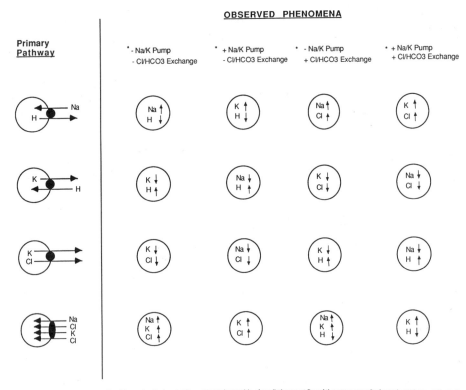

Figure 9. Cartoon depicting the interplay between various volume-sensitive flux pathways and the Na$^+$,K$^+$ pump and/or Cl$^-$/HCO$_3^-$ exchanger. The figures in the far left column depict the primary transport pathway. The figures in the first column to the right of the primary pathway depict the net effect of that pathway's operation. The figures in the three columns to the right depict the net effect of the primary pathway in parallel with the Na$^+$,K$^+$ pump and/or the Cl$^-$/HCO$_3^-$ exchanger. $^+$, the designated flux pathway is capable of mediating a net flux of the same magnitude as the primary pathway; $^-$, the designated flux pathway is incapable of mediating a flux of the magnitude of the primary pathway.

4. Cl$^-$/HCO$_3^-$ EXCHANGE: ANIONS AND pH

Dispersed through the above discussions of the various volume regulatory cation transport pathways are references to the consequences of parallel Cl$^-$/HCO$_3^-$ exchange function. Briefly, when operating in parallel with alkali metal/H$^+$ exchange the Cl$^-$/HCO$_3^-$ exchange "buffers" changes in pH$_i$ and mediates net Cl$^-$ flux. In contrast, when in parallel with alkali metal,Cl$^-$ cotransport or conductance pathways, the anion exchanger will buffer changes in [Cl$^-$]$_i$ and mediate changes in pH$_i$. Consequently, changes in pH$_i$ that occur in response to net cation flux and alkali metal : Cl$^-$ stoichiometries greater than 1 are not diagnostic of alkali metal/H$^+$ exchange nor is the absence of pH change and alkali metal : Cl$^-$ stoichiometry indicative of alkali metal cotransport or conductance (see Figs. 3, 4, 7, and 8). If net cation flux is via alkali metal/H^+ exchange and the anion exchanger is inhibited, then (1) net alkali metal flux will lead to pH changes that are predictable from a knowledge of net alkali metal ion flux and the system buffer power, and (2) the alkali metal : Cl^- stoichiometry will approach infinity. In contrast, if net cation flux is via alkali metal, Cl^- cotransport or conductance, then inhibition of parallel Cl^-/HCO_3^- exchange will (1) result in dissociation of pH changes from net alkali metal ion flux and (2) cause net alkali metal : Cl^- stoichiometry to approach its limiting value of unity (see Fig. 9).

5. THE Na⁺,K⁺ PUMP

While the above discussion has focused on the interaction of volume regulatory flux pathways and the Cl^-/HCO_3^- exchanger, we have omitted reference to another major player: the Na^+,K^+ pump. Since the volume regulatory pathways are dissipative, they ultimately depend on the energy stored in Na^+ or K^+ gradients generated by the Na^+,K^+ pump. The pump, as is true of the Cl^-/HCO_3^- exchanger, is also capable of obscuring the primary phenomenon. That is, if the pump is capable of transport rates approaching that of the volume regulatory pathways, it will have a major effect upon the net result of such transport (Fig. 9). If, for example, Na^+ enters the cell via Na^+/H^+ exchange, yet the Na^+,K^+ pump is capable of keeping pace with the Na^+/H^+ exchanger, the net result would be cell K^+ uptake and H^+ extrusion. If, in addition, the cell has a robust Cl^-/HCO_3^- exchanger, the net result would be cell K^+ and Cl^- uptake, a result with little resemblance to the initiating event.

ACKNOWLEDGMENTS. The author acknowledges the word processing skills of Leisa Miller, and the intellectual stimulation provided by Dr. Cris Lytle. Supported by USPHS Grant HL21179.

REFERENCES

Adorante, J. S., and Cala, P. M., 1987, Activation of electroneutral K flux in *Amphiuma* red blood cells by N-ethylmaleimide: Distinction between K/H exchange and KCl cotransport, *J. Gen. Physiol.* **90:**209–227.

Altamirano, A. A., and Russell, J. M., 1987, Coupled Na/K/Cl efflux: "Reverse" unidirectional fluxes in squid giant axons, *J. Gen. Physiol.* **89:**669–686.

Bourne, P. K., and Cossins, A. R., 1984, Sodium and potassium transport in trout (Salmo gairdneri) erythrocytes, *J. Physiol. (London)* **347:**361–375.

Cala, P. M., 1980, Volume regulation by *Amphiuma* blood cells. The membrane potential and its implication regarding the nature of the ion-flux pathways, *J. Gen. Physiol.* **76:**683–708.

Cala, P. M., 1983a, Volume regulation by red blood cells: Mechanism of ion transport, *Mol. Physiol.* **4:**33–52.

Cala, P. M., 1983b, Cell volume regulation by *Amphiuma* blood cells. The role of Ca as a modulator of alkali/H exchange, *J. Gen. Physiol.* **82:**761–784.

Cala, P. M., 1985, Volume regulation by *Amphiuma* red blood cells: Strategies for identifying alkali metal/H transport, *Fed. Proc.* **44:**2500–2507.

Cala, P. M., Mandel, L. J., and Murphy, E., 1986, Measurement of cytosolic free Ca during volume regulation in *Amphiuma* red blood cells, *Am. J. Physiol.* **19:**C423–C429.

Corcia, A., and Armstrong, W. M., 1983, KCl cotransport: A mechanism for basolateral chloride exit in *Necturus* gallbladder, *J. Membr. Biol.* **76:**173–182.

Duhm, J., 1987, Furosemide-sensitive K (Rb) transport in human erythrocytes: Modes of operation, dependence on extracellular and intracellular Na, kinetics, pH dependency, and the effect of cell volume and N-ethylmaleimide, *J. Membr. Biol.* **98:**15–32.

Duhm, J., and Gobel, B. O., 1984, Chloride-dependent, furosemide-sensitive K transport in human and rat erythrocytes. Dependence on external Na, internal Na, and cell volume, *Pfluegers Arch.* **402**(Suppl.): R11 (abstract).

Ellory, J. C., and Dunham, P. B., 1980, Volume-dependent passive potassium transport in LK sheep red cells, in: *Membrane Transport in Erythrocytes* (U. V. Lassen, H. H. Ussing, and J. O. Wieth, eds.), Alfred Benzon Symposium XIV, Munksgaard, Copenhagen, pp. 409–423.

Geck, P., and Heinz, E., 1986, The Na-K-2Cl cotransport system, *J. Membr. Biol.* **91:**97–105.

Geck, P., Pietrzyk, C., Burckhardt, B.-C., Pfeiffer, B., and Heinz, E., 1980, Electrically silent cotransport of Na^+, K^+, and Cl^- in Ehrlich cells, *Biochim. Biophys. Acta* **600:**432–447.

Grinstein, S., and Rothstein, A., 1986, Topical review: Mechanisms and regulation of the Na/H exchanger, *J. Membr. Biol.* **90:**1–12.

Grinstein, S., Clark, C. A., DuPre, A., and Rothstein, A., 1982, Volume-induced increase of anion permeability in human lymphocytes, *J. Gen. Physiol.* **80:**801–823.

Grinstein, S., Rothstein, A., Sarkadi, B., and Gelfand, E. W., 1984, Responses of lymphocytes to anisotonic media: Volume-regulating behavior, *Am. J. Physiol.* **246:**C204–C215.

Haas, M., and McManus, T. J., 1985, Effect of norepinephrine on swelling-induced potassium transport in duck red cells: Evidence against a volume-regulatory decrease under physiological conditions, *J. Gen. Physiol.* **85:**649–667.

Haas, M., Schmidt, W. F., and McManus, T. J., 1982, Catecholamine-stimulated ion transport in duck red cells. Gradient effects in electrically neutral [Na-K-2Cl] co-transport, *J. Gen. Physiol.* **80:**125–147.

Hazama, A., and Okada, Y., 1987, Electrophysiological evidence for independent activation of K and Cl conductances during regulatory volume decrease in cultured epithelial cells [abstract], *Eur. Soc. Comp. Physiol. Biochem.* 9th Conference.

Hoffmann, E. K., 1986, Anion transport systems in the plasma membrane of vertebrate cells, *Biochim. Biophys. Acta.* **864:**1–31.

Hoffmann, E. K., Simonsen, L. O., and Lambert, I. H., 1984, Volume-induced increase in K and Cl permeabilities in Ehrlich ascites tumor cells: Role for internal Ca, *J. Membr. Biol.* **78:**211–222.

Kaji, D., 1986, Volume-sensitive K transport in human erythrocytes, *J. Gen. Physiol.* **88:**719–738.

Knauf, P. A., 1979, Erythrocyte anion exchange and the band 3 protein. Transport kinetics and molecular structure, *Curr. Top. Membr. Transp.* **12:**249–363.

Kracke, G. R., Anatra, M. A., and Dunham, P. B., 1988, Asymmetry of Na-K-Cl cotransport in human erythrocytes, *Am. J. Physiol.* **254:**C243–C250.

Kramhoft, B., Lambert, I. H., Hoffmann, E. K., and Jorgensen, F., 1986, Activation of Cl-dependent K transport in Ehrlich ascites tumor cells, *Am. J. Physiol.* **251:**C369–C379.

Kregenow, F. M., 1971, The response of duck red cells to hypertonic media. Further evidence for a volume-controlling mechanism, *J. Gen. Physiol.* **58:**398–412.

Lauf, P. K., 1982, Evidence for chloride-dependent potassium and water transport induced by hyposmotic stress in erythrocytes of the marine teleost, *Opsanus tau, J. Comp. Physiol.* **146:**9–16.

Lauf, P. K., 1983, Thiol-dependent passive K/Cl transport in sheep red cells: I. Dependence on chloride and external K [Rb] ions, *J. Membr. Biol.* **73:**237–246.

Livne, A., Grinstein, S., and Rothstein, A., 1987, Volume-regulating behavior of human platelets, *J. Cell Physiol.* **131:**354–363.

Lytle, C., and McManus, T. J., 1986, A minimal model of [Na-K-2Cl] co-transport with ordered binding and glide symmetry, *J. Gen. Physiol.* **88:**36a (abstract).

Lytle, C., and McManus, T. J., 1987, Effect of loop diuretics and stilbene derivatives on swelling-induced [K-Cl] co-transport, *J. Gen. Physiol.* **90:**28a (abstract).

McManus, T. J., Haas, M., Starke, L. C., and Lytle, C., 1985, The duck red cell model of volume-sensitive chloride-dependent cation transport, *Ann. N.Y. Acad. Sci.* **456:**183–186.

O'Grady, S. M., Palfrey, H. C., and Field, M., 1987, Characteristics and function of Na-K-Cl cotransport in epithelial tissues, *Am. J. Physiol.* **253:**C177–C192.

Palfrey, H. C., Feit, P. W., and Greengard, P., 1980, cAMP-stimulated cation cotransport in avian erythrocytes: Inhibition by "loop" diuretics, *Am. J. Physiol.* **238:**C139–C148.

Parker, J. C., 1983, Hemolytic action of potassium salts on dog red blood cells, *Am. J. Physiol.* **244:**C313–C317.

Parker, J. C., 1984, Glutaraldehyde fixation of sodium transport in dog red blood cells, *J. Gen. Physiol.* **84:**789–803.

Parker, J. C., 1988, Na/H exchange and volume regulation in nonepithelial cells, in: *Na/H Exchange* (S. Grinstein, ed.), CRC Press, Boca Raton, pp. 179–190.

Parker, J. C., and Dunham, P. B., 1989, Passive cation transport, in: *Red Cell Membranes: Structure, Function and Clinical Aspects* (P. Agre and J. Parker, eds.), Dekker, New York.

Schmidt, W. F., III, and McManus, T. J., 1977, Ouabain-insensitive salt and water movements in duck red cells. I. Kinetics of cation transport under hypertonic conditions, *J. Gen. Physiol.* **70:**59–79.

Thornhill, W. B., and Laris, P. C., 1984, KCl loss and cell shrinkage in the Ehrlich ascites tumor cell induced by hypotonic media, 2-deoxyglucose, and propanolol, *Biochim. Biophys. Acta* **773:**207–218.

Tosteson, P. C., and Robertson, J. S., 1956, Potassium transport in duck red blood cells, *J. Cell. Comp. Physiol.* **47:**147–166.

Wieth, J. O., and Brahm, J., 1985, Cellular anion transport, in: *The Kidney: Physiology and Pathophysiology* (D. W. Seldin and G. Giebisch, eds.), Raven Press, New York, pp. 49–89.

Chloride Transport in the Squid Giant Axon

John M. Russell and Walter F. Boron

1. INTRODUCTION

The squid giant axon has been the object of study by neurobiologists for almost 50 years. It arises in the stellate ganglion as the syncytial fusion of several hundred small axons arising from separate ganglionic cells. Depending upon the squid species of origin, the diameter of the giant axon can vary from 200 μm to 1500 μm. The most commonly used species, *Loligo pealei* and *L. forbesi,* have axons with diameters of 400–900 μM. The large size, of the axon functionally serves to permit the rapid conduction of action potentials by reducing the electrical resistance of the axoplasm. Thus, the axon serves to rapidly transmit excitatory signals to the musculature of the mantle wall thereby permitting rapid escape or attack movements.

Chloride is, by far, the predominant free anion both in the axoplasm and in the fluid normally bathing the outside of the axon (hemolymph). Nevertheless, the study of Cl^- transport by this preparation has, until recently, lagged well behind the study of movements of Na^+, K^+, and Ca^{2+}. There are undoubtedly a number of reasons for this apparent neglect by neurobiologists. However, probably the single most important reason is the fact that, to electrophysiologists, the study of Cl^- transport across the axolemma has been relatively unrewarding. In fact, until fairly recently, there were no reports of Cl^- channels in squid axon (Inoue, 1985). Thus, the present interest in studying Cl^- transport by the giant axon stems largely from the role of Cl^- in axonal processes other than excitability. Figure 1 illustrates the documented mechanisms by which Cl^- is presently known to cross the axolemma. Two of the three mechanisms are electrically silent processes involving ion exchange (antiport) or ion cotransport (symport) processes. Each will be discussed separately in this chapter.

John M. Russell • Department of Physiology and Biophysics, University of Texas Medical Branch, Galveston, Texas 77550. *Walter F. Boron* • Department of Cellular and Molecular Physiology, Yale University, School of Medicine, New Haven, Connecticut 06510.

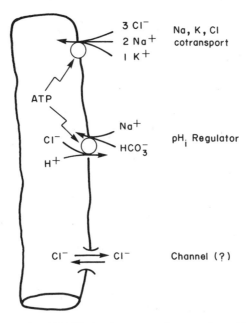

Figure 1. Mechanisms by which Cl⁻ has been reported to cross the squid axon cell membrane.

2. INTRAAXONAL AND EXTRACELLULAR [Cl⁻]

Before discussing the specific Cl⁻ transport processes, it is important to establish the baseline or physiological levels of Cl⁻ inside and outside the axon. The first reported measurements of total [Cl⁻] for giant axon axoplasm and for hemolymph were made in 1939 (Bear and Schmidt, 1939). They reported the [Cl⁻] of axoplasm to be 130 mM and in hemolymph, 474 mM. During the intervening 50 years, a number of other investigators have measured the [Cl⁻] of these two fluids with all but one obtaining essentially the same value (for review of this early literature see Russell, 1984). The one exception, however, had a rather far-reaching consequence. Steinbach (1941) reported a [Cl⁻] of 41 mM in the axoplasm from freshly dissected axons. Since such a value is close to that expected if Cl⁻ is in electrochemical equilibrium across the axolemma, it was often used by electrophysiologists as a typical value. Steinbach (1941) further reported that axoplasmic [Cl⁻] rapidly increased after dissection of the axon. At least nine other groups have measured axoplasmic [Cl⁻] and failed to find the low value reported by Steinbach (1941). Furthermore, Keynes (1963) made a very careful study of [Cl⁻] in freshly dissected axons and pointed out that the rate of increase of axoplasmic [Cl⁻] reported by Steinbach (1941) would require net Cl⁻ uptake rates of around 280 pmoles/cm²·sec. Such a rate is five to ten times higher than any reported unidirectional flux of Cl⁻ in the squid axon. Brinley and Mullins (1965) reported that [Cl⁻] in axoplasm obtained from squid that were taken in the winter was lower (71 mM) than from the same species of squid taken during the summer (111 mM). This observation required access to squid during winter and summer months. Such access is not currently possible and so no one has been able to confirm or extend this observation.

In summary, current evidence is that the free $[Cl^-]_i$ of squid axons is rather high (100–140 mM), which, given that $[Cl^-]_o = 490$ mM, yields an equilibrium potential for Cl^- of -29 to -37 mV (at 12°C). Since the *in situ* resting membrane potential of a squid axon is -60 to -70 mV, it is clear that there is a three- to fourfold higher $[Cl^-]_i$ than expected from simple electrochemical equilibrium conditions.

3. MECHANISMS OF TRANSAXOLEMMAL Cl⁻ MOVEMENTS

Early studies confirmed that Cl^- could cross the axolemma (Caldwell and Keynes, 1960). Thus, the high $[Cl^-]_i$ could not result from the impermeability of the axolemma to Cl^-.

3.1. Cl⁻ Conductance in the Squid Axon

There have been very few studies examining the Cl^- conductance of the axolemma. Adelman and Taylor (1961) found very little effect on the current–voltage relationship of replacing 80% of the external Cl^- with methylsulfate. Baker *et al.* (1962) reported that Cl^- conductance could account for no more than 20% of the leakage conductance. We have consistently observed that complete replacement of extracellular Cl^- with gluconate, sulfate, or methanesulfonate has very small effects on resting membrane potential. Similarly, varying the $[Cl^-]_i$ from 0 to 150 mM has only slight effects on resting membrane potential (Russell, 1976).

Recently, Inoue (1985) has reported a voltage-sensitive Cl^- conductance that could be blocked by intracellular application of SITS (extracellular application had only a slight inhibitory effect). Most of the study employed external and internal solutions that contained no Na^+ or K^+; Mg^{2+} or Ca^{2+} replaced the monovalents externally and TMA^+ or $TEMA^+$ was used in the internal fluid. The latter two organic cations also act to suppress voltage-sensitive K^+ channels. Tetrodotoxin was also added to suppress the voltage-sensitive Na^+ channel. Figure 2 illustrates the effect of varying $[Cl^-]_o$ on the current–voltage relationships in an axon perfused with a Cl^--free solution. At membrane potentials significantly more positive than normal resting potential, there developed an inward current which could be blocked by removal of external Cl^- (Cl^- replaced with phosphate) or by treatment with 100 μM SITS. Qualitatively similar results were obtained when the internal and external solutions were designed to be somewhat closer to a normal ionic condition. This voltage-sensitive current had essentially no time-dependent characteristics. By defining the Cl^- current as that current blocked by SITS, Inoue calculated the membrane permeability to Cl^- as a function of membrane potential. Figure 3 shows that the Cl^- permeability was activated at membrane potentials more positive than about 0 mV. It is of interest to note the relatively large current carried by this channel. At a membrane potential of -50 mV, the SITS-sensitive inward current was about 3 μA/cm². This is

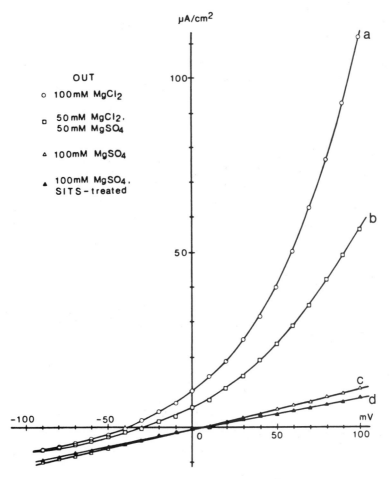

μA/cm²

OUT

○ 100 mM MgCl₂

□ 50 mM MgCl₂,
    50 mM MgSO₄

▵ 100 mM MgSO₄

▲ 100 mM MgSO₄,
    SITS-treated

Figure 2. Effects of varying $[Cl^-]_o$ and of internally applied SITS on the current–voltage relation of an internally perfused squid axon. The internal fluid contained 100 mM TEA and 0 mM Cl. (Reproduced from *The Journal of General Physiology*, 1985, Vol. 85, pp. 519–538 by copyright permission of the Rockefeller University Press.)

equivalent to a net efflux of around 30 pmoles/cm²·sec. This is larger than any reported unidirectional isotopic efflux. Whether this discrepancy is related to the different species of squid used in this electrical study or to methodological differences is unknown.

3.2. Coupled Na⁺,K⁺,Cl⁻ Cotransport

Na⁺,K⁺,Cl⁻ cotransport was first identified in Ehrlich ascites tumor cells (Geck *et al.*, 1980) and has since been identified in a variety of cell types (e.g., Chipperfield, 1986) including the squid giant axon (Russell, 1979, 1983; Altamirano and Russell,

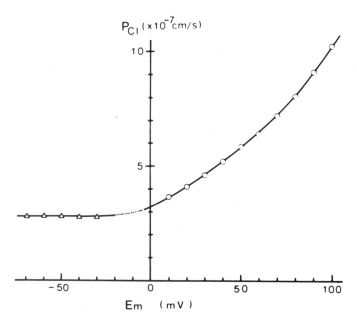

Figure 3. Membrane-voltage dependence of electrical Cl⁻ permeability (P_{Cl}). P_{Cl} was calculated using the Goldman–Hodgkin–Katz equation. (Reproduced from *The Journal of General Physiology*, 1985, Vol. 85, pp. 519–538 by copyright permission of the Rockefeller University Press.)

1987). The internally dialyzed squid giant axon provides an important advantage for the study of this cotransporter. It allows control of the solute conditions on both sides of the membrane. Furthermore, these controlled conditions can be maintained, unchanging, during any given experimental procedure. Such control permits rather more clear-cut interpretation of results than is possible with other preparations.

3.2.1. Basic Properties of the Axonal Na⁺,K⁺,Cl⁻ Cotransporter

These properties were identified several years ago (Russell, 1979, 1983) and will only be outlined here for the sake of completeness. A component of the unidirectional influxes (and effluxes, see below) of Cl⁻, Na⁺, and K⁺ are tightly coupled to one another with an absolute requirement that all three ions be present on the *cis* membrane side. The same component of the fluxes can be completely blocked by furosemide (0.3 mM) or by bumetanide (10 μM or less). The half-inhibition constant for bumetanide is about 0.1 μM. The disulfonic stilbene derivatives, SITS or DIDS, have no effect on these fluxes.

The preceding general properties are those found for the Na⁺,K⁺,Cl⁻ cotransporter in other preparations (e.g., Chipperfield, 1986). However, the squid axon cotransporter has several properties that have either not been observed in other cell types or have not been well characterized. A very characteristic property of the rate of Na⁺,K⁺,Cl⁻ cotransport by the squid axon is the inverse dependence on $[Cl^-]_i$.

Figure 4 shows an experiment in which both Cl^- and K^+ influxes were measured simultaneously into a single axon. At relatively high $[Cl^-]_i$ (150 mM), both influxes were very low. Removal of Cl_i^- by internal dialysis with a fluid in which Cl^- was replaced by glutamate stimulated both Cl^- and K^+ influxes. Returning $[Cl^-]_i$ to 150 mM resulted in inhibition of both fluxes. Treatment with 10 μM bumetanide reduced both fluxes to the level observed when $[Cl^-]_i$ was 150 mM (data not shown). Bumetanide-sensitive sodium influx is also inhibited by raising $[Cl^-]_i$. Thus, this cotransporter apparently cannot operate in the Cl^-/Cl^- exchange mode. Since K_i^+ was present throughout these experiments, it also follows that K-Cl/K-Cl exchange is not possible. It should be noted that $[Na^+]_i = 0$ for these experiments so any Na_i^+-activated modes of exchange fluxes, if present in squid axon, would not be observed (e.g., McManus, 1987; Duhm, 1987). Thus, we cannot yet say whether the glide symmetry model proposed by McManus (1987; see below) fits the Na^+,K^+,Cl^- cotransporter in squid axon.

3.2.2. Stoichiometry of the Process

An apparently major, and as yet unexplained, difference between Na^+,K^+,Cl^- cotransport in squid axon and that in most other preparations is the stoichiometry of the cotransport process. Net Na^+,K^+,Cl^- cotransport has been reported to occur with a stoichiometry of $1:1:2$ ($Na^+:K^+:Cl^-$) in Ehrlich ascites tumor cells (Geck et al., 1980), duck red blood cells (Haas et al., 1982), and MDCK cells (an epithelial cultured cell line; Aiton et al., 1981). In the squid axon, we have reported a stoichiometry of $2:1:3$ (Russell, 1983). A similar stoichiometry has been reported for ferret red blood cells (Hall and Ellory, 1985) although these data may have been contaminated with exchange fluxes (see below).

Bumetanide-sensitive exchange fluxes of Na^+ and K^+ have been identified in a

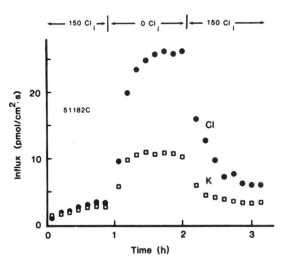

Figure 4. Effect of varying $[Cl^-]_i$ on unidirectional ^{36}Cl and ^{42}K influxes into an internally dialyzed squid axon. The dialysis fluid contained 4 mM ATP but no Na^+. Notice that decreasing $[Cl^-]_i$ from 100 mM to 0 mM increased both influxes and that the effect was reversible. (Reproduced from *The Journal of General Physiology*, 1983, Vol. 81, pp. 909–925 by copyright permission of the Rockefeller University Press.)

variety of preparations (e.g., see McManus, 1987, and Duhm, 1987). To the extent such fluxes occur, they could introduce an error in the determination of the stoichiometry of the cotransport process if one is measuring unidirectional isotopic fluxes. McManus and his co-workers have demonstrated both K^+/K^+ exchange (probably K-Cl/K-Cl exchange) and Na^+/Na^+ exchange under appropriate conditions in duck red cells. Indeed, these observations led McManus to propose the glide symmetry model of cotransport (McManus, 1987). This is a model of ordered binding in which the first ion bound is the first ion to debind, and so on. Figure 5 represents the reaction scheme for such a glide model which fits the data for erythrocytes (McManus, 1987; Duhm, 1987).

What may be seen from this particular model is that since Na^+ is first-on and first-off, exchange fluxes of all the ions require, as a minimum, the presence of intracellular Na^+ to permit the "reverse" movement to occur. Therefore, it is important to emphasize that our stoichiometry data were obtained in axons dialyzed with Na^+-free fluid thereby assuring that the axoplasmic $[Na^+]$ was very low. Furthermore, we have no evidence that any furosemide- or bumetanide-sensitive influxes of K^+ or Cl^- can occur in the absence of external Na^+. To determine stoichiometry we measured the influxes of ^{36}Cl, ^{22}Na, and ^{42}K which were inhibited either by furosemide or by increasing $[Cl^-]_i$ from 0 mM to 150 mM. All the experiments were performed as double-label experiments, using ^{42}K influx as the common label. We found that both treatments gave essentially the same stoichiometry. Further evidence against any exchange fluxes contaminating our results is the observation that raising *trans*-side $[Cl^-]$ (i.e., $[Cl^-]_i$) resulted in the inhibition of all three influxes, not their stimulation. (This interpretation must be tempered by the possibility that Cl_i^- might act at a site different from the transport sites to directly inhibit cotransport, i.e., a Cl^--sensitive modifier site may exist.)

We have not carefully studied the unidirectional efflux stoichiometry of the transporter but we have measured the bumetanide-sensitive effluxes of both ^{36}Cl and ^{22}Na under the same conditions and found their ratio to be $2.8:2$ ($Cl^- : Na^+$), which is reasonably close to the ratio of $3:2$ found for the influx stoichiometry (Altamirano and Russell, 1987).

It is important to point out explicitly that, if the stoichiometry in the squid axon is $2:1:3$ ($Na^+ : K^+ : Cl^-$), predictions based on the glide model proposed for red blood cells (which have a stoichiometry of $1:1:2$) might be in error. For example, if K^+ is the first-on, first-off ion in squid axon, then having intracellular K^+ would, in principle, permit exchange fluxes of K^+ (in the absence of intracellular Cl^- and Na^+). Clearly, much more work must be done to resolve this matter. The internally dialyzed squid axon provides a very useful model to test the glide symmetry hypothesis because one can independently and systematically vary the individual ion concentrations on either side of the membrane while maintaining steady-state conditions.

3.2.3. Reversibility and Asymmetry of the Process

We have demonstrated that the Na^+,K^+,Cl^- cotransporter in squid giant axon can mediate unidirectional fluxes in both directions across the axolemma. Such rever-

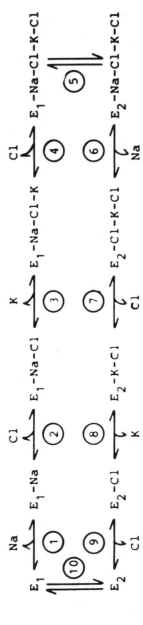

Figure 5. Reaction scheme for the glide symmetry model of Na^+, K^+, Cl^- cotransport in red blood cells as originally proposed by McManus (1987). E_1 = external-facing sites; E_2 = internal-facing sites. Partial reactions 3–8 would promote K^+/K^+ exchange; partial reactions 1–6 would promote Na^+/Na^+ exchange and partial reactions 1–7 would promote Cl^-/Cl^- exchange. Notice all three sets of partial reactions causing exchange fluxes require the presence of Na^+. (Reproduced, with permission, from Duhm, 1987.)

sibility would be expected for a secondary active transport process although, presumably, the physiologically dominant direction is inwards (see Section 3.2.4).

The fact that the transport process can operate in either direction should not be taken to mean that the properties of the cotransporter are perfectly symmetrical at the intra- and extracellular faces of the axolemma. One very important asymmetry is the absolute requirement for intracellular ATP (Russell, 1976, 1979, 1983; Altamirano and Russell, 1987) about which more will be said below (Section 3.2.4b). Another asymmetry appears to be the effect of *trans*-side Cl⁻. As already discussed (Sections 3.2.1 and 3.2.2), raising $[Cl^-]_i$ to around 150 mM reduces total Cl⁻, Na⁺, and K⁺ influxes by completely inhibiting all bumetanide-sensitive influxes of these ions. However, bumetanide-sensitive effluxes are relatively insensitive to $[Cl^-]_o$. Normal extracellular fluid for the axon has a very high [Cl⁻] (480–550 mM). Completely replacing Cl_o^- with methanesulfonate does result in a relatively small increase of Cl⁻ efflux (Altamirano and Russell, 1987) but it is obvious that the *trans*-side effect of Cl⁻ on efflux is small. Another difference that we have tentatively identified is for the $K_{0.5}$ for activation of bumetanide-sensitive Cl⁻ fluxes by Cl⁻. For efflux, the $K_{0.5}$ was found to be 53 mM with a V_{max} of only 10 pmoles/cm²·sec (Altamirano and Russell, 1987). Preliminary studies on Cl_o^- activation of bumetanide-sensitive Cl⁻ influx apparently revealed a very low affinity of the external site for Cl⁻ ($[Na^+]_o$ = 430 mM; $[K^+]_o$ = 10 mM). No clear sign of saturation was noted at $[Cl^-]_o$ = 560 mM; thus, no meaningful $K_{0.5}$ value could be calculated. However, it was clear that V_{max} for influx was much larger (at least 30–50 pmoles/cm²·sec) than we have noted for efflux. We must interpret this apparent kinetic asymmetry cautiously at this time since the ionic conditions were not the same on each side of the membrane for the influx and efflux experiments.

3.2.4. Energetics of Na⁺,K⁺,Cl⁻ Cotransport in Squid Axon

Table 1 lists the intracellular and extracellular concentrations of Na⁺, K⁺, Cl⁻, and HCO_3^- normally found in squid axoplasm and hemolymph, respectively. The theoretical equilibrium potentials (Nernst potential) are also presented for each ion. For

Table 1. Normal Transmembrane Ion Gradients
in Squid Giant Axon

	Intracellular	Extracellular	E_x (mV)
[Cl⁻] (mM)	120	490	−35
[Na⁺] (mM)	50	425	+54
[K⁺] (mM)	375	18	−77
pH	7.4	8.0	−35
[HCO₃⁻] (mM)	~2.5	~10	−35
	$E_m \cong -60$ to -70 mV		

none of these ions can the steady-state distribution be accounted for by a thermodynamically passive process or processes. Na^+ and K^+ distributions are the result of the ouabain-sensitive Na^+ pump (Na^+,K^+-ATPase) and we believe the Cl^- distribution is a result of the Na^+,K^+,Cl^- cotransporter. The nonequilibrium HCO_3^- distribution is believed to be the result of the Na^+-dependent Cl^-/HCO_3^- exchange (see Section 3.3).

3.2.4a. Is There Enough Energy in the Transmembrane Gradients of Na^+, K^+, and Cl^- to Account for the Observed $[Cl^-]_i$ via a Secondary Active Transport Mechanism?

If the higher-than-equilibrium $[Cl^-]_i$ results from Na^+, K^+,Cl^- cotransport, we must determine whether, in principle, there is enough energy in the relevant ion gradients to account for the observed Cl^- distribution. In order to determine this, we must calculate the free energy available for the operation of the cotransporter. And, in order to make this calculation, we must assume a stoichiometry for the net transport process. Since both reported stoichiometries are electroneutral, we can ignore the membrane potential. If we use the stoichiometry of $2:1:3$ ($Na^+:K^+:Cl^-$), then:

$$\Delta \dot{G}_{net} = RT \ln \left[\frac{(120)^3 \cdot (50)^2 \cdot 375}{(490)^3 \cdot (425)^2 \cdot 18} \right]$$

$$= -3.3 \text{ kcal}$$

At thermodynamic equilibrium this stoichiometry would support a $[Cl^-]_i$ of 740 mM. If we use the stoichiometry of $1:1:2$, $\Delta G_{net} = -1.1$ kcal. This latter stoichiometry would predict an equilibrium $[Cl^-]_i$ of about 315 mM. Thus, both stoichiometries could, in principle, easily support the observed Cl^- gradient, although the stoichiometry of $2:1:3$ has a considerable energetic advantage.

3.2.4b. What Is the Role of ATP in Na^+,K^+,Cl^- Cotransport?

Despite the fact that the preceding section shows that more than sufficient energy exists in the relevant ion gradients to account for the higher-than-equilibrium $[Cl^-]_i$ found in squid axoplasm, we have repeatedly demonstrated an absolute requirement for intracellular ATP by the cotransporter (Russell, 1976, 1979, 1983; Altamirano and Russell, 1987). Figure 6 shows that bumetanide-sensitive Cl^- influx is related to $[ATP]_i$ in a dose-dependent manner exhibiting a $K_{0.5}$ of 86 μM (Altamirano *et al.*, 1988). An ATP requirement for Na^+,K^+,Cl^- cotransport has been reported for the cotransporter in other cells as well (Chipperfield, 1986) yet again the requirement seems not to be on energetic grounds (Geck *et al.*, 1980).

What are the possible ways in which ATP might be utilized to promote activation of the Na^+,K^+,Cl^- cotransporter? At least three different mechanisms can be envisioned. One possibility is that the Na^+,K^+,Cl^- cotransporter is an ATPase, stoichiometrically consuming ATP during the operation of the cotransporter. Attempts by us to measure bumetanide-sensitive ATP utilization have been unsuccessful. Nevertheless, this possibility seems unlikely because vanadate, an inhibitor of ATPases, has

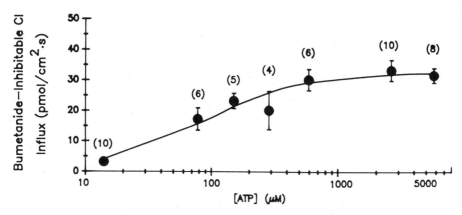

Figure 6. Dose–response relationship between axoplasmic [ATP] and bumetanide-sensitive Cl⁻ influx into the dialyzed squid giant axon. [ATP] was set by dialysis for 90–120 min with a dialysis fluid containing a given [ATP] plus an ATP buffer system consisting of 3 mM phosphoenolpyruvate and 3 mM arginine phosphate. The [ATP] of the extruded axoplasm was measured for every axon and corresponded with the [ATP] of the dialysis fluid. $[Cl^-]_i = 0$ mM; $[Na^+]_i = 0$ mM; $[K^+]_i = 400$ mM. Numbers in parentheses refer to the number of axons at each [ATP]. Solid line is fit by a nonlinear, least-squares version of the Hill equation [$K_{0.5} = 86$ μM; $V_{max} = 33$ pmoles/cm²•sec; Hill coefficient (n) = 0.97]. (Adapted, with permission, from Altamirano *et al.*, 1988.)

no effect on bumetanide-sensitive fluxes when ATP is present (Russell, 1983; and see below). Furthermore, as we have already shown, there are no obvious energetic reasons why an ATPase mechanism ought to be involved in Na^+, K^+, Cl^- cotransport.

Another possibility is that ATP might simply need to bind to the transporter or some essential cofactor in order to activate the cotransport function. Although we cannot completely rule out this possibility, we have some evidence that simple binding (without the possibility of hydrolysis) does not suffice to activate the cotransporter. Figure 7A shows an experiment in which Cl⁻ influx is measured while ATP is washed out of the axon. After ATP is washed out and Cl⁻ influx is stable at a low, baseline level, the so-called hydrolyzable ATP analogue, α,β-methylene-ATP, was added with no effect on Cl⁻ influx. Since this analogue is hydrolyzed at 1/1000th the rate of ATP in other systems (Yount, 1975), one cannot tell whether the lack of stimulation by this analogue was because the terminal phosphate group could not be hydrolyzed or whether it had no affinity for the ATP binding site. Figure 7B tests the latter possibility and shows that α,β-methylene-ATP apparently competes with ATP for the activation of Cl⁻ influx. This result strongly suggests that the analogue can bind to the cotransporter's ATP site, but not activate the cotransporter. Thus, ATP binding alone may not be sufficient to activate the cotransporter.

A third mechanism by which ATP could activate Na^+, K^+, Cl^- cotransport is by a phosphorylation/dephosphorylation mechanism utilizing a protein kinase and a protein phosphatase. According to such a model, phosphorylation of the cotransporter or an essential cofactor permits the cotransporter to mediate ion fluxes. Dephosphorylation would inactivate it. Recently, we have tested the following corollary of this hypothesis (Altamirano *et al.*, 1988); by reducing the rate of dephosphorylation, one

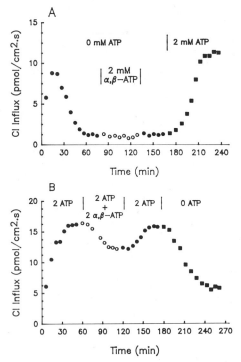

Figure 7. $^{36}Cl^-$ influx into internally dialyzed squid giant axons. Both axons were dialyzed with a fluid containing 50 mM Cl^- at 15°C; therefore, the Na^+,K^+,Cl^- cotransport rate was relatively low. (A) Effect of 4 mM α,β-methylene-ATP on Cl^- influx. Axon had been dialyzed with ATP-free dialysis fluid for 20 min prior to beginning the ^{36}Cl influx measure. As ATP was washed out, Cl^- influx declined. reaching a steady-state value of 1.2 pmoles/ cm^2·sec. Addition of α,β-methylene-ATP had no effect on Cl^- influx. However, when 4 mM ATP was presented to the axon, Cl^- influx increased to an average of 11.8 pmoles/cm^2·sec. (B) Effect of 2 mM α,β-methylene-ATP on Cl^- influx in the continuous presence of 2 mM ATP. After reaching a steady-state influx of 16.1 pmoles/ cm^2·sec in the presence of 2 mM ATP, 2 mM α,β-methylene-ATP was added to the dialysis fluid. This resulted in a fall of Cl^- influx to a steady state of 12.0 pmoles/cm^2·sec. This inhibitory effect was reversible upon removal of the α,β-methylene-ATP. Removal of ATP had its usual inhibitory effect resulting in a steady-state, ATP-insensitive Cl^- influx of about 5.5 pmoles/cm^2·sec. (From J. M. Russell, unpublished observations.)

should also reduce the rate of inactivation of cotransport caused by ATP depletion. Figure 8 shows the effects of two agents that are known to inhibit protein phosphatases. We compared the rate constants for inactivation of Cl^- influx caused by ATP depletion under three different conditions: (1) control conditions in which no protein phosphatase inhibitor was present, and when either (2) 40 μM vanadate or (3) 5 mM fluoride was presented intracellularly via the dialysis fluid. Neither vanadate nor fluoride had any statistically significant effect on Cl^- influx in the presence of ATP. (This result provides further evidence against the involvement of an ATPase.) However, both agents significantly slowed the rate of influx inactivation resulting from ATP depletion. Neither agent affected the rate of ATP hydrolysis by a squid brain homogenate (Altamirano *et al.*, 1988), a result that argues against the possibility that their effect on influx inactivation might be the result of a slower ATP depletion rate. Furthermore, at the end of one hour of ATP depletion, axons treated with either vanadate or fluoride had a significantly higher bumetanide-sensitive Cl^- influx remaining than did the control axons (cf. Fig. 8A,B,C). We interpret the preceding results to mean that the inactivation of bumetanide-sensitive Cl^- influx (presumably via Na^+,K^+,Cl^- cotransport) that occurs when ATP is depleted is a result of the dephosphorylation of the cotransporter or an essential cofactor. Vanadate and fluoride slow the inactivation by partially inhibiting this process. There are several possible reasons we did not observe complete prevention of the inactivation. Only one concentration of each of the inhibitors was used. Perhaps the protein phosphatase(s) was (were) incompletely inhib-

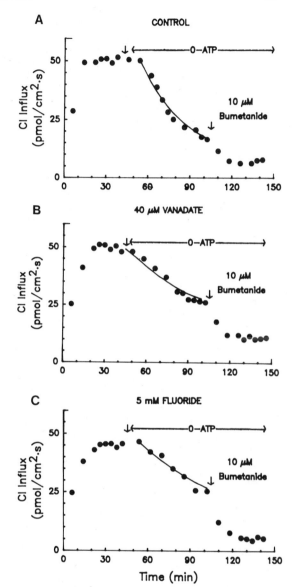

Figure 8. Inactivation of bumetanide-sensitive Cl⁻ influx into internally dialyzed axons during ATP depletion is slowed by intracellular vanadate or fluoride. (A) Control showing time course and degree of inactivation of bumetanide-sensitive Cl⁻ influx caused by 1 hr of ATP depletion prior to bumetanide treatment. The solid line through the data points represents a best exponential fit with a rate constant of -0.0306 min^{-1} ($t_{0.5}$ = 22.7 min). Axon diameter = 580 μm. (B) Effect of 40 μM vanadate present in the dialysis fluid throughout the experiment. Best exponential fit rate constant for inactivation = 0.0171 min^{-1} ($t_{0.5}$ = 40.5 min). Axon diameter = 560 μm. (C) Effect of 5 mm F⁻ present in the dialysis fluid throughout the experiment. Best exponential fit rate constant = 0.0137 min^{-1} ($t_{0.5}$ = 50.6 min). Axon diameter = 56 μm. (By permission, from Altamirano et al., 1988.)

ited. Perhaps more than one protein phosphatase can mediate the dephosphorylation. In fact, it is possible that vanadate and fluoride exert their effects on different phosphatases. A similar lack of complete inhibition of dephosphorylation by Phosphatase Inhibitor 2 has been reported in a study on ATP-dependent Ca²⁺ channels in heart muscle (Hescheler et al., 1987). Thus, whether this incomplete inhibitory effect represents spontaneous dephosphorylation or the presence of multiple protein phosphatases is unknown.

In summary, we believe our results are most consistent with Na^+, K^+, Cl^- cotransport being activated by phosphorylation and being inactivated by a protein phosphatase-mediated dephosphorylation mechanism. At present, we have no information about the presumed phosphorylation (activation) step. Such an activation/inactivation scheme implies that the Na^+, K^+, Cl^- cotransport rate is a regulated variable. However, at this time, the function of Na^+, K^+, Cl^- cotransport in the squid axon is unknown beyond being responsible for the high axoplasmic $[Cl^-]$. Why the axon requires a high $[Cl^-]_i$ is uncertain. One possibility is that the high $[Cl^-]_i$ is to permit the efficient functioning of the Na^+-dependent Cl^-/HCO_3^- exchange process which is responsible for regulating intracellular pH (see Section 3.3.2).

3.3. Na^+-Dependent Cl^-/HCO_3^- Exchange: The pH_i-Regulating Mechanism

Of the many animal cells tested, only red blood cells have a pH_i that is in electrochemical equilibrium with external pH and membrane potential (Roos and Boron, 1981). Table 1 lists the values for both intracellular and extracellular pH and the HCO_3^- concentrations inside the axon as well as in the fluid bathing the axon. As is the case with virtually all cells, the pH_i of the axon is more alkaline than expected if H^+ or HCO_3^- were in electrochemical equilibrium. Not only is the steady-state pH_i more alkaline than expected from passive thermodynamic considerations, but if the pH_i is made relatively acidic, there exists a mechanism in the axolemma to return pH_i to a normal value. We term the ion transport mechanisms responsible for the non-equilibrium behavior of pH_i, pH_i-regulating mechanisms. Given the near-universal presence of such acid-extruding mechanisms in animal cell membranes, it seems fair to conclude that the maintenance of pH_i at a relatively alkaline value rather far from electrochemical equilibrium is clearly one of the fundamental properties of animal cell membranes.

Interestingly, there appears to be more than one such pH_i-regulating mechanism used depending upon the specific animal cell being studied. At the moment the two about which the most is known are: (1) Na^+-dependent Cl^-/HCO_3^- exchange and (2) Na^+/H^+ exchange. Some cells use predominantly one or the other of these mechanisms, some use both, and for many cells the final answer as to the mechanism(s) used for pH_i regulation is still unknown.

The squid giant axon exclusively uses the Na^+-dependent Cl^-/HCO_3^- exchange mechanism to regulate pH_i. This fact simplifies the study of the transporter and, in fact, the squid axon has been one of the primary sources of information about the fundamental characteristics of this important transport process. For this review, we will focus on the role of Cl^- in this mechanism, but it is first necessary to present some of the fundamental general properties of the transporter.

3.3.1. General Properties of Na^+-Dependent Cl^-/HCO_3^- Exchange

Figure 9 illustrates one of the fundamental properties of the Na^+-dependent Cl^-/HCO_3^- exchanger. After the pH_i (measured with a longitudinally inserted glass micro-

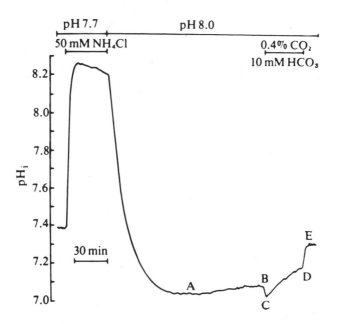

Figure 9. Demonstration of acid extrusion in a squid giant axon whose pH_i was lowered by prior exposure to NH_4^+. Washout of NH_4^+ caused the pH_i to fall to almost 7.0 whereas normal pH_i in the axon is about 7.4 (see trace prior to NH_4^+ application). In the nominal absence of HCO_3^-/CO_2, pH_i recovery was very slow. Addition of HCO_3^-/CO_2 caused an initial acidification (segment B–C) as a result of CO_2 entry and the subsequent formation and dissociation of carbonic acid. After the initial acidification, pH_i began to rise at a much faster rate (segment C–D) than when HCO_3^-/CO_2 were nominally absent. We term this recovery of pH_i, acid extrusion. The effect of HCO_3^-/CO_2 was fully reversible. (Reprinted by permission from *Nature*, Vol. 259, pp. 240–241, copyright © 1976, Macmillan Magazines Ltd.; Boron and DeWeer, 1976b.)

pH electrode) was acidified using the NH_4^+ prepulse technique (Boron and DeWeer, 1976a), the axon was initially bathed in a solution nominally free of CO_2. During this time (segment A–B) the recovery of pH_i was very slow (probably the result of ambient CO_2 causing a $[HCO_3^-] \cong 0.5$ mM at the pH of the external fluid, 8.0). However, when the external fluid was changed to one containing CO_2 and HCO_3^- (segment C–D of Fig. 9), the rate of pH_i recovery ("acid extrusion") was greatly accelerated. This acid extrusion process also requires that external Na^+ be present as well as intracellular ATP. The requirement for ATP does not seem to be on energetic grounds. Calculations like that in Section 3.2.4a. for the Na^+-dependent Cl^-/HCO_3^- exchanger reveal that the ion gradients contain sufficient energy to (in principle) maintain pH_i at about pH 8.0. Thus, we hypothesize that the role of ATP may be to modulate or regulate the activity of the exchanger (see Section 3.3.3). The process is completely inhibited by disulfonic acid stilbene derivatives such as SITS and DIDS, which are irreversible, or the reversible analogue, DNDS (Boron and Knakal, 1989).

All the aforementioned properties of the squid axon Na^+-dependent Cl^-/HCO_3^- exchanger, except one, are demonstrated by the other well-studied example of Na^+-dependent Cl^-/HCO_3^- exchange, the snail neuron. Thomas (1976) used mitochondrial poisons to reduce cellular [ATP] sufficiently to block Na^+ pump activity without inhibiting acid extrusion.

3.3.2. Role of Cl^- in the Acid Extrusion Process

In the absence of Cl_i^-, the squid axon cannot extrude an intracellular acid load. Figure 10 shows that removal of Cl_i^- by internal dialysis completely prevents recovery from an intracellular acid load (segment b–c). When the $[Cl^-]_i$ of the same axon was increased to 350 mM, the second exposure to CO_2/HCO_3^- resulted in a rapid rate of acid extrusion (segment f–g).

The dependence on Cl_i^- was found to be $[Cl^-]_i$-dependent (Boron and Russell, 1983). Figure 11 shows the different rates of acid extrusion in separate axons dialyzed with four different Cl^- concentrations. Data like these from 38 axons were fitted by a

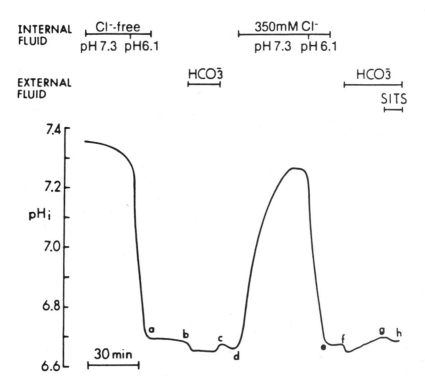

Figure 10. Demonstration of the dependence of acid extrusion on Cl_i^-. The axon was first dialyzed with a Cl^--free dialysis fluid (pH 6.1). The dialysis fluid flow was then turned off to return pH_i control to the axolemma. While Cl^--depleted, the axon could not extrude the acid load in the absence (segment a–b) or presence of (segment b–c) HCO_3^-/CO_2. After dialyzing the axon with a fluid containing 350 mM Cl^- and reacidifying the axoplasm, the dialysis fluid flow was again turned off and the pH_i monitored. With Cl_i^- present, acid extrusion occurred in the presence of HCO_3^-/CO_2. The acid extrusion was completely blocked by 0.5 mM SITS. (Reprinted by permission from *Nature*, Vol. 264, pp. 73–74, copyright © 1976, Macmillan Magazines Ltd.; Russell and Boron, 1976.)

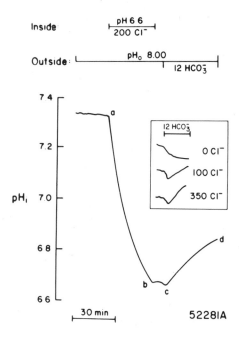

Figure 11. [Cl⁻]ᵢ-dependence of the rate of acid extrusion from dialyzed squid giant axons. At point a, dialysis was begun with a fluid containing 200 mM Cl⁻ at pH 6.6. Halting dialysis (point b) returned control of pHᵢ to the axon, but produced only a very slow pHᵢ recovery (b–c). The addition of 12 mm HCO₃⁻ to the SSW at a constant pH₀ of 8.00 caused pHᵢ to recover (c–d). The inset shows the results of similar experiments (comparable to segments b–c and c–d) on axons of approximately the same diameter. Although 12 mM HCO₃⁻ failed to stimulate pHᵢ recovery in the axon previously dialyzed with 0 mM Cl⁻ (top), the recovery rate was greater in axons dialyzed with 100 mM (middle) and 350 mM Cl⁻ (bottom). (Reproduced from *The Journal of General Physiology*, 1983, Vol. 81, pp. 373–399 by copyright permission of the Rockefeller University Press.)

nonlinear, least-squares fit to the Michaelis–Menten equation and yielded an apparent $K_m = 84 \pm 15$ mM for $[Cl^-]_i$. Thus, at least one function of the relatively high, normal $[Cl^-]_i$ of axoplasm (~ 120 mM) is to activate the pHᵢ regulating mechanism.

Not only is Cl_i^- required for the operation of the acid-extruding mechanism, but Cl^- is also transported out of the axon by this mechanism. Figure 12 shows the results of a unidirectional Cl^- efflux experiment in an internally dialyzed squid axon. The axon was first dialyzed with a fluid containing 150 mM Cl^- at a normal pH of 7.35. At this pHᵢ, there was no effect on Cl^- efflux when CO_2/HCO_3^- were presented externally. However, when the pHᵢ was lowered to 6.7 by internal dialysis, a second exposure to CO_2/HCO_3^- resulted in a reversible stimulation of Cl^- efflux (Russell and Boron, 1976; Boron and Russell, 1983). This HCO_3^--stimulated Cl^- efflux requires that pHᵢ be acidic, that intracellular ATP be present, and that Na^+ and HCO_3^- be present in the extracellular fluid. It is completely abolished by SITS. Extensive studies on the stoichiometry of the overall Na^+-dependent Cl^-/HCO_3^- exchange process revealed that one Cl^- is extruded from the axon, one Na^+ is taken up into the axon, while two acid equivalents are neutralized (Boron and Russell, 1983). Considerable uncertainty exists about the identity of the anion with which Cl^- actually exchanges. As shown in Fig. 1, it may be that Cl^- exchanges for an external HCO_3^- and that a H^+ is also extruded to give the stoichiometry of $1 \, Cl^- : 2 \, H^+$ neutralized. An alternative model also fits all available data (Boron and Russell, 1983; Boron and Knakal, 1989). In this latter model, Cl_i^- exchanges for the extracellular ion-pair $NaCO_3^-$, which upon entering the more acidic internal environment dissociates and neutralizes two acid equivalents. This model has the seductive, teleologic advantage of simply being an

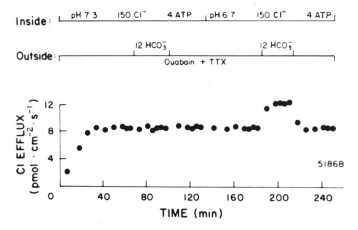

Figure 12. External HCO_3^- stimulates Cl^- efflux, but only when pH_i is acidic. At $pH_i = 7.3$, even though external Na^+ and internal ATP were present, there was no stimulation of Cl^- efflux when HCO_3^-/CO_2 was presented. However, when pH_i was reduced to 6.7 by continuous internal dialysis, exposure to HCO_3^-/CO_2 caused a prompt and reversible increase of Cl^- efflux. Such efflux could be blocked or prevented by treatment with SITS, removal of external Na^+, or removal of axoplasmic ATP. (Reproduced from *The Journal of General Physiology,* 1983, Vol. 81, pp. 373–399 by copyright permission of the Rockefeller University Press.)

anion exchanger and does not require discrete cation binding sites. Considerable further work will be required to reach a firm conclusion about which of the two models most accurately represents the actual mechanism.

3.3.3. The Role of ATP in the Operation of the Na^+-Dependent Cl^-/HCO_3^- Exchanger

One of the earliest observations about this transporter in the squid axon was that it could be inhibited by cyanide (Boron and DeWeer, 1976a) and had an absolute requirement for ATP (Russell and Boron, 1976). As is the case for the Na^+,K^+,Cl^- cotransporter (see Section 3.2.4a), there is more than enough potential energy in the relevant ion gradients. In this case, the gradients could maintain pH_i at a level slightly higher than 8.0. However, no evidence of activity by the ion transport mechanism can be detected at pH_i values higher than ~ 7.3. For example, no HCO_3^--dependent, SITS-inhibitable Cl^- efflux was noted at a pH_i of 7.35 (see Fig. 12). We take this kind of result to mean that the ion transport mechanism is inactivated at normal pH_i.

Recently, we have examined the possible role ATP might play in this pH_i activation of the ion transport process (Boron *et al.,* 1988). We have previously observed that axons depleted of ATP and dialyzed to an acidic pH_i could not extrude the imposed acid load even when Na^+ and HCO_3^- were present externally and Cl^- present internally. ATP-γ-S is an analogue of ATP that can phosphorylate or bind to proteins in a fashion that is only slowly reversible. Hence, it does not support ATPase functions (e.g., Yates and Duonce, 1976; Yasuoka *et al.,* 1982; DiPolo and Beauge, 1987), but does support some protein kinase-mediated functions (e.g., Gratecos and Fischer,

1974; Sun *et al.*, 1980). Figure 13 shows that dialyzing with ATP-γ-S for 60 min followed by 60 min of nucleotide-free dialysis resulted in an axon capable of extruding the imposed acid load when it was presented with HCO_3^-/CO_2. The inset shows that if the axon were dialyzed with ATP prior to the 60-min nucleotide-free dialysis period, subsequent exposure to HCO_3^-/CO_2 did not result in acid extrusion. Thus, ATP-γ-S presented to the axon at a pH of 6.7 activated the acid extrusion mechanism (i.e., the Na^+-dependent Cl^-/HCO_3^- exchanger) in a way that did not require the continual presence of the analogue. Control experiments using the luciferin–luciferase assay confirmed that 60 min of dialysis with nucleotide-free fluids reduced the nucleotide concentration to below 5 μM. We also showed that activation of acid extrusion by ATP-γ-S required that cellular Cl^- be present as well as extracellular Na^+. Furthermore, as Fig. 14 shows, ATP-γ-S can support HCO_3^--dependent Cl^- efflux. In this experiment, the axon was dialyzed with 8 mM ATP-γ-S/O mM ATP for 60 min prior to adding ³⁶Cl to measure unidirectional efflux. All during the efflux measurements the axon was dialyzed with a solution free of ATP-γ-S and ATP. Even 2 hr after washing away ATP-γ-S, the HCO_3^--dependent Cl^- efflux was still present. Axons treated with

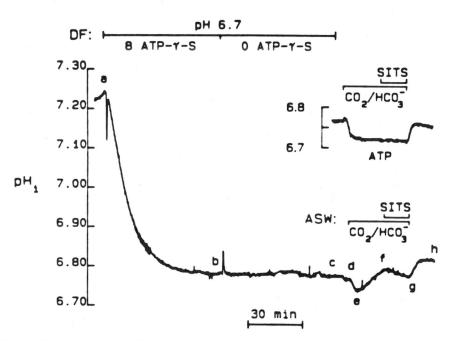

Figure 13. Pretreatment with ATP-γ-S supports acid extrusion. The axon was dialyzed for 1 hr with a pH 6.7 fluid containing 8 mM ATP-γ-S but no ATP (segment a–b), followed by another 60 min of dialysis with nucleotide-free dialysis fluid (segment b–c). After halting dialysis fluid, the axon was exposed to HCO_3^-/CO₂-containing external fluid, which caused a prompt acid extrusion (segment e–f) that was completely blocked by SITS (segment f–g). The inset shows part of an experiment identical in all respects except the axon was pretreated with 8 mM ATP rather than ATP-γ-S. Clearly, no acid extrusion was present after washing out ATP, confirming our earlier observations. (Reprinted by permission from *Nature*, Vol. 332, pp. 262–265, copyright © 1988, Macmillan Magazines Ltd.; Boron *et al.*, 1988.)

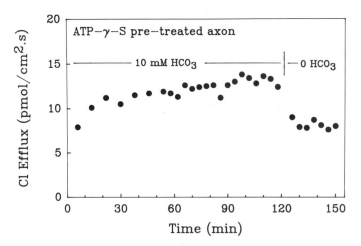

Figure 14. Pretreatment with ATP-γ-S stimulates HCO_3^- -dependent Cl^- efflux. This axon was dialyzed for 62 min prior to beginning to collect the efflux data with a fluid containing 8 mM ATP-γ-S, 0 mM ATP, and 150 mM Cl^-. In addition, the external fluid contained 0.5 mM CN throughout the entire experiment. At zero time, the dialysis fluid was changed to one containing ^{36}Cl but no ATP-γ-S or ATP. For the first 120 min of this experiment, the external fluid contained 10 mM HCO_3^- /0.4% CO_2 (pH 8.0). Then the external fluid was changed to a HCO_3^- /CO_2-free fluid (Cl^- replaced HCO_3^-). After 2 hr of dialysis with a fluid free of ATP-γ-S, this axon has a HCO_3^- -dependent Cl^- efflux of 5.1 pmoles/cm^2·sec. This value compares very well with data obtained from ATP-fueled axons under identical ionic conditions (Boron and Russell, 1983). At the end of this experiment, the axoplasm was extruded and analyzed for ATP content. It was found to contain 7 μM ATP.

100 μM DIDS in addition to the ATP-γ-S pretreatment did not have a HCO_3^- -dependent Cl^- efflux. Thus, it is clear that the ATP-γ-S was activating the Na^+-dependent Cl^-/HCO_3^- exchanger we have previously studied and characterized as ATP-requiring.

While performing these experiments, we noticed that if ATP-γ-S was presented to the axon at a pH of 7.35 and then washed out while reducing pH_i to 6.7, there was no HCO_3^-/CO_2-induced acid extrusion (Fig. 15). Thus, activation of the ion transport mechanism by ATP-γ-S requires that the pH_i be acidic. We examined this point further by applying ATP-γ-S at a pH_i of 6.7 and then during the nucleotide washout period, raising the axoplasmic pH to a value more alkaline than 7.3 for varying periods of time. Figure 16 illustrates the effect of this washout time at an alkaline pH_i on the acid extrusion rate by the ATP-γ-S-activated axons. Our results show that there is a time-dependent inactivation of the ion transporter with a half-time of about 11 min. Thus, we hypothesize that an alkaline pH_i causes either the stimulation of dephosphorylation (via a protein phosphatase?) or the debinding of the analogue.

The foregoing result is particularly exciting as it gives an insight into a possible mechanism for the pH_i activation/deactivation of the pH_i-regulating ion transport mechanism. It suggests that the pH_i sensitivity might reside in the ATP-dependent activation of the transporter. At a minimum it is consistent with dephosphorylation (or debinding) being stimulated as pH_i increases so that at normal pH_i, the ion transporter

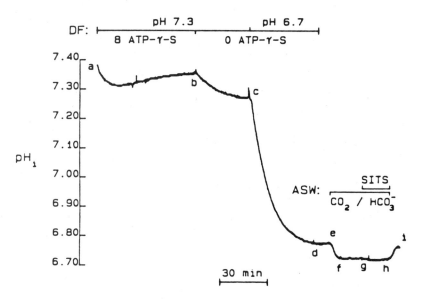

Figure 15. Activation of pH$_i$ regulation by ATP-γ-S depends on the pH$_i$. In this experiment, 8 mM ATP-γ-S was presented to the axon in a dialysis fluid at pH 7.3. Subsequently, the ATP-γ-S was washed out and the pH$_i$ dialyzed down to 6.75. Exposure to HCO$_3^-$/CO$_2$ had no activating effect on acid extrusion. (Reprinted by permission from *Nature,* Vol. 332, pp. 262–265, copyright © 1988, Macmillan Magazines Ltd.: Boron *et al.,* 1988.)

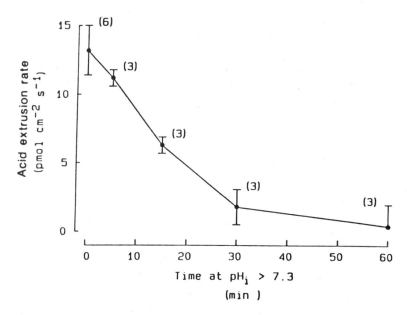

Figure 16. Time dependence of the inactivation of the ATP-γ-S supported acid extrusion caused by exposure to pH$_i$ > 7.3. Axons were treated with 8 mM ATP-γ-S at pH 6.7 for 60 min. A 60-min nucleotide-free washout period followed during which pH$_i$ was raised above 7.3 for periods ranging from 0 to 60 min. Summary of 18 experiments. (Reprinted by permission from *Nature,* Vol. 332, pp. 262–265, copyright © 1988, Macmillan Magazines Ltd.: Boron *et al.,* 1988.)

is completely inactivated. Whether or not the phosphorylation (or binding) step is stimulated by acidic pH_i is unknown.

4. SUMMARY

Cl^- transport in the squid axon is dominated by two, electrically silent, secondary active transport processes. The Na^+,K^+,Cl^- cotransport mechanism is probably responsible for the normal high $[Cl^-]_i$ typical of squid axon. At least one function of the high $[Cl^-]_i$ is apparently to activate the second Cl^- transport mechanism, Na^+-dependent Cl^-/HCO_3^- exchange. This latter mechanism clearly functions to adjust and maintain pH_i constant.

ACKNOWLEDGMENT. The authors' studies described herein were supported by NIH Grants NS-11946 (J.M.R.) and GM-06499 (W.F.B.).

REFERENCES

Adelman, W. J., and Taylor, R. E., 1961, Leakage current rectification in the squid giant axon, *Nature* **190:** 883–885.

Aiton, J. F., Chipperfield, A. R., Lamb, J. F., Ogden, P., and Simmons, N. L., 1981, Occurrence of passive furosemide-sensitive transmembrane potassium transport in cultured cells, *Biochim. Biophys. Acta* **646:** 389–398.

Altamirano, A. A., and Russell, J. M., 1987, Coupled Na/K/Cl efflux: "Reverse" unidirectional fluxes in squid axons, *J. Gen. Physiol.* **89:**669–686.

Altamirano, A. A., Breitwieser, G. E., and Russell, J. M., 1988, Vanadate and fluoride effects on Na^+-K^+-Cl^- cotransport in squid giant axon, *Am. J. Physiol.* **256:**C582–C586.

Baker, P. F., Hodgkin, A. L., and Shaw, T. T., 1962, The effects of changes of internal ionic concentrations on the electrical properties of perfused giant axons, *J. Physiol. (London)* **164:**355–374.

Bear, R. S., and Schmitt, F. O., 1939, Electrolytes in the axoplasm of the giant nerve fibers of the squid, *J. Cell Comp. Physiol.* **14:**205–215.

Boron, W. F., and DeWeer, P., 1976a, Intracellular pH transients in squid giant axons caused by CO_2, NH_3 and metabolic inhibitors, *J. Gen. Physiol.* **67:**91–112.

Boron, W. F., and DeWeer, P., 1976b, Active proton transport stimulation by CO_2/HCO_3^-, blocked by cyanide, *Nature* **259:**240–241.

Boron, W. F., and Knakal, R. C., 1989, Intracellular pH-regulating mechanism of the squid axon. Interaction between DNDS and extracellular Na^+ and HCO_3^-, *J. Gen. Physiol.* **93:**123–150.

Boron, W. F., and Knakal, R., 1989, Intracellular pH-regulating mechanism of the squid axon: Interaction between DNDS and extracellular Na^+ and HCO_3^-, *J. Gen. Physiol.* **93:**123–150.

Boron, W. F., and Russell, J.M., 1983, Stoichiometry and ion dependencies of the intracellular-pH-regulating mechanism in squid giant axons, *J. Gen. Physiol.* **81:**373–399.

Boron, W. F., Hogan, E., and Russell, J. M., 1988, pH-sensitive activation of the intracellular-pH regulation system in squid axons by ATP-γ-S, *Nature* **332:**262–265.

Brinley, F. J., Jr., and Mullins, L. J., 1965, Variations in the chloride content of isolated squid axons, *Physiologist* **8:**121.

Caldwell, P. C., and Keynes, R. D., 1960, The permeability of the squid giant axon to radioactive potassium and chloride ions, *J. Physiol. (London)* **154:**177–189.

Chipperfield, A. R., 1986, The $(Na^+$-K^+-$Cl^-)$ co-transport system, *Clin. Sci.* **71:**465–476.

DiPolo, R., and Beauge, L., 1987, In squid axons, ATP modulates Na^+-Ca^{2+} exchange by a Ca^{2+}_i-dependent phosphorylation, *Biochim. Biophys. Acta* **897**:347–354.

Duhm, J., 1987, Furosemide-sensitive K^+ (Rb^+) transport in human erythrocytes: Modes of operation, dependence on extracellular and intracellular Na^+, kinetics, pH dependency and the effect of cell volume and N-ethylmaleimide, *J. Membr. Biol.* **98**:15–32.

Geck, P., Pietrzy, K. C., Burckhardt, B. C., Pfieffer, B., and Heinz, E., 1980, Electrically silent cotransport of Na, K, and Cl in Ehrlich cells, *Biochim. Biophys. Acta* **600**:432–447.

Gratecos, D., and Fischer, E. H., 1974, Adenosine 5'-(3-thiotriphosphate) in the control of phosphorylase activity, *Biochem. Biophys. Res. Commun.* **58**:960–967.

Haas, M., Schmidt, W. F., III, and McManus, T. J., 1982, Catecholamine-stimulated ion transport in duck red blood cells. Gradient effects in electrically neutral (Na^+,K^+,2Cl) co-transport, *J. Gen. Physiol.* **80**:125–147.

Hall, A. C., and Ellory, J. C., 1985, Measurement and stoichiometry of bumetanide-sensitive (2Na:1K:3Cl) cotransport in ferret red cells, *J. Membr. Biol.* **85**:205–213.

Hescheler, J., Kameyama, M., Trautwein, W., Mieskes, G., and Soling, H.-D., 1987, Regulation of the cardiac calcium channel by protein phosphatases, *Eur. J. Biochem.* **165**:261–266.

Inoue, I., 1985, Voltage-dependent chloride conductance of the squid axon membrane and its blockade by some disulfonic stilbene derivatives, *J. Gen. Physiol.* **85**:519–538.

Keynes, R. D., 1963, Chloride in the squid giant axon, *J. Physiol. (London)* **168**:69–105.

McManus, T. J., 1987, Na,K,2Cl co-transport: Kinetics and mechanism, *Fed. Proc.* **46**:2378–2381.

Roos, A., and Boron, W. F., 1981, Intracellular pH, *Physiol. Rev.* **61**:296–434.

Russell, J. M., 1976, ATP-dependent chloride influx into internally dialyzed squid giant axons, *J. Membr. Biol.* **28**:335–349.

Russell, J. M., 1979, Chloride and sodium influx: A coupled uptake mechanism in the squid giant axon, *J. Gen. Physiol.* **73**:801–818.

Russell, J. M., 1983, Cation-coupled chloride influx in squid axon: Role of potassium and stoichiometry of the transport process, *J. Gen. Physiol.* **81**:909–925.

Russell, J. M., 1984, Chloride in the squid giant axon, *Curr. Top. Membr. Transp.* **22**:177–193.

Russell, J. M., and Boron, W. F., 1976, Role of chloride transport in regulation of intracellular pH, *Nature* **264**:73–74.

Steinbach, H. B., 1941, Chloride in the giant axons of the squid, *J. Cell Comp. Physiol.* **17**:57–64.

Sun, I. Y.-C., Johnson, E. M., and Allfrey, V. G., 1980, Affinity purification of newly phosphorylated protein molecules. Thiophosphorylation and recovery of histones H1, H2B and H3 and the high mobility group protein HMG-1 using adenosine 5'-O-(3-thiotriphosphate) and cyclic AMP-dependent protein kinase, *J. Biol. Chem.* **255**:742–749.

Thomas, R. C., 1976, Ionic mechanism of the H^+ pump in a snail neurone, *Nature* **262**:54–55.

Yasuoka, K., Kawakita, M., and Kaziro, Y., 1982, Interaction of adenosine-5'-O-(3-thiotriphosphate) with Ca^{2+},Mg^{2+}-adenosine triphosphatase of sarcoplasmic reticulum, *J. Biochem.* **91**:1629–1637.

Yates, D. W., and Duonce, V. C., 1976, The binding of nucleotides and bivalent cations to the calcium-and-magnesium ion-dependent adenosine triphosphatase from rabbit muscle sarcoplasmic reticulum, *Biochem. J.* **159**:719–728.

Yount, R. G., 1975, ATP analogs, *Adv. Enzymol.* **43**:1–56.

Intracellular Cl⁻ Regulation and Synaptic Inhibition in Vertebrate and Invertebrate Neurons

Francisco J. Alvarez-Leefmans

1. INTRODUCTION

Cl^- movements across plasma membrane channels and carriers play a central role in a number of mechanisms essential for neuronal function and survival. These include regulation and maintenance of intracellular pH (Boron, 1983; Thomas, 1984; see Russell and Boron, this volume), regulation and maintenance of cell volume (Ballanyi and Grafe, 1988; see Chapter 2, this volume), and modulation of neuronal excitability through anion channels activated by inhibitory neurotransmitters (Alger, 1985; Barker, 1985; Roberts, 1986; Siggins-Gruol, 1986), intracellular Ca^{2+} (see Mayer *et al.*, this volume), or transmembrane voltage (see Chesnoy-Marchais, this volume). Furthermore, Cl^- has recently been shown to exert modulatory effects on G proteins (Deterre *et al.*, 1983; Higashijima *et al.*, 1987). The latter are known to be an essential part of the intracellular messenger machinery coupling receptor binding of neurotransmitters (or hormones) to their specific cell responses. All the above considerations make evident the importance of understanding the mechanisms by which Cl^- is regulated and maintained in nerve cells. Given the wide spectrum of the subjects involved, the present account will be confined to considering Cl^- regulation in relation to inhibitory neurotransmitter actions.

2. Cl⁻ TRANSPORT AND SYNAPTIC INHIBITION

Cl^- is the predominant anion that carries current through neuronal membrane channels gated by inhibitory neurotransmitters such as GABA and glycine. The increase in Cl^- conductance (gCl) as well as the changes in transmembrane potential

Francisco J. Alvarez-Leefmans • Departamento de Farmacología y Toxicología, Centro de Investigacíon y de Estudios Avanzados del I. P. N., Mexico 07000, D. F., and Departamento de Neurobiología, Instituto Mexicano de Psiquiatría, Mexico 14370, D. F.

brought about by inhibitory transmitters are both thought to be determinants in generating postsynaptic (Llinás *et al.*, 1974; Dingledine and Langmoen, 1980; Thompson and Gähwiler, 1989a; see Section 4) as well as presynaptic inhibitory actions (see Section 5.1.3 and reviews by Davidoff and Hackman, 1985, and Alger, 1985). Both amplitude and direction of neurotransmitter-gated Cl^- currents and their associated transmembrane potential changes are mainly determined by the sign and magnitude of the difference between the resting potential (E_m) and the Cl^- equilibrium potential (E_{Cl}). The value of E_{Cl} is determined by the intracellular Cl^- activity, a_{Cl}^i, which is set by the relative contribution of various anion-transporting mechanisms present in the neuronal membrane. My objective in this chapter is to discuss the possible nature of these anion-transporting mechanisms, their role in the regulation of intracellular Cl^- levels, and their importance in the normal operation of *Cl^--dependent inhibitory synaptic processes.*

The importance of neuronal Cl^- regulatory mechanisms can be appreciated just by considering the functional significance of Cl^--mediated synaptic inhibition in the nervous system. Although the precise role of synaptic inhibition is not well understood, we know that suppressing it leads to epileptiform electrical activity (for review see Krnjević, 1983). GABAergic Cl^--dependent postsynaptic inhibition is known to be fragile and dependent on, among other factors, the effectiveness of anion transport mechanisms responsible for intracellular Cl^- homeostasis (e.g., Thompson and Gähwiler, 1989a,b). For instance, the depression of hyperpolarizing IPSPs that follows repetitive stimulation, or activity-dependent disinhibition (Ben-Ari *et al.*, 1979; McCarren and Alger, 1985), can be, in part, a result of intracellular Cl^- accumulation leading to decreases in IPSP driving force, i.e., the difference between E_m and E_{GABA} or E_{IPSP} (Huguenard and Alger, 1986; Korn *et al.*, 1987; Thompson and Gähwiler, 1989a). This results from the fact that the hyperpolarization associated with the IPSP (and not just the conductance change as previously thought; e.g., Eccles, 1964a) plays a significant inhibitory role (see above) and therefore intracellular Cl^- accumulation would be expected to result in significant disinhibition. As discussed below (Section 4.2), the mechanism of active Cl^- extrusion responsible for keeping an adequate driving force for the IPSP may be an outwardly directed K^+,Cl^- cotransport. Hence, increasing $[K^+]_o$ should increase $[Cl^-]_i$ (see Section 4.2.1). It is known that repetitive stimulation produces a transient elevation of $[K^+]_o$ to values as high as 12 mM (Lewis and Schuette, 1975; Heinemann and Lux, 1977; Somjen, 1979; Benninger *et al.*, 1980). Thus, the increase in $[K^+]_o$ may underlie intracellular Cl^- accumulation leading to the reduction in IPSP driving force and disinhibition following repetitive stimulation (Wong and Watkins, 1982; McCarren and Alger, 1985; Thompson and Gähwiler, 1989a,b). Activity-dependent disinhibition may be an important endogenous factor in the genesis of epilepsy (Korn *et al.*, 1987; Thompson and Gähwiler, 1989a).

Under normal conditions, postsynaptic as well as presynaptic inhibition appear to be involved in the processing of information in neuronal networks. Postsynaptic inhibitory action involves not only suppression of action potential generation on individual neurons in the classic sense (Eccles, 1964a, 1969) but also transient clamping of the transmembrane potential to a value set by the equilibrium potential of the anions (mainly Cl^-) to which the membrane becomes permeable. The latter transmembrane potential value seems to be suitable for the orchestrated expression of other ionic conductances that determine some neuronal firing patterns (e.g., Jahnsen and Llinás, 1984). This endows inhibitory synapses with the property of modulation of the intrinsic electrophysiological properties of neurons (Llinás, 1988). Presynaptic inhibition is less well understood than postsynaptic inhibition. Presynaptic inhibitory synapses selectively decrease or suppress transmitter release at individual sets of input fibers onto a postsynaptic neuron without affecting the electrical properties of the latter. Thus, presynaptic inhibitory synapses can be envisaged as selective filters of incoming information into a postsynaptic neuron. Presynaptic inhibition in the spinal cord is associated with primary afferent depolarization through axoaxonic synaptic contacts. Depolarization is due to the opening of $GABA_A$-activated Cl^- channels resulting in an *efflux* of Cl^- (see Section 5.1.3a). The depolarization *per se* as well as the increase in gCl are thought to be important in producing presynaptic inhibition in the spinal cord although their precise roles in modulating transmitter release for presynaptic inhibition is still uncertain (Section 5.1.3a). The presence of an inward active Cl^- transport in sensory neurons has long been suspected because of the depolarizing action of GABA in both the terminals and the cell bodies of primary sensory neurons (Levy, 1977; Nicoll and Alger, 1979; Davidoff and Hackman, 1985). The nature of this Cl^- transport mechanism is discussed below (Section 5.1).

Due to the three-dimensional organization of inhibitory interneurons, Cl^--dependent synaptic inhibition determines not only the excitability properties of individual pre- or postsynaptic elements but also those of entire neuronal networks. Thus, by shaping the firing pattern of neuronal networks in the time and space domains, synaptic inhibition plays a pivotal role in the processing of information in the nervous system. Examples of this principle emerging from the functional and geometric organization of inhibitory synapses, some of which can now be correlated to certain behavioral states, are found in thalamic and cortical neurons (e.g., see review by Steriade and Llinás, 1988). For instance, in primary visual (Sillito, 1984; Bolz and Gilbert, 1986) and somatosensory cortex (Dykes *et al.*, 1984), synaptic inhibition appears to shape many aspects of each neuron's receptive field. The latter constitutes the neuronal basis for the process of feature extraction from sensory information with which these cells are endowed. A clear example illustrating the role of intracortical Cl^--dependent GABAergic inhibition comes from the experiments of Sillito (1984). This author has studied the effect of iontophoretic application of bicuculline on the receptive field properties of visual cortical cells. Bicuculline is a powerful antagonist of Cl^--dependent GABAergic responses. The receptive field properties of "simple" and "complex" cells are drastically altered by suppressing Cl^--dependent inhibition. The changes include loss of directional and orientation selectivity (Sillito, 1984).

Other functional aspects of synaptic inhibition include its role in oscillatory behavior of neuronal networks (e.g., Traub *et al.*, 1989). Rhythmic oscillations of many neuronal networks are correlated with behaviors such as chewing, breathing, and walking. The amplitude and frequency of this rhythmic activity depend not only on intrinsic cellular properties (Llinás, 1988) but also on the connectivity and efficacy of excitatory and inhibitory synapses (Jahnsen and Llinás, 1984; Traub *et al.*, 1989). The role of synaptic inhibition in this kind of process and the functional implications of this type of activity on behavioral states have recently been reviewed (Steriade and Llinás, 1988; Traub *et al.*, 1989). Finally, the role of Cl^--dependent synaptic inhibition in preventing seizure activity in neuronal networks must be emphasized. Clearly, synaptic inhibition is essential for maintaining normal levels of neuronal excitability, its disruption in certain areas of the brain can lead to the synchronized discharges of epilepsy (Schwartzkroin and Wheal, 1984). Collapse of transmembrane Cl^- gradients disrupts inhibition; hence, the importance of understanding how such gradients are generated and maintained.

It is clear from all of the above considerations that intracellular Cl^- in nerve cells must be tightly regulated, hence the relevance of understanding the transport systems affecting transmembrane Cl^- distribution in neurons.

The Driving Force for Cl^--Dependent IPSPs Is Generated and Maintained by Active Transport Systems for Cl^-

Inhibitory transmitters such as GABA and glycine exert their action on receptors coupled to anion channels with relatively high selectivity toward Cl^- (see Chapters 8 and 12, this volume). GABA, the predominant inhibitory neurotransmitter in the CNS (Krnjević, 1981, 1984, Roberts, 1986), opens Cl^- channels when interacting with $GABA_A$ receptors thereby generating IPSPs. When acting on $GABA_B$ receptors, the same amino acid appears to open K^+ channels leading to relatively slow IPSPs (Nicoll, 1988; Connors *et al.*, 1988). Cl^--dependent postsynaptic potentials mediated by $GABA_A$ receptors can be either hyperpolarizing, as the classical IPSP, or depolarizing, as is the case in some nerve terminals, cell bodies, and dendrites (Simmonds, 1984; Alger, 1985; Davidoff and Hackman, 1985; Nicoll, 1988). The direction as well as the amplitude of Cl^- currents mediated by $GABA_A$ receptors are determined mainly by the sign and magnitude of the driving force $(E_m - E_{Cl})$ for the transmitter-activated Cl^- current. If E_{Cl} is kept more negative than E_m, the transmembrane Cl^- electrochemical gradient will be inwardly directed. Therefore, an increase in gCl brought by the transmitter will lead to an influx of Cl^- with the resulting membrane hyperpolarization. The Cl^--dependent IPSPs mediated by $GABA_A$ receptors located at or near the cell bodies of many vertebrate and invertebrate neurons constitute examples of this situation (e.g., Lux, 1971; Llinás and Baker, 1972; Krnjević, 1974; Barker and Ransom, 1978; Deisz and Lux, 1982; Connors *et al.*, 1988; Kaneda *et al.*, 1989; see Barker *et al.*, this volume).

There are cases in which E_{Cl} is more positive than E_m. In this situation the transmembrane Cl^- gradient is outwardly directed and therefore an increase in gCl will lead to an *efflux* of Cl^- with consequent membrane depolarization. The Cl^--dependent depolarizing potential mediated by $GABA_A$ receptors in the soma of peripheral ganglion neurons (Feltz and Rasminsky, 1974; Nishi *et al.*, 1974; Adams and Brown, 1975; Gallagher *et al.*, 1978; Hattori *et al.*, 1984; Ballanyi and Grafe, 1985; Alvarez-Leefmans *et al.*, 1988) and in the terminals of some intraspinal afferent fibers (Levy, 1977; Nicoll and Alger, 1979) illustrate this case.

The fact that in most nerve cells E_{Cl} has been consistently found to be either more positive or more negative than E_m implies that *Cl^- is not passively distributed across* *neuronal membranes*. Measurements of intracellular Cl^- activity (a^i_{Cl}) with liquid anion-exchanger microelectrodes in vertebrate and invertebrate neurons give support to this notion (see Section 3), which is in disagreement with the classical view originally proposed for mammalian motoneurons (Coombs *et al.*, 1955) and later extended to hippocampal pyramidal cells (Eccles *et al.*, 1977; Allen *et al.*, 1977). According to this view, Cl^- was thought to be in electrochemical equilibrium across the membrane.

To achieve and maintain a nonequilibrium transmembrane Cl^- distribution, neurons must be endowed with active transport mechanisms for Cl^-. These transport mechanisms are known in the neurobiological parlance as "Cl^- pumps," a term that is misleading since it implies *a priori a primary active transport system,* i.e., that whose energy is derived *directly* from the splitting of ATP. It turns out that most anion transport systems known to be present in animal cells are *secondary active transports,* i.e., the translocation of the substrate is coupled to the translocation of another substrate that uses the same carrier, in the opposite direction (countertransport or antiport) or in the same direction (cotransport or symport). The energy for the unequal distribution of the *driven* substrates is derived in this case from the energy present in the unequal distribution of the *driving* substrate (Stein, 1986). For these reasons, in the present account I prefer using the term *carrier* or even less committal the terms *active transport systems* or *mechanisms.* An outwardly directed Cl^- transport mechanism is needed to keep a^i_{Cl} at levels lower than expected for a passive distribution, maintaining E_{Cl} more negative than E_m. Similarly, to keep a^i_{Cl} at sufficiently high values to make E_{Cl} more positive than E_m requires an inwardly directed transport mechanism that actively accumulates Cl^-. In fact, both outwardly and inwardly directed Cl^- active transport mechanisms have been postulated to exist in nerve cells for some time (Boistel and Fatt, 1958; Keynes, 1963; Lux, 1971; Russell and Brown, 1972; Llinás *et al.,* 1974; Nishi *et al.,* 1974; Fatt, 1974; Ascher *et al.,* 1976). However, except for some of the work done on invertebrates the evidence provided for active Cl^- transport, particularly in vertebrate nerve cells, was indirect (see below).

The question arises as to the nature of these Cl^- transport systems. Work done on crayfish stretch receptor neurons has been interpreted as evidence for K^+,Cl^- cotransport as a possible mechanism for Cl^- extrusion in nerve cells (Aickin *et al.,* 1982, 1984; see Section 4.2.2). A similar mechanism may exist in vertebrate cortical neurons, although the evidence is indirect, and measurements of a^i_{Cl} are still lacking (Thompson *et al.,* 1988a,b; Thompson and Gähwiler, 1989a,b; Section 4.2.3). It has also been shown that invertebrate neurons can extrude Cl^- through a Na^+-dependent Cl^-/HCO_3^- exchange mechanism (Thomas, 1977, 1984; Boron and Russell, 1983; see Section 4.1 and Russell and Boron, this volume). Evidence for the presence of this transport system in vertebrate neurons is still lacking. Although this transporter has been proposed to act as a regulator of intracellular pH, it could also subserve the function of regulator of intracellular Cl^-, (Russell, 1980). Regarding possible mechanisms for uphill accumulation of Cl^-, a Na^+,K^+,Cl^- cotransport mechanism has been characterized in squid axons (Russell, 1983; see Russell and Boron, this volume). Only during the last few years has direct measurement of net Cl^- fluxes in vertebrate nerve cells become feasible. The nature of an inwardly directed Cl^- transport mechanism that maintains E_{Cl} more positive than E_m has started to be elucidated. It appears to be an electroneutral Na^+, K^+, Cl^- cotransport mechanism similar to that present in squid axons and other cell types (Alvarez-Leefmans *et al.*, 1988). A similar transport system seems to be present in sympathetic ganglion cells (Ballanyi and Grafe, 1985). The evidence for the presence of this transport mechanism in vertebrate neurons is discussed in Section 5.1.

3. EVIDENCE FOR NONPASSIVE Cl^- DISTRIBUTION ACROSS NEURONAL MEMBRANES

3.1. General Considerations

The Cl^- equilibrium potential, E_{Cl}, is defined by the Nernst equation:

$$E_{Cl} = \frac{RT}{F} \ln \frac{a^i_{Cl}}{a^o_{Cl}} \tag{1}$$

where a^i_{Cl} and a^o_{Cl} are the intracellular and extracellular Cl^- activities, and R, T, and F have their usual thermodynamic meaning. If Cl^- is passively distributed, E_m will have the same value as E_{Cl} and therefore

$$E_m = \frac{RT}{F} \ln \frac{a^i_{Cl}}{a^o_{Cl}} \tag{2}$$

The predicted intracellular Cl^- activity at electrochemical equilibrium $(a^i_{Cl})_{eq}$ can be obtained from Eq. (2):

$$(a^i_{Cl})_{eq} = a^o_{Cl} \exp \frac{E_m F}{RT} \tag{3}$$

Active Cl^- transport requires demonstrating that a^i_{Cl} is maintained at a level different from that predicted by Eq. (3). In other words, E_{Cl} and E_m will be different. However, equal values of E_m and E_{Cl} do not preclude the presence of active Cl^- transport as is the case when there is a substantial electrodiffusional Cl^- permeability, e.g., skeletal muscle (Harris and Betz, 1987; see also Aickin, this volume; Kaila and Voipio, this volume). As discussed in Chapter 1, so far the most reliable direct measurements of a^i_{Cl} and corresponding estimates of E_{Cl} in neurons are those carried out with microelectrodes containing liquid anion exchangers such as Corning 477315 and 477913. There are several examples of a^i_{Cl} measurements using the above approach, showing that in many nerve cells E_{Cl} is either more negative or less negative than E_m.

3.2. Evidence for a Lower a^i_{Cl} Value Than That Predicted for a Passive Cl^- Distribution across Neuronal Membranes

Examples of cases in which E_{Cl}, derived from direct liquid anion exchanger microelectrode measurements of a^i_{Cl}, has been found to be more negative than E_m include giant neurons (R2, R15, and L1 to L6) of the abdominal ganglion of *Aplysia californica* (Brown *et al.*, 1970; Kunze and Brown, 1971; Russell and Brown, 1972;

Brown and Kunze, 1974); medial pleural neurons also of *A. californica* (Ascher *et al.*, 1976); neurons with hyperpolarizing IPSPs of *Lymnaea stagnalis* (Gardner and Moreton, 1985); and stretch receptor neurons of the crayfish *Astacus fluviatilis* (Deisz and Lux, 1982). There are other examples of studies on molluscan neurons in which it was concluded that E_{Cl} was more negative than E_m (Kerkut and Meech, 1966a,b; Neild and Thomas, 1974; reviewed by Gardner and Moreton, 1985). However, a^i_{Cl} measurements were done either with metallic Ag microelectrodes (Kerkut and Meech, 1966a,b) or with Ag:AgCl microelectrodes (Neild and Thomas, 1974) and for the reasons discussed in Section 5.1 of Chapter 1, they need to be reassessed with more reliable intracellular probes. The finding that a^i_{Cl} in many invertebrate neurons is kept lower than the value predicted for a passive transmembrane distribution of Cl⁻ led to the proposal of an active Cl⁻ extrusion mechanism in some invertebrate neurons.

In vertebrate neurons a value of a^i_{Cl} *lower than predicted for equilibrium* distribution has not been *directly* demonstrated at this time. In fact, for many years it was postulated that in mammalian motoneurons and hippocampal pyramidal cells Cl⁻ was passively distributed across the membrane (Coombs *et al.*, 1955; Eccles, 1957; Allen *et al.*, 1977). According to this view, Cl⁻-dependent IPSPs resulted from inhibitory transmitter opening of channels that were permeable to both Cl⁻ and K⁺, and the normal hyperpolarizing nature of the IPSP was determined by the K⁺ equilibrium potential (E_K). Despite the strenuous experimental attempts made to demonstrate an involvement of K⁺ in the generation of Cl⁻-dependent IPSPs, no convincing evidence was produced (Coombs *et al.*, 1955; Eccles *et al.*, 1964a,b; Allen *et al.*, 1977). Furthermore, recent evidence with patch-clamp techniques shows that both glycine- and GABA$_A$-mediated responses are due to the opening of *anion*-selective channels (Bormann *et al.*, 1987; see Bormann, this volume, and Kaila and Voipio, this volume). An alternative model to that postulated by Eccles and co-workers to explain the ionic mechanisms underlying hyperpolarizing IPSPs was proposed by Lux, Llinás, and associates in the early 1970s (Lux, 1971; Llinás and Baker, 1972; Llinás *et al.*, 1974). The basic conclusions derived from their work on mammalian motoneurons were: (1) that the hyperpolarizing IPSP was generated by the influx of Cl⁻, (2) that the IPSP was normally hyperpolarizing because E_{Cl} was more negative than E_m, and (3) that the driving force for the IPSP $E_{Cl} - E_m$ was generated and maintained by an outwardly directed active Cl⁻ transport mechanism that was inhibited by exposure to the NH_3/NH_4^+ couple. Further indirect evidence for a^i_{Cl} (or $[Cl^-]_i$) values lower than predicted for passive distribution in mammalian neurons has grown considerably in the last two decades. Recent examples include rat spinal motoneurons (Forsythe and Redman, 1988), pyramidal neurons from guinea pig cingulate cortex (Thompson *et al.*, 1988a,b), and pyramidal neurons from rat hippocampus (Thompson and Gähwiler, 1989a,b). The conclusion is based on three main observations: (1) loading the cells with Cl⁻ produces inversion of polarity of IPSP due to a positive shift in the equilibrium potential for the IPSP (E_{IPSP}) or for the reversal potential of GABA-induced responses (E_{GABA}), which under normal conditions are more negative than E_m, (2) Cl⁻-loaded cells are able to extrude Cl⁻, as judged from return to control values of E_{IPSP}, through a mechanism which appears to be electroneutral, and (3) lowering $[Cl^-]_o$ leads to a positive shift in E_{GABA}. The underlying assumption in all these observations is that $E_{IPSP} = E_{Cl}$ or $E_{GABA} = E_{Cl}$.

Inasmuch as the anion channels opened by transmitters such as GABA and glycine are not perfectly selective for Cl^- but are also permeable to HCO_3^- (Kelly *et al.*, 1969; Bormann *et al.*, 1987; Kaila *et al.*, 1989), and since the HCO_3^- equilibrium potential (E_{HCO_3}) is normally held at values more positive than E_m, the common assumption that $E_{IPSP} = E_{Cl}$ (or $E_{Cl} = E_{GABA}$) is not tenable unless experimentally tested (see Kaila and Voipio, this volume). This objection, however, does not alter the conclusion that E_{Cl} is more negative than E_m in neurons showing Cl^--dependent hyperpolarizing IPSPs. On the contrary, if HCO_3^- also permeates through transmitter-gated anion channels and a membrane hyperpolarization is still produced, this implies that E_{IPSP} (or E_{GABA}) will actually be more positive than E_{Cl}. That is to say that if this consideration can be generalized, the value of E_{Cl} derived from E_{IPSP} or E_{GABA} has consistently been *underestimated* in the literature, as further discussed below (Section 4.2.3).

3.3. Evidence for a Higher a^i_{Cl} Value Than That Predicted for a Passive Cl^- Distribution across Some Neuronal Membranes

Many neurons from vertebrates and invertebrates have a higher intracellular chloride activity (a^i_{Cl}) than expected from a passive thermodynamic distribution. Among invertebrates perhaps the clearest example is the squid giant axon (Keynes, 1963; Russell, 1976) where an inwardly directed Cl^- transport mechanism has been extensively characterized (see Russell and Boron, this volume). Other invertebrate nerve cells include crayfish giant axons (Cornwall *et al.*, 1970) and motor giant neurons of crayfish abdominal ganglia (Moody, 1981). Early claims for some *Helix aspersa* neurons (D cells) having a higher a^i_{Cl} than that predicted for a passive distribution (Kerkut and Meech, 1966a,b) were contrasted by Neild and Thomas (1974). However, both groups of workers used solid-state silver-based ion-selective microelectrodes and therefore the actual a^i_{Cl} of either D or H cells in snail neurons still remains to be settled. In *L. stagnalis* neurons, Moreton and Gardner (1981; see also Gardner and Moreton, 1985) found some cells having E_{Cl} more positive than E_m. However, in their experiments E_m was artificially maintained under current-clamp conditions.

In vertebrates there are three clear examples of neurons in which E_{Cl} calculated from direct measurements of a^i_{Cl} and a^o_{Cl} with liquid anion exchanger microelectrodes has been found to be less negative than E_m. These are frog motoneurons (Bührle and Sonnhof, 1985), neurons from rat superior cervical ganglia (Ballanyi *et al.*, 1984; Ballanyi and Grafe, 1985), and frog dorsal root ganglion cells (Alvarez-Leefmans *et al.*, 1986, 1987, 1988). In all three cases, GABA produces a Cl^--mediated depolarization that has pharmacological properties consistent with activation of $GABA_A$ receptors. In the three cases, an inwardly directed Cl^- transport mechanism has been proposed to account for the higher a^i_{Cl} than that predicted for a passive distribution. The nature of this transport system is discussed in Section 5.

There are other instances in which $GABA_A$-mediated responses are depolarizing in mammalian nerve cells. This type of response is mainly found by focal application of GABA onto the dendrites of cultured motoneurons and in a variety of cortical cells from *in vitro* brain slices (Andersen *et al.*, 1980; Misgeld *et al.*, 1986; Nicoll, 1988;

Connors *et al.*, 1988; Janigro and Schwartzkroin, 1988; see also Barker *et al.*, this volume). The ionic mechanism of the dendritic depolarizing response to GABA is not well understood but it seems to involve an increase in Cl^- permeability (reviewed by Nicoll, 1988). It has been suggested that an inwardly directed Cl^- transport mechanism maintains E_{Cl} at a value less negative than E_m in the dendrites (Misgeld *et al.*, 1986; Müller *et al.*, 1989), although rigorous evidence for this statement is still rather meager. The issue is further discussed in Section 5.1.4.

4. NATURE OF ACTIVE Cl⁻ EXTRUSION MECHANISMS IN NEURONS

Early work in *Aplysia* giant neurons (Russell and Brown, 1972; Ascher *et al.*, 1976) showed that following an intracellular Cl^- load, cells were able to actively extrude Cl^-. Indirect evidence for active Cl^- extrusion was also obtained in cat motoneurons (Lux, 1971; Llinás *et al.*, 1974). The presence of this uphill Cl^- transport process would hold the value of E_{Cl} more negative than E_m and hence of the hyperpolarizing nature of Cl^--mediated IPSPs. There was also evidence suggesting that active Cl^- extrusion could be blocked by NH_3/NH_4^+ (a mixture of ammonium ions, NH_4^+ and ammonia gas, NH_3) in neurons of invertebrates (Meyer and Lux, 1974; Ascher *et al.*, 1976; Russell, 1978; Deisz and Lux, 1982; Aickin *et al.*, 1982) and vertebrates (Lux, 1971; Llinás *et al.*, 1974; Thompson *et al.*, 1988b). An example of the blocking effect of NH_3/NH_4^+ on possible Cl^- extrusion in cat trochlear motoneurons is shown in Fig. 1. NH_3/NH_4^+ blockage of the Cl^- extrusion process was accompanied by removal of the hyperpolarization of IPSPs without blocking the underlying increase in conductance (Lux *et al.*, 1970; Lux, 1971; Llinás *et al.*, 1974; Meyer and Lux, 1974; Raabe and Gumnit, 1975; Iles and Jack, 1980). Under these conditions the effectiveness of inhibition was considerably reduced, indicating that not only the conductance increase (ΔgCl) but also the hyperpolarization *per se* contributed to inhibitory action (Llinás *et al.*, 1974; Raabe and Gumnit, 1975; Iles and Jack, 1980). The finding that reducing or collapsing the Cl^- electrochemical gradient considerably reduces inhibition highlighted the role of active Cl^- extrusion for postsynaptic inhibition (see Fig. 2). It was proposed that the effects of NH_3/NH_4^+ were due to NH_4^+ rather than NH_3 (Ascher *et al.*, 1976; Aickin *et al.*, 1982). NH_3/NH_4^+ actions do not have a straightforward interpretation. This is because besides altering pH_i (Boron and De Weer, 1976; Roos and Boron, 1981; Boron, 1983), NH_4^+ can substitute for K^+ in a variety of membrane transport processes, including permeation through different kinds of ion channels and carriers (see review by Knepper *et al.*, 1989). Examples of the former include K^+ channels (Hille, 1975; Latorre and Miller, 1983; Rudy, 1988) and Na^+ channels (Edwards, 1982). Examples of the latter include the Na^+/K^+ pump, Na^+, K^+, Cl^- cotransport, and Na^+/H^+ exchange (for references see Knepper *et al.*, 1989). Perhaps because of its action in so many different membrane transport proteins, NH_4^+ has multiple effects in neuronal networks that make it difficult to provide a unique interpretation of its mechanism of action in Cl^- extrusion processes (Iles and Jack, 1980; Alger and Nicoll, 1983). Based on a series of experimental observations in *Aplysia* neurons and dialyzed squid axons, Russell (1980) pointed out that Cl^- efflux

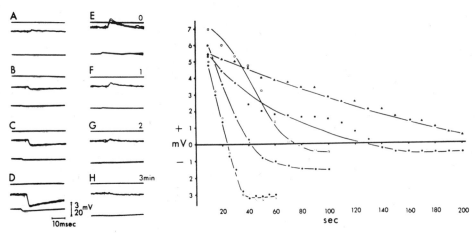

Figure 1. Chloride extrusion in cat trochlear motoneurons following an intravenous injection of ammonium acetate. IPSPs were evoked by ipsilateral vestibular nerve stimuli. (A–H) Intracellular records from a trochlear motoneuron 15 min after injecting 3 mmol ammonium acetate per kg of animal weight. (A) Vestibular nerve stimulus, 5 min after penetration, evoked only a small IPSP of reversed polarity. The IPSP was depolarizing because, due to ammonium acetate action on Cl^- extrusion, Cl^- accumulated inside the cell thereby shifting E_{IPSP} toward more positive values. In most cases, ammonium acetate eliminated all potential changes generated by vestibular nerve stimuli in trochlear motoneurons. Before ammonium acetate injection, a large, hyperpolarizing IPSP was normally recorded. However, after ammonium injection, Cl^- extrusion was partially inhibited leading to Cl_i^- accumulation and therefore alteration in the driving force for the IPSP, i.e., E_m was close to E_{Cl}. (B–D) Depolarizing current of increasing amplitude produced a proportionate reversal and amplitude increase in IPSP. The fact that the injection of a depolarizing DC current of increasing amplitude disclosed the IPSP indicated that the equilibrium potential for the IPSP rather than its underlying conductance was altered by ammonium acetate. The maneuver shown in B to D led to a further increase in $[Cl^-]_i$ as evident from the increase in size of the reversed IPSP when the polarizing current was suddenly removed (compare A with E). (E–H) Records obtained after the polarizing current shown in D was suddenly removed. (E) IPSP amplitude at moment of current removal; (F) 1 min, (G) 2 min, and (H) 3 min later. The time course of recovery in IPSP amplitude were interpreted as the time course of Cl^- extrusion. The graph on the right illustrates the time course of Cl^- extrusion before and after ammonium acetate injection. The open and closed circles indicated by an arrow, represent two sequential measurements of IPSP recovery after Cl^- injection before administration of 3 mM/kg ammonium acetate. The recovery plots were obtained from another motoneuron at 10 min (▲), 30 min (■), 45 min (□), and 60 min (●) after the ammonium injection. (Reproduced with permission from Llinás *et al.*, 1974.)

shared the same properties as acid extrusion. Therefore, he proposed that the Cl^- extrusion process could be the same as the acid extrusion mechanism found in invertebrate neurons, namely a Na^+-coupled Cl^-/HCO_3^- exchanger (see below). On the other hand, Aickin *et al.* (1982, 1984) and more recently Thompson *et al.* (1988a,b) and Thompson and Gähwiler (1989b) have presented evidence that, at first glance, appears incompatible with Russell's suggestion but favors the notion that a K^+,Cl^- symport is the primary mechanism for active Cl^- extrusion in vertebrate and invertebrate neurons (see below). In fact, the two proposed mechanisms are not mutually exclusive. At present, the evidence given for each of them as transport mechanisms for active Cl^- extrusion in neurons is still fragmentary and, in the case of the K^+,Cl^-

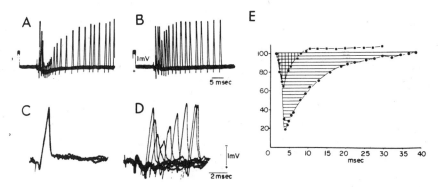

Figure 2. Interaction between field potentials evoked at the trochlear nucleus by vestibular and antidromic (IVth nerve) stimulation. The control antidromic field potential, shown in C, was elicited at increasing intervals after stimulation of the ipsilateral vestibular nerve. The slow positive wave (downward deflection) on which the antidromic field potentials are superimposed in A, represent the field potential evoked by vestibular nerve stimuli. The record in A was taken before, and those in B–D after, administration of 2.0 mmol/kg animal weight ammonium acetate. In A, note that the amplitude of the antidromic field potential was reduced for a period up to about 30 msec after vestibular nerve stimulation. Inhibition was maximal for an interval of 3–3.5 msec after the vestibular nerve stimulus, which corresponded with the peak of the extracellular positivity generated by vestibular nerve stimuli in the trochlear nucleus. This positive wave represented the IPSP recorded extracellularly. Following ammonium acetate injection, the positive field potential generated by vestibular nerve stimulation was abolished and its inhibitory action on the antidromic field markedly weakened (B). (E) Plot of the time course and amplitude of inhibition of the antidromic field potentials from a set of records similar to those shown in A and B. Dots indicate inhibition prior to, and squares inhibition following ammonium acetate injection; the horizontal and vertical lines show total area of inhibition under these two conditions, respectively. The inhibition was reduced to 10% of control, as calculated from the reduction of the total area of inhibition (horizontal lines) to that following ammonium acetate (vertical lines). (Reproduced with permission from Llinás *et al.*, 1974.)

cotransport, rather indirect, to be considered conclusive. In fact, in the crayfish stretch receptor neuron both transport processes may coexist in the same cell (Aickin *et al.*, 1982; Moser, 1985).

4.1. Na⁺-Coupled Cl⁻/HCO₃⁻ Exchanger as a Possible Mechanism for Active Cl⁻ Extrusion in Nerve Cells

Na⁺-dependent Cl⁻/HCO₃⁻ exchange was first described by Thomas (1977) in snail neurons. Subsequently, it was found in other invertebrate neurons including squid axons (Boron and Russell, 1983), crayfish central neurons (Moody, 1981), crayfish stretch receptor sensory neurons (Moser, 1985), and leech Retzius neurons (Schlue and Thomas, 1985; Schlue and Deitmer, 1988). There is no evidence for the presence of this transport system in vertebrate neurons (Chesler, 1986, 1987). Na⁺-dependent Cl⁻/HCO₃⁻ exchange has been considered primarily as pH$_i$ regulatory mechanism (Thomas, 1984). This system is inhibited by disulfonic stilbenes such as SITS and DIDS, by removing external Na⁺ or HCO₃⁻, or by depleting internal Cl⁻. The system has been characterized in great detail in squid giant axons (see Russell and Boron, this volume) and snail neurons (Thomas, 1984). Reviews dealing with this anion exchanger have appeared in the last decade (Roos and Boron, 1981; Russell and Boron, 1982; Boron, 1983; Thomas, 1984; see also Boron, 1985). This system couples the electroneutral, HCO_3^--dependent movement of one Na^+ and two equivalents of base in one direction, to the movement of one Cl^- in the opposite direction. Through the operation of this mechanism the inwardly directed chemical gradient for Na^+ can drive the uphill entry of HCO_3^- into the cell, coupled to the efflux of Cl^- from the cell. Thus, as already mentioned, it is not unreasonable to view this transport system as subserving the function of a pH_i regulator as well as that of a Cl_i^- regulator in nerve cells. Although there is some suggestive evidence supporting this dual function (Russell, 1980), further studies are needed to clarify this issue.

4.2. K^+,Cl^- Cotransport as a Possible Mechanism for Active Cl^- Extrusion in Nerve Cells

4.2.1. Diagnostic Criteria for K^+,Cl^- Cotransport

Carrier-mediated electroneutral K^+,Cl^- cotransport has been demonstrated in a variety of nonneuronal cell types (for references see Lauf, 1988; Reuss, 1989; Brugnara *et al.*, 1989). In this kind of transport system the movements of Cl^- and K^+ are directly coupled. It is a secondary active transport mechanism in which the uphill movement of one ion occurs with the driving force of the downhill movement of the partner ion. Under physiological conditions the chemical potential gradient for K^+, which is outwardly directed, can drive Cl^- uphill against its chemical potential gradient.

Proof of K^+,Cl^- cotransport requires demonstrating that: (1) Cl^- uphill transport is driven by downhill K^+ movement; (2) uphill transport of K^+ is driven by the downhill transport of Cl^-; (3) the Cl^- fluxes attributable to the cotransport do not occur by electrodiffusion and therefore are not dependent on the membrane potential; (4) the mechanism is Na^+-independent; and (5) the mechanism can be inhibited by sulfamoyl benzoic acid diuretics such as furosemide and bumetanide.

Experimental tests for the above criteria can be obtained by elevating $[K^+]_o$, keeping $[Cl^-]_o$ constant, and observing changes in a_{Cl}^i. If a K^+,Cl^- cotransport is present, elevating $[K^+]_o$ should produce an increase in a_{Cl}^i. It follows that decreasing $[K^+]_o$ should produce a fall in a_{Cl}^i (e.g., Reuss, 1983; Corcia and Armstrong, 1983). Usually it is possible to make alterations in $[K^+]_o$ within the range of 1 to 10 mM without substantial changes in E_m. It must be proved that Cl^- movements do not occur by an electrodiffusional pathway of any type, e.g., voltage dependent, passive leak, or transmitter gated. In this respect, it is important to keep in mind that in nervous tissues increasing $[K^+]_o$ might lead to leakage of inhibitory transmitters from nerve terminals and/or glial cells. Therefore, appropriate control experiments have to be made to exclude that possibility. Reversal of the mode of operation of the carrier can be achieved by experimental perturbation of the chemical gradients. For instance, reducing the outwardly directed K^+ gradient and simultaneously increasing the inwardly directed Cl^- gradient should produce an uphill increase in a_K^i concomitant with an increase in a_{Cl}^i (e.g., Reuss, 1983).

There are no specific pharmacologic blocking agents of K^+,Cl^- cotransport although recently [(dihydroindenyl)oxy]alkanoic acid has been introduced as a potent inhibitor of this transport system in red cells (Garay *et al.*, 1988; Vitoux *et al.*, 1989). Unfortunately, the above compound also blocks the Cl^-/HCO_3^- exchanger and at relatively high doses it induces a "nonspecific K^+ leak" (Garay *et al.*, 1988). Judicious use of derivatives of stilbene sulfonic acid and sulfamoylbenzoic acid in conjunction with experiments of the type described above should give clues on the nature of the Cl^- transporter. In duck red blood cells (Lytle and McManus, 1987), K^+,Cl^- cotransport is inhibited by sulfamoylbenzoic acid derivatives with the following order of potency: furosemide > bumetanide ≫ piretanide. Stilbene sulfonic acid derivatives such as SITS and DIDS also inhibit K^+,Cl^- cotransport with an IC_{50} similar to bumetanide (30 μM). It is important to bear in mind that Cl^- transport inhibitors can block not only carrier-mediated but also electrodiffusional-Cl^- transport pathways. For instance, furosemide blocks not only K^+,Cl^- cotransport (Kregenow, 1981; Lauf, 1988), but also Na^+,K^+,Cl^- cotransport (Ellory *et al.*, 1982; O'Grady *et al.*, 1987), Cl^-/HCO_3^- exchange (Brazy and Gunn, 1976), Na^+-dependent Cl^-/HCO_3^- exchange (Boron *et al.*, 1978; Russell, personal communication), and Cl^- channels (Nicoll, 1978; Cherksey and Zeuthen, 1987; Nakai *et al.*, 1988). Bumetanide is a widely used Na^+,K^+,Cl^- cotransport blocker because it inhibits this transporter with an IC_{50} of 0.1 to 0.3 μM (O'Grady *et al.*, 1987; Altamirano and Russell, 1987; Landry *et al.*, 1987). However, it inhibits Cl^- currents in epithelia with an IC_{50} of 30 μM (Landry *et al.*, 1987) and it is also capable of inhibiting Cl^-/HCO_3^- exchange (Gunn, 1985). All this emphasizes the need for caution in characterizing a Cl^- transport mechanism solely by its pharmacologic properties.

4.2.2. K^+,Cl^- Cotransport as a Mechanism for Cl^- Extrusion by Invertebrate Neurons

The crayfish stretch receptor neuron has been the subject of detailed studies to elucidate the nature of the Cl^- transport process responsible for Cl^- extrusion and its implications on the ionic mechanism of postsynaptic inhibition (Meyer and Lux, 1974; Deisz and Lux, 1982; Aickin *et al.*, 1982, 1984). The stretch receptor is a convenient preparation for these kinds of studies because the neurons are large enough to be impaled with ion-selective microelectrodes (e.g., Brown *et al.*, 1978; Deisz and Lux, 1982; Moser, 1985), the GABAergic inhibitory synapses are evenly distributed over the neuron and are not restricted to the dendrites (Peterson and Pepe, 1961), and their Cl^--dependent IPSPs are hyperpolarizing (for review see Roberts, 1986). Deisz and Lux (1982) measured a_{Cl}^i in stretch receptor neurons with liquid anion exchanger microelectrodes. Their preparations were maintained in nominally HCO_3^--free medium. Hence, Cl^-/HCO_3^- exchange (Moser, 1985) probably proceeded very slowly. They concluded that, after appropriate correction for a constant intracellular interference on Cl^--selective electrode readings, a_{Cl}^i was lower than predicted from a passive distribution and thus E_{Cl} was more negative than E_m. The reversal potential of the IPSP (E_{IPSP}) was approximately equal to E_{Cl}, implying the presence of an active Cl^- extrusion mechanism.

Complete replacement of Cl_o^- with isethionate and gluconate resulted in a transient decrease of E_m by up to 15 mV and a fall in a_{Cl}^i. Substitution of 50% of the Cl_o^- had no effect on E_m but produced a fall in a_{Cl}^i, and a slight increase in cell input resistance (R_{in}). Taken together these results suggest the presence of a significant electrodiffusive Cl^- permeability. The latter would not preclude the existence of an electroneutral Cl^- transport system but would certainly make its proof more difficult. Deisz and Lux (1982) also found that increasing $[K^+]_o$ to 20 mM produced a sustained depolarization concomitant with an increase in a_{Cl}^i. Complete removal of K_o^+ led to the opposite changes. The effects of $[K^+]_o$ on a_{Cl}^i were interpreted as being consistent with the presence of a K^+,Cl^- cotransport (Aickin *et al.*, 1982, 1984). However, the observations can also be interpreted as resulting from a membrane highly conductive to both Cl^- and K^+. Exposure to 5 mM NH_3/NH_4^+ (pH$_o$ 7.55) produced a depolarization of (1.5 to 4 mV) accompanied by an increase in a_{Cl}^i. E_{IPSP} was also shifted to more positive values although remaining more negative than E_m. R_{in} decreased by 20 $\pm$ 5.5% (see Hino, 1979), suggesting again that the increase in a_{Cl}^i could be accounted for by a high electrodiffusional permeability to Cl^-. The latter could represent passive-leak anion channels and/or transmitter-gated anion channels. Deisz and Lux (1982) tested for the second possibility by applying picrotoxin (PTX). Exposure to 10^{-4} M PTX produced a decrease in a_{Cl}^i, indicating a significant Cl^- influx via PTX-sensitive channels. However, application of 5 mM NH_4^+ in the presence of PTX caused a similar increase in a_{Cl}^i to that observed in the absence of PTX. This suggests that the NH_4^+-induced increase in a_{Cl}^i was not via an increase in the PTX-sensitive Cl^- conductance but did not exclude the possibility of a PTX-resistant Cl^- conductance. Furosemide (6 $\times 10^{-4}$ M) produced an increase in the apparent a_{Cl}^i. The effect could not be attributed to furosemide interference on the Cl^--selective microelectrode response since the change in a_{Cl}^i was accompanied by a shift in E_{IPSP} toward more positive values. Furosemide slightly depolarized the cells but R_{in} was not altered. These results suggest that the putative Cl^- extrusion mechanism is furosemide-sensitive.

In a subsequent study, Aickin *et al.* (1982) further assessed the effect of NH_4^+ and other cations and Cl^--transport inhibitors on the Cl^- extrusion mechanism of the stretch receptor neuron. They did not directly measure a_{Cl}^i but estimated it by measuring E_{IPSP} and the driving force for the IPSP, i.e., $E_{IPSP} - E_m$. Their basic assumption was that $E_{IPSP} = E_{Cl}$. This was a reasonable assumption considering that their experiments were conducted in the absence of HCO_3^- and that Deisz and Lux (1982) had demonstrated that $E_{IPSP} \simeq E_{Cl}$ by directly measuring a_{Cl}^i.

Aickin *et al.* (1982) found that application of NH_3/NH_4^+ induced a concentration-dependent reversible decrease in the driving force for the IPSP with an onset at about 0.2 mM. E_m and R_{in} also decreased in a concentration-dependent fashion. E_{IPSP} never became less negative than E_m. It is well known that exposure to NH_3/NH_4^+ induces an intracellular alkalinization and upon withdrawal of NH_3/NH_4^+ an internal acidification ensues (Boron and De Weer, 1976; Roos and Boron, 1981). However, Moser (1985) measuring pH$_i$ in the stretch receptor neuron found that during exposures to 20 mM

NH_4Cl (in HCO_3^--free media) there was only a small initial intracellular alkalinization (0.14 ± 0.05 pH unit), which peaked 1 or 2 min after the onset of the NH_4^+ application, and which was followed by a slow acidification (initial rate of 0.13 ± 0.02 pH unit/min) tending to a plateau at a pH_i of about 6.8. Removal of NH_4Cl from the media resulted in a fast acidification (see his Fig. 1). He concluded that the magnitude and time course of the pH_i response in the sensory neurons are in fact different from other preparations and may be explained by a large NH_4^+ permeability (see Hino, 1979; Moser, 1987) and a 90% reduction of membrane resistance in 20 mM NH_4Cl. In the experiments of Aickin *et al.* (1982), cells were exposed to NH_3/NH_4^+ for periods of about 10 min. Therefore, after 2 to 3 min of exposure to NH_3/NH_4^+ the cells probably started to become acidified rather than alkalinized. Considering the shape and magnitude of the observed pH_i response following NH_3/NH_4^+ exposure (Moser, 1985, 1987), the E_{IPSP} changes reported by Aickin *et al.* (1982) appear to be uncorrelated with pH_i. The pH_i-regulating system of the sensory cell requires Na^+ and Cl^- which according to Moser (1985) probably operates in a combined mechanism such as parallel Na^+/H^+–Cl^-/HCO_3^- exchange or tightly coupled Na^+-dependent Cl^-/HCO_3^- exchange similar to that present in other preparations (Thomas, 1984). The experiments of Aickin *et al.* (1982) were conducted in nominally HCO_3^--free medium. In addition, lowering pH_i by exposing the cells to acetate considerably reduced the driving force for the IPSP. Assuming that the GABA-activated anion channels were impermeable to acetate, this suggests that intracellular acidification was accompanied by an increase in a_{Cl}^i. These indirect observations are difficult to reconcile with the presence of Na^+-dependent Cl^-/HCO_3^- exchange operating under these conditions as a Cl^- extrusion mechanism. Taken together, the results suggest that the Cl^- extrusion mechanism studied by Aickin *et al.* (1982) is different from the pH_i-regulating system described by Moser in these cells. Exposure to NH_3/NH_4^+ at different pH_o values led Aickin and associates to conclude that the decrease in the driving force for the IPSP was due to the cation NH_4^+ interfering with the Cl^- extrusion mechanism (Aickin *et al.*, 1982, 1984). This concept was reinforced by the observation that other monovalent cations mimicked the effect of NH_4^+ in decreasing the driving force for the IPSP in the order: $Rb^+ > NH_4^+ > K^+ > Cs^+$. They further showed that furosemide (6×10^{-4} M) decreased the driving force for the IPSP, which was in agreement with the observations of Deisz and Lux (1982). DIDS (10^{-4} M) caused spontaneous firing and reduced E_{IPSP} to the threshold potential. R_{in} was also decreased. Aickin *et al.* (1982) finally concluded that the IPSP driving force is maintained in these neurons by a K^+,Cl^- cotransport. Measurement of K_i^+ (Brown *et al.*, 1978) indicates that the K^+ gradient is large enough to provide the energy for Cl^- extrusion. The K^+ site of the K^+,Cl^- cotransport mechanism exhibits, according to Aickin *et al.* (1982), the binding selectivity sequence: $Rb^+ > NH_4^+ > K^+ > Cs^+$. The mechanism is inhibited partially by furosemide and completely by DIDS. The possible presence of an electrodiffusional Cl^- leak was not completely ruled out. The above observations taken together with those of Moser (1985) allow a tentative conclusion, namely that under physiological conditions it is not unlikely that both a Na^+-dependent Cl^-/HCO_3^- exchange and a K^+,Cl^- cotransport may be involved in regulating a_{Cl}^i, although further studies are clearly needed to prove this hypothesis. Considering the criteria for K^+,Cl^-

cotransport that were outlined in Section 4.2.1, the existing evidence for this kind of transport in stretch receptor neurons is still fragmentary and incomplete.

4.2.3. A Mechanism for Cl⁻ Extrusion in Vertebrate Neurons: Evidence for K⁺,Cl⁻ Cotransport

It has recently been suggested that the mechanism of active Cl^- extrusion which normally maintains E_{Cl} more negative than E_m in mammalian cortical neurons is a K^+,Cl^- cotransport (Thompson et al., 1988a; Thompson and Gähwiler, 1989b). The experimental evidence backing up this hypothesis is indirect (see below) and can be summarized as follows: (1) Cl^- extrusion following a Cl_i^- load is inhibited by furosemide and bumetanide but not by SITS (the latter is in contrast to stretch receptor neurons in which the putative cotransporter is blocked by DIDS); (2) the mechanism is sensitive to $[K^+]_o$ in such a way that increasing $[K^+]_o$ leads to an increase in $[Cl^-]_i$; (3) Cl^- movements are not accompanied by changes in cell input resistance or E_m, suggesting that the mechanism is electroneutral; and (4) the mechanism transports SCN^- but not NO_3^-.

The experiments considered here were done in guinea pig cingulate cortical neurons of in vitro slices (Thompson et al., 1988a,b) and in CA_3 pyramidal cells in organotypic hippocampal slice cultures from neonatal rats (Thompson and Gähwiler, 1989a,b). Since a_{Cl}^i was not measured, $[Cl^-]_i$ and E_{Cl} were estimated indirectly from the reversal potentials (E_{IPSP}) of synaptically evoked GABAergic Cl^--dependent IPSPs, or from those of responses to perisomatic GABA application (E_{GABA}). The authors showed that $E_{GABA} \simeq E_{IPSP}$, and assumed that $E_{Cl} = E_{GABA}$ or E_{IPSP}. This assumption is valid only if Cl^- is the only permeant ion (see below). Shortly after neuronal impalement with 2 M KCl-filled microelectrodes, E_{IPSP} was shifted in the depolarizing direction until reaching a stable value. This was most likely due to diffusion of Cl^- from the microelectrode. Then Cl^- was iontophoresed intracellularly. This resulted in a further depolarizing shift in E_{IPSP} and the generation of a depolarizing IPSP at resting E_m. The depolarizing IPSP returned to control (preinjection) amplitude with a single exponential time course. Because the depolarizing shift in E_{IPSP} results from an increase in $[Cl^-]_i$, the rate of recovery of the amplitude of the IPSP was taken to reflect the rate of Cl^- extrusion from the cell. The recovery of the amplitude of the depolarizing IPSP after Cl^- iontophoresis occurred without changes in E_m or input resistance, suggesting that Cl^- efflux underlying IPSP recovery was electroneutral. The fact that the resting electrodiffusive P_{Cl} was negligible reinforced the latter interpretation. The rate of recovery of the amplitude of the depolarizing IPSP after iontophoretic Cl^- load was significantly reduced by furosemide (0.5–1.5 mM) and bumetanide (50 μM) but not by piretanide (50 μM). The lack of effect of piretanide was explained in terms of its relative potency (Fig. 3). However, higher doses of piretanide were not tested. High doses of SITS (0.5–1 mM) had no effect on recovery from Cl_i^- injection. Iontophoretic loading of the cells with NO_3^- or SCN^- resulted also in a depolarizing shift in E_{IPSP}. From the rates of recovery of IPSP amplitude after intracellular loading with either SCN^- or NO_3^-, it was concluded that the former but not the latter was transported by the anion extrusion mechanism (Fig. 3).

The K$^+$,Cl$^-$ cotransport hypothesis predicts that the [Cl$^-$]$_i$ and the rate of Cl$^-$ extrusion are both functions of $\Delta\mu_{K^+}$, the chemical potential for K$^+$. Accordingly, increasing [K$^+$]$_o$ should lead to a decrease in the rate of Cl$^-$ extrusion and an to increase in [Cl$^-$]$_i$. Thompson et al. (1988a,b) tested the effect of [K$^+$]$_o$ on the rate of recovery of IPSP amplitude following intracellular iontophoresis of Cl$^-$ and on E_{IPSP}. Increasing [K$^+$]$_o$ from 2.5 to 10 mM caused a reduction in the rate of recovery of the IPSP, suggesting a decrease in the rate of Cl$^-$ extrusion. They also showed that increasing [K$^+$]$_o$ in steps from 1 to 10 mM leads to a positive shift in E_{IPSP} (Fig. 5, filled circles). Assuming that E_{IPSP} is an indicator of [Cl$^-$]$_i$, as the authors did, the latter observation suggests that [Cl$^-$]$_i$ increased as [K$^+$]$_o$ was increased. They estimated [Cl$^-$]$_i$ from the observed values of E_{IPSP}, for each [K$^+$]$_o$, assuming that E_{IPSP} = E_{Cl}. These values are plotted in Fig. 4 (filled circles).

The values of [Cl$^-$]$_i$ calculated from E_{IPSP} are likely to be overestimates if K$^+$,Cl$^-$ cotransport is the only Cl$^-$ transport system and if the anion channels transiently opened by GABA are also permeable to HCO$_3^-$, as discussed below. Thompson et al. (1988b) showed that the resting P_{Cl} of cortical neurons is unmeasurably small. Assuming for the sake of argument that K$^+$,Cl$^-$ cotransport is the sole system transporting Cl$^-$, and that [K$^+$]$_i$ is held constant by the Na$^+$,K$^+$ pump, the driving force for the cotransport, $\Delta\mu_{K^+,Cl^-}$, is the sum of the chemical potential differences of K$^+$ and Cl$^-$ as expressed in the equation

$$\Delta\mu_{K^+,Cl^-} = \Delta\mu_{K^+} + \Delta\mu_{Cl^-} \qquad (4)$$

it follows that

$$\Delta\mu_{K^+,Cl^-} = RT \ln \frac{[K^+]_i}{[K^+]_o} + RT \ln \frac{[Cl^-]_i}{[Cl^-]_o} \qquad (5)$$

This system attains thermodynamic equilibrium when

$$\Delta\mu_{K^+,Cl^-} = 0 \qquad (6)$$

and therefore, for a 1:1 stoichiometry,

$$[K^+]_i \, [Cl^-]_i = [K^+]_o \, [Cl^-]_o \qquad (7)$$

and

$$[Cl^-]_i = \frac{[K^+]_o \, [Cl^-]_o}{[K^+]_i} \qquad (8)$$

It is important to mention that the above model neglects kinetic constraints, i.e., it is valid only when concentrations are well below saturation. For example, in the control saline used by Thompson et al. (1988b), [K$^+$]$_o$ was 5 mM, [Cl$^-$]$_o$ was 133 mM, and [K$^+$]$_i$ estimated from the reversal potential of the late IPSP, which is due to an increase

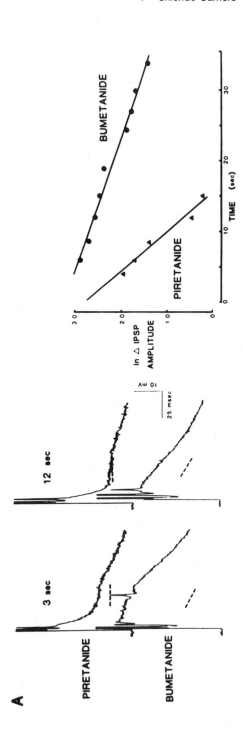

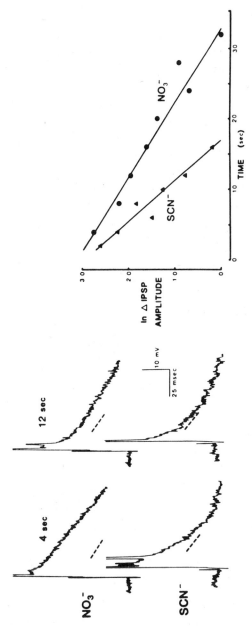

Figure 3. (A) Effect of bumetanide and piretanide on rate of recovery of IPSP amplitude following intracellular iontophoresis of Cl⁻ (2 min, 1.5-nA hyperpolarizing current) in guinea pig cingulate cortical neurons. Synaptic responses 3 and 12 sec after Cl⁻ injection are shown at the left; dashed lines indicate control amplitude in this and subsequent synaptic responses. On the right, the natural logarithm of the increase in IPSP amplitude over the control amplitude is plotted as a function of time after injection, for two representative cells exposed to 50 μM bumetanide or 50 μM piretanide. Time constant = 6.5 sec in piretanide, not significantly different from control, and 15 sec in bumetanide, noticeably shorter than control. (B) Comparison of the rates of extrusion of NO₃⁻ and SCN⁻ following intracellular iontophoresis (2 min, 1.0 nA) for two representative cells impaled with KNO₃- or KSCN-filled recording electrodes. Synaptic responses 4 and 12 sec after anion iontophoresis shown on the left. On the right, calculation of the time constant of Cl⁻ extrusion as in A. Time constant = 6 sec for SCN⁻, not significantly different from control Cl⁻ injections and 21 sec for NO₃, noticeably shorter than control Cl⁻ injections. Action potentials appear amputated due to high gain. (Modified from Thompson *et al.*, 1988.)

in gK, was about 120 mM. Substituting these values into Eq. (8), a value of about 5.5 mM is obtained for $[Cl^-]_i$. This is significantly lower than the value of 9.7 mM estimated by Thompson et al. (1988b) from E_{IPSP}, assuming that $E_{IPSP} = E_{Cl}$, i.e., that the neurotransmitter-gated anion channels are permeated only by Cl^-. Figure 4 shows $[Cl^-]_i$ estimated by Thompson et al. (1988b) for four different values of $[K^+]_o$, assuming that $E_{Cl} = E_{IPSP}$ (filled circles), as compared to those estimated using Eq. (8). If the simplifying assumptions for applying Eq. (8) are correct, the present arguments suggest that $[Cl^-]_i$ could be significantly lower than predicted from E_{IPSP} measurements, while the conclusion that $[Cl^-]_i$ is a function of $[K^+]_o$ could still be valid (see below). It is important to bear in mind that increasing $[K^+]_o$ could produce an increase in P_{Cl} with the consequent increase in $[Cl^-]_i$. This is because increasing $[K^+]_o$ could lead to an increase in release of transmitter from GABAergic synaptic terminals investing the neuron under observation (see Somjen, 1979). Obviously, this effect of $[K^+]_o$ on $[Cl^-]_i$ would have nothing to do with a K^+,Cl^- cotransport. Clearly, appropriate controls are needed to rule out this possibility.

In light of recent studies on the anion selectivity of GABA-gated Cl^- channels (Bormann et al., 1987; Kaila et al., 1989; Kaila and Voipio, this volume; see also Kelly et al., 1969), the practice of estimating E_{Cl} and $[Cl^-]_i$ (or a_{Cl}^i) from measurements of E_{IPSP} or E_{GABA} may lead to significant errors. This is because GABA-gated Cl^- channels are not perfectly selective for Cl^- but are also permeable to other anions among which the most important from a physiologic point of view is HCO_3^-. Under physiological conditions, HCO_3^- is present inside cells in significant concentrations. $[HCO_3^-]_i$ is determined primarily by pH_i and the internal CO_2 (Thomas, 1976). Since neurons maintain their pH_i at a level that is about one unit higher than expected from a passive distribution of H^+, the HCO_3^- equilibrium potential (E_{HCO_3}), which is the same as that for H^+, will be more positive than E_m (Roos and Boron, 1981; Thomas, 1984). When E_{Cl} is more negative than E_m, E_{IPSP} (or E_{GABA}) will have an intermediate value between E_{HCO_3} and E_{Cl}. The value of E_{IPSP} (or E_{GABA}) will be determined

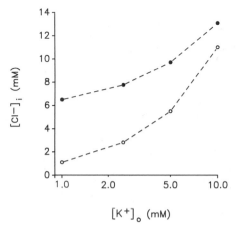

Figure 4. Estimated intracellular Cl^- concentration, $[Cl^-]_i$, as a function of external $[K^+]$ for a guinea pig cingulate cortical neuron of a brain slice maintained in vitro. Filled circles are estimates obtained from values of E_{IPSP} reported by Thompson et al. (1988b; their Fig. 1A). The actual E_{IPSP} values measured by Thompson et al. (1988b) are replotted in Fig. 5 (filled circles). $[Cl^-]_i$ was calculated with the Nernst equation assuming that $E_{IPSP} = E_{Cl}$. Open circles are the steady-state values predicted by Eq. (8), assuming that the only Cl^- transport system in the cell is a K^+,Cl^- cotransport with a 1:1 stoichiometry and that the membrane has a negligible resting electrodiffusive P_{Cl}. The value of $[K^+]_i$ used in the calculations was 120 mM. The assumptions of the model are further discussed in the text.

by the permeability ratio P_{HCO_3}/P_{Cl} for the anion channels opened by GABA; and by the external and internal concentrations (or more correctly activities) of Cl^- and HCO_3^-. Knowing the appropriate parameters, it is possible to calculate E_{IPSP} (or E_{GABA}) using the Goldman–Hodgkin–Katz equation:

$$E_{IPSP} = -\frac{RT}{F} \ln \frac{[Cl^-]_o + \alpha\,[HCO_3^-]_o}{[Cl^-]_i + \alpha\,[HCO_3^-]_i} \tag{9}$$

where $\alpha = P_{HCO_3}/P_{Cl}$; R, T, and F have their usual meaning; $[Cl^-]_o$ and $[HCO_3^-]_o$ are known; P_{HCO_3}/P_{Cl} has recently been determined for GABA-gated Cl^- channels in mouse cultured spinal neurons (Bormann *et al.*, 1987) and in crayfish muscle fibers (Kaila *et al.*, 1989). The measured values range from 0.18 to 0.42. Data obtained from simultaneous measurements of E_{GABA}, a^i_{Cl}, and pH_i in crayfish muscle gave a value for P_{HCO_3}/P_{Cl} of 0.33 (Kaila *et al.*, 1989). I have used this value for the calculations that follow. The other parameter needed to estimate E_{IPSP} under the ionic conditions described by Thompson *et al.* (1988b) is $[HCO_3^-]_i$. This can be estimated by knowing the difference between pH_o and pH_i. This is because CO_2 permeates the plasma membrane freely and at constant P_{CO_2}, for reasons already explained, E_{HCO_3} will match E_H. Thus,

$$E_H = 2.303\,\frac{RT}{zF}\,(pH_i - pH_o) \tag{10}$$

at 37°C and taking $pH_i = 7.2$ and $pH_o = 7.4$

$$E_H = 61.54 \text{ mV } (7.2 - 7.4)$$
$$\simeq -12 \text{ mV}$$

and thus

$$E_{HCO_3} = -12 \text{ mV}$$

From the Nernst equation it follows that

$$[HCO_3^-]_i = [HCO_3^-]_o \exp \frac{-12\,F}{RT} \tag{11}$$

For $[HCO_3^-]_o = 26$ mM, the calculated $[HCO_3^-]_i$ is about 16.4 mM. Another way of calculating $[HCO_3^-]_i$ is from the equilibrium distribution ratios as follows:

$$[HCO_3^-]_i = \frac{[H^+]_o}{[H^+]_i}\,[HCO_3^-]_o \tag{12}$$

which for the pH_o and $[HCO_3^-]_o$ used in Thompson's experiments and assuming that the pH_i was 7.2, gives a value of about 16.4 mM. Now, with all the above values, and knowing that $[Cl^-]_o = 133$ mM it is possible to estimate E_{IPSP} by using Eq. (9). Substituting the appropriate figures in this equation at 37°C and for the composition of the control saline used by Thompson *et al.* (1988b), which contains 5 mM K^+,

$$E_{IPSP} = -61.54 \log \frac{[133] + 0.33 \cdot [26]}{[5.5] + 0.33 \cdot [16.4]}$$

$$\approx -68.5 \text{ mV}$$

The value of -68.5 mV estimated for E_{IPSP} is close to that of -70 mV experimentally obtained by Thompson et al. (1988b). However, these estimated and observed values of E_{IPSP} differ considerably from the value of ≈ -85 mV, which would be obtained assuming that $E_{Cl} = E_{IPSP}$ and that the steady-state transmembrane Cl^- distribution is determined solely by K^+,Cl^- cotransport. Figure 5 shows E_{IPSP} values plotted against different $[K^+]_o$. The filled circles are the E_{IPSP} values measured by Thompson et al. (1988b), the open triangles represent E_{IPSP} estimated for a 1 : 1 K^+,Cl^- cotransport, assuming that $[K^+]_i = 120$ mM and that $E_{Cl} = E_{IPSP}$ and the open circles are estimated E_{IPSP} values assuming a 1 : 1 K^+,Cl^- cotransporter, $[K^+]_i = 120$ mM, $[HCO_3^-]_i = 16.4$ mM, and P_{HCO_3}/P_{Cl} for the GABA-gated channels is 0.33. The main conclusion that can be drawn from the experimental and theoretical data summarized in Fig. 5 is that the E_{IPSP} values measured by Thompson et al. (1988b) are consistent with the presence of a K^+,Cl^- cotransporter if it is assumed that the anion channels opened by GABA are also permeated by HCO_3^-. The model presented here is not consistent with the assumption that $E_{Cl} = E_{IPSP}$. If the model is correct, then the values for $[Cl]_i$ estimated from E_{IPSP}, assuming that $E_{IPSP} = E_{Cl}$ (Fig. 5, filled circles), are likely to be overestimates of the "true" $[Cl^-]_i$ (Fig. 4, open circles). Clearly, future experiments aimed to directly measure $[Cl^-]_i$ are needed to clarify the above issues and to confirm the presence of K^+,Cl^- cotransport in vertebrate nerve cells. Internal perfusion studies in isolated CNS neurons (e.g., Kaneda et al., 1989) in conjunction with patch-clamp studies (Bormann et al., 1987) should allow, in the near future, measurement of P_{HCO_3}/P_{Cl} for GABA-gated anion channels in other vertebrate neurons. Direct measurement of $[Cl^-]_i$ with liquid anion exchanger microelectrodes and/or Cl^--sensitive fluorescent indicators (Verkman et al., 1989) in brain slices, freshly isolated cells, or cell cultures will constitute powerful approaches to these problems.

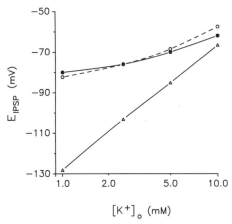

Figure 5. Equilibrium potential of the IPSP, E_{IPSP}, as a function of external $[K^+]$ for a guinea pig cingulate cortical neuron. Filled circles are values reported by Thompson et al. (1988b). Open triangles are values predicted by Eq. (9), assuming that $E_{IPSP} = E_{Cl}$, i.e. $\alpha = 0$, and that the only Cl^- transport system present in the cell is a K^+,Cl^- cotransport with a 1 : 1 stoichiometry. $[Cl^-]_o = 133$ mM as reported by Thompson et al. (1988b). The values used for $[Cl^-]_i$ are those shown by open circles in Fig. 4, i.e., those predicted by Eq. (8). Open circles: values of E_{IPSP} predicted by Eq. (9) for a cell having a 1 : 1 K^+,Cl^- cotransport, $[K^+]_i = 120$ mM, $[HCO_3^-]_i = 16.4$ mM, and $\alpha = 0.33$. The values used for $[Cl^-]_i$ are those shown by open circles in Fig. 4.

5. NATURE OF INWARDLY DIRECTED ACTIVE Cl⁻ TRANSPORT MECHANISM IN NERVE CELLS

Some vertebrate and invertebrate neurons keep their $[Cl^-]_i$ (or a^i_{Cl}) at levels higher than predicted from a passive distribution (Section 3.3). This implies that these nerve cells must be endowed with one or more inwardly directed active Cl^- transport mechanisms. The presence of a mechanism for uphill accumulation of Cl^- in nerve cells was first postulated by Keynes (1963) to explain the higher-than-passive intracellular free $[Cl^-]$ found in squid giant axons. The nature of the transport process responsible for active uptake of Cl^- by the axon has been elucidated by the work of Russell (1976, 1979, 1983, 1984; Altamirano and Russell, 1987; see Chapter 3, this volume). It is a secondary active transport process involving the electroneutral influx of Cl^- coupled to Na^+ and K^+, and the fluxes can be inhibited by furosemide and bumetanide, i.e., a Na^+, K^+, Cl^- cotransporter. This cotransport mechanism is not a unique feature of invertebrate nerve cells. In fact, there is recent evidence for the presence of a similar transport system in some vertebrate neurons such as those of frog dorsal root ganglia (Alvarez-Leefmans *et al.*, 1986, 1988) and probably those of rat sympathetic ganglion neurons (Ballanyi and Grafe, 1985). In addition, the system has been found and characterized in more or less detail in a variety of nonneuronal cells (e.g., Geck and Heinz, 1986; Chipperfield, 1986; O'Grady *et al.*, 1987; Lauf *et al.*, 1987; Haas, 1989). In fact, it was first identified almost 10 years ago by Geck *et al.* (1980) working in Ehrlich ascites tumor cells (see Geck and Heinz, 1986).

In nonneuronal cells there are other active transport systems capable of uphill accumulation of Cl^-. These include a Na^+-independent Cl^-/HCO_3^- exchanger, like the one present in heart Purkinje fibers and other cell types (Vaughan-Jones, 1988); a K^+-independent Na^+, Cl^- cotransporter, like the one characterized in canine tracheal epithelium (Widdicombe *et al.*, 1983); and perhaps also a primary active Cl^- transport, i.e., an ATP-driven Cl^- pump (Gerencser and Lee, 1983). At present it is not clear if K^+-independent Na^+, Cl^- cotransport is a distinct pathway for Cl^- transport or if it simply reflects an alternative mode of operation of Na^+, K^+, Cl^- cotransport. In fact, it has been suggested that Na^+, K^+, Cl^- cotransport may shift to Na^+, Cl^- cotransport depending on the physiological state of the cell (O'Grady *et al.*, 1987). Be that as it may, for the time being, it is not known whether Na^+-independent Cl^-/HCO_3^- exchange, K^+-independent Na^+, Cl^- cotransport, or a Cl^--stimulated ATPase are present in nerve cells.

There are other mechanisms capable of inward Cl^- transport across neuronal cell membranes like some uptake systems for amino acid transmitters which are coupled to Na^+ and Cl^- (Kanner and Schuldiner, 1987). One of the best studied transport systems of this kind is the one for GABA (Radian and Kanner, 1985; Radian *et al.*, 1986; Keynan and Kanner, 1988). It is a Na^+, Cl^-, GABA cotransporter whose primary function appears to be the uptake of GABA by glial cells, synaptic nerve terminals probably belonging to GABAergic inhibitory neurons, and cell bodies of non-GABAergic neurons like those of autonomic and sensory ganglia (Tanaka and Taniyama, 1986; Kanner and Schuldiner, 1987). The driving force for the process seems to be derived from the

Na$^+$ electrochemical gradient. So far the [Cl$^-$]$_i$ of GABAergic synaptic nerve terminals is not known. However, we know that glial cells as well as sympathetic and sensory neurons, all of which are endowed with this GABA uptake system, maintain [Cl$^-$]$_i$ at levels higher than predicted for a passive distribution (see Kimmelberg, this volume; Kettenmann, this volume). Therefore, it is likely that uptake of GABA by some neurons and glial cells is coupled to the downhill movement of Na$^+$ and the uphill movement of Cl$^-$. Whether a situation like this occurs in GABAergic nerve terminals remains to be elucidated. The GABA uptake system appears to be electrogenic. A stoichiometry of 2 Na$^+$: 1 Cl$^-$: 1 GABA has been proposed for the transport of this amino acid into brain membrane vesicles (Radian and Kanner, 1983). However, more complex stoichiometries cannot be excluded (Kanner and Schuldiner, 1987). Transport systems like the one considered here have been described for other amino acids in neuronal and nonneuronal tissues (for a recent example and references see Zelikovic et al., 1989; see also Yudilevich and Boyd, 1987; Kanner and Schuldiner, 1987). In nervous tissues these transport systems are thought to play an important role in the removal of transmitters from synaptic clefts thereby contributing to terminate the overall process of synaptic transmission (Iversen, 1975). In addition, they might subserve the function of assuring both constant and high levels of neurotransmitters in the neuron and low concentrations in the cleft. Furthermore, the reversal of this kind of carrier could contribute to the release of GABA and other amino acid transmitters (e.g., Nelson and Blaustein, 1982). The relative role played by amino acid, Na$^+$,Cl$^-$-coupled transport system in overall Cl$_i^-$ homeostasis is presently unknown. Activation of the GABA transporter may transiently alter the steady-state a_{Cl}^i during increases in inhibitory activity. It is possible that the GABA transporter may be activated under "basal" conditions thereby contributing to the steady-state a_{Cl}^i in some nerve terminals. Topical application of GABA to vertebrate neurons produces transient reduction in the steady-state a_{Cl}^i levels (Bührle and Sonnhof, 1985; Ballanyi and Grafe, 1985) as the result of opening of Cl$^-$ channels and consequent net Cl$^-$ efflux. If this kind of change occurs under physiologic conditions, powerful Cl$^-$ transport systems are needed to restore and then maintain a_{Cl}^i at steady-state levels. Ballanyi and Grafe (1985) have shown that a putative Na$^+$,K$^+$,Cl$^-$ cotransporter may be capable of restoring a_{Cl}^i to the levels found prior to GABA application in rat sympathetic ganglion neurons. Whether the GABA uptake system or other Cl$^-$ transport mechanisms are involved in Cl$^-$ replenishment in these or other nerve cells remains to be elucidated. Recovery of a_{Cl}^i levels after cellular Cl$^-$ depletion in dorsal root ganglion neurons appears to be achieved mainly through Na$^+$,K$^+$,Cl$^-$ cotransport (Alvarez-Leefmans et al., 1988), although other Cl$^-$ transport mechanisms have not been ruled out.

    In conclusion, Na$^+$,K$^+$,Cl$^-$ cotransport is the only mechanism that has been shown to be involved in keeping free [Cl$^-$]$_i$ at values higher than those predicted for a passive thermodynamic distribution in some nerve cells. Other transport systems like Na$^+$-independent Cl$^-$/HCO$_3^-$ exchange, Na$^+$,Cl$^-$ cotransport or Na$^+$,Cl$^-$ amino acid cotransport, or an ATP-driven Cl$^-$ pump might, in principle, be involved, but the evidence is still lacking. The study of Cl$_i^-$ homeostasis in vertebrate nerve cells is at a very early stage. Clearly more research is needed to answer the many questions that have emerged in the last few years and that I have briefly discussed in this section.

5.1. Na^+,K^+,Cl^- Cotransport as an Uphill Cl^- Accumulation System in Some Vertebrate Neurons

The identification and characterization of a Na^+,K^+,Cl^- cotransport system responsible for the uphill accumulation of Cl^- in invertebrate nerve cells is discussed by Russell and Boron (this volume). Here, I shall review and discuss the available evidence for the presence of a Na^+,K^+,Cl^- cotransport system in some vertebrate neurons.

5.1.1. The Need for an Inwardly Directed Active Cl^- Transport Mechanism in Some Vertebrate Nerve Cells

The presence of an inwardly directed active Cl^- transport system in some vertebrate neurons has long been suspected due to the depolarizing action of GABA in the cell bodies of primary sensory neurons (Nishi *et al.*, 1974; Deschenes *et al.*, 1976; Gallagher *et al.*, 1978, 1983) and sympathetic ganglion cells (Adams and Brown, 1975), as well as in the intraspinal terminals of primary afferent fibers (see reviews: Levy, 1977; Nicoll and Alger, 1979; Davidoff and Hackman, 1985). The depolarizing effect of GABA was proposed to result from opening of GABA-gated Cl^- channels, with the consequent efflux of Cl^- which transiently drives E_m toward E_{Cl}. This implied that the steady-state a^i_{Cl} must be higher than predicted from a passive distribution and, hence, the presence of an inwardly directed "Cl^- pump" was postulated (see below). The argument behind this hypothesis was that if GABA acts by selectively opening Cl^- channels, the GABA equilibrium potential, E_{GABA}, should be an indicator of the equilibrium potential for Cl^- (E_{Cl}). In spite of the difficulties involved in accurately determining E_{GABA}, its value was consistently found to be more positive than the resting membrane potential (E_m). If $E_{GABA} = E_{Cl}$, then the above neuronal types must be endowed with a so-called "inwardly directed chloride pump" (e.g., Nishi *et al.*, 1974; Adams and Brown, 1975; Deschenes *et al.*, 1976; Gallagher *et al.*, 1978, 1983). This "pump" would hold E_{Cl} above its equilibrium level, setting up an outwardly directed Cl^- electrical gradient. Until recently, no direct evidence was provided to support this hypothesis (see below and Alvarez-Leefmans *et al.*, 1988). Furthermore, pharmacologic agents known to interfere with Cl^- transport in other systems yielded conflicting interpretations (Nicoll, 1978; Wojtowicz and Nicoll, 1982; Gallagher *et al.*, 1983). The risks of assuming that E_{GABA} is a measure of E_{Cl} have already been discussed in relation to the possible permeability to HCO_3^- of the anion channels opened by GABA (see Section 4.2.3 and Kaila and Voipio, this volume). In the cases in which E_{Cl} was supposed to be more positive than E_m, the above assumption (i.e., that E_{GABA} is a measure of E_{Cl}) could be even more risky than in other cases so far discussed, since E_{HCO_3} is also expected to be more positive than E_m. For instance, consider the following extreme case. Suppose that $E_{Cl} = E_m$. Since E_{HCO_3} is more positive than E_m, current through GABA-opened channels would be carried only by HCO_3^-, a fact that would produce a depolarization. This could have been the case in early work since in most, if not all, experiments in which E_{GABA} was determined the preparations were superfused with CO_2/HCO_3^--buffered solutions. One important piece of evidence against the idea suggested in this example was that E_{GABA} was found to be dependent on $[Cl^-]_o$ in a manner predicted for a membrane permeable to Cl^- (e.g., Nishi et al., 1974; Adams and Brown, 1975; Gallagher et al., 1978). However, only on internally perfused DRG neurons has it been possible to show unequivocally that $E_{GABA} = E_{Cl}$ (Hattori et al., 1984). Unfortunately, in these experiments, neither the internal nor the external perfusion fluid contained HCO_3^-. Clearly a direct measurement of a_{Cl}^i was needed. This was provided by Ballanyi and co-workers for rat sympathetic ganglion cells (Ballanyi et al., 1984; Ballanyi and Grafe, 1985) and by Alvarez-Leefmans et al. (1986, 1987, 1988) for frog dorsal root ganglion (DRG) neurons. In both types of cells, a_{Cl}^i was measured using double-barreled liquid anion exchanger microelectrodes.

In sympathetic ganglion cells the reported a_{Cl}^i was 29.3 ± 4.4 mM (mean ± S.D., $n = 39$). The mean a_{Cl}^o was 94.5 mM and E_m values were in the range of −40 to −75 mV with a mean of −49.1 ± 5.4 mV. The measurements were made in CO_2/HCO_3^--buffered solutions at 30°C. No corrections were made in these a_{Cl}^i readings for possible intracellular anion interference on electrode response. Possible interferences included HCO_3^- or other endogenous intracellular anions, or anions leaking from the reference barrel (see Sections 5.2.1d and 5.2.1e of Chapter 1). However, even assuming that intracellular interferences accounted for as much as 5 mM (see Table 2 of Chapter 1) and assuming that there was no significant anion leakage from the reference barrel (see Section 5.2.1e of Chapter 1), a_{Cl}^i was 1.7 times higher than the value predicted for a passive transmembrane distribution of Cl^-. The a_{Cl}^i values corrected for an interference of about 5 mM are in fair agreement with predictions made from measurements of E_{GABA} (Adams and Brown, 1975) and electron microprobe measurements of total Cl^- content (Galvan et al., 1984), as discussed in Section 1.2 of Chapter 1. Similar results have been reported for frog motoneurons, where GABA also has a depolarizing action (Bührle and Sonnhof, 1983, 1985). The mean a_{Cl}^i values were 34.4 mM (Bührle and Sonnhof, 1983) and 32.3 mM (Bührle and Sonnhof, 1985). Again, the possibility of overestimation of a_{Cl}^i as a result of intracellular interfering anions was not experimentally ruled out by the authors. However, their reported values are in close agreement with measurements of total Cl^- content by electron probe X-ray microanalysis (Allakhverdov et al., 1980) assuming an intracellular activity coefficient for Cl^- of 0.76.

Measurement of a_{Cl}^i in DRG neurons, in which intracellular interferences with liquid anion exchanger microelectrode readings were estimated, constituted the first direct evidence for the notion that a_{Cl}^i is higher than expected for a passive distribution in these cells (Alvarez-Leefmans et al., 1988). The mean E_m was −57.9 ± 1.0 mV and a_{Cl}^i was 23.6 ± 1.0 mM ($n = 53$). The latter value was 2.6 times the activity expected for an equilibrium distribution and the difference between E_m and E_{Cl} was about 25 mV (see Fig. 6 and Table 1). The a_{Cl}^i values could not result from HCO_3^- interference with the microelectrode response since the measurements were conducted in the nominal absence of HCO_3^- in Ringer solution buffered with HEPES and equilibrated either with air or with O_2 and, therefore, the intracellular levels of HCO_3^- were likely to be extremely low. Measured a_{Cl}^i values were independent of the composition of the filling solution used in the reference barrel (see Fig. 8 of Chapter 1). These

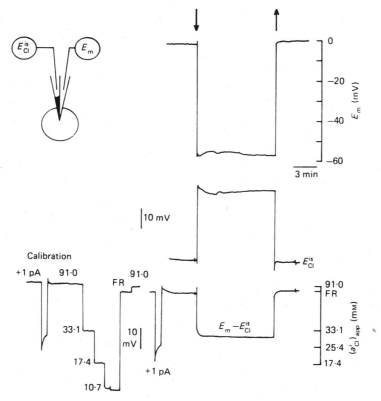

Figure 6. Measurement of a_{Cl}^i with a double-barrel Cl$^-$-selective microelectrode. The upper trace shows the potential recorded by the reference barrel (E_m) and the middle trace is the potential of the Cl$^-$-selective barrel (E_{Cl}^{is}). The bottom trace shows the differential signal ($E_m - E_{Cl}^{is}$) obtained by electronic subtraction. Arrows indicate electrode penetration and withdrawal from the cell. The calibration of the Cl$^-$-selective microelectrode before penetration is shown on the left of the bottom trace. FR denotes the standard bathing solution (frog Ringer solution) and the numbers indicate the activity of Cl$^-$ in pure solutions of KCl which were used for calibration. The scale on the right of the differential trace, labeled $(a_{Cl}^i)_{app}$, was drawn according to the calibration. Current pulses (1 pA) were passed through the ion-selective barrel to monitor its resistance which in this case was about 25 GΩ. The resistance of the reference barrel was 60 MΩ. (From Alvarez-Leefmans *et al.*, 1988.)

Table 1. Membrane Potential and Intracellular Cl$^-$ Activity in DRG Neurons

E_m (mV)	$(a_{Cl}^i)_{app}$ (mM)	$(a_{Cl}^i)_{oCl}$ (mM)	a_{Cl}^i (mM)	$(a_{Cl}^i)_{eq}$ (mM)	E_{Cl} (mV)	$E_m - E_{Cl}$ (mV)	No. cells/ No. animals
−57.7 ± 1.0	33.6 ± 0.9	10.7 ± 0.8	23.6 ± 1.0	9.2 ±0.3	−33.4 ± 1.4	−24.9 ± 1.7	53/26

Measurements done in standard frog Ringer solution, buffered with HEPES (5mM).
Values are means ± S.E.M.
$n = 53$ except for $(a_{Cl}^i)_{oCl}$ in which $n = 24$
$(a_{Cl}^i)_{app}$ = intracellular Cl$^-$ activity obtained by interpolation from the calibration scale.
$(a_{Cl}^i)_{oCl}$ = steady state apparent a_{Cl}^i read by the electrode in the nominal absence of external Cl$^-$
$a_{Cl}^i = (a_{Cl}^i)_{app} - (a_{Cl}^i)_{oCl}$
$(a_{Cl}^i)_{eq} = a_{Cl}^o \exp(E_m\,zF/RT)$
$E_{Cl} = (RT/F) \ln (a_{Cl}^o/a_{Cl}^i)$

reference filling solutions included 3 mM KCl, 4 M K-acetate, and the organic reference liquid ion exchanger of Thomas and Cohen (1981). These and other observations (see below) led to the conclusion that the high a_{Cl}^i found in DRG neurons did not result from cytosolic contamination with foreign interfering anions (see Section 5.2.1e of Chapter 1 and Alvarez-Leefmans and Noguerón, 1989).

5.1.2. Experimental Evidence for the Presence of Na$^+$,K$^+$,Cl$^-$ Cotransport as a Mechanism for Uphill Accumulation of Cl$^-$ in DRG Neurons

5.1.2a. Effects of Changes of Cl$_o^-$ on a_{Cl}^i. The experiments already discussed demonstrated that a_{Cl}^i in the soma of DRG neurons is two to three times higher than expected for an electrochemical equilibrium distribution, implying the presence of a mechanism for active inward transport of Cl$^-$. The question arises as to the mechanism by which this outwardly directed Cl$^-$ gradient is established and maintained. To answer this question, the effect of changes of [Cl$^-$]$_o$ on a_{Cl}^i was studied to test whether transmembrane Cl$^-$ movements were electrodiffusional or carrier-mediated. Figure 7 illustrates a typical experiment in which a_{Cl}^o was reduced nominally to 0 mM while a_{Cl}^i and E_m were continuously recorded. The apparent a_{Cl}^i measured by the microelectrode, $(a_{Cl}^i)_{app}$, decreased rapidly upon the removal of Cl$_o^-$, reaching a final steady-state level

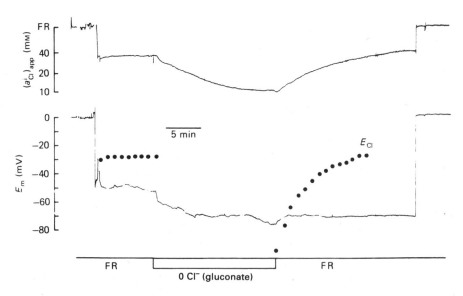

Figure 7. Effects of removing and readding external Cl$^-$ on a_{Cl}^i and E_m of a DRG neuron. The upper trace shows the differential signal ($E_m - E_{Cl}^{is}$) which indicated $(a_{Cl}^i)_{app}$. The lower trace is the membrane potential (E_m). At the periods indicated on the bottom, the bathing solution was changed from standard frog Ringer solution (FR) to a Cl$^-$-free solution in which Cl$^-$ was replaced by gluconate. The points superimposed on the E_m trace show the estimated E_{Cl} calculated from the value of a_{Cl}^i corrected for the recorded intracellular "interference" (10 mM) and for the measured a_{Cl}^o. The scale for $(a_{Cl}^i)_{app}$ was directly obtained from the calibration of the microelectrode and was not corrected for the recorded interference. The nature of the "interference" is discussed in the text. (From Alvarez-Leefmans *et al.*, 1988.)

of about 10 mM in about 10 min. This final steady-stage value, $(a_{Cl}^i)_{0Cl}$, was tentatively explained as interference from other anions with the response of the microelectrode, although part of it might well be Cl^- trapped inside the cell (see below). The recovery of $(a_{Cl}^i)_{app}$ started immediately after a_{Cl}^o was restored. The net Cl^- movements occurred without changes in E_m, i.e., they were electroneutral. The small rapid changes that can be observed in the E_m trace mostly resulted from measured liquid junction potentials at the reference (bath) electrode (see inset in Fig. 10). *These observations imply a low P_{Cl} of the DRG cell membrane* (see below) *and suggest that the movements of Cl^- occurred largely through a nonelectrodiffusional pathway, i.e., an electroneutral carrier mechanism.* The circles superimposed on the E_m trace in Fig. 7 correspond to the calculated values for E_{Cl} corrected for the measured "interference." It can be seen that at the beginning of the experiment E_{Cl} was well above its equilibrium value ($E_m - E_{Cl} = 28$ mV). Without correction for the above-mentioned "interference," E_{Cl} would be even more positive. On exposure to Cl^--free Ringer, the cell was depleted of Cl^- and immediately after restoring the external Cl^- activity to control levels, E_{Cl} was transiently below E_m. However, during most of the recovery period in Ringer solution, the movement of Cl^- occurred against its electrochemical potential difference across the cell membrane. This by itself constitutes direct evidence for the presence of an active process for accumulation of Cl^- in DRG neurons.

Both the decrease in a_{Cl}^i on exposure to Cl^--free Ringer as well as the recovery on returning to Cl^--containing Ringer, followed exponential time courses. Initial rates of decay and recovery of a_{Cl}^i, (da_{Cl}^i/dt), for seven cells were 4.1 ± 0.8 and 3.3 ± 0.3 mM/min, respectively. Calculated rate constants for decay and recovery corresponded to time constants of Cl_i^- emptying and reaccumulation of about 6 and 7 min, respectively. Assuming that the observed changes in a_{Cl}^i were only due to transmembrane fluxes of Cl^-, the net efflux of Cl^- can be calculated from the values of da_{Cl}^i/dt and the volume-to-surface ratio of the cells. For a cell of volume equivalent to a sphere of radius of 30×10^{-4} cm, the volume-to-surface ratio was 10^{-3} cm (see Section 6 of Chapter 1), and the Cl^- efflux was estimated to be 90×10^{-12} mole/cm² per sec. The latter was calculated by substituting the appropriate values in the following equation:

$$J_{Cl} = 1/\gamma_{Cl}^i [(da_{Cl}^i/dt) h + (dh/dt) a_{Cl}^i] \tag{13}$$

The general form of this equation has been discussed in Section 6 of Chapter 1. γ_{Cl}^i is the intracellular activity coefficient for Cl^- (taken as 0.76) and h is the volume-to-surface ratio of the cell. Since da_{Cl}^i/dt is the initial rate of change in a_{Cl}^i, it is expected that $dh/dt \simeq 0$. Assuming for the sake of argument that the calculated Cl^- efflux was purely electrodiffusive, the Cl^- permeability of the membrane, P_{Cl}, can be calculated using the constant-field theory (Goldman, 1943; Hodgkin and Katz, 1949), as discussed in Section 6.1 of Chapter 1. For $J_{Cl} = 90 \times 10^{-12}$ mole/cm² per sec and for the recorded mean values of E_m (-54 mV), a_{Cl}^i, and a_{Cl}^o (see Table 2 in Alvarez-Leefmans *et al.*, 1988), P_{Cl} was estimated to be 1.5×10^{-6} cm/sec. Having a value for P_{Cl}, the Cl^- conductance in the steady state, gCl, was also calculated. The corresponding gCl was 3.4×10^{-4} S/cm². A similar procedure gave values of 19.6×10^{-6} cm/sec and 4.4×10^{-4} S/cm² for P_{Cl} and gCl during reaccumulation of Cl^- in standard Ringer solution. The average specific membrane conductance of the DRG cell membrane, gm, had been estimated to be 2.1×10^{-4} S/cm^2 (Lopez and Alvarez-Leefmans, 1989). Comparison of the value of g_{Cl} calculated above with that of gm leads to an inconsistency, namely that gCl is larger than gm, the total membrane conductance. This *reductio ad absurdum* argument indicated to us that the assumption of electrodiffusion was incorrect and that a large fraction of the observed changes in a^i_{Cl} were due to an electroneutral carrier-mediated process.

Other observations supporting the notion that electrodiffusive P_{Cl} is low and, hence, that the Cl$^-$ movements cannot be explained in terms of electrodiffusion included the following: (1) steady-state measurements of a^i_{Cl} and E_m in many cells showed that E_m and E_{Cl} are very poorly correlated (see Fig. 8 of Chapter 1); (2) a^i_{Cl} did not change when cells were transiently depolarized by exposing them to high $[K^+]_o$; (3) neither cell input resistance (measured in the linear range of the I/V curve) nor the membrane potential was appreciably modified in nominally Cl$^-$-free solutions (Fig. 8 and Table 2).

The relation between a^i_{Cl} and a^o_{Cl} was further studied in experiments like the one shown in Fig. 9. After a long exposure to 0 Cl$^-$ to maximally reduce a^i_{Cl}, stepwise changes in external solutions having increasing concentrations of Cl$^-$ were tested. It was found that both the steady-state value of a^i_{Cl} as well as the rate at which a^i_{Cl} reached this value were proportional to a^o_{Cl}. As shown in Fig. 9B, the relation had a sigmoidal shape with a nonlinear behavior at low a^o_{Cl} and a tendency to saturate at a high, near-normal a^o_{Cl} activity. Half-maximal accumulation occurred at about 50 mM Cl^-_o. This observation suggests that under normal ionic conditions the inwardly directed transport system is saturated, i.e., under resting a^i_{Cl} the system is probably inhibited. Whether

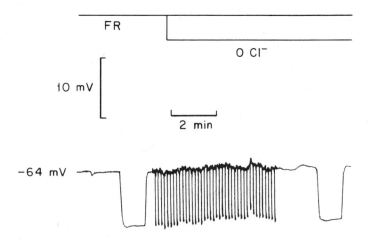

Figure 8. Input resistance and membrane potential of a DRG cell measured in standard frog Ringer solution (FR) and after external Cl$^-$ was replaced by gluconate (0 Cl$^-$) as indicated. Resting potential was −64 mV. Hyperpolarizing current pulses (0.25 nA, 500 msec) were injected through the microelectrode by means of a bridge circuit. First and last pulse shown were obtained at higher recorder speed to observe the shape of the transmembrane voltage transient. Note the absence of appreciable changes in E_m or input resistance in the Cl$^-$-free solution. (From Gamiño and Alvarez-Leefmans, unpublished observations.)

Table 2. Effect of External Cl⁻ on DRG Cell Input
Resistance (R_{in}) and Membrane Potential (E_m)

Control		0 Cl⁻		
E_m (mV)	R_{in} (mΩ)	ΔE_m (mV)	$(\Delta R_{in})_{t=0}$ (MΩ)	$(\Delta R_{in})_{t=\infty}$ (MΩ)
-62 ± 7	29 ± 3	2 ± 1	0	-0.1 ± 0.7

All values are means $\pm$ S.E.M.
t_0 = 1 to 5 min after solution change
t_∞ = 5 to 10 min after solution change
Fractional change in R_{in} for 0 Cl⁻ $\begin{cases} (\Delta R_{in})_{t=0}/R_{in} = 0 \\ (\Delta R_{in})_{t=\infty}/R_{in} = -0.01 \end{cases}$
$\Delta E_m = (E_m)_{FR} - (E_m)_{0Cl^-}$

this inhibition is due to kinetic or thermodynamic factors is unknown (see below). The shape of the relationship between a^i_{Cl} and a^o_{Cl} further suggested that the transport system has a complex stoichiometry.

5.1.2b. The Inwardly Directed Cl⁻ Transport System of DRG Cells Requires the Simultaneous Presence of Extracellular Na⁺ and K⁺.
As already mentioned, there are two likely transport systems that could be responsible for active accumulation of Cl⁻ in nerve cells. These are the Na⁺-independent anion exchanger and the Cl⁻ transport coupled to Na⁺ and K⁺. Since the experiments in DRG cells were conducted in the virtual absence of HCO_3^-, the anion exchanger, if present at all, should contribute very little to the observed accumulation of Cl⁻. Therefore, the dependence on extracellular Na⁺ and K⁺ of the active accumulation of Cl⁻ was explored.

The effects of Na⁺- or K⁺-free solutions on a^i_{Cl} were studied in the steady state and during the active reaccumulation from Cl⁻-free solutions. A typical experiment is illustrated in Fig. 10. Soon after impalement the cell was exposed for about 4 min to a solution in which Na⁺ was replaced by N-methyl-D-glucamine, keeping a^o_{Cl} at the same level as in the standard Ringer solution. This resulted in an immediate decrease in a^i_{Cl} accompanied by a hyperpolarization. Both E_m and a^i_{Cl} returned toward control levels upon readmission of standard Ringer solution to the bath. In the example shown, before a^i_{Cl} reached control levels, the cell was partially depleted of Cl⁻ by exposing it to a Cl⁻-free solution. Cl⁻ reaccumulation was then observed in the Na⁺-free solution and finally in standard Ringer solution. It was found that the rate of Cl⁻ reaccumulation in the Na⁺-free solution was slowed to about 15% of its value in standard Ringer solution. The concurrent changes in E_m and calculated values for E_{Cl} in the same experiment are shown in Fig. 11A, which illustrates more clearly the effects of Na_o^+ on the distribution of Cl⁻ across the cell membrane. Na⁺ removal reduced the basal intracellular accumulation of Cl⁻, as reflected by the reduction in the difference between E_m and E_{Cl}. In addition, Na⁺ removal after Cl⁻ depletion impeded the reaccumulation even though E_{Cl} was now more negative than E_m. *These experi-*

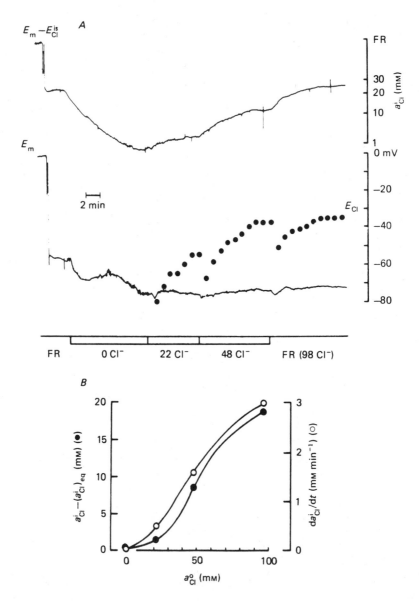

Figure 9. Relation between intracellular accumulation of Cl^- and a^o_{Cl} in a DRG neuron. (A) Pen recording of an experiment to test the effects of changing a^o_{Cl} on a^i_{Cl}. Upper trace shows a^i_{Cl}. The scale on the right of this trace was corrected for the intracellular interference recorded while the preparation was superfused with the Cl^--free solution. Lower trace is E_m. Filled circles superimposed on the E_m trace represent the position of E_{Cl} calculated from the measured a^o_{Cl} and a^i_{Cl}, at various times. Bottom marks indicate the periods at which the solutions were changed as well as their Cl^- activity (mM). After a brief period of recording in standard frog Ringer solution (FR), superfusion of the preparation with Cl^--free solution started, until a new steady state for a^i_{Cl} was reached. The cell was then exposed to solutions containing increasing Cl^- activities. Superfusion with each Cl^--containing solution lasted until a new steady state for a^i_{Cl} was reached. (B) Plot of values of the amount by which a^i_{Cl} was above equilibrium in the steady state, $(a^i_{Cl})_{eq}$ (●), against a^o_{Cl} for the experiment shown in A. The initial rate of increase in intracellular Cl^- activity, da^i_{Cl}/dt, for each external Cl^- activity, a^o_{Cl}, is also plotted (○). Lines were drawn by eye. (From Alvarez-Leefmans *et al.*, 1988.)

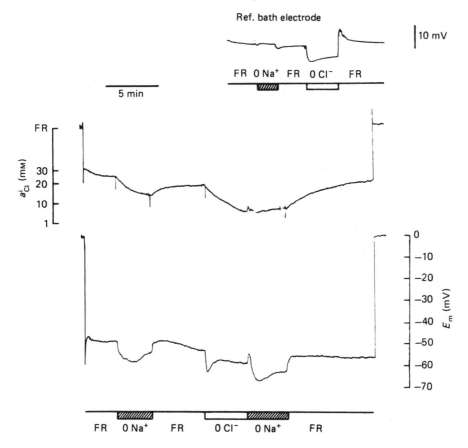

Figure 10. Effects of external Na⁺ on a^i_{Cl} in a DRG neuron. At the periods indicated at the bottom bars, the solution was changed from frog Ringer solution (FR) to a Na⁺-free solution (0 Na⁺) and back to FR. The cell was then exposed to Cl⁻-free solution (0 Cl⁻) and the reaccumulation of internal Cl⁻ was monitored in the absence of external Na⁺ (0 Na⁺) and in FR, i.e., in the presence of external Na⁺. The Na⁺-free solution was made by equimolar replacement of Na⁺ with *N*-methyl-D-glucamine. Inset at the top shows the liquid junction potentials of the reference (bath) electrode for each of the solutions used in the experiment. Note that the 0 Na⁺ solution generated a negligibly small junction potential in contrast to the Cl⁻-free solution. Therefore, the hyperpolarization recorded by the E_m barrel of the microelectrode when the cell was superfused with 0 Na⁺ solution was generated across the membrane while changes in the E_m trace produced when the 0 Cl⁻ solution was flushed in could be accounted for by a liquid junction potential at the reference bath electrode. (From Alvarez-Leefmans *et al.*, 1988.)

ments clearly indicate that Na⁺ is required for Cl⁻ reaccumulation after cell Cl⁻ depletion, and for the maintenance of a^i_{Cl} in the steady state. Therefore, it can be concluded that Na^+_o is required to accumulate Cl⁻ above its equilibrium distribution and for keeping it out of equilibrium in basal conditions.

The effects of K^+_o on a^i_{Cl} were similar to those observed for Na⁺. Figure 11B shows a plot of E_m and E_{Cl} from an experiment similar to that shown in Fig. 10 but in

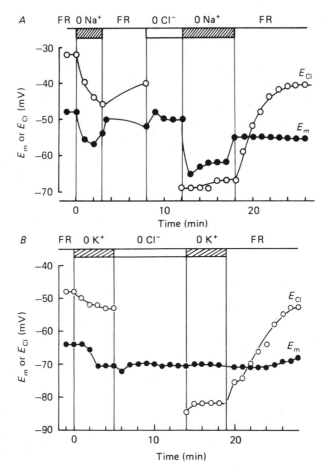

Figure 11. (A) Effects of external Na$^+$ on the intracellular accumulation of Cl$^-$. Membrane potential (E_m, ●) and the calculated Cl$^-$ equilibrium potential (E_{Cl}, ○) from measured a^i_{Cl} and a^o_{Cl} are plotted on the same time scale, using data obtained from the experiment illustrated in Fig. 10. (B) Effects of external K$^+$ on the intracellular accumulation of Cl$^-$. Symbols as in A. Data were obtained from an experiment similar to that shown in Fig. 10 but in which external K$^+$ was removed instead of external Na$^+$. (From Alvarez-Leefmans *et al.*, 1988.)

which external K$^+$ was removed during the periods indicated. Removal of K^+_o was accompanied by an immediate reduction in a^i_{Cl} revealed in the graph by the change in E_{Cl} toward more negative values. In the absence of K^+_o the reaccumulation of Cl$^-$ in standard Ringer solution was impeded.

A similar dependence of a^i_{Cl} reaccumulation on external Na$^+$ and K$^+$ was found in rat sympathetic ganglion cells by Ballanyi and Grafe (1985). They studied a^i_{Cl} reaccumulation after the cells had been partially depleted of Cl^-_i by exposing them to GABA, a maneuver that in addition to opening Cl$^-$ channels produces a decrease in a^i_K, without measurable changes in a_{Na}^i. They found that the post-GABA a_{Cl}^i recovery was considerably slowed in the absence of either external Na^+ or K^+. However, the interpretation of this finding is complicated because post-GABA a_{Cl}^i recovery occurred without the expected changes in a_{Na}^i.

Summarizing the above results show that the steady-state a_{Cl}^i levels in DRG cells depend on the simultaneous presence of extracellular Na^+ and K^+. Similarly, the active reaccumulation of Cl^- after Cl_i^- depletion requires the simultaneous presence of Na^+ and K^+ in the bathing solution. The latter was observed in DRG cells and also in autonomic neurons, although the methods used for depleting the cells of Cl^- were different. Altogether these findings suggest the presence of a Na^+,K^+,Cl^- cotransporter in the soma of DRG cells and probably in those of sympathetic ganglion cells.

Still lacking for vertebrate neurons is the demonstration that the fluxes of Cl^- are tightly coupled to those of the two co-ions, Na^+ and K^+. Future experiments are clearly needed to elucidate this issue.

5.1.2c. Inwardly Directed Cl⁻ Transport System of DRG Neurons Is Sensitive to "Loop" Diuretics.

Some sulfamoyl benzoic acid derivatives such as furosemide and bumetanide are known to block Na^+,K^+,Cl^- cotransport in a variety of tissues whether excitable or not (Ellory and Stewart, 1982; Geck and Heinz, 1986; O'Grady *et al.*, 1987). In view of the Na^+- and K^+-dependence of a_{Cl}^i, we studied the effect of furosemide and bumetanide on net Cl^- fluxes in DRG cells. First we tested the effect of furosemide (0.5–1 mM) on the recovery of a_{Cl}^i after the cells had been depleted of Cl^- by exposure to Cl^--free solutions. Experiments of this type revealed that reaccumulation of Cl_i^- was reversibly slowed by furosemide. On average, the rate of recovery of a_{Cl}^i in the presence of 1 mM furosemide was 0.37 mM/min ($n = 4$), which was about 10% the rate of recovery in standard Ringer solution.

Bumetanide was also an efficient inhibitor of Cl^- reaccumulation at lower concentrations (10^{-5} M) than furosemide. This is illustrated in the experiment of Fig. 12. Bumetanide was first applied during the last part of the a_{Cl}^i decay in Cl^--free solution and then during 4 min with standard Ringer solution. It can be observed that bumetanide sharply reduced the slope of the a_{Cl}^i decay and also inhibited the recovery of a_{Cl}^i in standard Ringer (compare with Fig. 7). The fact that bumetanide inhibited the decay of a_{Cl}^i in Cl^--free solution is independent evidence for the transport-mediated nature of a substantial fraction of the efflux of Cl^- when Cl_o^- is removed. The "washout" of the drug appeared to be very slow as judged from the rate of recovery of a_{Cl}^i well after its removal.

Taken together with the observations discussed in Sections 5.1.2a–d, the inhibitory effect of bumetanide on the efflux of Cl^- suggests that an important fraction of the *outwardly directed* movement of Cl^- in the Cl^--free solutions could reflect the operation of Na^+,K^+,Cl^- cotransport in the "reverse" mode. Reversibility is a property inherent in this kind of secondary active transport system which by definition can be fueled by the energy stored within the combined thermodynamic gradients of the transported solutes. That this system can operate in the reverse mode has been shown in internally dialyzed squid axons by Altamirano and Russell (1987).

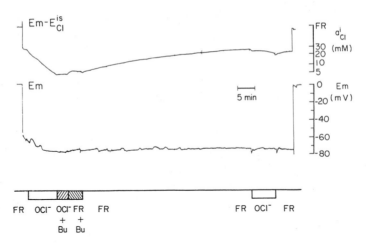

Figure 12. Effect of bumetanide on intracellular Cl$^-$ depletion and accumulation in a DRG neuron. Pen recording of a^i_{Cl} (upper trace) and E_m (lower trace). Bathing solutions were changed as indicated at the bottom. After a brief period of recording in standard Ringer (FR), external Cl$^-$ was removed and a^i_{Cl} decayed to a level of about 5 mM. Then bumetanide (Bu) addition to the Cl$^-$-free solution stopped further decline in a^i_{Cl}. The bathing medium was then switched to FR containing Bu and recovery of a^i_{Cl} was observed for a few minutes. Besides changing the slope of the a^i_{Cl} decay, bumetanide significantly slowed the recovery of a^i_{Cl} in FR (compare with Fig. 7). A similar sequence of solution changes was repeated about 45 min later, but eliminating the bumetanide-containing solution. It can be observed that both inward and outward net Cl$^-$ movements were largely inhibited. a^i_{Cl} was corrected for an interference of 10.7 mM. Gluconate was used as Cl$^-$ substitute. Bumetanide was added to both Cl$^-$-free solution and FR to a final concentration of 10^{-5} M. (Alvarez-Leefmans, Gamiño, and Giraldez, unpublished observations.)

Studies by Gallagher *et al.* (1983) showed that GABA-induced depolarizations in cat DRG cell bodies were depressed by SITS, furosemide, and bumetanide. They assumed that if these agents affected an inwardly directed Cl$^-$ transport process, E_{GABA} would be shifted towards E_m. Since the depression of GABA responses was not associated with changes in E_{GABA}, Gallagher *et al.* (1983) suggested that these Cl$^-$ transport inhibitors exerted their action by blocking Cl$^-$ channels or some other site associated with the GABA receptor complex. Based on this indirect evidence, they concluded that the inwardly directed Cl$^-$ transport system present in cat DRG cells was insensitive to bumetanide, furosemide, and other agents, and therefore that the mechanism responsible for establishing and maintaining the Cl$^-$ gradient in these cells was "unique." Similar results had been reported by Nicoll (1978) who found that furosemide (1×10^{-3} M) had little effect on the extrapolated reversal potential for the GABA response recorded in frog DRG cells, although the response itself was depressed by the drug. He concluded that furosemide acts primarily by blocking the conductance increase elicited by GABA rather than disrupting the ionic gradient responsible for the depolarizing action of the amino acid. The conclusions derived from our results (Alvarez-Leefmans *et al.*, 1988) are in contrast to those of Nicoll (1978) and Gallagher *et al.* (1983). The evidence discussed here clearly shows that the Cl$^-$ transport process present in DRG cell membranes is inhibitable by loop diuretics (furosemide and bumetanide), a fact that gives further support to the notion that the so-called "chloride pump" is a cation-coupled transport system. On the other hand, our results are consistent with those of Wojtowicz and Nicoll (1982) showing that piretanide, another 5-sulfamoyl benzoic acid derivative, changed E_{GABA} in DRG cells toward the resting potential by an average of 6 mV, 15 to 60 min following its application.

Experiments designed to study the action of loop diuretics or other agents on E_{GABA}, using the latter as an indicator of E_{Cl}, have to be interpreted with caution for the following reasons: (1) E_{GABA}, whether extrapolated or not, cannot be accurately determined on intact DRG cells because of strong voltage- and time-dependent rectification present in their membrane (e.g., Ito, 1957; Kostyuk *et al.*, 1981; Brown *et al.*, 1981; Mayer and Westbrook, 1983). (2) In addition to their action on Cl⁻ transport, loop diuretics behave as GABA-gated Cl⁻ channel blockers (Nicoll, 1978; Wojtowicz and Nicoll, 1982; Gallagher *et al.*, 1983), which implies a reduction in the signal-to-noise ratio for determination of the apparent reversal of GABA-induced responses. (3) Since resting P_{Cl} is very low in DRG cells, inhibition of steady-state Cl⁻ transport might go undetected for long periods of time; for example, a decrease in a_{Cl}^i of 10 mM would shift E_{GABA} by about 9 mV in the hyperpolarizing direction. This might well be within the error of the technique (Gallagher *et al.*, 1978). (4) So far there is no experimental validation of the assumption that $E_{GABA} = E_{Cl}$ in sensory neurons.

5.1.2d. Thermodynamics of Cl⁻ Transport and [Cl⁻]ᵢ in DRG Neurons.

The $[Cl^-]_i$ of DRG neurons (or any cell) at steady state will be set by the relative contribution of the various anion-transporting systems present in its membrane. The electroconductive pathways of DRG cells include GABA-activated Cl⁻ channels (Hattori *et al.*, 1984) and Ca^{2+}-activated Cl⁻ channels (see Mayer *et al.*, this volume). However, in the steady state when these conductances are not activated, P_{Cl} is, as discussed, extremely low. The concentrations of Na^+ and K^+ in these cells can be assumed to be determined by the Na^+,K^+ pump. The only carrier-mediated pathway for Cl⁻ that has been shown to be present in DRG cells is Na^+,K^+,Cl^- cotransport, which is electroneutral. Other Cl⁻ transport systems may be present but, so far, they have not been investigated. In squid axons, it has been shown that this transport system moves 2 Na^+, 1 K^+, and 3 Cl⁻ into the axon (Russell, 1983, 1984). However, in most vertebrate cells the stoichiometry is 1 Na^+ : 1 K^+ : 2 Cl⁻ (Geck and Heinz, 1986; Haas, 1989). If by analogy with other vertebrate cells we assume that in DRG neurons the system has a 1 : 1 : 2 stoichiometry, it follows that the driving force for this system will be the sum of the chemical potential differences of Na^+ and K^+ plus twice the Cl⁻ potential difference. The system will attain thermodynamic equilibrium when

$$[Na^+]_i \, [K^+]_i \, [Cl^-]_i^2 = [Na^+]_o \, [K^+]_o \, [Cl^-]_o^2 \tag{14}$$

Since the transport system is electroneutral, thermodynamic equilibrium will not be affected by E_m. Assuming for the sake of argument that this is the only Cl⁻ transport system operating in the steady state in DRG neurons, the expected $[Cl^-]_i$ when the guinea pig hippocampal slices. The rationale for this experiment was that GABA, by activating a Cl⁻ conductance, should induce changes in a_{Cl}^o secondary to Cl⁻ influx or transport reaches thermodynamic equilibrium (assuming no kinetic constraints) is given by

$$[Cl^-]_i = \left(\frac{[Na^+]_o \, [K^+]_o \, [Cl^-]_o^2}{[Na^+]_i \, [K^+]_i} \right)^{1/2} \qquad (15)$$

For $[Na^+]_o = 115$ mM, $[K^+]_o = 2.5$ mM, $[Cl^-]_o = 122$ mM, $[Na^+]_i = 25$ mM, and $[K^+]_i = 105$ mM, $[Cl^-]_i$ will be about 40.4 mM. Taking 0.76 as the intracellular activity coefficient for Cl^-, the predicted a^i_{Cl} is 30.7 mM, which is about 7 mM more than the reported a^i_{Cl} value of 23.6 mM (Alvarez-Leefmans *et al.*, 1988).

There are reasons to suspect that our value of 23.6 mM for a^i_{Cl} is an *underestimate*. These reasons are: (1) that the mean interference level used to correct the a^i_{Cl} readings was 10.7 mM, which is about twice the mean level found by other workers in a variety of other tissues (see Table 2 of Chapter 1), and (2) for the Na^+,K^+,Cl^- cotransport system to run in the reverse direction, it will need a minimum Cl^-_i concentration which might be determined by kinetic factors (Altamirano *et al.*, 1989). The fact that both the steady-state value of a^i_{Cl} and the rate at which a^i_{Cl} reached this value were proportional to a^o_{Cl} in such a way that the relation had a sigmoidal shape with a nonlinear behavior at low a^o_{Cl} (and presumably therefore at low a^i_{Cl}) favors this hypothesis. If it is assumed that half of the "interference" level, i.e., the final steady-state reading of a^i_{Cl} in Cl^--free solution, is Cl^- trapped in the cell, then mean a^i_{Cl} in DRG neurons will be about 28.6 mM, which means that $[Cl^-]_i \simeq 37.6$ mM, a value close to the 40.4 mM predicted by Eq. (15). It remains to be tested whether half or more of the so-called intracellular interference is indeed trapped Cl^-. Finally, the various Cl^- transport systems that are known to be present in the cell bodies and probably in the terminals of DRG cells are illustrated in Fig. 13.

5.1.3. Functional Significance of Na⁺,K⁺,Cl⁻ Cotransport in Nerve Cells

Na^+,K^+,Cl^- cotransport has been shown to be present in a wide variety of animal tissues, and it has been postulated to serve a number of physiological functions. The latter include maintenance and regulation of cell volume (Geck and Heinz, 1986; Hoffman and Simonsen, 1989), and transport of salt and water across absorptive and secretory epithelia (O'Grady *et al.*, 1987; Haas, 1989).

Due to the characteristic electrical activity of neurons, which involves synaptic and action potentials in diverse spatiotemporal patterns, the constancy of the intracellular ionic environment of nerve cells is continuously challenged. Upsetting the steady-state intracellular ion concentrations results in alterations of transmembrane electrochemical and osmotic gradients. Under physiological conditions, nerve cells are able to restore and maintain ionic gradients that determine their resting membrane potential and volume. This implies the presence in neuronal membranes of powerful ion transport mechanisms responsible for intracellular ion homeostasis. It has been proposed that the Na^+,K^+,Cl^- cotransporter may be one of these transport systems. For instance, it has been shown that GABA-induced depolarizations in sympathetic ganglion neurons involve a transient reduction in a^i_{Cl} and a^i_K. Restoration of a^i_{Cl} and a^i_K to steady-state values seems to be achieved by a system with pharmacology and ionic

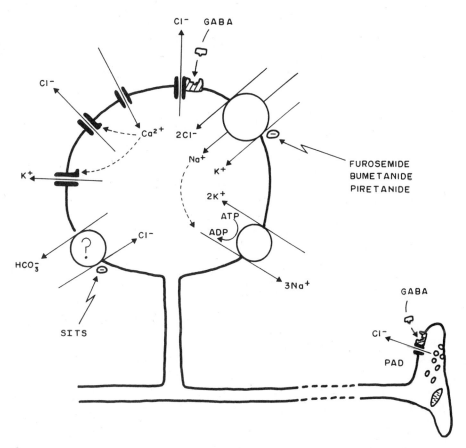

Figure 13. Schematic representation of the various channels and carriers present in the cell body and intraspinal terminals of a DRG cell. The only ion channel depicted in the terminal is a GABA-gated Cl^- channel which when activated generates primary afferent depolarization (PAD). In the cell body there are Ca^{2+} channels, Ca^{2+}-activated Cl^- and K^+ channels (Na^+ channels not shown), and GABA-gated Cl^- channels. Internal Cl^- is maintained at a level higher than predicted for passive distribution by the operation of the $Na^+,K^+,2Cl^-$ transport system. It is not known whether Cl^-/HCO_3^- exchange is present in these cells. Na^+ and K^+ gradients are maintained by the Na^+,K^+ pump.

dependence similar to the Na^+,K^+,Cl^- cotransporter (Ballanyi *et al.*, 1984; Ballanyi and Grafe, 1985). Amino acid-induced outward K^+ and Cl^- movements in frog motoneurons are accompanied by measurable decreases in cell volume (Serve *et al.*, 1988). The mechanism of cell volume restoration is unknown but it has been proposed that it may be the Na^+,K^+,Cl^- cotransporter which might be stimulated by cell shrinkage (Ballanyi and Grafe, 1988). Therefore, it has been postulated for neurons that, just as it has been postulated for many nonexcitable tissues (see reviews by Reuss, 1988; Hoffman and Simonsen, 1989), Na^+,K^+,Cl^- cotransport may be involved in regulation and maintenance of cell volume. However, at present, this suggestion can only be regarded as an interesting working hypothesis.

Inwardly Directed Cl⁻ Transport and Presynaptic Inhibition in Vertebrates. Presynaptic inhibition was discovered in 1957 by Frank and Fuortes. Recording intracellularly in extensor motoneurons, they found that EPSPs in response to primary afferent stimulation could be reduced by conditioning volleys to other afferent axons (from flexor muscles). The latter were postulated to establish axoaxonic contacts with the primary afferent terminals (see Davidoff and Hackman, 1984). The hypothesis is that activation of the presynaptic axon reduces the amount of transmitter released by the arrival of the action potential in the terminal of the primary afferent axon. Stimulation of presynaptic axons produces depolarization of the primary afferent terminal, which is thought to partially inactivate the Na^+ conductance mechanism, leading to a smaller presynaptic action potential with the consequent reduction in transmitter release (Eccles, 1964b). Other factors causing presynaptic inhibition have been postulated that are not mutually exclusive; they include block of action potential invasion to the terminals due to the transmitter-induced membrane "shunt," as well as reduction in the Ca^{2+} current through activation of $GABA_B$ receptors (for review see Davidoff and Hackman, 1985). The hypothesis of Eccles (1964b) has been extensively tested in the last 25 years. Some of its provisions have been fully accepted; some are still in dispute (cf. Burke and Rudomín, 1977; Nicoll and Alger, 1979; Davidoff and Hackman, 1984, 1985).

Since presynaptic inhibition and primary afferent depolarization (PAD) are both reduced by the GABA antagonists picrotoxin and bicuculline, it has been suggested that GABA is the neurotransmitter mediating PAD and presynaptic inhibition (Schmidt, 1963; Levy, 1977; Nicoll and Alger, 1979; Davidoff and Hackman, 1985). The exact mechanism by which PAD is produced and its precise role in modulating transmitter release for presynaptic inhibition are still not certain. Given the striking resemblance in the pharmacological properties of the GABA receptors present in the intraspinal terminals and in the cell bodies of primary sensory neurons of amphibians (Curtis *et al.*, 1961; Nishi *et al.*, 1974; Barker *et al.*, 1975; Akaike *et al.*, 1985) and mammals (DeGroat, 1972; Gallagher *et al.*, 1978; Desarmenien *et al.*, 1984; see also review by Levy, 1977). The receptors involved in both cases seem to be of the $GABA_A$ type, i.e., GABA–barbiturate–benzodiazepine–receptor complexes coupled to Cl^- channels (Olsen and Leeb-Lundberg, 1981; Bowery *et al.*, 1984; Olsen and Venter, 1986). Activation of these receptors by GABA produces depolarization in the terminals as well as in the cell bodies of primary sensory neurons. Based on these similarities, it is reasonable to postulate the existence in the intraspinal terminals of primary sensory neurons of an inwardly directed Cl^- transport mechanism similar to that present in their parent cell bodies. This transport system would serve the purpose of generating and maintaining an outwardly directed Cl^- gradient, and therefore restoring a_{Cl}^i after GABA action. This higher-than-passive a_{Cl}^i would make possible the depolarization of the axon terminals (PAD) that is associated with presynaptic inhibition (see Fig. 13). This hypothesis would receive further support if the ionic mechanism of GABA-induced depolarization in the terminals was the same as in their cell bodies. There is evidence that GABA selectively increases the Cl^- permeability of DRG cell bodies (Nishi *et al.*, 1974; Deschenes *et al.*, 1976; Gallagher *et al.*, 1978; Hattori *et al.*, 1984). Furthermore, in DRG cells, Nishi *et al.* (1974) found that reducing the $[Na^+]_o$ to a tenth of its normal value shifted E_{GABA} by about 5 mV toward E_m, and eventually reduced the amplitude of the response to below control values. These observations can now be understood, in the light of our present results, as being the consequence of a reduction in a_{Cl}^i following partial removal of Na_o^+. Interestingly, Wojtowicz and Nicoll (1982) found that piretanide, a sulfamoyl benzoic acid derivative that blocks Na^+,K^+,Cl^- cotransport, depressed primary afferent GABA responses, and presynaptic inhibition in the frog spinal cord.

Agreement is still lacking about the ionic mechanism underlying GABA-induced depolarization of central terminals of primary sensory neurons in the isolated spinal cord. There is evidence suggesting that Cl^- ions mediate GABA-induced depolarizations, although the contribution of Na^+ ions is still uncertain. GABA responses are depressed when Cl_o^- is reduced or removed (Barker and Nicoll, 1972, 1973; Kudo *et al.*, 1975; Otsuka and Konishi, 1976). However, some data indicate that the GABA response in the terminals is abolished after prolonged superfusion of the cord with Na^+-free Ringer solution (Barker and Nicoll, 1972, 1973; Padjen *et al.*, 1973; Otsuka and Konishi, 1976; Constanti and Nistri, 1976). The latter finding has been put forward by some authors (e.g., Nistri, 1983) as evidence for an increase in Na^+ conductance induced by GABA, which would partly explain its depolarizing action. Other authors conclude that it is unclear if the process involves an increased permeability to Na^+, Cl^-, or both ions (Nicoll and Alger, 1979; reviewed by Davidoff and Hackman, 1985). Although this important issue still needs further experimental analysis, our results in DRG cells (Alvarez-Leefmans *et al.*, 1988) suggest that block of GABA-induced responses by prolonged exposure to Na^+-free solutions could be the consequence of Cl_i^- depletion of the terminals. This would be sufficient to explain the depression of GABA-induced responses in the terminals without the need for invoking GABA-induced opening of Na^+ channels. This interpretation would also agree with the wealth of evidence showing that GABA acts by selectively opening Cl^- channels in nerve cells (Krnjević, 1974; Bormann *et al.*, 1987).

5.1.4. Depolarizing Action of GABA in the Dendrites of Vertebrate Neurons. An Inwardly Directed Cl⁻ Transport in the Dendrites?

In mammalian CNS neurons kept *in vitro*, focal application of GABA produces hyperpolarizing responses when applied at or near the cell body, and depolarizing responses when applied in the dendrites (Barker and Ransom, 1978; Andersen *et al.*, 1980; Alger and Nicoll, 1982; Wong and Watkins, 1982; Misgeld *et al.*, 1986; Connors *et al.*, 1988; Janigro and Schwartzkroin, 1988). In hippocampal pyramidal cells, both depolarizing and hyperpolarizing responses were found to be Cl^--dependent (reviewed by Nicoll, 1988). Furthermore, somatic and dendritic responses were blocked by $GABA_A$ antagonists. This led to the suggestion (Newberry and Nicoll, 1985; Misgeld *et al.*, 1986) that a_{Cl}^i is significantly higher in the dendrites than in the soma of the same cell and that the Cl_i^- of each neuron is regulated by two transport systems. According to this view, an outwardly directed Cl^- transport mechanism maintains an inward Cl^- gradient in the pyramidal cell soma and an inwardly directed Cl^- transport system maintains an outward Cl^- gradient in the dendrites (Misgeld *et al.*, 1986). More recently, Müller *et al.* (1989) measured regional changes in Cl_o^- with ion-selective microelectrodes in efflux depending on the sign and magnitude of the driving force for Cl^-. Focal application of GABA to the pyramidal cell body layer of CA_3 or CA_1 induced a decrease of a_{Cl}^o whereas application to the dendritic layer of CA_3 or CA_1 induced an increase in a_{Cl}^o. The changes in external Cl^- persisted in the presence of tetrodotoxin and were blocked by bicuculline. The authors concluded that a_{Cl}^i in the soma is maintained lower than predicted for a passive distribution due to an outwardly directed Cl^- transport system that sets up an inwardly directed Cl^- gradient. In the dendrites of the same cells, a_{Cl}^i is higher than predicted for a passive distribution, due to the presence of an inwardly directed Cl^- transport system. This implies a nonuniform Cl_i^- distribution in pyramidal cells, i.e., there are Cl_i^- gradients. An alternative explanation is that Cl_i^- is uniformly distributed along pyramidal cells and therefore E_{Cl} would be constant with respect to space. What is different is the value of E_m along the cells, i.e., pyramidal cells of *in vitro* slices are not necessarily isopotential due, for example, to removal of tonic excitation in the dendrites. This means that E_m could be more negative in the dendrites than in the cell body even though E_{Cl} would be the same. a_{Cl}^i would be uniformly distributed along the neuron due to the presence of an electroneutral Cl^- transport system, which, by definition, would be insensitive to E_m. Clearly more detailed studies are needed to clarify this issue. Another alternative explanation to the above findings is that the channels coupled to the GABA receptors have different ion selectivities in the dendrites and in the soma.

6. CONCLUSIONS

The study of ion transport mechanisms in vertebrate nerve cells is at a very early stage. The case of Cl^- is just an example of this general statement. It is ironic that in this era of recombinant DNA technology, one of the greatest achievements of human brains, we know so little about how vertebrate nerve cells regulate their intracellular ionic environment. Compared with what is known about ion transport in epithelial cells, our knowledge in vertebrate neurons is primitive. It is expected that the miniaturization of ion-selective microelectrodes achieved during the last few years, as well as the availability of new fluorescent probes for measuring intracellular ion activities, combined with various cell isolation procedures and dialysis techniques will produce a revolution in the field of ion transport in vertebrate neurons.

ACKNOWLEDGMENTS. I thank Drs. J. M. Russell and R. Llinás for many stimulating discussions. I am also greatly indebted to Dr. L. Reuss who offered numerous valuable criticisms and suggestions on the manuscript. I also thank Dr. S. Gamiño for allowing me to reproduce Fig. 8 from our unpublished results. I am also grateful to Ms. Lynette Durant for excellent secretarial assistance. This work was partially supported by CONACYT Grant PO28CCOX891556.

REFERENCES

Adams, P. R., and Brown, D. A., 1975, Actions of γ-aminobutyric acid on sympathetic ganglion cells, *J. Physiol. (London)* **250**:85–120.

Aickin, C. C., Deisz, R. A., and Lux, H. D., 1982, Ammonium action on postsynaptic inhibition in crayfish neurones: Implications for the mechanism of chloride extrusion, *J. Physiol. (London)* **329**:319–339.

Aickin, C. C., Deisz, R. A., and Lux, H. D., 1984, Mechanisms of chloride transport in crayfish stretch receptor neurones and guinea pig vas deferens: Implications for inhibition mediated by GABA, *Neurosci. Lett.* **97**:239–244.

Akaike, N., Hattori, K., Inomata, N., and Oomura, Y., 1985, γ-Aminobutyric-acid- and pentobarbitone-gated chloride currents in internally perfused frog sensory neurones, *J. Physiol. (London)* **360**:367–386.

Alger, B. E., 1985, GABA and glycine: Postsynaptic actions, in: *Neurotransmitter Actions in the Vertebrate Nervous System* (M. A. Rogawski and J. L. Barker, eds.), pp. 33–69, Plenum Press, New York.

Alger, B. E., and Nicoll, R. A., 1982, Pharmacological evidence for two kinds of GABA receptor on rat hippocampal pyramidal cells studied *in vitro*, *J. Physiol. (London)* **328**:125–141.

Alger, B. E., and Nicoll, R. A., 1983, Ammonia does not selectively block IPSPs in rat hippocampal pyramidal cells, *J. Neurophysiol.* **49**:1381–1391.

Allakhverdov, B. L., Burovina, I. V., Chmykhova, N. M., and Shapovalov, A. I., 1980, Electron probe x-ray microanalysis of intracellular sodium, potassium and chlorine contents in amphibian motoneurones, *Neuroscience* **5**:2023–2031.

Allen, G. I., Eccles, J., Nicoll, R. A., Oshima, T., and Rubia, F. J., 1977, The ionic mechanisms concerned in generating the IPSPs of hippocampal pyramidal cells, *Proc. R. Soc. London Ser. B* **198**:363–384.

Altamirano, A. A., and Russell, J. M., 1987, Coupled Na/K/Cl efflux. "Reverse" unidirectional fluxes in squid giant axons, *J. Gen. Physiol.* **89**:669–686.

Altamirano, A. A., Breitwieser, G. E., and Russell, J. M., 1989, Na⁺,K⁺,Cl⁻ coupled transport in squid axon, *Acta Physiol. Scand.* **136**(Suppl. 582):16.

Alvarez-Leefmans, F. J., and Noguerón, I., 1989, Intracellular chloride homeostasis in vertebrate nerve cells, *Acta Physiol. Scand.* **136**(Suppl. 582):17.

Alvarez-Leefmans, F. J., Gamiño, S. M., and Giraldez, F., 1986, Direct demonstration that chloride ions are not passively distributed across the membrane of dorsal root ganglion cells of the frog: Preliminary studies on the nature of the chloride pump, *Biophys. J.* **49**:413a.

Alvarez-Leefmans, F. J., Giraldez, F., and Gamiño, S. M., 1987, Intracellular chloride regulation in vertebrate sensory neurones, *Neuroscience* **22**(Suppl.):S200.

Alvarez-Leefmans, F. J., Gamiño, S. M., Giraldez, F., and Noguerón, I., 1988, Intracellular chloride regulation in amphibian dorsal root ganglion neurones studied with ion-selective microelectrodes, *J. Physiol. (London)* **406**:225–246.

Andersen, P., Dingledine, R., Gjerstad, L., Langmoen, I. A., and Laursen, A., 1980, Two different responses of hippocampal pyramidal cells to application of gamma-amino butyric acid, *J. Physiol. (London)* **305**:279–296.

Ascher, P., Kunze, D., and Neild, T. O., 1976, Chloride distribution in *Aplysia* neurons, *J. Physiol. (London)* **256**:441–464.

Ballanyi, K., and Grafe, P., 1985, An intracellular analysis of γ-aminobutyric-acid-associated ion movements in rat sympathetic neurons, *J. Physiol. (London)* **365**:41–58.

Ballanyi, K., and Grafe, P., 1988, Cell volume regulation in the nervous system, *Renal Physiol. Biochem.* **3–5**:142–157.

Ballanyi, K., Grafe, P., Reddy, M. M., and Ten Bruggencate, G., 1984, Different types of potassium transport linked to carbachol and γ-aminobutyric acid actions in rat sympathetic neurons, *Neuroscience* **12**:917–927.

Barker, J. L., 1985, GABA and glycine: Ion channel mechanisms, in: *Neurotransmitter Actions in the Vertebrate Nervous System* (M. A. Rogawski and J. L. Barker, eds.), pp. 71–100, Plenum Press, New York.

Barker, J. L., and Nicoll, R. A., 1972, Gamma-aminobutyric acid. Role in primary afferent depolarization, *Science* **176**:1043–1045.

Barker, J. L., and Nicoll, R. A. 1973, The pharmacology and ionic dependency of amino acid responses in the frog spinal cord, *J. Physiol. (London)* **228**:259–277.

Barker, J. L., and Ransom, B. R., 1978, Amino acid pharmacology of mammalian central neurones grown in tissue culture, *J. Physiol. (London)* **280**:331–354.

Barker, J. L., Nicoll, R. A., and Padjen, A., 1975, Studies on convulsants in the isolated frog spinal cord. I. Antagonism of amino acid responses, *J. Physiol. (London)* **245:**521–536.

Ben-Ari, Y., Krnjević, K., and Reinhardt, W., 1979, Hippocampal seizures and failure of inhibition, *Can. J. Physiol. Pharmacol.* **57:**1462–1466.

Benninger, C., Kadis, J., and Prince, D. A., 1980, Extracellular calcium and potassium changes in hippocampal slices, *Brain Res.* **187:**165–182.

Boistel, J., and Fatt, P., 1958, Membrane permeability change during inhibitory transmitter action in crustacean muscle, *J. Physiol. (London)* **144:**176–191.

Bolz, J., and Gilbert, C. D., 1986, Generation of end-inhibition in the visual cortex via interlaminar connections, *Nature* **320:**362–364.

Bormann, J., Hamill, O. P., and Sakmann, B., 1987, Mechanism of anion permeation through channels gated by glycine and γ-aminobutyric acid in mouse cultured spinal neurones, *J. Physiol. (London)* **385:** 243–286.

Boron, W. F., 1983, Transport of H^+ and of ionic weak acids and bases, *J. Membr. Biol.* **72:**1–16.

Boron, W. F., 1985, Intracellular pH-regulating mechanism of the squid axon, *J. Gen. Physiol.* **85:**325–345.

Boron, W. F., and De Weer, P., 1976, Intracellular pH transients in squid giant axons caused by CO_2, NH_3 and metabolic inhibitors, *J. Gen. Physiol.* **67:**91–112.

Boron, W. F., and Russell, J. M., 1983, Stoichiometry and ion dependencies of the intracellular-pH-regulating mechanism in squid giant axons, *J. Gen. Physiol.* **81:**373–399.

Boron, W. F., Russell, J. M., Brodwick, M. S., Keifer, D. W., and Roos, A., 1978, Influence of cyclic AMP on intracellular pH regulation and chloride fluxes in barnacle muscle fibers, *Nature* **276:**511–513.

Bowery, N. G., Hill, D. R., Hudson, A. L., Price, G. W., Turnbull, M. J., and Wilkin, G. P., 1984, Heterogeneity of mammalian GABA receptors, in: *Actions and Interactions of GABA and Benzodiazepines* (N. G. Bowery, ed.), pp. 81–108, Raven Press, New York.

Brazy, P., and Gunn, R. B., 1976, Furosemide inhibition of Cl transport in human red blood cells, *J. Gen. Physiol.* **68:**583–599.

Brown, A. M., and Kunze, D. L., 1974, Ionic activities in identifiable *Aplysia* neurons, in: *Ion-Selective Microelectrodes* (H. J. Berman and N. C. Hebert, eds.), pp. 57–73, Plenum Press, New York.

Brown, A. M., Walker, J. L., and Sutton, R. B., 1970, Increased chloride conductance as the proximate cause of hydrogen ion concentration effects in *Aplysia* neurons, *J. Gen. Physiol.* **56:**559–582.

Brown, H. M., Ottoson, D., and Rydquist, B., 1978, Crayfish stretch receptor: An investigation with voltage-clamp and ion-sensitive electrodes, *J. Physiol. (London)* **284:**155–180.

Brown, T. H., Perkel, D. H., Norris, J. C., and Peacock, J. N., 1981, Electrotonic structure and specific membrane properties of mouse dorsal root ganglion neurons, *J. Neurophysiol.* **45:**1–15.

Brugnara, C., Thuong, V. H., and Tosteson, D. C., 1989, Role of chloride in potassium transport through a K-Cl cotransport system in human red blood cells, *Am. J. Physiol.* **256:**994–1003.

Bührle, C. P., and Sonnhof, U., 1983, Intracellular ion activities and equilibrium potentials in motoneurones and glia cells of the frog spinal cord, *Pfluegers Arch.* **396:**144–153.

Bührle, C. P., and Sonnhof, U., 1985, The ionic mechanism of postsynaptic inhibition in motoneurons of the frog spinal cord, *Neuroscience* **14:**581–592.

Burke, R. E., and Rudomin, P., 1977, Spinal neurons and synapses, in: *Handbook of Physiology,* Section 1, *The Nervous System,* Volume 1, *Cellular Biology of Neurons,* Part 2 (E. R. Kandel, ed.), pp. 877–944, American Physiological Society, Bethesda.

Cherksey, B. D., and Zeuthen, T., 1987, A membrane protein with a K^+ and a Cl^- channel, *Acta Physiol. Scand.* **129:**137–138.

Chesler, M., 1986, Regulation of intracellular pH in reticulospinal neurones of the lamprey, *Petromyzon marinus, J. Physiol. (London)* **381:**241–261.

Chesler, M., 1987, pH regulation in the vertebrate central nervous system: Microelectrode studies in the brain stem of the lamprey, *Can. J. Physiol. Pharmacol.* **65:**986–993.

Chipperfield, A. R., 1986, The (Na^+-K^+-Cl^-) cotransport system, *Clin. Sci.* **71:**465–467.

Connors, B. W., Malenka, R. C., and Silva, L. R., 1988, Two inhibitory postsynaptic potentials and GABA$_A$ and GABA$_B$ receptor-mediated responses in neocortex of rat and cat, *J. Physiol. (London)* **406:**443–468.

Constanti, A., and Nistri, A., 1976, A comparative study of the action of γ-aminobutyric acid and piperazine on the lobster muscle fiber and the spinal cord, *Br. J. Pharmacol.* **57:**347–358.

Coombs, J. S., Eccles, J. C., and Fatt, P., 1955, The specific ionic conductances and the ionic movements across the motoneuronal membrane that produce the inhibitory postsynaptic potential, *J. Physiol. (London)* **130:**326–373.

Corcia, A., and Armstrong, W. M., 1983, KCl cotransport: A mechanism for basolateral chloride exit in *Necturus* gallbladder, *J. Membr. Biol.* **76:**173–182.

Cornwall, M. C., Peterson, D. F., Kunze, D. L., Walker, J. L., and Brown, A. M., 1970, Intracellular potassium and chloride activities measured with liquid ion exchanger microelectrodes, *Brain Res.* **23:** 433–436.

Curtis, D. R., Phillis, J. W., and Watkins, J. C., 1961, Actions of amino acids on the isolated hemisected spinal cord of the toad, *Br. J. Pharmacol. Chemother.* **16:**262–283.

Davidoff, R. A., and Hackman, J. C., 1984, Spinal inhibition, in: *Handbook of the Spinal Cord,* Volume 2, *Anatomy and Physiology of the Spinal Cord,* (R. A. Davidoff, ed.), pp. 385–459, Dekker, New York.

Davidoff, R. A., and Hackman, J. C., 1985, GABA: Presynaptic actions, in: *Neurotransmitter Actions in the Vertebrate Nervous System* (M. A. Rogawski and J. L. Barker, eds.), pp. 3–32, Plenum Press, New York.

DeGroat, W. C., 1972, GABA-depolarization of a sensory ganglion: Antagonism by picrotoxin and bicuculline, *Brain Res.* **38:**429–439.

Deisz, R. A., and Lux, H. D., 1982, The role of intracellular chloride in hyperpolarizing postsynaptic inhibition of crayfish stretch receptor neurones, *J. Physiol. (London)* **326:**123–138.

Desarmenien, M., Feltz, P., Occhipinti, G., Santangelo, F., and Schlichter, R., 1984, Coexistence of GABA_A and GABA_B receptors on A and C primary afferents, *Br. J. Pharmacol.* **81:**327–333.

Deschenes, M., Feltz, P., and Lamour, Y., 1976, A model for an estimate *in vivo* of the ionic basis of presynaptic inhibition: An intracellular analysis of the GABA-induced depolarization in rat dorsal root ganglia, *Brain Res.* **118:**486–493.

Deterre, P., Gozlan, H., and Bockaert, J., 1983, GTP-dependent anion-sensitive adenylate cyclase in snail ganglia potentiation of neurotransmitter effects, *J. Biol. Chem.* **258:**1467–1473.

Dingledine, R., and Langmoen, I. A., 1980, Conductance changes and inhibitory actions of hippocampal recurrent IPSPs, *Brain Res.* **185:**277–287.

Dykes, R. W., Landry, P., Metherate, R., and Hicks, T. P., 1984, Functional role of GABA in cat primary somatosensory cortex: Shaping receptive fields of cortical neurons, *J. Neurophysiol.* **52:**1066–1093.

Eccles, J. C., 1957, *The Physiology of Nerve Cells,* Johns Hopkins Press, Baltimore.

Eccles, J. C., 1964a, *The Physiology of Synapses,* Springer, Berlin.

Eccles, J. C., 1964b, Presynaptic inhibition in the spinal cord, *Prog. Brain Res.* **12:**65–89.

Eccles, J. C., 1969, *The Inhibitory Pathways of the Central Nervous System,* Thomas, Springfield, Ill.

Eccles, J. C., Eccles, R. M., and Ito, M., 1964a, Effects of intracellular potassium and sodium injection on the inhibitory postsynaptic potential, *Proc. R. Soc. London Ser. B* **160:**181–196.

Eccles, J. C., Eccles, R. M., and Ito, M., 1964b, Effects produced on inhibitory postsynaptic potentials by the coupled injections of cations and anions into motoneurones, *Proc. R. Soc. London Ser. B* **160:**197–210.

Eccles, J., Nicoll, R. A., Oshima, T., and Rubia, F. J., 1977, The anionic permeability of the inhibitory postsynaptic membrane of hippocampal pyramidal cells, *Proc. R. Soc. London Ser. B* **198:**345–361.

Edwards, C., 1982, The selectivity of ion channels in nerve and muscles, *Neuroscience* **7:**1355–1366.

Ellory, J. C., and Stewart, G. W., 1982, The human erythrocyte Cl⁻-dependent Na–K cotransport system as a possible model for studying the action of loop diuretics, *Br. J. Pharmacol.* **75:**183–188.

Ellory, J. C., Dunham, P. B., Logue, P. J., and Stewart, G. W., 1982, Anion-dependent cation transport in erythrocytes, *Philos. Trans. R. Soc. London Ser. B* **299:**483–495.

Fatt, P., 1974, Postsynaptic cell characteristics determining membrane potential changes, in: *Lecture Notes in Biomathematics,* Volume 4, *Physics and Mathematics of the Nervous System* (M. Conrad, W. Güttinger, and M. Dal Cin, eds.), pp. 150–170, Springer-Verlag, Berlin.

Feltz, P., and Rasminsky, M., 1974, A model for the mode of action of GABA on primary afferent terminals: Depolarizing effects of GABA applied iontophoretically to neurones of mammalian dorsal root ganglia, *Neuropharmacology* **13:**553–563.

Forsythe, I. D., and Redman, S. J., 1988, The dependence of motoneuron membrane potential on extracellular ion concentrations studied in isolated rat spinal cord, *J. Physiol. (London)* **404:**83–99.

Frank, K., and Fuortes, M. G. F., 1957, Presynaptic and postsynaptic inhibition of monosynaptic reflexes, *Fed. Proc.* **16:**39–40.

Gallagher, J. P., Higashi, H., and Nishi, S., 1978, Characterization and ionic basis of GABA-induced depolarizations recorded *in vitro* from cat primary afferent neurones, *J. Physiol. (London)* **275:**263–282.

Gallagher, J. P., Nakamura, J., and Shinnick-Gallagher, P., 1983, The effects of temperature, pH and Cl^- pump inhibitors on GABA responses recorded from cat dorsal root ganglia, *Brain Res.* **267:**249–259.

Galvan, M., Dörge, A., Beck, F., and Rick, R., 1984, Intracellular electrolyte concentrations in rat sympathetic neurones measured with an electron microprobe, *Pfluegers Arch.* **400:**274–279.

Garay, R. P., Nazaret, C., Hannaert, P. A., and Cragoe, E. J., 1988, Demonstration of a $[K^+,Cl^-]$-cotransport system in human red cells by its sensitivity to [(dihydroindenyl)oxy]alkanoic acids: Regulation of cell swelling and distinction from the bumetanide-sensitive $[Na^+,K^+,Cl^-]$-cotransport system, *Mol. Pharmacol.* **33:**696–701.

Gardner, D. R., and Moreton, R. B., 1985, Intracellular chloride in molluscan neurons, *Comp. Biochem. Physiol.* **80A:**461–467.

Geck, P., and Heinz, E., 1986, The Na-K-2Cl cotransport system, *J. Membr. Biol.* **91:**97–105.

Gerencser, G. A., and Lee, S. H., 1983, Cl^--stimulated adenosine triphosphatase: Existence, location and function, *J. Exp. Biol.* **106:**143–161.

Gold, M. R., and Martin, A. R., 1982, Intracellular Cl^- accumulation reduces Cl^- conductance in inhibitory synaptic channels, *Nature* **299:**828–830.

Goldman, D. E., 1943, Potential, impedance and rectification in membranes, *J. Gen. Physiol.* **27:**37–60.

Gunn, R. B., 1985, Bumetanide inhibition of anion exchange in human red blood cells, *Biophys. J.* **47:**326a.

Haas, M., 1989, Properties and diversity of (Na-K-Cl) cotransporters, *Annu. Rev. Physiol.* **51:**443–457.

Harris, G. L., and Betz, W. J., 1987, Evidence for active chloride accumulation in normal and denervated rat lumbrical muscle, *J. Gen. Physiol.* **90:**127–144.

Hattori, K., Akaike, N., Oomura, Y., and Kuraoka, S., 1984, Internal perfusion studies demonstrating GABA-induced chloride responses in frog primary afferent neurons, *Am. J. Physiol.* **246:**C259–C265.

Heinemann, V., and Lux, H. D., 1977, Ceiling of stimulus induced rises in extracellular potassium concentration in the cerebral cortex of the cat, *Brain Res.* **120:**231–249.

Higashijima, T., Ferguson, K. M., and Sternweis, P. C., 1987, Regulation of hormone-sensitive GTP-dependent regulatory proteins by chloride, *J. Biol. Chem.* **262:**3597–3602.

Hille, B., 1975, Ionic selectivity of Na and K channels of nerve membranes, in: *Membranes: A Series of Advances,* Volume 3 (G. Eisenman, ed.), pp. 255–323, Dekker, New York.

Hino, N., 1979, Action of ammonium ions on the resting membrane of crayfish stretch receptor neuron, *Jpn. J. Physiol.* **29:**99–102.

Hodgkin, A. L., and Katz, B., 1949, The effect of sodium ions on the electrical activity of the giant axon of the squid, *J. Physiol. (London)* **108:**37–77.

Hoffman, E. K., and Simonsen, L. O., 1989, Membrane mechanisms in volume and pH regulation in vertebrate cells, *Physiol. Rev.* **69:**315–382.

Huguenard, J. R., and Alger, B. E., 1986, Whole cell voltage-clamp study of the fading of GABA-activated currents in acutely dissociated hippocampal neurons, *J. Neurophysiol.* **56:**1–18.

Iles, J. F., and Jack, J. J. B., 1980, Ammonia: Assessment of its action on postsynaptic inhibition as a cause of convulsions, *Brain* **103:**555–578.

Ito, M., 1957, The electrical activity of spinal ganglion cells investigated with intracellular microelectrodes, *Jpn. J. Physiol.* **7:**297–323.

Iversen, L. L., 1975, Uptake processes for biogenic amines, in: *Handbook of Psychopharmacology,* Volume 3 (L. L. Iversen, ed.), pp. 381–442, Plenum Press, New York.

Jahnsen, H., and Llinás, R., 1984, Ionic basis for the electroresponsiveness and oscillatory properties of guinea-pig thalamic neurons *in vitro, J. Physiol. (London)* **349:**227–247.

Janigro, D., and Schwartzkroin, P. A., 1988, Effects of GABA on CA3 pyramidal cell dendrites in rabbit hippocampal slices, *Brain Res.* **453:**265–274.

Kaila, K., Pasternack, M., Saarikoski, J., and Voipio, J., 1989, Influence of GABA-gated bicarbonate conductance on membrane potential, current and intracellular chloride in crayfish muscle fibres, *J. Physiol. (London)* **416:**161–181.

Kaneda, M., Wakamori, M., and Akaike, N., 1989, GABA-induced chloride current in rat isolated Purkinje cells, *Am. J. Physiol.* **256:**C1153–C1159.

Kanner, B. I., and Schuldiner, S., 1987, Mechanism of transport and storage of neurotransmitters, *CRC Crit. Rev. Biochem.* **22:**1–38.

Kelly, J. S., Krnjević, K., Morris, M. E., and Yim, G. K. W., 1969, Anionic permeability of cortical neurons, *Exp. Brain Res.* **7:**11–31.

Kerkut, G. A., and Meech, R. W., 1966a, The internal chloride concentration of H and D cells in the snail brain, *Comp. Biochem. Physiol.* **19:**819–832.

Kerkut, G. A., and Meech, R. W., 1966b, Microelectrode determination of the intracellular chloride concentration in nerve cells, *Life Sci.* **5:**453–456.

Keynan, S., and Kanner, B. I., 1988, γ-Aminobutyric acid transport in reconstituted preparations from rat brain: Coupled sodium and chloride fluxes, *Biochemistry* **27:**12–17.

Keynes, R. D., 1963, Chloride in the squid giant axon, *J. Physiol. (London)* **169:**690–705.

Knepper, M. A., Packer, R., and Good, D. W., 1989, Ammonium transport in the kidney, *Physiol. Rev.* **69:** 179–249.

Korn, S. J., Giacchino, J. L., Chamberlin, N. L., and Dingledine, R., 1987, Epileptiform burst activity induced by potassium in the hippocampus and its regulation by GABA-mediated inhibition, *J. Neurophysiol.* **57:**325–340.

Kostyuk, P. G., Veselovsky, N. S., Fedulova, S. A., and Tsyndrenko, A. Y., 1981, Ionic currents in the somatic membrane of rat dorsal root ganglion neurons. III. Potassium currents, *Neuroscience* **6:**2439–2444.

Kregenow, F. M., 1981, Osmoregulatory salt transporting mechanisms: Control of cell volume in anisosmotic media, *Ann. Rev. Physiol.* **43:**493–505.

Krnjević, K., 1974, Chemical nature of synaptic transmission in vertebrates, *Physiol. Rev.* **54:**418–540.

Krnjević, K., 1981, Transmitters in motor systems, in: *Handbook of Physiology,* Section 2, *The Nervous System,* Volume II, *Motor Control* (V. B. Brooks, ed.), pp. 107–154, American Physiological Society, Baltimore.

Krnjević, K., 1983, GABA-mediated inhibitory mechanisms in relation to epileptic discharges, in: *Basic Mechanisms of Neuronal Hyperexcitability* (H. Jasper and N. van Gelder, eds.), pp. 249–280, Liss, New York.

Krnjević, K., 1984, Neurotransmitters in cerebral cortex: A general account, in: *Cerebral Cortex,* Volume 2, *Functional Properties of Cortical Cells* (E. G. Jones and A. Peters, eds.), pp. 39–61, Plenum Press, New York.

Kudo, Y., Abe, N., Goto, S., and Fukuda, H., 1975, The chloride-dependent depression by GABA in the frog spinal cord, *Eur. J. Pharmacol.* **32:**251–259.

Kunze, D. L., and Brown, A. M., 1971, Internal potassium and chloride activities and the effects of acetylcholine on identifiable *Aplysia* neurons, *Nature (London)* **229:**229–231.

Landry, D. W., Reitman, M., Cragoe, E. J., and Al-Awqati, Q., 1987, Epithelial chloride channel, *J. Gen. Physiol.* **90:**779–798.

Latorre, R., and Miller, C., 1983, Conduction and selectivity in potassium channels, *J. Membr. Biol.* **71:** 11–30.

Lauf, P. K., 1988, K : Cl cotransport: Emerging Molecular aspects of a ouabain-resistant, volume-responsive transport system in red blood cells, *Renal Physiol. Biochem.* **3–5:**248–259.

Lauf, P. K., McManus, T. J., Haas, M., Forbush, B., Duhm, J., Flatman, P. W., Saier, M. H., and Russell, J. M., 1987, Physiology and biophysics of chloride and cation cotransport across cell membranes, *Fed. Proc.* **46:**2377–2394.

Levy, R. A., 1977, The role of GABA in primary afferent depolarization, *Prog. Neurobiol. (Oxford)* **9:**211–267.

Lewis, D. V., and Schuette, W. H., 1975, NADH fluorescence and $[K^+]_o$ changes during hippocampal electrical stimulation, *J. Neurophysiol.* **38:**405–417.

Llinás, R., 1988, The intrinsic electrophysiological properties of mammalian neurons: Insights into central nervous function, *Science* **242:**1654–1664.

Llinás, R., and Baker, R., 1972, A chloride-dependent inhibitory postsynaptic potential in cat trochlear motoneurons, *J. Neurophysiol.* **35**:484–492.

Llinás, R., Baker, R., and Precht, W., 1974, Blockage of inhibition by ammonium acetate action on chloride pump in cat trochlear motoneurons, *J. Neurophysiol.* **37**:522–532.

Lopez, R., and Alvarez-Leefmans, F. J., 1984, Electrotonic structure and specific membrane properties of frog dorsal root ganglion neurons maintained *in vitro, Soc. Neurosci. Abstr.* **10**(1):429.

Lux, H. D., 1971, Ammonium and chloride extrusion: Hyperpolarizing synaptic inhibition in spinal motoneurons, *Science* **173**:555–557.

Lux, H. D., Loracher, C., and Neher, E., 1970, The action of ammonium on postsynaptic inhibition of cat spinal motoneurons, *Exp. Brain Res.* **11**:431–447.

Lytle, C., and McManus, T. J., 1987, Effect of loop diuretics and stilbene derivatives on swelling-induced KCl cotransport, *J. Gen. Physiol.* **90**:28a.

McCarren, M., and Alger, B. E., 1985, Use-dependent depression of IPSPs in rat hippocampal pyramidal cells *in vitro, J. Neurophysiol.* **53**:557–571.

Mayer, M. L., and Westbrook, G. L., 1983, A voltage-clamp analysis of inward (anomalous) rectification in mouse spinal sensory ganglion neurons, *J. Physiol. (London)* **340**;19–45.

Meyer, H., and Lux, H. D., 1974, Action of ammonium on a chloride pump, *Pfluegers Arch.* **350**:185–195.

Misgeld, U., Deisz, R. A., Dodt, H. V., and Lux, H. D., 1986, The role of chloride transport in postsynaptic inhibition of hippocampal neurons, *Science* **232**:1413–1415.

Moody, W. J., 1981, The ionic mechanisms of intracellular pH regulation in crayfish neurons, *J. Physiol. (London)* **316**:293–308.

Moreton, R. B., and Gardner, D. R., 1981, Increased intracellular chloride activity produced by the molluscicide, N-(triphenylmethyl)morpholine (Frescon), in *Lymnaea stagnalis* neurons, *Pestic. Biochem. Physiol.* **15**:1–9.

Moser, H., 1985, Intracellular pH regulation in the sensory neuron of the stretch receptor of the crayfish (*Astacus fluviatilis*), *J. Physiol. (London)* **362**:23–38.

Moser, H., 1987, Electrophysiological evidence for ammonium as a substitute for potassium in activating the sodium pump in a crayfish sensory neuron, *Can. J. Physiol. Pharmacol.* **65**:141–145.

Müller, W., Misgeld, U., and Lux, H. D., 1989, γ-Aminobutyric acid-induced ion movements in the guinea pig hippocampal slice, *Brain Res.* **484**:184–191.

Nakai, K., Sasaki, K., Matsumoto, M., and Takashima, K., 1988, Effects of furosemide on the resting membrane potentials and the transmitter-induced responses of the *Aplysia* ganglion cells, *Tohoku J. Exp. Med.* **156**:79–90.

Neild, T. O., and Thomas, R. C., 1974, Intracellular chloride activity and the effects of acetylcholine in snail neurones, *J. Physiol. (London)* **242**:453–470.

Nelson, M. T., and Blaustein, M. P., 1982, GABA efflux from synaptosomes: Effects of membrane potential, and external GABA and cations, *J. Membr. Biol.* **69**:213–223.

Newberry, N. R., and Nicoll, R. A., 1985, Comparison of the action of baclofen with γ-aminobutyric acid on rat hippocampal pyramidal cells *in vitro, J. Physiol. (London)* **360**:161–185.

Nicoll, R. A., 1978, The blockade of GABA mediated responses in the frog spinal cord by ammonium ions and furosemide, *J. Physiol. (London)* **283**:121–132.

Nicoll, R. A., 1988, The coupling of neurotransmitter receptors to ion channels in the brain, *Science* **241**:545–551.

Nicoll, R. A., and Alger, B. E., 1979, Presynaptic inhibition: Transmitter and ionic mechanisms, *Int. Rev. Neurobiol.* **21**:217–258.

Nishi, S., Minota, S., and Karczmar, A. G., 1974, Primary afferent neurones: The ionic mechanism of GABA-mediated depolarization, *Neuropharmacology* **13**:215–219.

Nistri, A., 1983, Spinal cord pharmacology of GABA and chemically related amino acids, in: *Handbook of the Spinal Cord,* Volume 1, *Spinal Cord Pharmacology* (R. A. Davidoff, ed.), pp. 45–104, Dekker, New York.

O'Grady, S. M., Palfrey, H. C., and Field, M., 1987, Characteristics and functions of Na-K-Cl cotransport in epithelial tissues, *Am. J. Physiol.* **253**:177–192.

Olsen, R. W., and Leeb-Lundberg, F., 1981, Convulsant and anticonvulsant drug binding sites related to

GABA-regulated chloride ion channels, in: *GABA and Benzodiazepine Receptors* (E. Costa, G. Di-Chiara, and G. L. Gessa, eds.), pp. 93–103, Raven Press, New York.

Olsen, R. W., and Venter, J. C., 1986, *Benzodiazepine/GABA Receptors and Chloride Channels: Structural and Functional Properties,* Liss, New York.

Otsuka, M., and Konishi, S., 1976, GABA in the spinal cord, in: *GABA in Nervous System Function* (E. Roberts, T. N. Chase, and D. B. Tower, eds.), pp. 197–202, Raven Press, New York.

Padjen, A., Nicoll, R., and Barker, J. L., 1973, Synaptic potentials in the isolated frog spinal cord studied using sucrose gap, *J. Gen. Physiol.* **61:**270–271.

Peterson, R. P., and Pepe, I. A., 1961, The fine structure of inhibitory synapses in the crayfish, *J. Biophys. Biochem. Cytol.* **11:**159–169.

Raabe, W., and Gumnit, R. J., 1975, Disinhibition in cat motor cortex by ammonia, *J. Neurophysiol.* **38:** 347–355.

Radian, R., and Kanner, B. I., 1983, Stoichiometry of sodium- and chloride-coupled γ-aminobutyric acid transport by synaptic plasma membrane vesicles isolated from rat brain, *Biochemistry* **22:**1236–1241.

Radian, R., and Kanner, B. I., 1985, Reconstitution and purification of the sodium and chloride-coupled γ-aminobutyric acid transporter from rat brain, *J. Biol. Chem.* **260:**11859–11865.

Radian, R., Bendahan, A., and Kanner, B. I., 1986, Purification and identification of the functional sodium- and chloride-coupled γ-aminobutyric acid transport glycoprotein from rat brain, *J. Biol. Chem.* **261:** 15437–15441.

Reuss, L., 1983, Basolateral KCl cotransport in a NaCl-absorbing epithelium, *Nature* **305:**723–726.

Reuss, L., 1988, Cell volume regulation in nonrenal epithelia, *Renal Physiol. Biochem.* **3–5:**187–201.

Reuss, L., 1989, Ion transport across gallbladder epithelium, *Physiol. Rev.* **69:**503–545.

Roberts, E., 1986, GABA: The road to neurotransmitter status, in: *Benzodiazepine/GABA Receptors and Chloride Channels: Structural and Functional Properties,* pp. 1–39, Liss, New York.

Roos, A., and Boron, W. F., 1981, Intracellular pH, *Physiol. Rev.* **61:**296–434.

Rudy, B., 1988, Diversity and ubiquity of K channels, *Neuroscience* **25:**729–749.

Russell, J. M., 1976, ATP-dependent chloride influx into squid giant axon, *J. Membr. Biol.* **28:**335–350.

Russell, J. M., 1978, Effects of ammonium and bicarbonate-CO₂ on intracellular chloride levels in *Aplysia* neurons, *Biophys. J.* **22:**131–137.

Russell, J. M., 1979, Chloride and sodium influx: A coupled uptake mechanism in the squid giant axon, *J. Gen. Physiol.* **73:**801–818.

Russell, J. M., 1980, Anion transport mechanisms in neurons, *Ann. N.Y. Acad. Sci.* **341:**510–523.

Russell, J. M., 1983, Cation-coupled chloride influx in squid axon: Role of potassium and stoichiometry of the transport process, *J. Gen. Physiol.* **81:**909–925.

Russell, J. M., 1984, Chloride in the squid giant axon, *Curr. Top. Membr. Transp.* **22:**177–193.

Russell, J. M., and Boron, W. F., 1982, Intracellular pH regulation in squid giant axons, in: *Intracellular pH: Its Measurement, Regulation and Utilization in Cellular Functions,* pp. 221–237, Liss, New York.

Russell, J. M., and Brown, A. M., 1972, Active transport of chloride by the giant neuron of the *Aplysia* abdominal ganglion, *J. Gen. Physiol.* **60:**499–518.

Schlue, W.-R., and Deitmer, J. W., 1988, Ionic mechanisms of intracellular pH regulation in the nervous system, *Ciba Found. Symp.* **139:**49–69.

Schlue, W.-R., and Thomas, R. C., 1985, A dual mechanism for intracellular pH regulation by leech neurons, *J. Physiol. (London)* **364:**327–338.

Schmidt, R. F., 1963, Pharmacological studies on the primary afferent depolarization of the toad spinal cord, *Pfluegers Arch.* **277:**325–346.

Schwartzkroin, P. A., and Wheal, H. V., 1984, *Electrophysiology of Epilepsy,* Academic Press, New York.

Serve, G., Endres, W., and Grafe, P., 1988, Continuous electrophysiological measurements of changes in cell volume of motoneurons in the isolated frog spinal cord, *Pfluegers Arch.* **411:**410–415.

Siggins, G. R., and Gruol, D. L., 1986, Mechanisms of transmitter action in the vertebrate central nervous system, in: *Handbook of Physiology,* Section 1, *The Nervous System,* Volume 14, (T. E. Bloom, ed.), pp. 1–114, American Physiological Society, Bethesda.

Sillito, A. M., 1984, Functional considerations of the operation of GABAergic inhibitory processes in the visual cortex, in: *Cerebral Cortex,* Volume 2, *Functional Properties of Cortical Cells* (E. G. Jones and A. Peters, eds.), pp. 91–117, Plenum Press, New York.

Simmonds, M. A., 1984, Physiological and pharmacological characterization of the actions of GABA, in: *Actions and Interactions of GABA and Benzodiazepines* (N. G. Bowery, ed.), pp. 27–40, Raven Press, New York.

Somjen, G. G., 1979, Extracellular potassium in the mammalian central nervous system, *Annu. Rev. Physiol.* **41:**159–177.

Stein, W. D., 1986, *Transport and Diffusion across Cell Membranes,* Academic Press, New York.

Steriade, M., and Llinás, R., 1988, The functional states of the thalamus and the associated neuronal interplay, *Physiol. Rev.* **68:**649–736.

Tanaka, C., and Taniyama, K., 1986, GABA transport in peripheral tissues: Uptake and efflux, in: *GABAergic Mechanisms in the Mammalian Periphery* (S. L. Erdö and N. G. Bowery, eds.), pp. 57–72, Raven Press, New York.

Thomas, R. C., 1976, The effect of carbon dioxide on the intracellular pH and buffering power of snail neurones, *J. Physiol. (London)* **255:**715–735.

Thomas, R. C., 1977, The role of bicarbonate, chloride and sodium ions in the regulation of intracellular pH in snail neurones, *J. Physiol. (London)* **273;**317–338.

Thomas, R. C., 1984, Experimental displacement of intracellular pH and the mechanism of its subsequent recovery, *J. Physiol. (London)* **354:**3P–22P.

Thomas, R. C., and Cohen, C. J., 1981, A liquid ion-exchanger alternative to KCl for filling intracellular reference microelectrodes, *Pfluegers Arch.* **390:**96–98.

Thompson, S. M., and Gähwiler, B. H., 1989a, Activity-dependent disinhibition. I. Repetitive stimulation reduces IPSP driving force and conductance in the hippocampus *in vitro, J. Neurophysiol.* **61:**501–511.

Thompson, S. M., and Gähwiler, B. H., 1989b, Activity-dependent disinhibition. II. Effects of extracellular potassium, furosemide, and membrane potential on E_{Cl^-} in hippocampal CA3 neurons, *J. Neurophysiol.* **61:**512–523.

Thompson, S. M., Deisz, R. A., and Prince, D. A., 1988a, Outward chloride/cation cotransport in mammalian cortical neurons, *Neurosci. Lett.* **89:**49–54.

Thompson, S. M., Deisz, R. A., and Prince, D. A., 1988b, Relative contributions of passive equilibrium and active transport to the distribution of chloride in mammalian cortical neurons, *J. Neurophysiol.* **60:**105–124.

Traub, R. D., Miles, R., and Wong, R. K. S., 1989, Model of the origin of rhythmic population oscillations in the hippocampal slice, *Science* **243:**1319–1325.

Vaughan-Jones, R. D., 1988, Regulation of intracellular pH in cardiac muscle, *Ciba Found. Symp.* **139:**23–46.

Verkman, A. S., Sellers, M. C., Chao, A. C., Leung, T., and Ketcham, R., 1989, Synthesis and characterization of improved chloride sensitive fluorescent indicators for biological applications, *Anal. Biochem.* **178:**355–361.

Vitoux, D., Oliviero, O., Garay, R. P., Cragoe, E. J., Galacteros, F., and Benzard, Y., 1989, Inhibition of K^+ efflux and dehydration of sickle cells by [(dihydroindenyl)oxy]alkanoic acid: An inhibitor of the K^+,Cl^- cotransport system, *Proc. Natl. Acad. Sci. USA* **86:**4273–4276.

Widdicombe, J. H., Nathanson, I. T., and Highland, E., 1983, Effects of loop diuretics on ion transport by dog tracheal epithelium, *Am. J. Physiol.* **245:**C388–C396.

Wojtowicz, J. M., and Nicoll, R. A., 1982, Selective action of piretamide on primary afferent GABA responses in the frog spinal cord, *Brain Res.* **236:**173–181.

Wong, R. K. S., and Watkins, D. J., 1982, Cellular factors influencing GABA response in hippocampal pyramidal cells, *J. Neurophysiol.* **48:**938–951.

Yudilevich, D. L., and Boyd, C. A. R., 1987, *Amino Acid Transport in Animal Cells,* Physiological Society Study Guides, No. 2, Manchester University Press, Great Britain.

Zelikovic, I., Stejskal-Lorenz, E., Lohstroh, P., Budreau, A., and Chesney, R. W., 1989, Anion dependence of taurine transport by rat renal brush border membrane vesicles, *Am. J. Physiol.* **256:**646–655.

Chloride Transport across Glial Membranes

H. K. Kimelberg

1. INTRODUCTION

Ions flow across cell membranes by two major classes of transport mechanisms. One class is represented by ion channels, which are best viewed as water-filled pathways through specific proteins embedded in the membrane lipid bilayer (Hille, 1984). Such channels show selectivity for different ions according to size and charge. The rate of movement of ions through these channels is determined by the electrochemical driving force on the ion and the individual conductances of the channels. Characteristically, channel-mediated fluxes are very large, in excess of 10^6 ions/sec per channel (Hille, 1984). The number of channels per unit area, the rate at which these channels open, and the duration of time they are open are important determinants of channel-mediated ion fluxes. Channel opening can be modified by transmembrane voltage, specific ligands, or when tension is applied to the membrane (Sachs, 1988). In the case of the various chloride channels, all these forces seem to operate as we shall learn later in this chapter as well as in other chapters in this volume.

Chloride can also be transported by an entirely different class of transport processes, which are usually characterized as being electrically neutral since they can exchange Cl^- for Cl^- or another anion, or transport an equal number of anionic plus cationic charges in the same direction (Tosteson, 1981). The best example of an anion exchange process is the Cl^-/HCO_3^- exchanger, which is present at a particularly high concentration in red blood cells where it is responsible for mediating Cl^-/HCO_3^- exchange during the ion transport movements associated with respiration (Lowe and Lambert, 1983). Perhaps the most intensively studied example of a cotransport system is the system that catalyzes movement of Na^+, K^+, and Cl^-, and can result in active accumulation of Cl^- utilizing the inward driving force of the Na^+ chemical gradient (Geck and Heinz, 1986). This system will enable the cell to accumulate KCl, since the Na^+ taken up will be continuously pumped out by the Na^+ pump (Na^+,K^+-ATPase). The Na^+ pump is a primary active transport system (Stein, 1986), since it directly utilizes energy derived from the hydrolysis of ATP to actively accumulate K^+ and

---

H. K. Kimelberg • Division of Neurosurgery, Departments of Biochemistry and Pharmacology/Toxicology, and Program in Neuroscience, Albany Medical College, Albany, New York 12208.

extrude Na^+. There is also a Na^+,Cl^- cotransport system that functions to accumulate NaCl, and a K^+,Cl^- cotransport system that usually functions in the efflux mode (Warnock et al., 1983).

In this chapter I will discuss the Cl^- channels and Cl^- exchange or cotransport systems that have been described as present in glial cells. These cells include astrocytes, oligodendrocytes, and Schwann cells, although the bulk of the work done so far has been on astrocytes. The chapter will be organized under headings arranged according to the different transport systems described above or in relation to a specific function (Sections 7 and 8) or from a historical perspective (Section 2).

2. EARLY ELECTROPHYSIOLOGICAL STUDIES OF GLIA

Before proceeding to discuss more recent studies on anion transport in glia it is worth mentioning the first electrophysiological studies of glial cells done by Kuffler, Nicholls, and Orkand, which provided some of the first information on the ion transport properties of glial cells (Nicholls and Kuffler, 1964; Kuffler et al., 1966; Orkand, 1977). These workers initially used the leech central nervous system and later the optic nerve of an amphibian, the mud puppy Necturus maculosus, since in both preparations glial cells were readily accessible to impalement by microelectrodes. The most detailed studies were done in Necturus where it was found that a population of glial cells could be obtained that had a mean membrane resting potential of -89 mV. These potentials responded to changing $[K^+]_o$ in a perfectly Nernstian manner, i.e., a plot of membrane potential versus the logarithm of $[K^+]_o$ at 23°C gave a slope of 59 mV per tenfold change in $[K^+]_o$ over a 1.5 mM to 75 mM range of $[K^+]_o$ according to the Nernst equation:

$$E_m \ (\mathrm{mV}) = 2.303 \ RT/F \ \log \ [K^+]_o/[K^+]_i \qquad (1)$$

When such behavior is found, it also indicates that $[K^+]_i$ remains constant while $[K^+]_o$ is being varied. When E_m is zero, then the equation reduces to $[K^+]_i = [K^+]_o$. In the case of the Necturus optic nerve, $[K^+]_i$ was found to be 99 mM. It is worth noting that for the K^+ dependence studies, Kuffler et al. (1966) rejected any glial cells that gave a potential less negative than -85 mV, because of presumed imperfect impalements or damaged cells. This Nernstian behavior with regard to K^+ implies that the Cl^- conductance is at least about a 100-fold less than the K^+ conductance and thus Cl^- does not contribute to the membrane potential. Alternatively, the Cl^- conductance could be large enough that the $[Cl^-]_i/[Cl^-]_o$ ratio rapidly equilibrates with the changing membrane potential due to changing $[K^+]_o$. This in fact was the behavior found by Hodgkin and Horowicz (1959) for frog skeletal muscle, where the permeabilities for K^+ and Cl^- were 1–2 and 4×10^6 cm/sec, respectively. Also, Hodgkin and Horowicz (1959) did find transient deviations from a Nernstian response when $[K^+]_o$ was rapidly changed without altering $[Cl^-]_o$. They attributed this to the time needed for Cl^- to passively reequilibrate with the new membrane potential. Such transient changes were not examined in the early studies on glial cells (Kuffler *et al.*, 1966). More recent studies have detected measurable Cl⁻ conductance in some glial cells. Also, the ion channels that are responsible for passive ion conductances can be directly measured using patch-clamp techniques and Cl⁻ channels have been identified by this method in cultured glial cells, as will be discussed in Section 4 (see also Chapter 6).

3. CHLORIDE CONTENT OF GLIAL CELLS

3.1. $[Cl^-]_i$ Estimates in Intact Nervous System

Following the points raised above, it is clearly important to know if the intracellular free chloride concentrations of glial cells in the brain are actually in equilibrium with the membrane potentials of the respective cells, since this will immediately tell us if we need only postulate Cl⁻ channels or whether some active transport of Cl⁻ needs to be considered. In the intact nervous system, it is only possible to obtain a value for $[Cl^-]_i$ in single cells using ion-specific microelectrodes (see Section 3.2) or X-ray microprobe analysis. However, Smith *et al.* (1981) did calculate a range for glial cell $[Cl^-]_i$ of 36–46 mM in rat cerebral cortex based on a number of assumptions and an average measured value for total Cl_i^- of 22 mmoles/kg water, with an extracellular space of 12–14%. The assumptions were that, on the average, Cl⁻ passively equilibrates in neurons with a $[Cl^-]_i$ of 10 mM, and a 2:1 relative neuron–glial volume.

3.2. Use of Ion-Selective Microelectrodes

A more satisfactory approach is to try to directly measure the a^iCl, or intracellular Cl⁻ activity, by impalement of glial cells in intact brain or slices using ion-selective microelectrodes (ISMs). Because of the small size of glial cells and the difficulties of positively identifying which cells are impaled, this is a formidable task. However, Grafe and Ballanyi (1987) and Ballanyi *et al.* (1987) have recently used ISMs to measure a^iCl and other ion activities in glia of tissue slices from the olfactory cortex of guinea pigs. They reported values for a^iK of 66 mM, a^iNa of 25 mM, and a^iCl of 6 mM, with an E_m of −84 mV. Stimulation of the lateral olfactory tract caused a rise of external K⁺ activity, (a^oK) from 2.2 to around 7 mM. Concomitantly, the glial cell depolarized and both a^iK and a^iCl rose. It was concluded that a^iCl in these glial cells was always in equilibrium with E_m and the rise in a^iK and a^iCl was largely driven by Donnan forces, as in frog skeletal muscle (Boyle and Conway, 1941; Hodgkin and Horowicz, 1959). Grafe and Ballanyi (1987) also showed that K⁺ uptake into glial cells could be driven by the Na⁺ pump in response to a rise in a^oK from 2.2 to 6.5 mM, including when the K⁺ channels were blocked with 0.5 mM Ba^{2+}. An increased Na⁺ pump activity in glial cells in response to a rise in $[K^+]_o$ from 3 to 9 mM is predicted from studies on isolated glia and cultured astrocytes where the Na⁺ pump, as in other cells (Stein, 1986), was found to saturate at $[K^+]_o$ values of 10–12 mM and the $K_{0.5}$ for activation by K^+ was around 1 mM (Kimelberg _et al.,_ 1978a,b). However, the high Cl^- permeability of glial cells in olfactory cortex is not found in cultured astrocytes. Rodent astrocytes in primary culture are relatively impermeable to Cl^- and E_{Cl} is more positive than E_m, although this behavior may change when the cells become depolarized at $[K^+]_o \geq 20$ mM (see Sections 3.3 and 5.2). In fact, the behavior of the olfactory cortex glial cells closely corresponds to that described by Kettenmann (1987) for cultured oligodendrocytes from mouse spinal cord (also see below), raising the question of whether astrocytes or oligodendrocytes were being studied in the olfactory cortex. Also, if transmembrane Cl^- distribution is in equilibrium with the membrane potential of glial cells, it is not clear where the excess Cl_i^- determined in cortical tissue _in situ_ (Smith _et al.,_ 1981; see Section 3.1) is located.

3.3. Use of Glial Cell Cultures

An alternative and widely used approach to solving the severe methodological problem of studying glial cells _in vivo_ is to use well-defined primary cultures, and cultures of astroglial, oligodendroglial, and Schwann cells can all be made with reasonable-to-excellent purity. Measuring ^{36}Cl steady-state levels in primary astrocyte cultures, Kimelberg (1981) reported that $[Cl^-]_i$ was 31 to 34 mM. Such values are three to five times higher than predicted from equilibration with the average membrane potentials of -70 mV, measured for the same cells (Kimelberg _et al.,_ 1979b), i.e., Cl^- was apparently being actively accumulated. These values are surprisingly close to those estimated by Smith _et al._ (1981) for glial cells _in situ_ (see Section 3.1). These results were later confirmed by Kettenmann and colleagues (Kettenmann, 1987; Kettenmann _et al.,_ 1987) using ISMs to determine a^iCl in primary astrocyte cultures from neonatal rat cerebral cortex. It was found that a^iCl was 20–40 mM, giving an E_{Cl} of -30 to -48 mV. Kettenmann (1987) also found that the average E_{Cl} of cultured, mouse spinal cord oligodendrocytes was -61 ± 10 mV. This was closer to, but still more positive than predicted for a passive distribution since the average resting membrane potentials were -64 mV. When $[K^+]_o$ was increased from 5.4 to 50 mM, a^iCl rose from about 10 to 50 mM in oligodendrocytes, and this rise in Cl_i^- was insensitive to 0.1 mM furosemide. There was considerable variability in the behavior of the cultured oligodendrocytes and a^iCl ranged from 7 to 30 mM and E_m minus E_{Cl} between $+22$ and -13 mV for individual cells, the mean difference being $+3 \pm 9$ mV. The author suggested that this variability might be due to different populations of oligodendrocytes. Also, the K^+ conductance must still predominate in these cells since when $[Cl^-]_o$ was reduced from 125 to 10 mM there was no immediate depolarization, which might have been expected since E_{Cl} would now be, at least transiently, around 0 mV. Alternatively, it was suggested that $[Cl^-]_i$ might change as rapidly as $[Cl^-]_o$ could be exchanged. Thus, at present it seems that Cl^- transport in cultured oligodendrocytes derived from mouse spinal cord is predominantly through Cl^- channel pathways. However, when $[Cl^-]_i$ was reduced by a different method, namely by reducing $[Cl^-]_o$, the reaccumulation of Cl_i when Cl_o was restored to its normal value occurred via the Na^+,K^+,Cl^- cotransport mechanism (see Kettenmann, this volume). Presumably the differences are related to the K^+-induced depolarization.

4. STUDIES ON Cl⁻ CONDUCTANCES AND CHANNELS IN GLIA

4.1. Low or High Cl⁻ Conductance?

As discussed in the preceding two sections, it has been assumed since the pioneering work of Kuffler and his colleagues (Kuffler *et al.*, 1966; Nicholls and Kuffler, 1964) that the Cl^- conductance of glial cells was insignificant compared to the overwhelming K^+ conductance, since changes in $[Cl^-]_o$ did not affect the glial membrane potential. As noted above, however, this behavior does not hold for other preparations and in *Necturus* and the leech could have been due either to a very low Cl^- conductance or to high Cl^- permeability resulting in rapid equilibration of Cl^- within the time periods before the first measurements were made. Continuous membrane potential recordings in glia in a *Necturus* optic nerve preparation (Bracho *et al.*, 1975) and in cultured astrocytes (Kimelberg *et al.*, 1982b, 1986; Walz *et al.*, 1984) have *usually* shown no effect of removing Cl_o^-. Again either the Cl^- conductance is small compared to those of other ions, or $[Cl^-]_i$ has equilibrated very rapidly indeed, i.e., within the perfusion time taken to exchange all the Cl_o^-, since otherwise transient effects should have been detected. In a few cases, however, approximately in 10% of the total cells in primary cultures of cerebral cortical astrocytes that we examined, a depolarization was observed when Cl_o^- was removed (Kimelberg *et al.*, 1982b). This behavior is shown in Fig. 1 for one of these cells, where we were also studying the effect of $(-)$ norepinephrine (NE). As can be seen in the second trace from the top, removal of all Cl^- from the medium, by replacement with isethionate, resulted in a 10-mV depolarization. Although in this experiment Cl^- removal was preceded by addition of 10^{-5} M NE, such depolarizing responses were also seen in cells not previously exposed to NE. It can be seen that in Cl^--depleted cells (second and third trace from top) the depolarizing response to NE was reduced compared to the response in the top tracing, while the depolarizing response to NE in Cl^--loaded cells when Cl_o^- was also removed, was increased (bottom trace). These data suggested that part of the depolarizing response due to NE could involve the opening of Cl^- channels. Further evidence of this is given in Section 6 and shown in Fig. 8. This behavior of a few of the cultured astrocytes is similar to observations in leech neuropil glia by Walz and Schlue (1982), who reported that there was a significant depolarization in these cells when Cl_o^- was removed. This depolarization was followed by a hyperpolarization, presumably as the cell loses Cl^- and the membrane potential approaches the K^+ equilibrium potential. We never saw such a hyperpolarization for the astrocyte cultures, possibly because even when the Cl^- conductance was appreciable it was still significantly less than the K^+ conductance. Walz and Schlue (1982) reported a 20% decrease in conductance that paralleled the presumed loss of Cl^- as indicated by the secondary hyperpolarization. Kettenmann (1987) also reported a 27% increase in input resistance in cultured oligodendrocytes when all the bath Cl^- was removed.

The more common behavior for primary astrocyte cultures is shown in Fig. 2 where removal of Cl_o^- had no effect on the membrane potential, as shown in the middle trace where three additions of Cl^--free medium, with gluconate or isethionate serving as the replacement anion, showed no changes in resting membrane potential. The

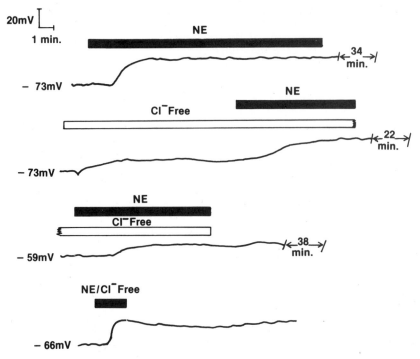

Figure 1. Depolarization of membrane potentials of rat brain astrocytes in primary culture due to omission of Cl$^-$ from the medium. At the beginning of the first trace, cells were in normal medium (122 mM NaCl, 3 mM KCl, 1.3 mM CaCl$_2$, 0.4 mM MgSO$_4$, 1.2 mM KH$_2$PO$_4$, 20 mM HEPES pH 7.4, 10 mM NaHCO$_3$, and 10 mM glucose). Cells were incubated in this solution for 1–2 hr and then the recordings made using standard electrophysiological techniques (Hirata *et al.*, 1983; Kimelberg *et al.*, 1979b). NE refers to 10^{-5} M (−) norepinephrine and was applied as indicated by the closed bars. The times of exposure to Cl$^-$-free medium are indicated by open bars. In "Cl$^-$-free" medium, NaCl and KCl were replaced with the respective isethionate salts and CaCl$_2$ was replaced with Ca(NO$_3$)$_2$. The times at the end of each trace represent a break in the trace of the duration indicated. The depolarization of the resting membrane potential by 14 mV seen at the beginning of the third trace seems to be principally associated with the long exposure of Cl$^-$-free medium and consequent depletion of cell Cl$^-$. Temperature of the experiment was 35°C; cells were 4 weeks old.

bottom trace shows that a high concentration of the anion cotransport inhibitor, furosemide, had no effect on the potential, consistent with the Na$^+$,K$^+$,Cl$^-$ cotransport system carrying no net charge in the steady state. This will be discussed in Section 5.2 in relation to Cl$^-$ cotransport. Clearly, since different cells in primary astrocyte cultures can give different results in response to removal of Cl$_o^-$, there appears to be some heterogeneity among the cells as to the density of Cl$^-$ channels relative to K$^+$ channels. However, in Fig. 2 the cultures were treated for about 2 hr with 1 mM dibutyryl cAMP to cause them to round up to facilitate impalement, while in Fig. 1 the cells were allowed to stand for at least 1 hr in salt solution (see Fig. 1), which resulted in a smaller proportion of the cells rounding up. Thus, it is possible that treatment with

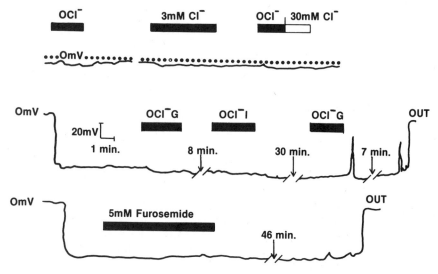

Figure 2. (Upper trace) Effect of changes in $[Cl^-]_o$ (as indicated) on electrode potential when both electrodes are out of the cell. (Middle trace) Effect of Cl^--free medium [replaced with gluconate (G) or isethionate (I) as indicated] on membrane potential of cell. (Lower trace) Effect of perfusion with furosemide (5 mM). Two-week-old cells treated with 1 mM dibutyryl cAMP for approximately 2 hr in HCO_3^--buffered medium (same medium as in Fig. 1 except 25 mM HCO_3^- and no HEPES was present). Solution was continuously bubbled with a 5% CO_2/95% air atmosphere before and during the recording. Temperature was 35–37°C. 4.5 mM K^+ on tracings indicates return to control medium. (From Kimelberg *et al.*, 1986, reproduced with permission.)

dibutyryl cAMP is responsible for the different behavior seen, i.e., dibutyryl cAMP may close Cl^- channels.

4.2. Patch-Clamp Studies

Recent work with primary astrocyte cultures using patch-clamp recordings in the whole-cell mode has shown the existence of voltage-gated anion conductances (Bevan *et al.*, 1985; Gray and Ritchie, 1986). These normally outward currents mediated by Cl^- moving inwardly, are turned on at membrane potentials of -40 mV or more positive and turned off at $+30$ mV, and are blocked by 20–100 μM of the disulfonate stilbene inhibitors DIDS or SITS. The Cl^- channels have a selectivity of chloride = bromide > methylsulfate > sulfate > isethionate > acetate. Gluconate appears to be virtually impermeant. Using patch-clamp techniques, Sonnhof (1987) also demonstrated anion-selective channels with large conductances of around 400 pS in primary astrocyte cultures. Ninety percent of these channels showed a strong voltage dependence, while the remaining 10% showed no voltage dependence. Nowak *et al.* (1987) also found a group of large anion channels with a mean conductance of 385 $\pm$ 27 pS which, as in the whole-cell voltage-clamp studies of Bevan *et al.* (1985) and Gray and

Ritchie (1986), were open between -40 and $+30$ mV. Nowak *et al.* (1987) also found a population of small Cl^- conductance channels (5 pS) opened by membrane hyperpolarization. It also appears that Cl^- channels in cultured astrocytes can be ligand-gated, as will be discussed in Section 6. Also see Kettenmann (this volume) for a full discussion of Cl^- channels in cultured glial cells.

5. ANION EXCHANGE AND COTRANSPORT IN GLIA

Work in nonneural tissue has shown that there are numerous pathways for Cl^- transport across cell membranes other than channels, and that these pathways are very important physiologically. Prominent among these is the anion exchange system for mediating Cl^-/HCO_3^- exchange, first described in the red blood cell membrane where it is necessary for rapid transport of HCO_3^- secondary to hydration and dehydration of CO_2 in respiration (Lowe and Lambert, 1983). The gene for this transport system has been cloned and the amino acid sequence determined (Kopito *et al.*, 1987). Thus, it will be of considerable interest to establish localization of the Cl^-/HCO_3^- exchange system in the different cell types of the nervous system using *in situ* hybridization, as well as by using antibodies to the membrane protein (Kay *et al.*, 1983). Another important system is the Na^+,K^+,Cl^- cotransport system involved in active intracellular accumulation of Cl^- and consequently in cell volume regulation and transcellular epithelial transport (Geck and Heinz, 1986). These systems have been identified in glial cells, and also in neurons as discussed in Chapters 3 and 4.

5.1. Anion Exchange Transport

We initiated studies on anion exchange in astrocytes some years ago (Bourke *et al.*, 1975; Kimelberg *et al.*, 1979a). The stimulus for this came from early observations that a considerable proportion of the high $[K^+]$-induced swelling of brain slices seemed to involve astrocytes (Tower and Bourke, 1966; Franck and Schoffeniels, 1972; Schousboe, 1972; Bourke *et al.*, 1975; Moller *et al.*, 1974). Since we found that a proportion of this swelling was dependent on HCO_3^-, was specifically inhibited by a number of anion exchange inhibitors, and could be mimicked under conditions of low $[K^+]_o$ by addition of a number of transmitters (Kimelberg and Bourke, 1982; Kimelberg *et al.*, 1982a; Bourke *et al.*, 1983), it appeared that it could not simply be explained by Donnan forces due to the increased $[K^+]_o \times [Cl^-]_o$ product. Lund-Anderson and Hertz (1970) also found K^+-induced swelling of brain slices when $[K^+]_o$ was increased, but in this case KCl was added to the solution making it hyperosmotic. This also suggested an active process of some kind since, if we assume that about as much KCl would be taken up by the cell by simple diffusion as was added to the medium, the increased osmolarity of the medium would offset any tendency of the cell to swell. Since astrocytic swelling is a common pathological response of the CNS to a variety of insults (Kimelberg and Ransom, 1986), the mechanism of this high $[K^+]$ model of cytotoxic edema (Klatzo *et al.*, 1984) was of considerable interest. The fact that a portion of the swelling in brain slices was in fact dependent on the presence of HCO_3^-/CO_2 buffer, and this same portion was specifically inhibited by anion exchange inhibitors such as SITS, strongly implicated the anion exchange system in this process (Bourke *et al.*, 1975, 1983; Kimelberg, 1979).

These processes were also studied in primary astrocyte cultures. Primary cultures are generally chosen in preference to studies on transformed glial cell lines since there is always uncertainty as to how closely the properties of such lines resemble those of normal astrocytes *in situ*. Thus, cells of the C_6 rat glioma line have average membrane potentials of -36 mV (Kukes *et al.*, 1976), which are quite different from the -70 to -90 mV potentials reported for astroglia *in situ* (Orkand, 1977). In contrast, the primary astroglial cultures we and others have used show many of the characteristics of normal astroglia. These include staining for the astrocyte-specific marker glial fibril-lary acidic protein (GFAP) comprising the intracellular intermediate filaments, and such GFAP-positive cells constitute at least 90% of the cells present in the primary cultures (Kimelberg, 1983). The cells have average membrane potentials of -70 mV and show a near-Nernstian response to varying $[K^+]_o$ similar to the properties of characterized mammalian astroglia *in situ*. Their morphology resembles that of astro-glia *in situ* when induced to form processes by treatment with dibutyryl cAMP or NE. Also, they contain astrocyte-specific enzymes such as glutamine synthetase (Haller-meyer *et al.*, 1981). There is currently a dramatic increase in work using primary astrocyte cultures which is showing similarities between these cultures and the proper-ties of astrocytes *in situ*. I have not chosen to list these references here since the list would always be incomplete and would cover a lot of work unrelated to ion transport studies. However, all this work tends to support the use of primary cultures as models for astrocyte function *in situ* (also see Federoff and Vernadakis, 1986).

Such primary astrocyte cultures show SITS-sensitive Cl^- transport and saturation kinetics for Cl^- with an apparent $K_{0.5}$ for uptake of ^{36}Cl of 36 mM Cl_o^- and the efflux of ^{36}Cl from preloaded cells is stimulated by the presence of extracellular Cl^- or HCO_3^- (Kimelberg, 1981). Similar properties have also been described for a glioma cell line (LRM 55) (Wolpaw and Martin, 1984). As noted in Section 3, $[Cl^-]_i$ in primary astrocyte cultures is severalfold higher than expected from electrochemical equilibrium. The mechanisms for active transport of Cl^- in cultured astrocytes seem to principally involve the Na^+,K^+,Cl^- cotransport system, but there is some effect due to the involvement of the anion exchange system (see Section 5.2). SITS, an inhibitor of the anion exchange system, inhibits the initial rate of uptake of Cl^- (Fig. 3), and to a varying extent, decreases the steady-state concentration of Cl^- (Kimelberg *et al.*, 1979a). Reduction of temperature also leads to reduction in the rate of uptake and the final steady-state levels of ^{36}Cl (Kimelberg *et al.*, 1978c), as also shown in Fig. 3. It can be seen that at 20°C, 1 mM SITS had only a minimal effect on uptake of ^{36}Cl compared to its effect at 37°C, and the control uptake of ^{36}Cl at 20°C was very similar to the uptake seen at 37°C in the presence of 1 mM SITS. Thus, as in red blood cells (Lowe and Lambert, 1983), the anion exchange system in astrocytes is very tem-perature-sensitive. Furosemide has a greater inhibitory effect than SITS on ^{36}Cl uptake by primary astrocyte cultures (Kimelberg and Frangakis, 1985) and the relative contri-bution of the Na^+,K^+,Cl^- cotransport and the anion exchange systems to overall Cl^- transport in astrocytes will be discussed in the following section. Based on an intra-

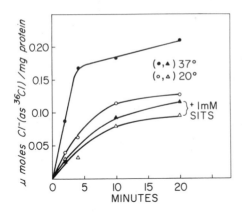

Figure 3. Effect of SITS and temperature on ^{36}Cl uptake in primary astrocyte cultures. Uptake was measured in cells growing on plastic tissue culture petri dishes in the same solution as in Fig. 1 except no HCO_3^- was present. SITS, at a final concentration of 1 mM, was added 5 min before addition of ^{36}Cl (3 μCi). After the times indicated, the cells were rapidly washed and aliquots taken for scintillation counting and protein determinations. (From Kimelberg *et al.*, 1978c, reproduced with permission.)

cellular volume of 4 μl/mg protein, the approximate maximum value of 200 nmoles Cl^-/mg protein seen in Fig. 3 gives an $[Cl^-]_i$ of 50 mM, close to the values of 31 to 43 mM previously reported (Kimelberg, 1981; see also Section 3). Such high intracellular levels of Cl^- may be a means of reducing intracellular acidity by promoting net Cl^- efflux for HCO_3^- influx on the exchange carrier, and addition of a large dose of furosemide (5 mM) does result in internal acidification in cultured astrocytes (Kimelberg *et al.*, 1982a). *In situ*, the exchange carrier could also function in the reverse direction under conditions such as increased extracellular acidity, with uptake of Cl^- and efflux of HCO_3^- to control pH_o (Kraig *et al.*, 1985; Kimelberg *et al.*, 1986). There is evidence that the anion exchange system is present in the CNS *in situ*, although its localization to glia has not been demonstrated (Ahmad and Loeschcke, 1983). Indeed, an exclusive localization of this ubiquitous transport system to astrocytes or glia seems unlikely, and there is clear evidence for SITS-sensitive and HCO_3^--dependent restoration of pH_i after an acid load in both invertebrate and lamprey neurons (Russell and Boron, 1976; Thomas, 1977; Chesler, 1987).

5.2. Anion Cotransport

The other major electrically neutral, carrier-mediated anion transport systems seen in eukaryotic cells are the cation-coupled Cl^- cotransport systems. The most intensively studied and characterized of these systems is that catalyzing the coupled uptake of K^+, Na^+, and Cl^- often, but not always, in the ration of 1 : 1 : 2. A major site where this system is seen in mammals is in the thick ascending loop of Henle (TAL) in the kidney, where it is responsible for active reabsorption of Cl^- (Greger, 1985). It appears it can either function in a number of other modes or there are separate systems catalyzing different cotransport stoichiometries such as $Na^+ + Cl^-$ and $K^+ + Cl^-$ cotransport (Warnock *et al.*, 1983). These systems are inhibited by a variety of pharmacological agents, some of which are well known as loop diuretics. Some of these inhibitors are quite specific, such as bumetanide, while a common inhibitor of the $Na^+,K^+,2Cl^-$ system, furosemide, also inhibits the anion exchanger (Brazy and Gunn, 1976). However, the fact that furosemide inhibits the uptake of K^+ as well as Cl^- in various glial cultures supports the postulate of the existence of a cotransport as well as the anion exchange system in primary cultures of rat (Kimelberg and Frangakis, 1985) and mouse astrocytes (Walz and Hertz, 1984), and a rat glioma cell line (C_6) (Johnson *et al.*, 1982). Bumetanide also inhibits K^+ and Cl^- uptake in primary cultures of rat astrocytes (Kimelberg and Frangakis, 1985), and furosemide inhibits high $[K^+]_o$-induced swelling in cat brain slices (Bourke *et al.*, 1983). The Tamm–Horsfall protein, which is thought to represent or to be associated with the K^+,Na^+,$2Cl^-$ cotransport system in kidney (Greven *et al.*, 1984), has been localized to astrocytes (radial Bergmann glial fibers) in rat cerebellum by immunocytochemistry (Zalc *et al.*, 1984), showing that the existence of this transport system in astrocyte cultures reflects its localization *in situ*. This localization is illustrated in Fig. 4 where it can also be seen that it colocalized with sulfogalactosylceramide, which has been implicated in the functioning of the Na^+,K^+-ATPase (Karlsson *et al.*, 1971).

In transporting epithelia, the cotransport system functioning in the $Na^+ + K^+ + 2 Cl^-$ mode has a major role in transepithelial transport of Cl^-, since by actively accumulating this ion, say from the luminal compartment because of specific localization of the transport system on the lumen-facing membrane, it can raise $[Cl^-]_i$ above its electrochemical equilibrium. Receptor-mediated opening of Cl^- channels on the contraluminal face can then lead to Cl^- efflux down its electrochemical gradient, with Na^+ being pumped out of the cell by the Na^+ pump and K^+ diffusing out through conductance channels (Frizzell *et al.*, 1979, 1986; Greger, 1985). In nonepithelial cells the anion cotransport systems have a major role in volume regulation through net uptake of K^+, Na^+, and Cl^- in shrunken cells, or efflux of KCl in swollen cells (Grinstein *et al.*, 1984; Hoffman, 1985, 1987). In the uptake mode the cotransport mechanism is a very efficient multiplier system since with the Na^+ pump operating in a $3 Na^+$ ₒᵤₜ for $2 K^+$ ᵢₙ exchange per ATP hydrolyzed, the two systems working in parallel can lead to the uptake of $3 Na^+ + 3 K^+ + 6 Cl^-$ per ATP (assuming a $1 Na^+ : 1 K^+ : 2 Cl^-$ stoichiometry for the cotransporter). Since the Na^+,K^+ pump also accumulates $2 K^+$, 5 KCl can be accumulated per ATP (Geck and Heinz, 1986). The additional Cl^-, or another anion, will leave the cell with the additional $1 Na^+$ pumped out that is not exchanged for K^+. Functioning in the $K^+ + Cl^-$ mode, the cotransport or a related system catalyzes KCl efflux to effect loss of KCl from the cell, such as occurs in volume regulatory decrease when the swollen cell is volume-regulating back to its normal volume. This topic of volume regulation is dealt with in more detail in Section 8.

The kinetic properties of the cotransporter system for different ions are of importance in the brain because during pathological conditions such as ischemia, hypoxia, or spreading depression, $[K^+]_o$ in the nervous system rises to very high levels (30–80 mM; Siesjo, 1984; Hansen, 1985). It appears that in primary astrocyte cultures, as in other cells (Greger, 1985; Stein, 1986), this system has a high affinity for K^+. Work done in the author's laboratory on the kinetics of this system for K^+ using primary astrocyte cultures is shown in Fig. 5 (Kimelberg, 1987). The cotransport system was measured as the bumetanide-sensitive component of ^{86}Rb influx. This system showed saturation kinetics with an apparent $K_{0.5}$ of about 2 mM K^+, for both the mannitol-stimulated and the control influx (see Fig. 5B). The large increase in the rate of uptake

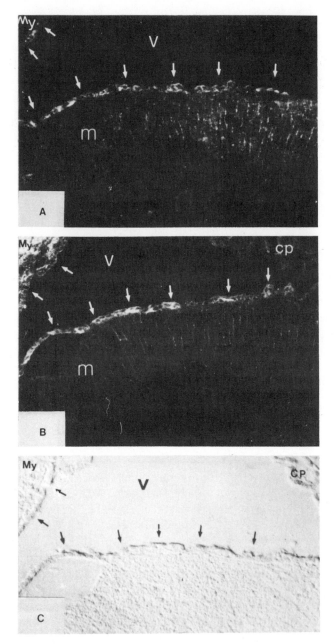

Figure 4. Immunofluorescence localization of (A) Tamm–Horsfall (TH) glycoprotein and (B) the acidic glycosphingolipid sulfogalactosylceramide (SGC) on adjacent sections of rat cerebellum. The field shows a part of the roof of the fourth ventricle. In the molecular layer (m) the astrocytic Bergmann cell fibers are positive for both TH (A) and SGC (B). Ependymal cells lining the roof of the fourth ventricle (V) also contain both TH and SGC. A small part of the brain stem forming the floor of the ventricle can also be seen in the upper left corner, and this myelinated bundle (My) is SGC-positive (B) but TH-negative (A). cp = choroid plexus, which appears to be negative for both TH and SGC. (C) Same field as B as seen with interference contrast optics × 100. (From Zalc *et al.,* 1984, reproduced with permission.)

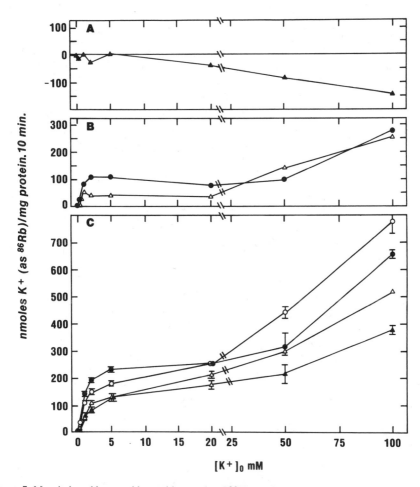

Figure 5. Mannitol- and bumetanide-sensitive uptake of ^{86}Rb as a function of varying $[K^+]_o$. $[Na^+]$ in the medium was reduced proportional to the increase in $[K^+]_o$ to maintain isotonicity. Cells were preincubated for 30 min with the various media indicated ± 200 mM mannitol ± 0.1 mM bumetanide, before being replaced with identical media but containing ^{86}Rb. The cultures, growing in plastic 12-well trays, were then incubated for a further 10 min and then rapidly washed, solubilized in 1 N NaOH, and aliquots taken for scintillation counting and protein determinations. Influx of K^+ was then calculated from the ^{86}Rb taken up and expressed per mg protein. Total uptake is shown in panel C. Panel B shows the bumetanide-sensitive influxes with and without mannitol, while A represents the effect of mannitol on the bumetanide-insensitive fluxes. A, bumetanide-insensitive flux given by (mannitol + bumetanide) − (control + bumetanide): B, bumetanide-sensitive fluxes, filled circles = mannitol − (mannitol + bumetanide), open triangles = control − (control + bumetanide): C, open circles = control; filled circles + mannitol; open triangles control + bumetanide; filled triangles, mannitol + bumetanide. Data shown are means of *n* = 3 wells ± S.E.M., 26-day-old cultures. (From Kimelberg, 1987, reproduced with permission.)

at $[K^+]_o > 20$ mM may be partly due to depolarization-induced opening of voltage-gated K^+ channels, although, as shown in Fig. 5B, a portion of this increase was also sensitive to bumetanide. The amount of mannitol used (200 mM) was sufficient to make the cells shrink to about 50% of their normal volume (Kimelberg and Frangakis, 1985). It is of interest that the bumetanide-sensitive transport system was only preferentially stimulated by such shrinkage at concentrations of $K^+ < 20$ mM, since at higher $[K^+]_o$ the bumetanide-sensitive component was equally stimulated in the presence or absence of mannitol (Fig. 5B). As shown in Fig. 5A the bumetanide-insensitive component actually decreased in the presence of mannitol at $[K^+]_o > 5$ mM; this may reflect inhibitory effects of cell shrinkage on K^+ channels. Figure 6 shows that ^{36}Cl uptake is also stimulated by addition of 200 mM mannitol, and this uptake is inhibited by furosemide. In this respect Cl^- behaves in the same way as ^{86}Rb. These studies were done after 30 min exposure to mannitol. This protocol was used because the stimulation of ion transport by mannitol does not begin until after 5–10 min and is maximal after 20–30 min exposure, suggesting the involvement of a second-messenger system. This is shown for ^{86}Rb in Fig. 7 where 200 mM mannitol and ^{86}Rb were added at time 0 on the abscissa.

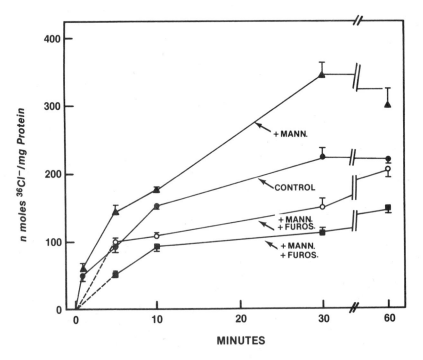

Figure 6. Hypertonic-induced stimulation of ^{36}Cl uptake due to addition of 200 mM mannitol and the effects of furosemide. Cells were preincubated with 1 mM furosemide in HCO_3^--buffered reaction media (see Fig. 2) with or without 200 mM mannitol for 30 min. This was then replaced with 0.5 ml of the identical medium but containing 2 μCi ^{36}Cl. Cultures were 26 days old. Each point represents the mean of 3 wells ± S.E.M. (From Kimelberg, 1987, reproduced with permission.)

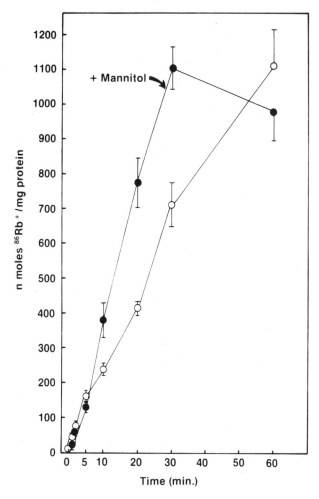

Figure 7. Time course of the uptake of ⁸⁶Rb by primary astrocyte cultures in the presence (filled circles) or absence (open circles) of 200 mM mannitol. Fresh HCO₃⁻ -buffered reaction medium (Fig. 2) containing 0 or 200 mM mannitol and 2 μCi of ⁸⁶Rb was added at 0 time. Twenty-five-day-old cultures, n = mean of 4 wells ± S.E.M.

Shrinkage of the cells upon exposure to hypertonic media has been reported to specifically activate bumetanide-sensitive K⁺ uptake in a variety of cell types, such as avian red blood cells (Kregenow, 1981; Hoffman, 1986) and cultured chick heart cells (Frelin *et al.,* 1986). The same activation in cultured astrocytes implies that they should show regulatory volume increase (RVI) since such net accumulation of Na⁺,K⁺, and 2Cl⁻ should result in uptake of osmotically obligated water. However, using [¹⁴C]3-*O* -methyl-D-glucose (Kletzien *et al.,* 1975) we were not able to detect any volume increase (Kimelberg and Frangakis, 1986), leaving the fate of the Na⁺, K⁺, and Cl⁻ taken up an open question. Perhaps it is lost passively, or the [¹⁴C]3-*O*-methyl-D-glucose method is not sensitive enough to reflect volume changes under these conditions. Stimulation of the rate of bumetanide-sensitive ^{86}Rb uptake by hypertonic conditions is seen in the astrocyte cultures at all concentrations of $K_o^+ < 20$ mM (Fig. 5). In this respect the astrocytes behave differently from duck erythrocytes that have been exposed to hypertonic media, where the cotransport system and RVI require $[K^+]_o$ of > 7 mM or more (Schmidt and McManus, 1977).

It is likely that the cotransport system will be activated at the high $[K^+]_o$ (e.g., 30–80 mM) often encountered in pathological states, such as ischemia (Hansen, 1985; Siesjo, 1984; Kimelberg and Ransom, 1986). At higher $[K^+]_o$ of 50 or 100 mM, there was a similar increase in bumetanide-sensitive influx into the astrocyte cultures under both control and hypertonic conditions and also an increase in bumetanide-insensitive fluxes under both control and hypertonic conditions. However, under hypertonic conditions the increase in bumetanide-insensitive uptake was less relative to the same uptake under isotonic conditions, and thus this component actually decreased in the presence of mannitol at high $[K^+]_o$ as compared to isotonic conditions (see Fig. 5A). The increased bumetanide-insensitive fluxes at $[K^+]_o > 20$ mM are not likely to represent uptake of K^+ by the Na^+ pump since this latter uptake saturates at $[K^+]_o$ of around 10 mM in astrocyte cultures (Kimelberg et al., 1978a,b). However, we did not do these experiments in the presence of ouabain to conclusively rule out any contribution of the Na^+ pump. These fluxes may then in part represent voltage-gated K^+ channels, which have been described in similar cultures as opening at membrane potentials of -40 mV or more positive (Bevan and Raff, 1985; Bevan et al., 1985). At 20 mM $[K^+]_o$ these cells depolarize to -50 to -40 mV (Kimelberg et al., 1979b, 1982a), which could thus lead to the opening of voltage-gated K^+, and also Cl^- channels (Bevan et al., 1985; Nowak et al., 1987), which in turn would lead to influx of KCl and cell swelling. Indeed, we have seen swelling of primary astrocyte cultures at 50 mM $[K^+]_o$ using [^{14}C]urea to measure the intracellular space (see Fig. 4 in Kimelberg and Frangakis, 1986). Such uptake into astrocytes could be an additional mechanism for reducing pathologically high $[K^+]_o$ (Bevan et al., 1985). It is interesting that at high $[K^+]_o$ in the presence of 200 mM mannitol, total ^{86}Rb uptake is lower due to reduced bumetanide-insensitive ^{86}Rb uptake, suggesting that under these conditions some of the K^+ channels have closed, perhaps to prevent loss of K_i^+ and further shrinkage.

Recent work has identified another anion cotransport system in glial cells. This is the electrogenic cotransport system first described by Boron and Boulpaep in salamander kidney cells operating in a Na^+-HCO_3^- ratio of $1:2$ (Boron and Boulpaep, 1985). Schlue and Deitmer (1988) found this system in neuropile glial cells of the leech, but not in the Retzius neurons located in the same ganglia. It was identified in these cells when a change from a HEPES buffered to a CO_2/HCO_3^- buffered media caused an intracellular alkalinization, rather than the expected acidification due to intracellular hydration of freely permeable CO_2. It was also associated with an approximately 5 mV hyperpolarization of the membrane potential, was dependent on medium Na^+, and was blocked by 0.5 mM DIDS. A SITS-sensitive Na^+- and HCO_3^--dependent hyperpolarization has been reported for glial cells from Necturus optic nerve (Astion and Orkand, 1988).

6. TRANSMITTER-INDUCED CHANGES IN Cl⁻ TRANSPORT

6.1. Transmitter-Induced Cl⁻ Conductances

Electrophysiological studies have implicated increased Cl⁻ conductance in the depolarization of primary astrocyte cultures due to the receptor-mediated action of at least two transmitters. Kettenmann and co-workers (Kettenmann and Schachner, 1985; Kettenmann et al., 1987) reported a GABA-induced depolarization that appears to be mediated via a GABA$_A$ receptor since it was inhibited by bicuculline and picrotoxin and mimicked by the GABA$_A$ receptor agonist muscimol. Furthermore, the reversal potential was close to −40 mV, a value that was close to E_{Cl} as determined for the same cells using Cl⁻-specific ISMs. The depolarization required 0.1 to 1 mM GABA, was greater at lower temperatures, and rapidly desensitized. Both desensitization and ED$_{50}$ values for GABA of 30–60 μM have been reported for rat brain GABA receptor responses (Houamed et al., 1984).

Kimelberg et al., (1986) reported that the NE-induced depolarization of primary astrocyte cultures was sensitive to changes in [Cl⁻]$_o$. In Fig. 8 we show that when the astrocyte membrane potential was clamped at around −10 mV by increasing [K⁺]$_o$, the normal depolarizing response to NE was converted to a hyperpolarizing response (Fig. 8, lower trace). However, when [Cl⁻]$_o$ was reduced to 5 mM to change E_{Cl} to around +50 mV, addition of NE again caused a depolarization (Fig. 8, middle trace). This result suggested that the effect of NE was at least partly mediated by a change in Cl⁻ conductance. This NE-induced depolarization had previously been shown to be an α-adrenergic-receptor-mediated response (Hirata et al., 1983), and we have recently shown that it is specifically α$_1$-mediated (Bowman and Kimelberg, 1987). Because [Cl⁻]$_i$ in astrocytes in primary cultures appears to be at a higher concentration than expected from electrochemical equilibrium (see Section 3), the effect of increased Cl⁻ conductance is normally to depolarize the cell, since there will be a net outward movement of Cl⁻, i.e., an inward depolarizing current. Also, neither the anion exchange nor cotransport systems seem to be involved since 0.01 mM bumetanide or 1 mM furosemide had no effect on the NE-induced depolarization (Kimelberg et al., 1986).

6.2. Cl⁻-Dependent Transmitter Uptake

The uptake of several transmitters by cultured glial cells also appears to depend on Cl⁻. For instance, we found that the Na⁺-dependent uptake of serotonin was completely dependent on Cl⁻ (Katz and Kimelberg, 1985), mimicking the behavior found in platelets (Lingjaerde, 1971). It has also recently been reported that in an LRM glioma cell, which has some astrocytic properties, the uptake of L-glutamate is dependent on Cl⁻ in both the presence and absence of Na⁺ (Waniewski and Martin, 1984). We have found that only Na⁺-independent uptake of L-[³H]glutamate in primary astrocyte cultures is actually dependent on medium Cl⁻ (Kimelberg and Pang, 1987). Since this component is only about 5% of the total uptake, with about 95% of the total uptake being Na⁺-dependent, the Cl⁻-dependent component does not seem as though

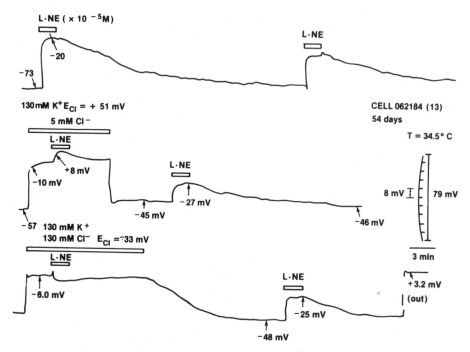

Figure 8. Effect of alterations in $[Cl^-]_o$ at different K^+-clamped values of the membrane potential (E_m) on the $(-)$ NE-induced depolarization of astrocytes. (Top recording) The first and second depolarizing response to 10^{-5} M NE of a "naive" cell, i.e., in a culture not previously exposed to NE. Initial E_m was -73 mV, and this depolarized to -20 mV with a 2-min exposure to 10^{-5} M NE. The second response to 10^{-5} M NE was about one-half that of the first, indicating considerable desensitization. The medium was buffered with 25 mM HEPES and did not contain HCO_3^- (see Fig. 1) with $[K^+]_o = 4.2$ mM. (Middle recording) Subsequent exposure of the same cell to NE when the medium was changed to one with high K^+ composition (130 mM K^+, 14.3 mM Na^+) in which the $[Cl^-]_o$ was reduced to 5 mM (Cl^- replaced with methanesulfonate), i.e., E_{Cl} is made positive, $\sim +50$ mV. Addition of NE still gave a small depolarizing response. When reexposed to normal K^+ medium ($K^+ = 4.5$ mM) with normal $[Cl^-]_o$, the cell rapidly repolarized and showed a normal depolarizing response to addition of 10^{-5} M NE. (Lower recording) When K^+ was increased to 130 mM to again "clamp" the cell at -6 mV and $[Cl^-]_o$ was kept constant at 130 mM, addition of 10^{-5} M NE caused a hyperpolarization. When reexposed to 4.5 mM K^+ medium, the cell slowly repolarized in contrast to the rapid repolarization seen when $[Cl^-]_o$ had been reduced. The calculated Cl^- equilibrium potentials (E_{Cl}) based on a $[Cl^-]_i$ of 40 mM (Kimelberg, 1981; Kettenmann, 1987) are shown when $[K^+]_o$ was increased to 130 mM. (From Kimelberg *et al.*, 1986, reproduced with permission.)

it is likely to normally play a significant role in glutamate uptake. Addition of L-glutamate to primary astrocyte cultures also results in stimulation of ^{36}Cl uptake (Kimelberg and Pang, 1987; Kimelberg *et al.*, 1989a). As discussed above, this does not seem likely to be involved in Na^+-dependent L-[^3H]glutamate uptake and may reflect increased uptake of Cl^- via channels, since we know that L-glutamate and kainic acid depolarize astrocytes in primary culture (Bowman and Kimelberg, 1984; Kettenmann and Schachner, 1985). The Cl^- could be entering via channels that are already open, to accompany Na$^+$, or via Cl$^-$ channels opened by the induced de-polarization (see Section 4).

7. Cl⁻ TRANSPORT AND VOLUME REGULATION

7.1. Cell Volume-Regulating Mechanisms in Other Cells

Many cells are able to regulate their volumes back toward control levels when they have been swollen or shrunken due to exposure to anisotonic media (Kregenow, 1981; Siebens, 1985; Grinstein et al., 1984; Hoffman, 1985, 1986, 1987; Gilles, 1987; Chamberlin and Strange 1989; see Cala, this volume). This phenomenon is termed regulatory volume decrease (RVD) when cells are first swollen due to exposure to hypotonic media or regulatory volume increase (RVI) when cells are first shrunken due to exposure to hypertonic media.

In different cell types, both RVI and RVD have been shown to involve a variety of mechanisms. Thus, RVI can occur by coupled Cl$^-$/HCO$_3^-$ and Na$^+$/H$^+$ exchange to mediate the net uptake of NaCl, or by uptake of NaCl by Na$^+$,Cl$^-$ cotransport. In both cases, Na$^+$ subsequently exchanges for K$^+$ on the Na$^+$ pump to effect net uptake of KCl. Alternatively, RVI can occur by Na$^+$,K$^+$,2Cl$^-$ cotransport, again with subsequent exchange of Na$^+$ for K$^+$. RVD can occur by separate K$^+$ and Cl$^-$ channels, K$^+$,Cl$^-$ cotransport, or coupled K$^+$/H$^+$ and Cl$^-$/HCO$_3^-$ or OH$^-$ exchange (Cala, 1980; Kregenow, 1981; Grinstein et al., 1984; Sarkadi et al., 1984; Hoffman, 1985, 1987; Siebens, 1985). All these volume regulatory systems involve Cl$^-$ in some way. There are a number of inhibitors that can be used to dissect out the contributions of the different systems. Na$^+$,K$^+$,2Cl$^-$ and Na$^+$,Cl$^-$ or K$^+$,Cl$^-$ cotransport are all inhibited by bumetanide or furosemide (Hoffman, 1986; Siebens, 1985), and thus may be variants of a single system or different systems that have a similar site, such as for Cl$^-$, on which the inhibitors bind. All the systems involving Na$^+$ are involved in net uptake of ions for restoring cell volume in shrunken cells, since they can use the inward driving force of the Na$^+$ gradient to drive the uptake of K$^+$ and Cl$^-$. While the other cotransport systems can function in the steady state, it is thought that the Na$^+$,Cl$^-$ uptake system is activated by prior hypotonic swelling and then resuspension of the cells in isotonic medium (Hoffman, 1987; Siebens, 1985; Grinstein et al., 1984). The K$^+$,Cl$^-$ system transports K$^+$ and Cl$^-$ at a 1 : 1 ratio and is sensitive to furosemide and bumetanide. It is thought to be a variant of the Na$^+$,K$^+$,2Cl$^-$ cotransport system. It mediates the net efflux of KCl in RVD. Finally, separate K$^+$ and Cl$^-$ conductance channels have also been shown to be involved in RVD. The K$^+$ channels are thought to be activated by increased influx of Ca^{2+} (Grinstein et al., 1984). Similarly, the Cl$^-$ channels may also be activated by Ca^{2+} (Hoffman, 1986; see Mayer et al., this volume). These processes might be activated via the Ca^{2+} binding protein calmodulin, since the anticalmodulin drug pimozide inhibits the KCl loss induced by swelling, or addition of the Ca^{2+} ionophore A23187 stimulates KCl loss in lymphocytes or Ehrlich ascites tumor cells (Hoffman, 1987). In addition to being stimulated by volume changes, these transport systems can also be affected by transmitters. In some cells, such as avian erythrocytes, the K^+, Na^+, $2\,Cl^-$ system can be activated by catecholamines via increased intracellular cAMP, presumably through cAMP-mediated phosphorylation (Geck and Heinz, 1986).

7.2. Effects of Hypertonic Media on Cultured Astrocytes

As discussed in Section 6, stimulation of the K^+,Na^+,$2Cl^-$ cotransport system in primary astrocyte cultures by exposure of the cells to hypertonic media is a response seemingly designed to effect RVI in shrunken cells. However, when we measured cell volume with [^{14}C]3-O-methyl-D-glucose under the same conditions we found no evidence for restoration of cell volume to control levels, so that the net uptake of ions by this system seems to be insufficient to cause such volume restoration, or else the tracer method was not sensitive enough to measure any effect. This lack of RVI behavior upon shrinkage alone, however, is similar to observation in C_6 glioma cells, rat hepatocytes, human peripheral blood lymphocytes, and Ehrlich ascites tumor cells where recovery of volume after hypertonic shock was not seen over the time course of the experiment and in these studies the total size of the cells was measured directly using a Coulter counter (Hoffman, 1987; Siebens, 1985). In Ehrlich ascites tumor cells and cultured chick cardiac cells, as in primary astrocyte cultures, an up to 20-fold increase in the rate of furosemide-sensitive ^{86}Rb influx was seen when the cells were shrunken in hypertonic medium (Frelin *et al.*, 1986; Geck and Heinz, 1986).

7.3. Effects of Hypotonic Media on Cultured Astrocytes

7.3.1. RVD

When exposed to hypotonic media, primary astrocyte cultures, like the other cell types discussed above, initially swell in proportion to the hypotonicity of the medium (up to three-fold when 100 mM NaCl has been removed) and then show RVD which is complete over a subsequent period of 10 to 30 min. We found no significant effect of 1.0 mM furosemide or 0.01 mM bumetanide on RVD (Kimelberg and Frangakis, 1985; Kimelberg and Goderie, 1988), and if RVD is principally due to loss of KCl this may be because it occurs through separate K^+ and Cl^- conductance channels. Alternatively, RVD may be occurring by loss of organic compounds, e.g., amino acids or amines such as taurine (Gilles, 1987; Hoffman, 1987). Taurine is rapidly released from primary astrocyte cultures when they are swollen in hypotonic medium (Kimelberg *et al.*, 1989b; Morales and Schousboe, 1988); also, taurine has been suggested to affect volume regulation in astrocytes via an effect on CO_2 (Van Gelder, 1983). As mentioned above, when the astrocytes shrink in response to exposure to medium made hypertonic by addition of 200 mM mannitol they do not show RVI but remain at this shrunken volume for at least 90 min (Kimelberg and Frangakis, 1985); this behavior is similar to that seen for many other cells. However, RVI has been reported in Ehrlich ascites tumor cells or lymphocytes after exposure to hypotonic media and then subsequent shrinkage upon reexposure to isotonic media (Grinstein *et al.*, 1984; Hoffman *et al.*,

1983). It has been suggested that this is due to loss of KCl from the cell so that when the cells are reexposed to isotonic media there is an increased driving force for inward transport of KCl on the cotransport uptake system (Geck and Heinz, 1986). We have found the same shrinkage of our cells after exposure to isotonic medium following a previous exposure to hypotonic medium and have also observed a reswelling back to control levels. This reswelling was inhibited by furosemide or bumetanide (Fig. 9). This suggests the involvement of the cation-coupled Cl^- transport system in this RVI response. Amiloride, which inhibits the Na^+/H^+ exchange system in our cells (Kimelberg and Ricard, 1982), had no effect, suggesting that NaCl uptake due to simultaneous functioning of Na^+/H^+ plus Cl^-/HCO_3^- exchangers, as has been reported to occur for RVI in *Amphiuma* red blood cells (Cala, 1980 and Chapter 2, this volume), is not involved.

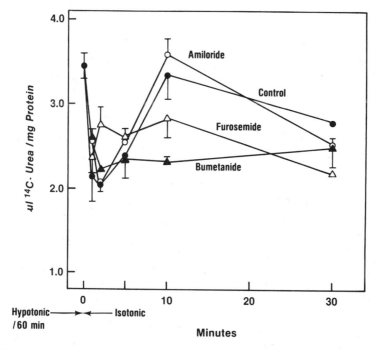

Figure 9. Effects of transport inhibitors on the regulatory volume increase (RVI) due to exposure to isotonic medium after a prior regulatory volume decrease (RVD) cycle for 60 min in hypotonic medium. Volume was measured as the [^{14}C]urea space (expressed as μl/mg protein). The inhibitors were added when the cells were exposed to isotonic medium after the initial 60-min exposure to hypotonic medium. The 0 time point represents the volume at the end of the RVD cycle, just before addition of isotonic medium. [^{14}C]urea was present during the 60-min exposure to hypotonic medium and was present in the isotonic medium at the same radioactivity per milliliter as during the RVD cycle. HCO_3^--buffered medium was used throughout. Media were made hypotonic by omission of 100 mM NaCl. The inhibitor concentrations used were: amiloride and furosemide 1 mM; bumetanide 0.01 mM. $n = 3$ wells ± S.E.M. (From Kimelberg and Frangakis, 1986, reproduced with permission.)

7.3.2. Increased Membrane Permeability

Recently, we have found that when cells were swollen in hypotonic medium, there was increased uptake of [^{14}C] mannitol, which we had been using as a marker for any residual extracellular fluid left by our washing procedure. The amount of [^{14}C]mannitol taken up was proportional to the degree of hypotonicity of the medium (Kimelberg and Goderie, 1988). This suggested the opening of large pores to non-electrolytes, in contrast to the view that during RVD only specific Cl$^-$ or K$^+$ channels are opened. Also there is a marked increase in uptake or release of taurine under hypotonic conditions (Morales and Schousboe, 1988; Kimelberg et al., 1989b). We have also found that swelling of the cells causes an immediate depolarization, followed by a gradual repolarization which corresponds in its time course to RVD. A rapid return to preswelling membrane potentials of -70 to -80 mV occurs when isotonic conditions are restored (Kimelberg and O'Connor, 1988). These data suggest the opening of either cation channels permeable to Na$^+$ or the opening of Cl$^-$ channels, and presumably their progressive closure during RVD. Such ion channels are likely to belong to the class of stretch-activated channels which have now been described in a number of cells as being activated by pressure applied through patch pipettes (Kullberg, 1987; Sachs, 1988), or by swelling the intact cell and measuring currents in the cell-attached patch mode (Christensen, 1987). These channels are usually cation-selective in mammalian cells, but in bacteria they are anion-selective (Kullberg, 1987). They are assumed to contribute to RVD. As mentioned above, RVD is thought to be mediated in some cells by activation of separate conductive K$^+$ and Cl$^-$ channels, in other cells by activation of furosemide- and bumetanide-sensitive K$^+$,Cl$^-$ cotransport, while in still other cells the net efflux of organic compounds such as amino acids occurs. All these processes need to be examined in detail in glial cells, especially astrocytes, since they characteristically swell in a number of pathological states, as discussed in the next section. If channels for ions and transport systems for organic compounds are indeed being opened in astrocytes when they swell in situ, as implied by our in vitro studies described above, these cells are likely to lose not only ions such as Cl$^-$ and K$^+$ which will affect the properties of neighboring neurons, but perhaps other neuroactive compounds such as glutamate, which are normally taken up and concentrated in astrocytes.

8. Cl$^-$ TRANSPORT IN GLIA IN PATHOLOGICAL STATES

In previous sections the point has been made that glial cells are likely to possess many of the Cl$^-$ channels and carriers now known to exist in other cells. While most of this evidence is from studies on glial cultures, there is some direct immunocytochemical proof for the existence of the Na$^+$,K$^+$,2Cl$^-$ cotransport system on astroglial cells in situ (see Fig. 4 and Section 5.1). There is evidence for the presence of Cl$^-$ channels in invertèbrate glial cells and glia in slices of guinea pig olfactory cortex in situ, and considerable evidence for their existence in glial cultures, as discussed in Section 4 (see also Kettenmann, this volume). There are several likely consequences of the activation of Cl⁻ channels on astroglial cells. One of these appears to be a depolarization of the astroglial membrane potential. This depolarization could then lead to activation of voltage-dependent channels, e.g., Ca^{2+}, with numerous secondary effects. One major role for such a Cl⁻ channel in pathological conditions is likely to be to mediate efflux of KCl for restoration of cell volume during RVD, as discussed in the preceding section. In addition, when astrocytes are swollen under conditions of high $[K^+]_o$, i.e., ≥ 20 mM, the opening of both K^+ and Cl⁻ channels will lead to swelling via net uptake of KCl. However, these considerations are all speculative and in only one case has a systematic experimental attempt been made to relate a pathological condition to Cl⁻ transport in astrocytes. This was in studies conducted by Robert S. Bourke, the present author, and colleagues on the relation between the astrocytic swelling seen in head injury, and Cl⁻ transport on the anion exchange system (see below). Administration of an inhibitor of the anion exchange system reduced astrocytic swelling as determined ultrastructurally. It is of particular interest that treatment with this drug also caused a significant decrease in mortality and the impaired neurological behavior associated with experimental head injury (Cragoe, 1987; Cragoe *et al.*, 1982, 1986; Kimelberg *et al.*, 1987).

The head injury model used involves a brief period of rapid acceleration–deceleration forces applied to the intact heads of anesthetized cats. Forty minutes after this, a period of hypoxia lasting 1 hr is imposed on the ventilated animals (Nelson *et al.*, 1982). Forty minutes after the injury a maximum amount of astrocytic swelling was seen, which in the case of the perivascular astrocytes increased their cross-sectional area about two-fold as determined by thin-section electron microscopy (see Table 1 and Barron *et al.*, 1988). Administration of the anion transport inhibitor L-644,711 at 20 min after injury, applied either intravenously, or intracisternally at about a 200-fold lower dose, improved the percent mortality of treated animals over control by up to 40% (Cragoe *et al.*, 1986; Kimelberg *et al.*, 1987). These data are shown in Table 2. The ability of the drug to inhibit perivascular astrocytic swelling occurring 40 min after trauma alone, is shown in the bottom row of Table 1.

We originally suggested that the astroglial swelling induced in cat brain slices was due to the coupled effects of Cl^-/HCO_3^- and Na^+/H^+ exchange transport systems operating in astroglia leading to net uptake of NaCl (Kimelberg, 1979). Such transport systems were found to be present in rat primary astrocyte cultures (Kimelberg *et al.*, 1979a, 1982a). In support of this mechanism, recent studies have shown that modification of indane diuretics to ostensibly nondiuretic congeners of the L-644,711 type which are potent inhibitors of astroglial swelling *in situ* (see above), converts them from inhibitors of cation and Cl⁻ cotransport to effective inhibitors of the Cl^-/HCO_3^- ion exchange system in human erythrocytes (Garay *et al.*, 1986). This model of astroglial swelling was also supported by the studies of Ransom *et al.* (1985) who reported that stimulation of the isolated rat optic nerve caused a decrease in the extracellular space (ECS). Developmental studies suggested that this might be due to activity-dependent swelling of glial cells and this shrinkage of the ECS was abolished in Cl⁻-free medium and reduced when SITS (1 mM) (see Fig. 3) or furosemide (10 mM) was present. Since transmitters will not be released in this purely axonal–glial preparation, the authors suggested that acid–base or CO_2 changes coupled with in-

Table 1. Effects of Experimental Head Injury on Capillary Luminal Area
and Perivascular Astrocytic Area and Glycogen Content[a]

Conditions (No. of cats)[b]	Capillary luminal area (μm^2)	Perivascular neuropil area (μm^2)	Astrocytic area as as % perivascular cytoplasm	No. glycogen granules/100 μm^2 astrocytic area
Control (4)	33.9 ± 12.0[c]	122.2 ± 47.6	11.4 ± 5.8	4.0 ± 2.0
40 min after trauma (4)	29.2 ± 10.4*	112.8 ± 48.7	22.0 ± 12.0*	1.2 ± 1.0*
100 min after trauma (2)	31.3 ± 12.7	119.1 ± 45.7	14.1 ± 4.6	5.5 ± 1.9
100 min after trauma + hypoxia/60 min (2)	31.4 ± 10.7	118.2 ± 46.1	18.1 ± 5.0	2.1 ± 1.4
Hypoxia/60 min only	42.8 ± 12.2*	150.0 ± 43.2	11.9 ± 3.3	3.0 ± 1.5
40 min after trauma + L-644,711 (4)[d]	33.4 ± 10.5	112.3 ± 45.7	10.9 ± 3.2	3.8 ± 2.9

[a]From Barron *et al.* (1988).
[b]See text for description of injury model. Trauma (67 sec) was given 30 min after anesthesia was induced and, when given, hypoxia (6% O_2/94% N_2) was imposed 40 min after trauma for a further 60 min.
[c]The values listed in this table are the means of 50–100 observations (25 per cat) ± S.D.
[d]When L-644,711 was given, cats were injected intravenously 20 min after trauma at a dose of 10 mg/kg (see Table 2).
*Different from control values with $p = 0.01$ or less.

creased transport of Cl^- and HCO_3^- on anion transport systems resulting in swelling of glial cells could be responsible for the ECS shrinkage.

Increased lactic acid is known to be associated with a poor outcome in traumatic brain injury (Plum, 1983). Other workers have shown the existence of Cl^-/HCO_3^- anion exchangers in the CNS and emphasized the importance of these systems in glial cells in the control of brain pH (Ahmad and Loeschcke, 1983; Kraig *et al.*, 1985). However, some resolution is needed of the recent findings of alkalinizing Na^+-HCO_3^- inward cotransport (Schlue and Deitmer, 1988) or neuronal-activity-induced alkaliniza-

Table 2. Effect of L-644,711 on the Mortality of Cats Subjected
to an Acceleration–Deceleration Plus Hypoxia Head Injury[a,b]

Mode of injection	Dose ($\mu g/kg$)	Treated animals		Control animals		Δ%[c]
		Deaths/total	% mortality	Deaths/total	% mortality	
Intracisternal	0.57	7/15	47	9/18	50	3
	57	4/19**	21	17/28	61	40**
Intravenous	1,000	12/25	48	14/25	56	8
	2,500	5/21	24	8/19	42	18
	5,000	1/19*	5	7/22	33	27**
	10,000	6/16**	38	15/19	79	42**

[a]Treated animals were given the dose of drugs shown, either by direct injection into the cisterna magna or intravenously through a cannula into the femoral vein 20 min after trauma and 20 min before hypoxia was imposed. Control animals were injected with vehicle only. The experiments were always run as pairs of drug-treated and control animals and the animal for each category was selected by a coin toss. (Data from Cragoe *et al.*, 1986, and Kimelberg *et al.*, 1987.)
[b]Treated versus control level of significance: *, $p < 0.05$; **, $p < 0.025$.
[c]Δ% = % mortality-control minus % mortality-treated.

tion (Chesler and Kraig, 1987, 1989) in glial cells and studies that have shown that the astrocyte seems to be a very acidic intracellular compartment during cerebral ischemia (Kraig and Nicholson, 1987). It is also of interest that Becker (1985) has reported preliminary studies showing that addition of a base (Tris) improved survival in cats subjected to a severe fluid percussion brain injury, further supporting a critical role for acid–base balance in brain trauma. Prevention of swelling and accumulation of Na^+ within the astrocyte is likely to preserve active uptake of transmitters by these cells which usually depend on an inwardly directed Na^+ gradient (Fonnum *et al.*, 1980). Thus, the excitotoxic effects of glutamate resulting in the death of glutamate-sensitive neurons (Rothman and Olney, 1987) may be partly prevented by its active accumulation within astrocytes, a process that could be inhibited in swollen astrocytes with dissipated Na^+ gradients. In addition, the swelling of astrocytes may open large pores which allow glutamate to leave the cell (see Section 8), also promoting excitotoxic effects (Kimelberg *et al.*, 1989b).

9. CONCLUSIONS

Movement of Cl^- across the plasma membranes of glial cells occurs through channels and carriers which are similar to those encountered for other cell types. Some of these routes for transmembrane Cl^- movements thought to be present in astrocytes are illustrated in Fig. 10. To this we should now add the electrogenic Na^+-HCO_3^- cotransport system. The bumetanide- and furosemide-sensitive $Na^+,K^+,2Cl^-$ cotransport system seems to be responsible for active transport of Cl^- into these cells (Kimelberg and Frangakis, 1985). The SITS-sensitive anion exchanger mediates exchange of Cl^- and HCO_3^- and will equilibrate the intra- to extracellular ratios of these ions in the absence of other competing fluxes. Since the Na^+ chemical gradient provides the driving force for the cotransport system, Na^+ will tend to accumulate inside the cell and will have to be pumped out by the Na^+ pump. We depict the latter pump as operating in the 3 Na^+ out : 2 K^+ mode since we have evidence for this transport system being electrogenic in astrocytes (Bowman and Kimelberg, 1984). Na^+ may also enter on the anion exchanger as $NaCO_3^-$ in exchange for Cl^- (Garay *et al.*, 1986; Boron, 1986), while an electrogenic Na^+ + 2 or 3 HCO_3^- cotransport system (Boron and Boulpaep, 1983; Boron, 1986) has recently been shown to be present in leech neuropile and *Necturus* optic nerve glia (Astion and Orkand, 1988; Schlue and Deitmer, 1988). As discussed in this chapter, there are a number of other systems that may or may not be present in astrocytes. One of these is the furosemide- or bumetanide-sensitive K^+,Cl^- efflux system which normally catalyzes KCl efflux in RVD. However, it appears that the RVD which occurs in glia upon swelling does not involve K^+,Cl^- cotransport because neither furosemide nor bumetanide affects RVD in these cells (Kimelberg and Frangakis, 1985; Kimelberg and Goderie, 1988). The RVI which occurs in astrocytes after exposure to hypotonic and a subsequent exposure to isotonic medium is sensitive to both bumetanide and furosemide and thus it is likely to involve the operation of the $Na^+,K^+,2Cl^-$ cotransport system (see Fig. 9 and Kimelberg and Frangakis, 1985, 1986).

Figure 10 also illustrates the postulated α_1-adrenergic-receptor-mediated opening

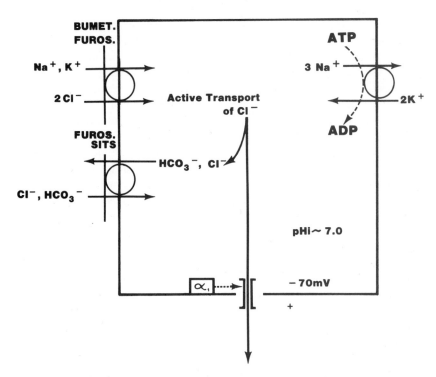

Figure 10. Diagram of different anion transport systems thought to be present in glial cells. See text for discussion. (From Kimelberg *et al.*, 1986, reproduced with permission.)

of Cl^- channels. Since Cl^- is actively accumulated in primary astrocyte cultures and E_{Cl} is more positive than E_m, opening of Cl^- channels will drive E_m toward E_{Cl}, which in these cells is between -30 and -50 mV (normal $E_m = -70$ mV) (Kimelberg *et al.*, 1986; Kettenmann, 1987) resulting in a depolarization. Alternatively, it is quite possible that the depolarization could also be partly due to the opening of chemically gated Na^+ or Ca^{2+} channels or the closure of K^+ channels (Akerman *et al.*, 1988), and this point will have to be clarified in future studies. Kettenmann *et al.* (1987) have shown that there are $GABA_A$ receptors linked to Cl^- channels in primary astrocyte cultures, so that this is another example of receptors being linked to Cl^- channels in astrocytes in primary culture (see also Chapter 6, this volume).

For technical reasons (Kimelberg, 1983), much of the work so far discussed has been done in primary cultures. The question remains as to whether these cells possess the same properties as astrocytes *in situ*. Indeed, as discussed in Section 3, Grafe and Ballanyi (1987) and Ballanyi *et al.* (1987) found a low a^iCl in glial cells in guinea pig olfactory cortex slices using ISMs and concluded that Cl^- is in electrochemical equilibrium. These data are in contrast to the results found for astrocytes in primary cultures, but very similar to the characteristics of oligodendrocytes in primary culture (Kettenmann, 1987). Thus, it will be of interest to precisely identify the type of glial cells impaled in the guinea pig olfactory slice. As far as the electrically neutral Cl^-

transport systems are concerned, Zalc *et al.* (1984) have identified the $Na^+,K^+,2Cl^-$ cotransport system in astrocytic Bergmann glial cells *in situ* by immunocytochemistry (see Fig. 4). Such techniques, combined with *in situ* hybridization to detect synthesis of specific mRNAs, in addition to antibodies to detect the protein product, seem to be a way in which some of these questions on the localization of transport processes in glial cells *in situ* can be resolved. These findings have important implications for the functioning of these cells. The electrically silent Cl^-/HCO_3^- exchange or electrogenic $Na^+-HCO_3^-$ cotransport systems could control intraglial and extracellular pH while the cation-coupled Cl^- cotransport systems can clearly function in volume control, as they appear to do in other cells, and would also remove K^+ and Cl^- from the extracellular space by net uptake. These carrier-mediated anion transport systems could thus cause large changes in the ion content of astrocytes and other glial cells without affecting their membrane potentials, and have up to now been largely neglected in discussions of anion transport in the mammalian CNS (but see Kimelberg and Bourke, 1982). In addition, the Cl^- conductance in these cells may be regulated by membrane voltage and transmitters such as GABA and NE. The opening of K^+ or Cl^- channels during astroglial swelling may contribute to RVD in these cells, and conversely contribute to swelling of these cells under pathological conditions when $[K^+]_o$ reaches levels of 20 mM or greater.

Whether these anion transport systems occur in all glial cells uniformly or are regionally specialized is also an important question. The concept of a generalized glial cell which has one or two basic properties, such as K^+ spatial buffering, has been a reassuringly simplistic view that many neuroscientists still have. In light of the experimental evidence summarized in this chapter and growing evidence of diversity in other glial properties, such as recently reviewed for astrocytes (Federoff and Vernadakis, 1986), it no longer seems possible to hold such a limited view. Specialization of function in terms of different brain regions also seems probable; indeed there is already evidence for it in terms of glial structure, antigenic markers, and glial-specific enzymes. Also, within an individual glial cell, certain regions of the surface membrane may have membrane specializations, such as the assemblies of intramembrane particles preferentially localized to the regions of the plasma membranes of astrocytes facing blood vessels or the pia mater (Landis and Reese, 1981), which may represent, for example, specific ion transport processes (Kimelberg, 1983). Although we are starting to have a fair idea about Cl^- transport in astroglial cells, the nature of Cl^- transport in other glial cells, namely oligodendroglia and microglia in the CNS and Schwann cells in the peripheral nervous system, is less known. Further elucidation of the properties and nature of these Cl^- transport systems and whether the Cl^- transport systems described in this chapter vary for glial cells from different regions of the nervous system or different membrane regions of the same glial cell, and how this is related to glial function are questions that seem worth investigating and will provide research topics for many years to come.

ACKNOWLEDGMENTS. Work from the author's laboratory was supported by grants BNS 8213873 from NSF and NS 23750 from NIH. I thank all the authors and publishers for permission to reproduce published work and Mrs. E. P. Graham for word processing.

REFERENCES

Ahmad, H. R., and Loeschcke, H. H., 1983, Evidence for a carrier mediated exchange diffusion of HCO_3^- against Cl^- at the interphases of the central nervous system, in: *Central Neurone Environment* (M. W. Schlafke, H. P. Loepchen, and W. R. See, eds.), Springer-Verlag, Berlin, pp. 13–21.

Akerman, K. E. O., Enkvist, M. O. K., and Holopainen, I., 1988, Activators of protein kinase C and phenylephrine depolarize the astrocyte membrane by reducing the K^+ permeability, *Neurosci. Lett.* **92:** 265–269.

Astion, M. L., and Orkand, R. K., 1988, Electrogenic Na^+/HCO_3^- cotransport in neuroglia, *Glia* **1:**355–357.

Ballanyi, K., Grafe, P., and Bruggencate, G. T., 1987, Ion activities and potassium uptake mechanisms of glial cells in guinea-pig olfactory cortex slices, *J. Physiol.* **382:**159–174.

Barron, K. D., Dentinger, M. P., Kimelberg, H. K., Nelson, L. R., Bourke, R. S., Keagan, S., Mankes, R., and Cragoe, E. J., Jr., 1988, Ultrastructural features of a brain injury model in cat, *Acta Neuropathol.* **76:**295–307.

Becker, D. P., 1985, Brain acidosis in head injury: A clinical trial, in: *Central Nervous System Trauma—Status Report* (D. P. Becker and J. T. Povlishock, eds.), NINCDS, pp. 229–242.

Bevan, S., and Raff, M., 1985, Voltage-dependent potassium currents in cultured astrocytes, *Nature* **315:** 229–232.

Bevan, S., Chiu, S. Y., Gray, P. T. A., and Ritchie, J. M., 1985, The presence of voltage-gated sodium, potassium and chloride channels in rat cultured astrocytes, *Proc. R. Soc. London Ser. B* **225:**229–313.

Boron, W. F., 1986, Intracellular pH in epithelial cells, *Annu. Rev. Physiol.* **48:**377–388.

Boron, W. F., and Boulpaep, E. L., 1983, Intracellular pH regulation in the renal proximal tubule of the salamander. Basolateral HCO_3^- transport, *J. Gen. Physiol.* **81:**53–94.

Bourke, R. S., Kimelberg, H. K., West, C. R., and Bremer, A. M., 1975, The effect of HCO_3^- on the swelling and ion uptake of monkey cerebral cortex under conditions of raised extracellular potassium, *J. Neurochem.* **25:**323–328.

Bourke, R. S., Kimelberg, H. K., Daze, M., and Church, G., 1983, Swelling and ion uptake in cat cerebrocortical slices: Control by neurotransmitters and ion transport mechanisms, *Neurochem. Res.* **8:** 5–24.

Bowman, C. L., and Kimelberg, H. K., 1984, Excitatory amino acids depolarize rat brain astrocytes in primary culture, *Nature* **311:**656–659.

Bowman, C. L., and Kimelberg, H. K., 1987, Pharmacological properties of the norepinephrine-induced depolarization of astrocytes in primary culture: Evidence for the involvement of an alpha₁-adrenergic receptor, *Brain Res.* **423:**403–407.

Boyle, P. J., and Conway, E. J., 1941, Potassium accumulation in muscle and associated changes, *J. Physiol. (London)* **100:**1–63.

Bracho, H., Orkand, P. M., and Orkand, R. K., 1975, A further study of the fine structure and membrane properties of neuroglia in the optic nerve of *Necturus, J. Neurobiol.* **6:**395–410.

Brazy, P. C., and Gunn, R. B., 1976, Furosemide inhibition of chloride transport in human red blood cells, *J. Gen. Physiol.* **68:**583–599.

Cala, P. M., 1980, Volume regulation by *Amphiuma* red blood cells, *J. Gen. Physiol.* **76:**683–708.

Chamberlin, M. E., and Strange, K., 1989, Anisosmotic cell volume regulation: A comparative view, *Am. J. Physiol.* **257:**C159–C173.

Chesler, M., 1987, pH regulation in the vertebrate central nervous system: Microelectrodes studies in the brain stem of the lamprey, *Can. J. Physiol. Pharmacol.* **65:**986–993.

Chesler, M., and Kraig, R. P., 1987, Intracellular pH of astrocytes increases rapidly with cortical stimulation, *Am. J. Physiol.* **253:**R666–670.

Chesler, M., and Kraig, R. P., 1989, Intracellular pH transients of mammalian astrocytes, *J. Neurosci.* **9:** 2011–2019.

Christensen, O., 1987, Mediation of cell volume regulation by Ca^{2+} influx through stretch-activated channels, *Nature* **330:**66–68.

Cragoe, E. J., Jr., 1987, Drugs for the treatment of traumatic brain injury, *Med. Res. Rev.* **7:**271–305.

Cragoe, E. J., Gould, N. P., Woltersdorf, O. W., Ziegler, C., Bourke, R. S., Nelson, L. R., Kimelberg, H. K., Waldman, J. B., Popp, A. J., and Sedransk, N., 1982, Agents for the treatment of brain injury. 1. (Aryloxy)alkanoic acids, *J. Med. Chem.* **25**:567–579.

Cragoe, E. J., Jr., Woltersdorf, O. W., Jr., Gould, N. P., Pietruszkiewicz, A. M., Ziegler, C., Sakurai, Y., Stokker, G. E., Anderson, P. S., Bourke, R. S., Kimelberg, H. K., Nelson, L. R., Barron, K. D., Rose, J. R., Szarowski, D., Popp, A. J., and Waldman, J. B., 1986, Agents for the treatment of brain edema. 2. [(2,3,9,9a-tetrahydro-3-oxo-9a-substituted-1H-fluoren-7-yl)oxy] alkanoic acids and some of their analogs, *J. Med. Chem.* **29**:825–841.

Fedoroff, S., and Vernadakis, A., eds., 1986, *Astrocytes,* Volumes 1–3, Academic Press, New York.

Fonnum, F., Karlsen, R. L., Malthe-Sorenssen, D., Sterri, S., and Walaas, I., 1980, High affinity transport systems and their role in transmitter action, in: *The Cell Surface and Neuronal Function* (C. W. Cotman, G. Poste, and G. L. Nicolson, eds.), North-Holland, Amsterdam, pp. 455–504.

Franck, G., and Schoffeniels, E., 1972, Cationic composition of rat cerebral cortex slices. Comparative study during development, *J. Neurochem.* **19**:395–402.

Frelin, C., Chassande, O., and Lazdunski, M., 1986, Biochemical characterization of the $Na^+/K^+/Cl^-$ co-transport in chick cardiac cells, *Biochem. Biophys. Res. Commun.* **134**:326–331.

Frizzell, R. A., Field, M., and Schultz, S. G., 1979, Sodium-chloride transport by epithelial tissues, *Am. J. Physiol.* **236**:F1–F8.

Frizzell, R. A., Halm, D. R., Rechkemmer, G., and Shoemaker, R. L., 1986, Chloride channel regulation in secretory epithelia, *Fed. Proc.* **45**:2727–2731.

Garay, R. P., Hannaert, P. R., Nazaret, C., and Cragoe, E. J., Jr., 1986, The significance of the relative effects of loop diuretics and anti-brain edema agents on the Na^+, K^+, Cl^- co-transport system and the $Cl^-/NaCO_3^-$ anion exchanger, *Naunyn-Schmiedeberg's Arch. Pharmacol.* **334**:202–209.

Geck, P., and Heinz, E., 1986, The Na-K-2Cl cotransport system, *J. Membr. Biol.* **91**:97–105.

Gilles, R., 1987, Regulation in cells of euryhaline invertebrates, *Curr. Top. Membr. Transp.* **30**:205–247.

Grafe, P., and Ballanyi, K., 1987, Cellular mechanisms of potassium homeostasis in the mammalian nervous system, *Can. J. Physiol. Pharmacol.* **65**:1038–1042.

Gray, P. T. A., and Ritchie, J. M., 1986, A voltage-gated chloride conductance in rat cultured astrocytes, *Proc. R. Soc. London Ser. B* **228**:267–288.

Greger, R., 1985, Ion transport mechanisms in thick ascending limb of Henle's loop of mammalian nephron, *Physiol. Rev.* **65**:760–797.

Greven, J., Kolling, B., Bronewski-Schwarzer, B. V., Junker, M., Neffgen, B., and Nilius, R. M., 1984, Evidence of a role of Tamm-Horsfall protein in the tubular action of furosemide-like loop diuretics, in: *Diuretics* (J. B. Puschett, ed.), Elsevier, Amsterdam, pp. 203–209.

Grinstein, S., Rothstein, A., Sarkadi, B., and Gelfand, E. W., 1984, Responses of lymphocytes to anisotonic media: Volume-regulating behavior, *Am. J. Physiol.* **246**:C204–C215.

Hallermeyer, K., Harmening, C., and Hamprecht, B., 1981, Cellular localization and regulation of glutamine synthetase in primary cultures of brain cells from newborn mice, *J. Neurochem.* **37**:43–52.

Hansen, A. J., 1985, Effect of anoxia on ion distribution in the brain, *Physiol. Rev.* **65**:101–148.

Hille, B., 1984, *Ionic Channels of Excitable Membranes,* Sinauer Associates, Sunderland, Mass., pp. 1.

Hirata, H., Slater, N. T., and Kimelberg, H. K., 1983, Alpha-adrenergic receptor-mediated depolarization of rat neocortical astrocytes in primary culture, *Brain Res.* **270**:358–362.

Hodgkin, A. L., and Horowicz, P., 1959, The influence of potassium and chloride on the membrane potential of single muscle fibres, *J. Physiol. (London)* **148**:127–160.

Hoffman, E. K., 1985, Role of separate K^+ and Cl^- channels and of Na^+/Cl^- cotransport in volume regulation in Ehrlich cells, *Fed. Proc.* **44**:2513–2519.

Hoffman, E. K., 1986, Anion transport systems in the plasma membrane of vertebrate cells, *Biochim. Biophys. Acta* **864**:1–31.

Hoffman, E. K., 1987, Volume regulation in cultured cells, *Curr. Top. Membr. Transp.* **30**:125–180.

Hoffman, E. K., Sjoholm, C., and Simonsen, L. O., 1983, Na^+, Cl^- cotransport in Ehrlich ascites tumor cells activated during volume regulation (regulatory volume increase), *J. Membr. Biol.* **76**:269–280.

Houamed, K. M., Bilbe, G., Smart, T. G., Constanti, A., Brown, D. A., Barnard, E. A., and Richards, B. M., 1984, Expression of functional GABA, glycine and glutamate receptors in *Xenopus* oocytes injected with rat brain mRNA, *Nature* **310**:318–321.

Johnson, J. H., Dunn, D. P., and Rosenberg, R. N., 1982, Furosemide-sensitive K^+ channel in glioma cells but not neuroblastoma cells in culture, *Biochem. Biophys. Res. Commun.* **109:**100–105.

Karlsson, K. A., Samuelsson, B. E., and Steen, G. O., 1971, Lipid pattern and Na^+-K^+-dependent adenosine triphosphatase activity in the salt gland of duck before and after adaptation in hypertonic saline, *J. Membr. Biol.* **5:**169–184.

Katz, D., and Kimelberg, H. K., 1985, Kinetics and autoradiography of high affinity uptake of serotonin by primary astrocyte cultures, *J. Neurosci.* **5:**1901–1908.

Kay, M. M. B., Tracey, C. M., Goodman, J. R., Cone, J. C., and Bassel, P. S., 1983, Polypeptides immunologically related to band 3 are present in nucleated somatic cells, *Proc. Natl. Acad. Sci. USA* **80:**6882–6886.

Kettenmann, H., 1987, K^+ and Cl^- uptake by cultured oligodendrocytes, *Can. J. Physiol. Pharmacol.* **65:** 1033–1037.

Kettenmann, H., and Schachner, M., 1985, Pharmacological properties of gamma-aminobutyric-acid-, glutamate-, and aspartate-induced depolarizations in cultured astrocytes, *J. Neurosci.* **5:**3295–3301.

Kettenmann, H., Backus, K. H., and Schachner, M., 1987, Gamma-aminobutyric-acid opens Cl^- channels in cultured astrocytes, *Brain Res.* **404:**1–9.

Kimelberg, H. K., 1979, Glial enzymes and ion transport in brain swelling, in: *Neural Trauma* (A. J. Popp, R. S. Bourke, L. R. Nelson, and H. K. Kimelberg, eds.), Raven Press, New York, pp. 137–153.

Kimelberg, H. K., 1981, Active accumulation and exchange transport of chloride in astroglial cells in culture, *Biochim. Biophys. Acta* **646:**179–184.

Kimelberg, H. K., 1983, Primary astrocyte cultures—a key to astrocyte function, *Cell Mol. Neurobiol.* **3:**1–6.

Kimelberg, H. K., 1987, Anisotonic media and glutamate-induced ion transport and volume responses in primary astrocyte culture, *J. Physiol. (London)* **82:**294–303.

Kimelberg, H. K., and Bourke, R. S., 1982, Anion transport in the nervous system, in: *Handbook of Neurochemistry,* 2nd ed. (A. Lajtha, ed.), Plenum Press, New York, pp. 31–67.

Kimelberg, H. K., and Frangakis, M. V., 1985, Furosemide- and bumetanide-sensitive ion transport and volume control in primary astrocyte cultures from rat brain, *Brain Res.* **361:**125–134.

Kimelberg, H. K., and Frangakis, M. V., 1986, Volume regulation in primary astrocyte cultures, *Adv. Biosci.* **61:**177–186.

Kimelberg, H. K., and Goderie, S. K., 1988, Volume regulation after swelling in primary astrocyte cultures, in: *The Biochemical Pathology of Astrocytes* (M. D. Norenberg, A. Schousboe, and L. Hertz, eds.), Liss, New York, pp. 299–311.

Kimelberg, H. K., and O'Connor, E. R., 1988, Swelling of astrocytes causes membrane potential depolarization, *Glia* **1:**219–224.

Kimelberg, H. K., and Pang, S., 1987, Effects of L-glutamate on ion transport processes and swelling in primary astrocyte cultures, *Soc. Neurosci. Abstr.* **13:**195.

Kimelberg, H. K., and Ransom, B. R., 1986, Physiological and pathological aspects of astrocytic swelling, in: *Astrocytes,* Volume 3 (S. Fedoroff and A. Vernadakis, eds.), Academic Press, New York, pp. 129–166.

Kimelberg, H. K., and Ricard, C., 1982, Control of intracellular pH in primary astrocyte cultures by external Na^+, *Trans. Am. Soc. Neurochem.* **13:**112.

Kimelberg, H. K., Narumi, S., Biddlecome, S., and Bourke, R. S., 1978a, $(Na^+ + K^+)$ ATPase, $^{86}Rb^+$ transport and carbonic anhydrase activity in isolated brain cells and cultured astrocytes, in: *Dynamic Properties of Glia Cells* (G. Franck, L. Hertz, E. Schoffeniels, and D. B. Tower, eds.), Pergamon Press, Elmsford, N.Y., pp. 347–357.

Kimelberg, H. K., Biddlecome, S., Narumi, S., and Bourke, R. S., 1978b, ATPase and carbonic anhydrase activities of bulk-isolated neuron, glia and synaptosome fractions from rat brain, *Brain Res.* **141:**305–323.

Kimelberg, H. K., Biddlecome, S., Bourke, R. S., and Bowman, C. L., 1978c, Membrane potential and Cl^- transport properties of primary glial cultures from rat brain, in: *Frontiers of Biological Energetics, I* (P. L. Dutton, J. S. Leigh, and A. Scarpa, eds.), Academic Press, New York, pp. 563–572.

Kimelberg, H. K., Biddlecome, S., and Bourke, R. S., 1979a, SITS-inhibitable Cl^- transport and Na^+-dependent H^+ production in primary astroglial cultures, *Brain Res.* **173:**111–124.

Kimelberg, H. K., Bowman, C. L., Biddlecome, S., and Bourke, R. S., 1979b, Cation transport and membrane potential properties of primary astroglial cultures from neonatal rat brains, *Brain Res.* **177:** 533–550.

Kimelberg, H. K., Bourke, R. S., Stieg, P., Barron, K. D., Hirata, H., Pelton, E. W., and Nelson, L. R., 1982a, Swelling of astroglia after injury to the central nervous system: Mechanisms and consequences, in: *Head Injury: Basic and Clinical Aspects* (R. G. Grossman and P. L. Gildenberg, eds.), Raven Press, New York, pp. 31–44.

Kimelberg, H. K., Hirata, H., Bowman, C., and Mazurkiewicz, J., 1982b, Effects of K⁺, Na⁺ and Cl⁻ on membrane potentials and I–V curves of primary astrocyte cultures, *Soc. Neurosci. Abstr.* **8:**238.

Kimelberg, H. K., Bowman, C. L., and Hirata, H., 1986, Anion transport in astrocytes, *Ann. N.Y. Acad. Sci.* **481:**334–353.

Kimelberg, H. K., Cragoe, E. J., Jr., Nelson, L. R., Popp, A. J., Szarowski, D., Rose, J. W., Woltersdorf, O. W., Jr., and Pietruszkiewicz, A. M., 1987, Improved recovery from a traumatic-hypoxic brain injury in cats by intracisternal injection of an anion transport inhibitor, *Central Nervous System Trauma* **4:**3–14.

Kimelberg, H. K., Pang, S., and Treble, D. H., 1989a, Excitatory amino acid-stimulated uptake of ²²Na⁺ in primary astrocyte cultures, *J. Neurosci.* **9:**1141–1149.

Kimelberg, H. K., Goderie, S., and Waniewski, R., 1989b, Hypoosmotic media-induced release of amino acids from astrocytes, *Soc. Neurosci. Abst.* **15:**353.

Klatzo, I., Suzuki, R., Orzi, F., Schuier, F., and Nitsch, C., 1984, Pathomechanisms of ischemic brain edema, in: *Recent Progress in the Study and Therapy of Brain Edema* (T. G. Go and A. Baethmann, eds.), Plenum Press, New York, pp. 1–10.

Kletzien, R. F., Pariza, M. W., Becker, J. E., and Potter, V. R., 1975, A method using 3-O-methyl-D-glucose and phloretin for the determination of intracellular water space of cells in monolayer cultures, *Anal. Biochem.* **68:**537–544.

Kopito, R. R., Andersson, M., and Lodish, H. F., 1987, Structure and organization of the murine band 3 gene, *J. Biol. Chem.* **262:**8035–8040.

Kraig, R. P., and Nicholson, C., 1987, Profound acidosis of presumed glial during ischemia, in: *Cerebrovascular Diseases* (M. E. Raichle and W. J. Powers, eds.), New York, Raven Press, pp. 97–102.

Kraig, R. P., Pulsinelli, W. A., and Plum, F., 1985, Hydrogen ion buffering during complete brain ischemia, *Brain Res.* **342:**281–290.

Kregenow, F. M., 1981, Osmoregulatory salt transporting mechanisms: Control of cell volume in anisotonic media, *Annu. Rev. Physiol.* **43:**493–505.

Kuffler, S. W., Nicholls, J. G., and Orkand, R. K., 1966, Physiological properties of glial cells in the central nervous system of amphibia, *J. Neurophysiol.* **29:**768–787.

Kukes, G., Elul, R., and de Vellis, J., 1976, The ionic basis of the membrane potential in a rat glial cell line, *Brain Res.* **104:**71–92.

Kullberg, R., 1987, Stretch-activated ion channels in bacteria and animal cell membranes, *Trends Neurosci.* **10:**38–39.

Landis, D. M. D., and Reese, T. S., 1981, Membrane structure in mammalian astrocytes: A review of freeze-fracture studies on adult, developing, reactive and cultured astrocytes, *J. Exp. Biol.* **95:**35–48.

Lingjaerde, O., Jr., 1971, Uptake of serotonin in blood platelets in vitro. I. The effects of chloride, *Acta Physiol. Scand.* **81:**75–83.

Lowe, A. G., and Lambert, A., 1983, Chloride–bicarbonate exchange and related transport processes, *Biochim. Biophys. Acta* **694:**353–374.

Lund-Anderson, H., and Hertz, L., 1970, Effects of potassium and of glutamate on swelling and on sodium and potassium content in brain cortex slices from adult rats, *Exp. Brain Res.* **11:**199–212.

Moller, M., Mollgard, K., Lund-Anderson, H., and Hertz, L., 1974, Concordance between morphological and biochemical estimates of fluid spaces in rat brain cortex slices, *Exp. Brain Res.* **22:**299–314.

Morales, H. P., and Schousboe, A., 1988, Volume regulation in astrocytes: A role for taurine as an osmoeffector, *J. Neurosci. Res.* **29:**505–509.

Nelson, L. R., Auen, E. L., Bourke, R. S., Barron, K. D., Malik, A. B., Cragoe, E. J., Jr., Popp, A. J., Waldman, J. B., Kimelberg, H. K., Foster, V. V., Creel, W., and Schuster, L., 1982, A comparison of animal head injury models developed for treatment modality evaluation, in: *Head Injury: Basic and Clinical Aspects* (R. G. Grossman and P. L. Gildenberg, eds.), Raven Press, New York, pp. 117–127.

Nicholls, J. G., and Kuffler, S. W., 1964, Extracellular space as a pathway for exchange between blood and neurons in the central nervous system of the leech: Ionic composition of glial cells and neurons, *J. Neurophysiol.* **27**:645–671.

Nowak, L., Ascher, P., and Berwald-Netter, Y., 1987, Ionic channels in mouse astrocytes in culture, *J. Neurosci.* **7**:101–109.

Orkand, R. K., 1977, Glial cells, in: *Handbook of Physiology—The Nervous System*, Volume I, Part 2 (E. R. Kandel, ed.), American Physiological Society, Bethesda, pp. 855–875.

Plum, F., 1983, What causes infarction in ischemic brain? *Neurology* **33**:222–233.

Ransom, B. R., Yamate, G. L., and Connors, B. W., 1985, Activity-dependent shrinkage of extracellular space in rat optic nerve: A developmental study, *J. Neurosci.* **5**:532–535.

Rothman, S. M., and Olney, J. M., 1987, Excitotoxicity and the NMDA receptor, *Trends Neurosci.* **10**:299–302.

Russell, J. M., and Boron, W. F., 1976, Role of chloride transport in regulation of intracellular pH, *Nature* **264**:73–74.

Sachs, F., 1988, Mechanical transduction in biological systems. *CRC Crit. Rev. Biomed. Eng.* **16**:141–149.

Sarkadi, B., Attisano, L., Grinstein, S., Buchwald, M., and Rothstein, A., 1984, Volume regulation of Chinese hamster ovary cells in anisoosmotic media, *Biochim. Biophys. Acta* **774**:159–168.

Schlue, W.-R., and Deitmer, J. W., 1988, Ionic mechanisms of intracellular pH regulation in the nervous system, in: *Proton Passage Across Cell Membranes*, CIBA Foundation Symposium 139, Wiley, Chichester, England, pp. 47–69.

Schmidt, W. F., III, and McManus, T. J., 1977, Ouabain-insensitive salt and water movements in duck red cells, *J. Gen. Physiol.* **70**:59–79.

Schousboe, A., 1972, Development of potassium effects on ion concentrations and indicator spaces in rat brain cortex slices during postnatal ontogenesis, *Exp. Brain Res.* **15**:521–531.

Siebens, A. W., 1985, Cellular volume control, in: *The Kidney: Physiology and Pathophysiology* (D. W. Seldin and G. Giebisch, eds.), Raven Press, New York, pp. 91–115.

Siesjo, B. K., 1984, Cerebral circulation and metabolism, *J. Neurosurg.* **60**:883–908.

Smith, Q. R., Johanson, C. E., and Woodbury, D. M., 1981, Uptake of $^{36}Cl^-$ and $^{22}Na^+$ by the brain–cerebrospinal fluid system: Comparison of the permeability of the blood–brain and blood–cerebrospinal fluid barriers, *J. Neurochem.* **37**:117–124.

Sonnhof, U., 1987, Single voltage-dependent K^+ and Cl^- channels in cultured rat astrocytes, *Can. J. Physiol. Pharmacol.* **65**:1043–1050.

Stein, W. D., 1986, *Transport and Diffusion Across Cell Membranes*, Academic Press, New York, pp. 425–474.

Thomas, R. C., 1977, The role of bicarbonate, chloride and sodium ions in the regulation of intracellular pH in snail neurones, *J. Physiol. (London)* **273**:317–338.

Tosteson, D. C., 1981, Cation countertransport and cotransport in human red cells, *Fed. Proc.* **40**:1429–1433.

Tower, D. B., and Bourke, R. S., 1966, Fluid compartmentation and electrolytes of cat cerebral cortex in vitro. III. Ontogenetic and comparative aspects, *J. Neurochem.* **13**:1119–1137.

Van Gelder, N. M., 1983, A central mechanism of action for taurine: Osmoregulation, bivalent cations, and excitation threshold, *Neurochem. Res.* **8**:687–699.

Walz, W., and Hertz, L., 1984, Intense furosemide-sensitive potassium accumulation in the presence of pathologically high extracellular potassium levels, *J. Cerebral Blood Flow Metab.* **4**:301–304.

Walz, W., and Schlue, W. R., 1982, External ions and membrane potential of leech neuropil glial cells, *Brain Res.* **239**:119–138.

Walz, W., Wuttke, W., and Hertz, L., 1984, Astrocytes in primary cultures: Membrane potential characteristics reveal exclusive potassium conductance and potassium accumulator properties, *Brain Res.* **292**:367–374.

Waniewski, R. A., and Martin, D. L., 1984, Characterization of L-glutamic acid transport by glioma cells in culture: Evidence for sodium-independent, chloride-dependent high affinity influx, *J. Neurosci.* **4**:2237–2246.

Warnock, D. G., Greger, R., Dunham, P. B., Benjamin, M. A., Frizzell, R. A., Field, M., Spring, K. R., Ives, H. E., Aronson, P. S., and Seifter, J., 1983, Ion transport processes in apical membrane of epithelia, *Fed. Proc.* **43:**2473–2487.

Wolpaw, E. W., and Martin, D. L., 1984, Cl⁻ transport in a glioma cell line: Evidence for two transport mechanisms, *Brain Res.* **297:**317–327.

Zalc, B., Collet, A., Monge, M., Ollier-Hartman, M. P., Jacque, C., Hartman, L., and Baumann, N. A., 1984, Tamm-Horsfall protein, a kidney marker is expressed on brain sulfogalactosylceramide-positive astroglial structures, *Brain Res.* **291:**182–187.

Chloride Channels and Carriers in Cultured Glial Cells

H. Kettenmann

1. INTRODUCTION

Since their discovery, glial cells have generally been considered to be passive elements with only a few of their properties being recognized as of great importance for the proper functioning of the nervous system, e.g., the formation of myelin by oligo-dendrocytes and Schwann cells (for reviews see Morell and Norton, 1980), the clearance of K^+ from the extracellular space by astrocytes (Orkand, 1977, 1980), the guidance of neurons during development (Rakic, 1981), and the uptake of neurotransmitters (Hertz, 1979). Neurons and glial cells are separated by the extracellular space and communication between these two cell populations requires that signals travel across the space. Release of K^+ into the extracellular space during neuronal activity and the response of glial cells, which take up the excess K^+ to regulate extracellular concentrations, is an example of such a signal between neurons and glial cells (e.g., Salem et al., 1975; Sykova and Orkand, 1980; Walz and Hertz, 1983b). The potassium uptake by glial cells is mediated by passive processes, namely spatial buffering and passive KCl uptake, and/or by stimulation of the Na^+/K^+-ATPase (Kettenmann, 1987a). The efficiency of spatial buffering seems to be determined by the density and distribution of K^+ channels (Newman, 1985a,b, 1986; Orkand, 1977), that of KCl uptake by the relative density of K^+ and Cl^- channels (Ballanyi et al., 1986; Kettenmann, 1987b). Thus, expression of Cl^- channels in glia can play a functional role in K^+ homeostasis. Recent observations indicate that not only K^+ undergoes changes in the extracellular space during neuronal activity, but also Na^+, Ca^{2+}, H^+, and Cl^- (Chesler, 1987; Dietzel et al., 1982; Nicholson, 1980a,b). Thus, glial cells may be involved in controlling the free concentration of other physiologically relevant ions including H^+ and Cl^-. This would not be surprising, since most ion transport systems across cell membranes function as co- or countertransporters. A well-known example is the countertransport of Na^+ and K^+ by the Na^+/K^+-ATPase, which is present in glial cells (Orkand, 1977). Other transport systems present in glial cells include Na^+/H^+ and Cl^-/HCO_3^- exchangers, Na^+/HCO_3^- and K^+/Cl^- cotransporters

H. Kettenmann • Department of Neurobiology, University of Heidelberg, D-6900 Heidelberg, Federal Republic of Germany.

(Hoppe and Kettenmann, 1989a; Kettenmann and Schlue, 1988; Kimelberg *et al.*, 1979). The combined activity of these carriers and the Na^+, K^+, Cl^-, and HCO_3^- channels which can be expressed by glial cells (Bevan *et al.*, 1986; Gray *et al.*, 1986; Kettenmann, 1987b; Astion *et al.*, 1987; Tang *et al.*, 1979) are likely to strongly influence nervous tissue extracellular ionic microenvironment.

This review will summarize present knowledge about Cl^- channels and carriers in glial cells to clarify their role in ionic regulation in the extracellular compartment and will discuss the implications this has for the functional role of glial cells.

2. Cl^- CHANNELS IN GLIAL CELLS

The early electrophysiological characterization of glial cells described their membrane as being exclusively permeable to K^+ (e.g., Kuffler *et al.*, 1966). The possibility of recording from small uncoupled glial cells with the patch-clamp technique uncovered channels permeable to other ions such as Na^+ (Chiu *et al.*, 1984; Shrager *et al.*, 1985) and Cl^- (Bevan *et al.*, 1984). More recent investigations carried out on cultured mammalian astrocytes, oligodendrocytes, and Schwann cells (Fig. 1) have shown the presence of at least three different types of channels permeable to Cl^- in these cells. The first type is in fact an anion-selective channel having a large unitary conductance and is usually found only in isolated membrane patches. The second type are low-conductance Cl^- channels some of which seem to be voltage-gated. The last is activated by the neurotransmitter GABA and is the glial $GABA_A$ receptor-linked Cl^- channel. The following sections (2.1, 2.2, 2.3) briefly review the features of the three types of Cl^- channels in glial cells.

2.1. The High-Conductance Anion-Selective Channel

This channel has been found in rat Schwann cells (Bevan *et al.*, 1984; Gray and Ritchie, 1985), and in cultured astrocytes from rat cerebral cortex (Fig. 2), but not in oligodendrocytes (Bevan *et al.*, 1985, 1986; Nowak *et al.*, 1987; Sonnhof, 1987). The single-channel conductance was between 385 pS (Nowak *et al.*, 1987) and 456 pS (Sonnhof, 1987) under conditions where both sides of the membrane were exposed to 140 mM NaCl. Large anions such as methyl sulfate and isethionate, but not ascorbate or glutamate, could traverse this channel. The channel was also partially permeable to Na^+, implying an imperfect selectivity against cations. The permeability of the channel to Na^+ was 20% of that of Cl^- (Gray *et al.*, 1986). The channel was rarely found in cell-attached patches, but was spontaneously active after isolation of the membrane to form inside-out patches. It has been suggested that as yet unidentified internal factors control the activity of this channel, keeping it in the closed configuration under physiological conditions (Gray *et al.*, 1986). Isolation of the channel from the cell resulted in its activation since the control factors are assumed to be no longer present at the inner side of the membrane. As in other cells in which it has been found to be present (e.g., Schwarze and Kolb, 1984), the activity of the glial anion channel was strongly dependent on membrane potential (Sonnhof, 1987). At potentials close to 0

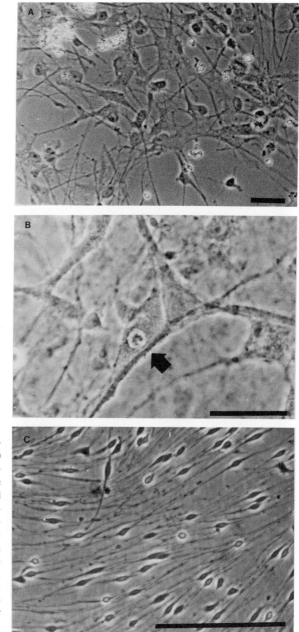

Figure 1. Cultured astrocytes, oligodendrocytes, and Schwann cells. (A) Astrocytes from rat cerebral hemispheres were cultured according to the methods developed by McCarthy and de Vellis (1980) as modified by Keilhauer *et al.* (1985). (B) Oligodendrocyte (arrow) from mouse spinal cord maintained for 4 weeks in culture (Kettenmann *et al.*, 1983). (C) Schwann cells from mouse nervus ischiadicus cultured for 4 days according to Seilheimer and Schachner (1987). Bars = 100 μm. (Courtesy of B. Seilheimer.)

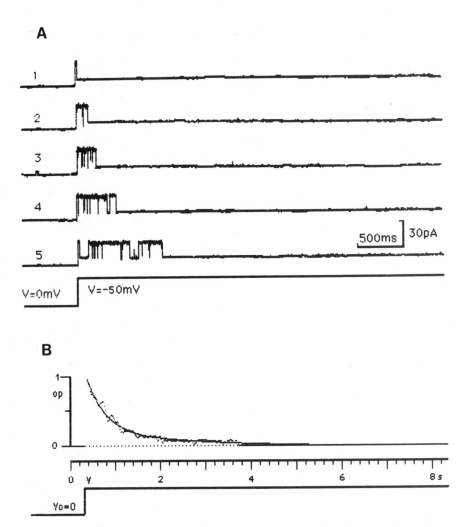

Figure 2. High-conductance anion channel. Current fluctuations of a single anion channel from cultured rat astrocytes recorded with the patch-clamp technique in the inside-out configuration. The potential was repeatedly stepped from 0 to -50 mV. The Cl^- concentration was approximately equal on both sides of the membrane. Channel openings to five voltage steps have been arranged according to their duration. Openings are interrupted by closures of varying durations. After being open for a period that varies stochastically, the channel shuts to an inactive state from which it can no longer reach the open state, unless the potential is reset to zero for some time. (B) Conditional average of the openings of single Cl-channels from a series of 79 voltage steps from 0 to -50 mV. Five of these ensembles are shown in A. The average current is normalized to drive the conditional open probability (op). The open probability is 0.98 immediately after the voltage step and then slowly relaxes to zero. When fitted with an exponential function, the corresponding time constant is 375 msec. (Modified from Sonnhof, 1987.)

mV, the channel has the highest probability of being open (close to 1), while open probability rapidly declines with polarization to either negative or positive potentials (Gray *et al.*, 1986). Anion-selective channels of large unitary conductance have also been identified in cell membranes of cultured rat skeletal muscle (Blatz and Magleby, 1983; see Blatz, this volume), chicken skeletal muscle, FO cells, thymocytes, freshly isolated peritoneal macrophages, Kupffer cells (Schwarze and Kolb, 1984), apical membranes of various epithelial cell lines (Nelson *et al.*, 1984; Kolb *et al.*, 1985), cultured pulmonary alveolar cells (Schneider *et al.*, 1985; Krouse *et al.*, 1986), and rat basophilic leukemia cells (Lindau and Fernandez, 1986).

The physiological role of this channel in glial cells is unknown. It has been suggested that it might represent half a gap junction (Gray *et al.*, 1986; Sonnhof, 1987). Evidence in favor of this hypothesis can be derived from comparison of corresponding single-channel properties. The serial conductance of two associated anion channels would agree with a conductance of about 150 pS as measured for single cell-to-cell channels in gap junctions (Neyton and Trautmann, 1985). The pronounced voltage-dependent kinetics fits well with the behaviour of voltage-gated gap junctions (Schwarze and Kolb, 1984) and a similar weak anion selectivity has been observed for the proposed gap junctional channel in lens cells (Zampighi and Hall, 1985). However, the glial channel is distinctive, compared to gap junctions, for its inability to pass glutamate (Sonnhof, 1987). It is conceivable, however, that the channel undergoes a conformational change when a pair of them form a gap junction. As an alternative to being half a gap junction, the large anion-selective channel could serve to allow the passage of small molecules and metabolites across the cell membrane (Gray *et al.*, 1986).

2.2. Low-Conductance Cl⁻ Channels

The resting membrane of glial cells appears to be permeable not only to K⁺, but to a smaller extent also to Cl⁻ (Kettenmann, 1987b). In addition, astrocytes have a Cl⁻ conductance that is activated by depolarizing potentials. Bevan *et al.* (1985) working in rat cultured astrocytes found in whole-cell patch recordings that when stepping the membrane potential from -70 to $+120$ mV an outward current was evoked that resulted from an inward Cl⁻ movement. The current was partially blocked when Cl⁻ was replaced by glutamate and as shown in Fig. 3, it was completely blocked when replaced by ascorbate (Bevan *et al.*, 1985). In cultured mouse oligodendrocytes, a resting Cl⁻ conductance seems to be present. This is suggested by cell input resistance measurements in Cl⁻-containing solution compared with those in solutions where Cl⁻ was substituted by isethionate (Kettenmann, 1987b). The increase in input resistance by 27% in a low-Cl⁻ solution suggests that part of the resting membrane conductance is Cl⁻ dependent (Fig. 4).

Nowak *et al.* (1987) observed single Cl⁻ channels in outside-out patches from mouse astrocytes in culture. The open probability of these channels increased when the patch was hyperpolarized from -40 to -100 mV; therefore, it is likely that these channels are active at physiological resting potentials. The single-channel conductance was about 5 pS, but the channel also showed a half-conductance substate (Nowak *et*

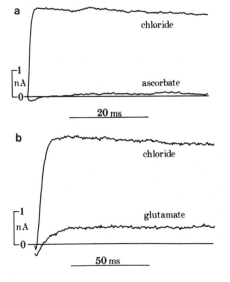

Figure 3. Cl⁻ current recorded from a cultured rat astrocyte. Currents evoked by stepping membrane potential from -70 mV holding potential to $+120$ mV with either Cl⁻ (top trace) or ascorbate (lower trace) as external anion. Pipette contained 150 mM CsCl to block K⁺ currents. (With permission from Bevan *et al.*, 1985.)

al., 1987). U. Sonnhof (personal communication) has found Cl⁻ channels active at the resting membrane potential with a conductance of 30 pS in cultured rat astrocytes. It is reasonable to think that these channels are responsible for the Cl⁻ conductance of the resting cell membrane as described above.

2.3. The Glial GABA$_A$ Receptor-Coupled Cl⁻ Channel

The third type of Cl⁻ channel found in glial cells has been extensively studied in astrocytes (Backus *et al.*, 1988; Bormann and Kettenmann, 1988; Kettenmann *et al.*,

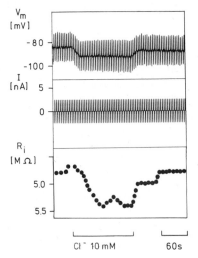

Figure 4. Input resistance of cultured mouse oligodendrocytes in low $[Cl^-]_o$. The input resistance was determined by injection of current pulses (I) with one electrode (middle trace) while recording the resulting change in membrane potential (V_m, upper trace) with a second, separate electrode. Cell input resistance (R_i) was calculated and displayed in the graph at the bottom. $[Cl^-]_o$ was reduced from 126 to 10 mM in the superfusate during time indicated by bar. The apparent hyperpolarization of the membrane potential in low $[Cl^-]_o$ was caused by liquid junction potential of the low-Cl⁻ solution with the reference electrode; a similar polarization could be recorded outside in the bath fluid. (Modified from Kettenmann, 1987b.)

1984a, 1987, 1988a,b; Kettenmann and Schachner, 1985; Fig. 5) and, to a smaller degree, in oligodendrocytes (Gilbert *et al.,* 1984; Kettenmann *et al.,* 1984b). The action of GABA on Schwann cells remains to be tested. The opening of this Cl⁻ channel is triggered by GABA, the major inhibitory neurotransmitter in the central nervous system, and leads to an efflux of Cl⁻ and thus to a depolarization of the cell (Figs. 5 and 7). A decrease in [Cl⁻]$_i$ upon application of GABA has been recorded in cultured oligodendrocytes (Hoppe and Kettenmann, 1989b). The channel is pharmacologically similar to that of the neuronal GABA$_A$ receptor. The depolarizing response elicited by GABA is mimicked by muscimol and antagonized by picrotoxin and bicuculline as in neurons (Kettenmann *et al.,* 1988a,b; Kettenmann and Schachner, 1985). Further similarities to the neuronal GABA$_A$ receptor are the responses of the glial cells to barbiturate and benzodiazepine receptor agonists, namely an amplification of the GABA response. As in neurons, antagonists of the benzodiazepine receptor block the action of benzodiazepine agonists (Backus *et al.,* 1988). The action of "inverse" benzodiazepine receptor agonists which reduce the Cl⁻ current in neurons revealed differences between neurons and astrocytes: at low [GABA] the GABA-induced depolarization is increased in the presence of the inverse agonists DMCM and Ro 22-7497 (Backus *et al.,* 1988; Bormann and Kettenmann, 1988).

The properties of the single Cl⁻ channel triggered by GABA were analyzed using the patch-clamp technique. As in neurons, the single-channel conductance of the main state is 30 pS (Fig. 6), while three different substates can be distinguished, at 13, 21, and 44 pS. The kinetic behavior of the channel is similar to that described for neuronal GABA-activated Cl⁻ channels (Bormann and Kettenmann, 1988). Thus, astrocytes, and probably also oligodendrocytes, express GABA receptors linked to Cl⁻ channels which share many properties with the neuronal GABA$_A$ receptor. The small, but significant pharmacological difference implies that the molecules forming the receptor might not be identical to those found in neurons.

At present, there is no evidence for a function of GABA receptors in glial cells. To understand their functional role, the cellular distribution and subcellular location of GABA receptors on glial cells *in vivo* have to be known. Since astrocytic processes can be in close proximity to synapses, glial GABA receptors could be exposed to GABA released into synaptic clefts. Thus, astrocytes could "sense" the activity of adjacent GABAergic synapses. The consequences of this information transfer from neuron to glia remains speculative. A recent model predicts that the glial GABA Cl⁻ channels might serve to buffer [Cl⁻]$_o$ in the vicinity of GABAergic synapses. This is likely since glial processes can be found in close vicinity to synaptic regions and since activation of glial GABA receptors induces an efflux of Cl⁻ from the glial cell while the activation of neuronal receptors leads, in most cases, to an influx of Cl⁻ into the neuron (Bormann and Kettenmann, 1988). The glial GABA receptor has recently been detected in astrocytes *in situ,* indicating that the expression of the receptor is not a tissue culture artifact (MacVicar *et al.,* 1989). Modulation of many cellular properties, including metabolic and enzymatic activities, receptor expression, channel properties, activity of carrier systems, and intercellular communication, may also be involved. Clearly more needs to be learned about the functions of glial cells, before the implications of a GABA-mediated signal from neuron to glia can be understood.

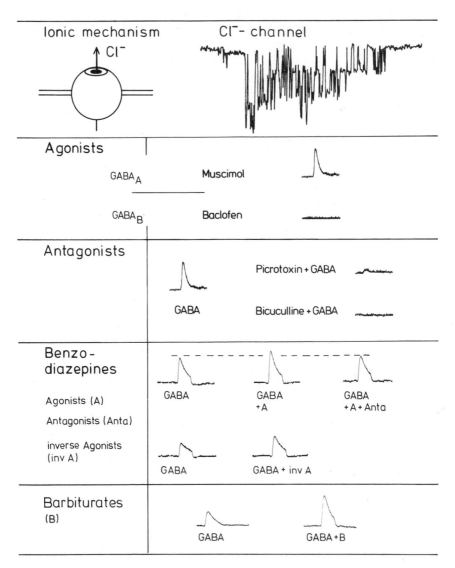

Figure 5. Properties of GABA receptors in cultured rat astrocytes. The scheme summarizes the ionic and pharmacological properties of the GABA receptor in astrocytes. GABA opens Cl^- channels via a receptor that is similar to the neuronal $GABA_A$ receptor. The agonist of the neuronal $GABA_A$ receptor, muscimol, is effective in depolarizing the astrocyte in contrast to the agonist of the neuronal $GABA_B$ receptor, baclofen. The antagonists of the neuronal $GABA_A$ receptor, picrotoxin and bicuculline, block the GABA-induced depolarization. Benzodiazepine receptor agonists (A) enhance the GABA response (compare GABA, GABA + A). The enhancement can be blocked by benzodiazepine receptor antagonists (Anta; compare GABA, GABA + A + Anta). Inverse agonists (inv A) of the benzodiazepine receptor show differences between neurons and glial cells. In contrast to neurons, the GABA response of astrocytes is enhanced and not decreased. The effect of barbiturates (B) is similar as in neurons, namely results in an increase of the GABA response.

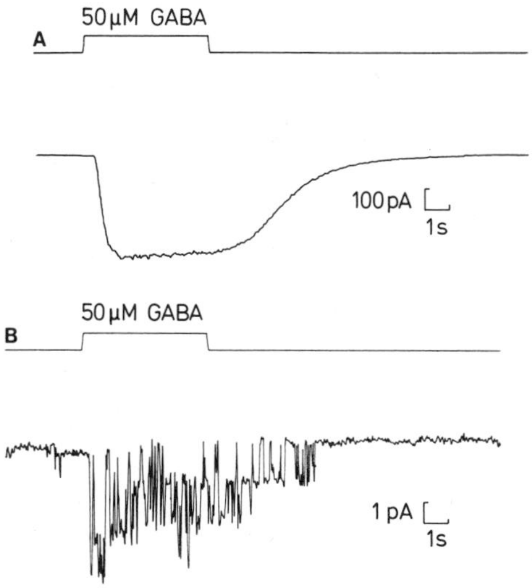

Figure 6. GABA-activated Cl⁻ currents. (A) Membrane currents from a cultured rat astrocyte were recorded using the patch-clamp technique in the whole-cell recording configuration. (B) From the same cell, currents of single Cl⁻ channels were recorded using the patch-clamp technique in the outside-out configuration. Membrane potential was clamped at -70 mV, [Cl⁻] was 120 mM on both sides of the membrane. GABA was applied as indicated by bar. (Modified from Bormann and Kettenmann, 1988.)

3. INTRACELLULAR Cl⁻ ACTIVITY IN GLIAL CELLS

In several glial preparations the levels of Cl_i^- have been determined using ion-selective microelectrodes. Cultured astrocytes from rat cerebral cortex had the highest [Cl⁻]$_i$ (30–50 mM) of all vertebrate glial cells. This corresponds to a Cl⁻ equilibrium potential (E_{Cl}) of -30 to -47 mV (Kettenmann *et al.*, 1987). Since these cells had membrane potentials as negative as -70 to -90 mV, the driving force for Cl⁻ through an open channel could be as high as -60 mV. This finding explains the large depolarizations of up to 50 mV elicited by GABA. The high [Cl⁻]$_i$ in astrocytes, far above electrochemical equilibrium, were originally suggested by Kimelberg (1981) who estimated [Cl⁻]$_i$ from flux measurements to be 40 mM. In cultured oligodendrocytes from mouse spinal cord, [Cl⁻]$_i$ was found to be closer to, but still higher than predicted for passive distribution (Kettenmann, 1987b). Figure 7 shows a penetration of an oligodendrocyte from mouse spinal cord with a Cl⁻-sensitive microelectrode. These cells showed a large heterogeneity in the difference between E_{Cl} and E_m ranging from 0 to 30 mV with a mean of 10 mV (Hoppe and Kettenmann, 1989a). This potential difference in which E_{Cl} was more positive than E_m, accounted for the driving force of the GABA-induced depolarization assuming that, as in astrocytes, GABA selectively opens Cl⁻ channels. Figure 8 illustrates a depolarization of an oligodendrocyte's membrane potential by GABA and the concomitant decrease in [Cl⁻]$_i$ toward the level that would exist if Cl⁻ was passively distributed. It indicates that GABA triggered an outward movement of Cl⁻ and implies that oligodendrocytes express GABA-activated Cl⁻ channels. After the depolarizing response, [Cl⁻]$_i$ is restored within a few minutes (Hoppe and Kettenmann, 1989b).

In contrast to these findings, unidentified glial cells from guinea pig olfactory cortex (Ballanyi *et al.*, 1986) and frog spinal cord (Bührle and Sonnhof, 1983) showed a passive Cl⁻ distribution with E_{Cl} within a few millivolts of the resting membrane

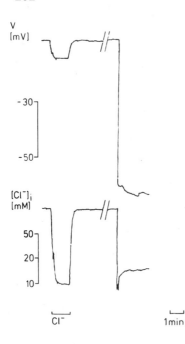

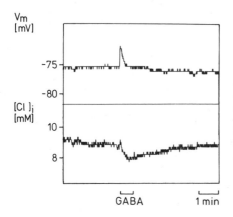

Figure 7. Penetration of an oligodendrocyte with a Cl⁻-sensitive microelectrode. A double-barreled microelectrode (Corning resin 477913) was used to measure $[Cl^-]_i$ of a cultured oligodendrocyte from mouse spinal cord. The electrode was calibrated in the bath by decreasing $[Cl^-]_o$ from 126 to 10 mM (bar at beginning of trace). The change in $[Cl^-]_o$ caused a liquid junction potential at the membrane potential (V) electrode. After impalement of an oligodendrocyte, resting membrane potential stabilized at about −65 mV, $[Cl^-]_i$ at 12 mM. The bathing solution contained 26 mM HCO_3^-. Temperature was between 25 and 30°C. (From Hoppe and Kettenmann, 1989a.)

potential. High values for $[Cl^-]_i$ of 167 mM in squid Schwann cells were reported by Villegas *et al.* (1965). Their measurements were obtained by soaking the tissue in sucrose solution and analyzing changes in $[Cl^-]$. They also reported $[K^+]_i$ of 220 mM and $[Na^+]_i$ of 312 mM, which are quite different from values found in other glial cells.

From the foregoing results it can be concluded that some glial cells, namely cultured astrocytes and oligodendrocytes, must possess active transport mechanism(s) that move Cl^- into the cell while other glial cells may not possess such active transport mechanisms since they have an electrochemically passive Cl^- distribution, or if present, such transport mechanisms would be hidden by a high resting Cl^- permeability (P_{Cl}).

Figure 8. Change in $[Cl^-]_i$ induced by GABA. $[Cl^-]_i$ and membrane potential were measured from an oligodendrocyte with a Cl⁻-sensitive microelectrode. Application of GABA (10^{-3} M) at the time indicated by the bar resulted in a transient depolarization of the membrane potential and a decrease in $[Cl^-]_i$. The delay between the peak of the depolarization and the change in $[Cl^-]_i$ is possibly caused by the time required for Cl^- to diffuse from the membrane to the tip of the electrode. (From Hoppe and Kettenmann, 1989b.)

4. Cl⁻ CARRIERS

Cl⁻ transporters can be studied by analyzing fluxes of radioactively labeled Cl⁻ or by depleting cells of Cl⁻ and measuring Cl⁻ reuptake with ion-selective microelectrodes. Using radioactive tracers, Kimelberg and colleagues have studied Cl⁻ movements in astrocytes and found that Cl⁻ is transported either by exchange for HCO_3^- or in combination with K^+ (Bourke *et al.*, 1977; Kimelberg, 1981; Kimelberg *et al.*, 1979). The uptake of Cl⁻ could be blocked by SITS, furosemide, and bumetanide (Kimelberg and Frangakis, 1985; Kimelberg *et al.*, 1979; for more detail see Kimelberg, this volume).

Cultured oligodendrocytes from mouse spinal cord are well suited for impalement with ion-selective microelectrodes (Kettenmann *et al.*, 1983; Kettenmann, 1988; Fig. 7). Thus, by directly monitoring changes in $[Cl^-]_i$, the dynamics of Cl⁻ carriers can be analyzed. Cl⁻ uptake was studied by experimentally decreasing $[Cl^-]_i$; this was achieved by maintaining cells for 2 to 5 min in low (10 mM) Cl⁻ solutions (Kettenmann, 1987b; Hoppe and Kettenmann, 1989a), which decreased $[Cl^-]_i$ by about 5 mM. Upon return to normal $[Cl^-]_o$, $[Cl^-]_i$ recovered to its resting value. This uptake of Cl⁻ could be blocked by furosemide (Hoppe and Kettenmann, 1989a), which has been reported to interfere with Cl⁻ transport mechanisms. In the presence of furosemide, $[Cl^-]_i$ approached the value expected for passive distribution (Fig. 9). Removal of furosemide restored the resting $[Cl^-]_i$. The blocker of Na^+,K^+,Cl^- cotransport, bumetanide, or application of Na^+-free bathing solution, also inhibited the uptake of Cl⁻. Cl⁻ accumulation is not blocked in HCO_3^--free solution indicating that Cl^-/HCO_3^- exchange is not involved in Cl⁻ uptake by oligodendrocytes. This is substantiated by the finding that SITS and DIDS, which specifically block Cl^-/HCO_3^- exchange, did not interfere with Cl⁻ uptake in oligodendrocytes (Hoppe and Kettenmann, 1989a). It can be concluded that Cl⁻ uptake comes about by the activity of a Na^+,K^+-coupled Cl⁻ cotransport and that such cells do not express a Cl^-/HCO_3^- exchanger (Hoppe and Kettenmann, 1989a).

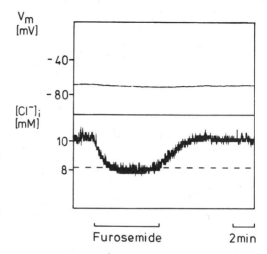

Figure 9. Effect of furosemide on aCl_i^- (intracellular Cl⁻ activity). A double-barreled Cl⁻-selective microelectrode served to record membrane potential (V_m) and aCl_i^- from a cultured oligodendrocyte. Furosemide (10^{-3} M) was applied as indicated by bar. The dashed line is the expected passive aCl_i^- for the recorded membrane potential. (Modified from Hoppe and Kettenmann, 1989s.)

The mechanisms regulating $[Cl^-]_i$ in cultured oligodendrocytes are therefore, in contrast to all other neural cells so far studied, independent from that which regulates pH_i. In oligodendrocytes, two carriers mediate the extrusion of acid, Na^+/H^+ exchange and Na^+,HCO_3^- cotransport (Kettenmann and Schlue, 1988). All neurons so far described express a Cl^-,HCO_3^--coupled exchange system. The strict separation of Cl^- and H^+ regulation in oligodendrocytes suggests a special functional relevance, but elucidation of this awaits further studies.

5. KCl UPTAKE BY GLIAL CELLS

Glial cells respond to an increase of K_o^+ by a combined uptake of K^+ and Cl^-. It is one of the mechanisms by which glial cells contribute to K^+ homeostasis of the extracellular space (Ballanyi *et al.*, 1986; Kettenmann, 1987b).

5.1. Uptake of Cl⁻ Induced by Elevated [K⁺]ₒ in Oligodendrocytes

An increase of $[K^+]_o$ in the extracellular space has been shown to lead to an increase in $[Cl^-]_i$ (Ballanyi *et al.*, 1986; Kettenmann, 1987b). In oligodendrocytes the increase of $[Cl^-]_i$ showed a large variability. In some cells $[Cl^-]_i$ was almost unchanged, in others it increased up to 30 mM, when $[K^+]_o$ was elevated from 5.4 to 50 mM (Kettenmann, 1987b; Hoppe and Kettenmann, 1989a; Fig. 10). This uptake could be mediated either by the activity of a carrier or by passive entry through Cl^- channels. These two mechanisms can be distinguished by comparing the uptake in the presence and absence of furosemide, which blocks the activity of Cl^- carriers. In oligodendrocytes and glial cells in olfactory cortex slices, the uptake of Cl^- by the increase in $[K^+]_o$ was only slightly affected by furosemide, indicating that this uptake is not mediated by Cl^- carriers (Ballanyi *et al.*, 1986; Kettenmann, 1987a).

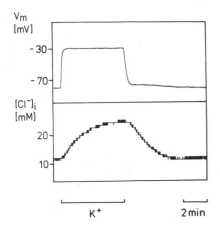

Figure 10. Cl^- uptake in oligodendrocytes induced by elevated $[K^+]_o$. A double-barreled Cl^--sensitive microelectrode served to record membrane potential (V_m) and $[Cl^-]_i$. $[K^+]$ was elevated in the bath from 5.4 to 50 mM as indicated by bar. The membrane potential was depolarized and $[Cl^-]_i$ increased from about 11 to 25 mM.

5.2. Mechanism of KCl Uptake in Oligodendrocytes

The following hypothesis can explain the KCl uptake. Glial cells are depolarized by an increase in $[K^+]_o$. The difference between the depolarized membrane potential and the Cl^- equilibrium potential causes the net entry of Cl^- through Cl^- channels, presumably one of the Cl^- channels open at the resting membrane potential. This leads to the observed increase in $[Cl^-]_i$. The influx of negative charge induces the influx of K^+ into the cell. It follows that the uptake of K^+ is linked to the uptake of Cl^-. The amount of K^+ increase would be determined by the number and conductivity of Cl^- channels present in the membrane. Thus, passive KCl uptake could be efficiently regulated by the density and distribution of K^+ and Cl^- channels.

5.3. Mechanism of KCl Uptake in Astrocytes

Cl^- uptake in astrocytes triggered by an increase in $[K^+]_o$ was less efficient when compared to oligodendrocytes (Kettenmann, 1987b). This could be due to the presence of different mechanisms of uptake. Based on flux measurements, Kimelberg (1981), Walz and Hertz (1983a), and Walz and Hinks (1985) found that astrocytic Cl^- uptake was mainly furosemide sensitive and therefore mainly accomplished by K^+,Cl^- cotransport and not by passive KCl uptake as found for oligodendrocytes (for details see Kimelberg, this volume). The functional significance of these differences in the mechanism of KCl uptake between astrocytes and oligodendrocytes remains to be elucidated.

5.4. Comparison of KCl Uptake in Astrocytes and Oligodendrocytes

Figure 11 schematically summarizes Cl^- carriers and channels thus far characterized in cultured oligodendrocytes and astrocytes. Oligodendrocytes either possess an inwardly directed Cl^- active transport process which is less active than that in astrocytes or alternatively the resting Cl^- permeability is higher than in astrocytes, shunting the oligodendrocytic Cl^- pump. The rapid Cl^- movement induced by K^+ in oligodendrocytes as compared to astrocytes favors the latter hypothesis (Kettenmann,

ASTROCYTE   OLIGODENDROCYTE

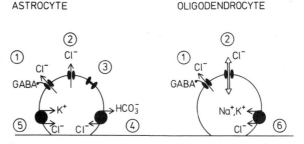

Figure 11. Cl^- channels and carriers in cultured astrocytes and oligodendrocytes. Astrocytes (left scheme) possess GABA-activated Cl^- channels (1), resting Cl^- channels (2), and high-conductive anion channels (3). The carriers responsible for the high $[Cl^-]_i$ consist of Cl^-/HCO_3^- exchange and K^+,Cl^- cotransport. Oligodendrocytes (right scheme) express Cl^- resting channels (2) and a carrier cotransporting Cl^- with K^+ and/or Na^+ (6). A subpopulation of oligodendrocytes express GABA-activated Cl^- channels (1).

1987b). In oligodendrocytes, $[Cl^-]_i$ can rapidly change through passive movements resulting in a larger capacity of these cells to take up KCl. In astrocytes, KCl uptake is mainly restricted to the activation of a K^+, Cl^- cotransport that results in a significantly slower rate of KCl uptake.

These different properties could be of importance for the mechanisms of K^+ uptake in the two glial cell populations. Oligodendrocytes are only weakly electrically coupled as compared to astrocytes and passive K^+ regulation via spatial buffer currents within the syncytium is restricted to astrocytes (Kettenmann and Ransom, 1988). In contrast, oligodendrocytes can control $[K^+]_o$ within the borders of a single cell by passive KCl uptake and their membrane properties are specialized for this task.

ACKNOWLEDGMENTS. I thank Dr. M. Schachner for discussion, help, and comments on the manuscript, K. H. Backus, D. Hoppe, B. R. Ransom, and H. Sontheimer for discussion, and J. Magin and J. Maier for technical assistance. This work was supported by Bundesministerium für Forschung und Technologie and DFG (SFB 317).

REFERENCES

Astion, M. L., Coles, J. A., and Orkand, R. K., 1987, Effects of bicarbonate on glial membrane potential in Necturus optic nerve, *Neurosci. Lett.* **76:**47–52.

Backus, K. H., Kettenmann, H., and Schachner, M., 1988, Effect of benzodiazepines and pentobarbital on the GABA-induced depolarization in cultured astrocytes, *Glia* **1:**132–140.

Ballanyi, K., Grafe, P., and ten Bruggencate, G., 1986, Ion activities and potassium uptake mechanisms of glial cells in guinea-pig olfactory cortex slices, *J. Physiol. (London)* **382:**159–174.

Bevan, S., Gray, P. T. A., and Ritchie, J. M., 1984, A high conductance anion-selective channel in rat Schwann cells, *J. Physiol. (London)* **348:**18P.

Bevan, S., Chiu, S. Y., Gray, P. T. A., and Ritchie, J. M., 1985, The presence of voltage-gated sodium, potassium and chloride channels in rat cultured astrocytes, *Proc. R. Soc. London Ser. B* **225:**299–313.

Bevan, S., Chiu, S. Y., Gray, P. T. A., and Ritchie, J. M., 1986, Voltage-gated ion channels in rat cultured astrocytes, in: *Ion Channels in Neural Membranes,* (R. J. Ritchie and R. D. Keynes, eds.), Liss, New York, pp. 159–174.

Blatz, A. L., and Magleby, K. L., 1983, Single voltage-dependent chloride-sensitive channels of large conductance in cultured rat muscle, *Biophys. J.* **43:**237–241.

Bormann, J., and Kettenmann, H., 1988, Patch clamp study of GABA receptor Cl^- channels in rat cultured astrocytes, *Proc. Natl. Acad. Sci. USA* **85:**9336–9340.

Bourke, R. S., Daze, M. A., and Kimelberg, H. K., 1977, Chloride transport in mammalian astroglia, in: *Dynamic Properties of Glial Cells* (E. Schoffeniels, G. Franck, L. Hertz, and D. B. Tower, eds.), Pergamon Press, Elmsford, N.Y., pp. 337–346.

Bührle, C. P., and Sonnhof, U., 1983, Intracellular ion activities and equilibrium potential in motoneurones and glial cells of the frog spinal cord, *Pfluegers Arch.* **396:**144–153.

Chesler, M., 1987, pH regulation in the vertebrate central nervous system: Microelectrode studies in the brain stem of the lamprey, *Can. J. Physiol. Pharmacol.* **65:**986–993.

Chiu, S. Y., Shrager, P., and Ritchie, J. M., 1984, Neuronal-type Na^+ and K^+ channels in rabbit cultured Schwann cells, *Nature* **311:**156–157.

Dietzel, I., Heinemann, U., Hofmeier, G., and Lux, H. D., 1982, Stimulus induced changes in extracellular Na^+- and Cl^--concentration in relation to changes in the size of the extracellular space, *Exp. Brain Res.* **40:**432–439.

Gilbert, P., Kettenmann, H., and Schachner, M., 1984, Gamma-aminobutyric acid directly depolarizes cultured oligodendrocytes, *J. Neurosci.* **4:**561–569.

Gray, P. T. A., and Ritchie, J. M., 1985, Ion channels in Schwann and glial cells, *Trends Neurosci.* **8:**411–415.

Gray, P. T. A., Bevan, S., Chiu, S. Y., Shrager, P., and Ritchie, J. M., 1986, Ionic conductances in mammalian Schwann cells, in: *Ion Channels in Neural Membranes,* (R. J. Ritchie and R. D. Keynes, eds.), Liss, New York, pp. 145–157.

Hertz, L., 1979, Functional interactions between neurons and astrocytes. I. Turnover and metabolism of putative amino acid transmitters, *Prog. Neurobiol.* **13:**277–323.

Hoppe, D., and Kettenmann, H., 1989a, Carrier-mediated Cl⁻ transport in cultured oligodendrocytes, *J. Neurosci. Res.* **22:**274–282.

Hoppe, D., and Kettenmann, H., 1989b, GABA triggers a Cl⁻ efflux from cultured oligodendrocytes, *Neurosci. Lett.* **97:**334–339.

Keilhauer, G., Meier, D. H., Kuhlmann-Krieg, S., Nieke, J., and Schachner, M., 1985, Astrocytes support incomplete differentiation of an oligodendrocyte precursor cell, *EMBO J* **44:**2499–2504.

Kettenmann, H., 1987a, Oligodendrocytes control extracellular potassium by active uptake and spatial buffering, in: *Dynamic Properties of Glia Cells II* (T. Grisar, G. Franck, L. Hertz, W. T. Norton, M. Sensenbrenner, and D. M. Woodbury, eds.), Pergamon Press, Elmsford, N.Y., pp. 155–164.

Kettenmann, H., 1987b, K⁺ and Cl⁻ uptake by cultured glial cells, *Can. J. Physiol. Pharmacol.* **65:**1033–1037.

Kettenmann, H., 1988, Electrophysiological methods applied in nervous system cultures, in: *Neuromethods,* Volume 10, *Neurochemistry VIII* (A. A. Boulton, G. B. Baker, and W. Walz, eds.), Humana Press, Clifton, N.J., pp. 493–544.

Kettenmann, H., and Ranson, R. R., 1988, Electrical coupling between astrocytes and between oligodendrocytes in mammalian cell cultures, *Glia* **1:**64–73.

Kettenmann, H., and Schachner, M., 1985, Pharmacological properties of GABA, glutamate and aspartate induced depolarizations in cultured astrocytes, *J. Neurosci.* **5:**3295–3301.

Kettenmann, H., and Schlue, W. R., 1988, pH regulation in cultured oligodendrocytes, *J. Physiol. (London)* **406:**147–162.

Kettenmann, H., Sonnhof, U., and Schachner, M., 1983, Exclusive potassium dependence of the membrane potential in cultured mouse oligodendrocytes, *J. Neurosci.* **3:**500–505.

Kettenmann, H., Backus, K. H., and Schachner, M., 1984a, Aspartate, glutamate and gamma-aminobutyric acid depolarize cultured astrocytes, *Neurosci. Lett.* **52:**25–29.

Kettenmann, H., Gilbert, P., and Schachner, M., 1984b, Depolarization of cultured oligodendrocytes by glutamate and GABA, *Neurosci. Lett.* **47:**271–276.

Kettenmann, H., Backus, K. H., and Schachner, M., 1987, GABA opens Cl⁻ channels in cultured astrocytes, *Brain Res.,* **404:**1–9.

Kettenmann, H., Backus, K. H., and Schachner, M., 1988a, GABA receptors on cultured astrocytes, in: *Glial Cell Receptors* (H. K. Kimelberg, ed.), Raven Press, New York, pp. 95–106.

Kettenmann, H., Backus, K. H., and Schachner, M., 1988b, Glial GABA receptors, in: *Biochemical Pathology of Astrocytes* (M. Norenberg, L. Hertz, and A. Schousboe, eds.), Liss, New York, pp. 587–598.

Kimelberg, H. K., 1981, Active accumulation and exchange transport of chloride in astroglial cells in culture, *Biochim. Biophys. Acta* **464:**179–184.

Kimelberg, H. K., and Frangakis, M. V., 1985, Furosemide and bumetanide sensitive ion transport and volume control in primary astrocyte cultures from rat brain, *Brain Res.* **361:**125–134.

Kimelberg, H. K., Biddlecome, S., and Bourke, R. S., 1979, SITS-inhibitable Cl⁻ transport and Na⁺-dependent H⁺ production in primary astroglial cultures, *Brain Res.* **173:**111–124.

Kolb, H. A., Brown, C. D. A., and Murer, H., 1985, Identification of a voltage-dependent anion channel in the apical membrane of a Cl⁻ secretory epithelium (MDCK), *Pfluegers Arch.* **403:**262–265.

Krouse, M. E., Schneider, G. T., and Gage, P. W., 1986, A large anion-selective channel has seven conductance levels, *Nature* **319:**58–60.

Kuffler, S. W., Nicholls, J. G., and Orkand, R. K., 1966, Physiological properties of glial cells in the central nervous system of amphibia, *J. Neurophysiol.* **29:**768–787.

Lindau, M., and Fernandez, J. M., 1986, A patch-clamp study of histamine-secreting cells, *J. Gen. Physiol.* **88:**349–368.

McCarthy, K. D., and de Vellis, J., 1980, Preparation of separate astroglial and oligodendroglial cell cultures from rat cerebral tissue, *J. Cell Biol.* **85:**890–902.

MacVicar, B. A., Tse, F. W. Y., Crackton, S. E., and Kettenmann, H., 1989, GABA activated Cl⁻ channels in astrocytes of hippocampal slices, *J. Neurosci.* **9:**3577–3583.

Morell, P., and Norton, W. T., 1980, Myelin, *Sci. Am.* **242:**74–89.

Nelson, D. J., Tang, J. M., and Palmer, L. G., 1984, Single channel recordings of apical membrane chloride conductance in A6 epithelial cells, *J. Membr. Biol.* **80:**81–89.

Newman, E. A., 1985a, Regulation of potassium levels by glial cells in the retina, *Trends Neurosci.* **8:**156–157.

Newman, E. A., 1985b, Membrane physiology of retinal glial (Muller) cells, *J. Neurosci.* **5:**2225–2239.

Newman, E. A., 1986, High potassium conductance in astrocyte endfeet, *Science* **233:**453–454.

Neyton, J., and Trautmann, A., 1985, Single channel currents of an intracellular junction, *Nature* **317:**331–335.

Nicholson, C., 1980a, Measurement of extracellular ions in the brain, *Trends Neurosci.* **3:**216–218.

Nicholson, C., 1980b, Dynamics of the brain cell microenvironment, *Neurosci. Res. Progr. Bull.* **18:**180–322.

Nowak, L., Ascher, P., and Berwald-Netter, Y., 1987, Ionic channels in mouse astrocytes, *J. Neurosci.* **7:**101–109.

Orkand, R. K., 1977, Glial cells, in: *Handbook of Physiology,* Section 1, Volume 1 (E. R. Kandel, ed.), American Physiological Society, Bethesda, pp. 855–873.

Orkand, R. K., 1980, Extracellular potassium accumulation in the nervous system, *Fed. Proc.* **39:**1515–1518.

Rakic, P., 1981, Neuronal–glial interaction during brain development, *Trends Neurosci.* **4:**184–187.

Salem, R. D., Hammerschlag, R., Bracho, H., and Orkand, R. K., 1975, Influence of potassium ions on accumulation and metabolism of (¹⁴C) glucose by glial cells, *Brain Res.* **86:**499–503.

Schneider, G. T., Cook, D. I., Gage, P. W., and Young, J. A., 1985, Voltage-sensitive, high conductance chloride channels in the luminal membranes of cultured pulmonary alveolar (type II) cells, *Pfluegers Arch.* **404:**354–357.

Schwarze, W., and Kolb, H., 1984, Voltage dependent kinetics of an anionic channel or large unit conductance in macrophages and myotube membranes, *Pfluegers Arch.* **402:**281–291.

Seilheimer, B., and Schachner, M., 1987, Regulation of neural cell adhesion molecule expression on cultured mouse Schwann cells by nerve growth factor, *EMBO J.* **6:**1611–1616.

Shrager, P., Chiu, S. Y., and Ritchie, J. M., 1985, Voltage-dependent sodium and potassium channels in mammalian cultured Schwann cells, *Proc. Natl. Acad. Sci. USA* **82:**948–952.

Sonnhof, U., 1987, Single voltage-dependent K⁺ and Cl⁻ channels in cultured rat astrocytes, *Can. J. Physiol. Pharmacol.* **65:**1043–1050.

Sykova, E., and Orkand, R. K., 1980, Extracellular potassium accumulation and transmission in frog spinal cord, *Neurosci.* **5:**1421–1428.

Tang, C. M., Strichartz, G. R., and Orkand, R. K., 1979, Sodium channels in axons and glial cells of the optic nerve of Necturus maculosa, *J. Gen. Physiol.* **74:**629–642.

Villegas, J., Villegas, L., and Villegas, R., 1965, Sodium, potassium, and chloride concentrations in the Schwann cell and axon of the squid nerve fiber, *J. Gen. Physiol.* **49:**1–7.

Walz, W., and Hertz, L., 1983a, Intracellular ion changes of astrocytes in response to extracellular potassium, *J. Neurosci. Res.* **10:**411–423.

Walz, W., and Hertz, L., 1983b, Functional interactions between neurons and astrocytes. II. Potassium homeostasis at the cellular level, *Prog. Neurobiol.* **20:**133–183.

Walz, W., and Hinks, E. C., 1985, Carrier-mediated KCl accumulation accompanied by water movements involved in the control of physiological K⁺ levels by astrocytes, *Brain Res.* **343:**44–51.

Zampighi, G. A., and Hall, J. E., 1985, Purified lens junctional protein form channels in planar lipid films, *Proc. Natl. Acad. Sci. USA* **82:**8468–8472.

Chloride Transport across the Sarcolemma of Vertebrate Smooth and Skeletal Muscle

C. Claire Aickin

1. INTRODUCTION

Our knowledge of the handling of Cl^- by muscle cells has increased immensely since the classic work of Boyle and Conway (1941) which revolutionized concepts about the permeability of the sarcolemma to Cl^-. In a complete turnaround from being considered impermeant, Cl^- ions were deemed to have free passage across the membrane. The majority view was that Cl^- permeability was so high that the transmembrane distribution was assumed to be determined entirely by the membrane potential, i.e., Cl^- was distributed at electrochemical equilibrium. All of interest that could possibly happen would be a reduction in the dominating Cl^- conductance causing a large decrease in the resting membrane conductance which would therefore seriously affect the electrical stability of the cell—the generation of myotonic activity (Lipicky and Bryant, 1966; Adrian and Marshall, 1976). As with all successful revolutions, the new regime dominated ideas, even in other cell types, for several years.

Recent research has changed this view. In the case of smooth and cardiac muscle, Cl^- is certainly not passively distributed across the sarcolemma nor is the membrane permeability to Cl^- high. Skeletal muscle poses more of a problem since here the permeability undoubtedly is very high. Nevertheless, there is clear evidence from a number of sources that Cl^- is actively transported into the sarcoplasm of this muscle type as well. This chapter will consider the advances in our understanding of the transmembrane Cl^- distribution, and the transport mechanisms and passive leaks by which it is regulated, in vertebrate smooth and skeletal muscle.

C. Claire Aickin • The University Department of Pharmacology, Oxford OX1 3QT, United Kingdom.

2. SMOOTH MUSCLE

2.1. Cl⁻ Distribution and Sarcolemmal Cl⁻ Permeability

The classical picture for Cl^- ions in smooth muscle is one of a high intracellular level, considerably higher than predicted for a passive distribution, and a relatively large membrane permeability (see Prosser, 1974; Casteels, 1981). Determination of the $[Cl^-]$ by diverse methods has given values within the range of 39–107 mM, equivalent to a Cl^- equilibrium potential (E_{Cl}) of -33 to -6 mV (see Table 1). Estimation of the intracellular level of extrapolation of radioisotope efflux agrees with these values (see Table 1), yet the magnitude of such effluxes could indicate a resting permeability to Cl^- (P_{Cl}) of the same order as that to K^+ if all the Cl^- efflux was via an electrodiffusive pathway. These two conclusions seem mutually exclusive, since together they imply fundamentally unlikely conditions. First, they would require a considerable degree of active inward transport of Cl^- to balance the necessarily large outward leak. Second, such a large electrical permeability to an ion with a significantly depolarizing equilibrium potential would tend to drive the resting potential to a relatively depolarized level, since E_{Cl} is at least 30 mV less negative than the E_m actually measured. Such a discrepancy could be resolved by substantial contribution of electrogenic mechanisms to the resting potential (see Casteels, 1969a). Both of those requirements would impose a large metabolic load on the cell. Intuitively then, one or other of the classical conclusions must be wrong.

Table 1. Intracellular Chloride Concentration in Smooth Muscle

Preparation	Method	$[Cl^-]_i$ (mM)	Authors
Guinea pig vas deferens	Ion analysis	57	Casteels (1969b)
	Ion analysis	50	Aickin and Brading (1982a)
Guinea pig taenia coli	Ion analysis	75	Brading and Widdicombe (1977)
	Ion analysis	55	Casteels (1969a)
Rat uterus	Ion analysis	39	Hamon et al. (1976)
Rat portal vein	Ion analysis	90	Wahlström (1973a)
Rat aorta	Ion analysis	56	Jones (1974)
Dog carotid artery	Ion analysis	96	Jones and Karreman (1969)
Rabbit pulmonary artery	Ion analysis	107	Jones et al. (1973)
Rat aorta	Ion analysis	141	Jones et al. (1973)
Rabbit portal vein	Ion analysis	50	Jones et al. (1973)
Rabbit taenia coli	Ion analysis	85	Jones et al. (1973)
Guinea pig taenia coli	³⁶Cl flux	58	Casteels (1969a)
Guinea pig vas deferens	³⁶Cl flux	51	Casteels (1969b)
	³⁶Cl flux	52	Aickin and Brading (1982a)
Rat portal vein	³⁶Cl flux	64	Wahlström (1973a)
Rat uterus	³⁶Cl flux	39	Hamon et al. (1976)
Rabbit portal anterior mesenteric vein	Electron probe	78	Somlyo et al. (1979)

Table 2. Intracellular Chloride Ion Activity in Smooth Muscle[a]

Preparation	a_{Cl}^i (mM)	E_m (mV)	E_{Cl} (mV)	Authors
Guinea pig vas deferens	41.2 ± 6.7 (79)	−67.6 ± 7.8 (79)	−24.3	Aickin and Brading (1982a)
Guinea pig ureter	51.1 ± 4.0 (13)	−48.7 ± 5.4 (13)	−18.6	Aickin and Vermuë (1983)
Cat proximal colon	—	−59.6 ± 1.8	−38.0 ± 3.3	Shearin and Walker (1983)

[a]Values are mean ± S.D. of an observation. Number of observations given in parentheses.

2.1.1. Is [Cl⁻]$_i$ Really So High?

Taking these conclusions in turn, are the high values for $[Cl^-]_i$ a real reflection of the cytosolic $[Cl^-]$? Each of the methods used to obtain the data in Table 1 measures total cellular content, including any compartmentalized or bound ions, and thus they could overestimate the free concentration. However, it seems unlikely that the high values are the result of sequestration of Cl^- within intracellular organelles. If there was significant intracellular binding of Cl^- as there is apparently of Na^+ (see Aickin, 1987a), this would require an apparent intracellular activity coefficient of around 0.2, which is within the range quoted for Na^+ in a variety of preparations (e.g., Lev, 1964; Lee and Fozzard, 1975). However, there is no evidence of Cl^- binding significantly to proteins or lipids.

This problem was resolved by the direct measurement of Cl_i^- activity (a_{Cl}^i) using Cl^--sensitive microelectrodes. In every case, a_{Cl}^i has been found to be considerably higher than predicted by the membrane potential (E_m) and the $[Cl^-]_o$ (see Table 2). As a final worry, Cl^--sensitive electrodes, based on the Corning ion exchanger (477315), are far from being perfectly selective for Cl^-, but show significant sensitivity to many other anions (see Chapter 1). Intracellular interference therefore could possibly be responsible for the high apparent level. This can be checked by equilibration of the preparation in Cl^--free media and under these conditions a_{Cl}^i falls to an apparent 3.1 ± 0.8 mM (mean ± S.D. of an observation, $n = 33$) in guinea pig vas deferens (Aickin and Brading, 1982a) and 3.3 ± 0.9 mM ($n = 8$) in guinea pig ureter (Aickin and Vermuë, 1983). Since values of 41.2 ± 6.7 mM ($n = 79$) and 51.1 ± 4.0 mM ($n = 13$) were recorded in vas deferens and ureter respectively, when 8 and 16.4 mM would have been consistent with a passive distribution, there can be no doubt that under normal conditions a_{Cl}^i is far above the equilibrium level with E_{Cl} considerably depolarized with respect to the resting potential.

2.1.2. Is P_{Cl} Really as High as P_K?

Several basic observations are difficult to reconcile with P_{Cl} being as large as P_K. First, membrane potential in most smooth muscle cells is scarcely affected by changes in $[Cl^-]_o$ (e.g., Holman, 1958; Kuriyama, 1963; Hirst and van Helden, 1982; Aickin and Brading, 1982a) whereas large, transient changes would be expected, as are observed in skeletal muscle (Hodgkin and Horowicz, 1959). Second, $[Cl^-]_i$ is independent of membrane potential (Casteels and Kuriyama, 1966; Aickin and Brading, 1982a, 1983) as is ^{36}Cl efflux (Casteels and Kuriyama, 1966; Gerstheimer et al., 1987). Third, total replacement of Na_o^+ with K^+ causes very little cellular swelling (Jones et al., 1973) and no measurable change in total tissue water content (Casteels and Kuriyama, 1966; Brading and Tomita, 1968). Considerable swelling would be expected if Cl^- were relatively permeant. It would therefore seem likely that radio-isotope efflux measurements overestimate the electrical conductance. This was initially intimated by Casteels (1981); indeed, precedent had already been set in both red blood cells (Hunter, 1977; Knauf et al., 1977) and barnacle muscle (Russell and Brodwick, 1976, 1979; Ashley et al., 1978) where a substantial portion of the Cl^- flux was shown to be carrier-mediated and the Cl^- electrical permeability very low.

2.1.2a. Studies Using Net Cl^- Fluxes.

Studies of net Cl^- movements using ion-selective microelectrodes also are consistent with electricaly silent, carrier-mediated transport. Thus, investigation of the changes in a_{Cl}^i observed following the alteration of $[Cl^-]_o$ revealed that not only is the active accumulation of Cl^- carrier-mediated but so too is at least part of the loss of Cl_i^- seen on removal of Cl_o^- (Aickin and Brading, 1982b, 1983; Aickin and Vermuë, 1983; see also Section 2.2.1). As shown in Fig. 1, the rate of fall of a_{Cl}^i in Cl^--free solution in the guinea pig vas deferens is considerably slowed by the nominal absence of CO_2 and HCO_3^- (mean half-time 11.2 ± 0.9 min, $n = 20$ versus 3.2 ± 0.2 min, $n = 16$ in the presence of 3% CO_2) and profoundly slowed by the presence of the stilbene disulfonate derivative DIDS (half-time 44.8 ± 1.9 min, $n = 4$; Aickin and Brading, 1984). Similarly, ^{36}Cl efflux into Cl^--free solution is slowed by the nominal absence of CO_2 and HCO_3^- and greatly slowed by the presence of DIDS (Aickin and Brading, 1982b, 1984). Steady-state efflux of ^{36}Cl, i.e., efflux into normal, Cl^--containing Krebs solution, has more commonly been used for determination of P_{Cl}. Efflux is not greatly different whether it be continually into Cl^--containing or Cl^--free solution with either permeant or non-permeant anion substitutes (Casteels, 1971; Aickin and Brading, 1983). However, the fractional loss of ^{36}Cl is markedly *increased* by readdition of Cl_o^- (Aickin and Brading, 1983). These results are readily explained if the majority of the Cl^- movements are carrier-mediated. Since efflux into Cl^--containing solution is unaffected by the presence or nominal absence of CO_2, it would appear that the anion exchanger then predominantly operates in a Cl^- self-exchange mode (Aickin and Brading, 1983). This efflux is also greatly inhibited by the presence of DIDS. Unfortunately, there is increasing evidence that the stilbene disulfonate derivates not only inhibit anion exchange but may also block movements of Cl^- which carry charge (e.g., Knauf et al., 1977; Vaughan and Fong, 1978; White and Miller, 1979; Xie et al., 1983; Nelson et al., 1984; Inoue, 1985; Montrose et al., 1987; Miller and Richard, this volume). Thus, calculation of P_{Cl} from any loss of Cl_i^- in the presence of DIDS may yield an underestimate. Inhibition of anion exchange caused by the absence of CO_2 and HCO_3^- from the bathing medium during net loss of Cl_i^- is equally unsatisfactory, since

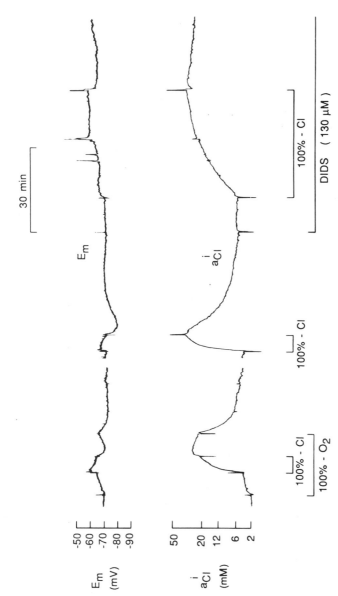

Figure 1. Pen recordings of an experiment to investigate the role of anion exchange in the accumulation and loss of Cl_i^- in guinea pig vas deferens upon addition and removal of Cl_O^-. Continuous measurements of the Cl_i^- activity (a_{Cl}^i) and membrane potential (E_m) were made with a double-barreled, Cl^--sensitive microelectrode (Aickin, 1981) filled with the Corning Cl^- ion-exchanger (477315) in one barrel and the reference liquid ion-exchanger (RLIE; Thomas and Cohen, 1981) in the other. The preparation was superfused with Cl^--free solution, equilibrated with 3% CO_2, 97% O_2 at pH 7.38 except for the intervals indicated, and was maintained at 35°C. Nominally CO_2-free solutions were equilibrated with 100% O_2 and buffered to pH 7.38 with Na-HEPES. (Reproduced from Aickin and Brading, 1984.)

metabolic production of CO_2 allows some turnover of the exchange (Aickin and Brading, 1984). In addition to the anion exchanger, there is also a Na^+,K^+,Cl^- cotransport mechanism involved in transmembrane movements of Cl^- (see Section 2.2.2), including the efflux of ^{36}Cl into Cl^--containing solution (Kreye *et al.*, 1981). It is therefore apparent that, until we can be certain of specific and complete inhibition of transport processes, neither net nor steady-state efflux of Cl^- can reflect the true electrical permeability. However, one point is certain: total, uninhibited efflux must yield a substantial overestimate of P_{Cl}.

2.1.2b. Studies Measuring Cl⁻-Dependent Changes in Membrane Conductance.

A seemingly more direct method for determination of P_{Cl} is to measure the total membrane conductance in Cl^--containing, and Cl^--deficient solutions. The difference between these measures is then assumed to equal the Cl^- conductance. Obviously this demands that no other conductance is affected by the alteration in Cl_o^-. This may be true in electrically uncoupled cells, but is unlikely to be so in the electrically coupled cells of smooth muscle, where the total conductance is dependent upon the conductance of junctional membranes. Alteration of $[Cl^-]$ has significant effects on pH_i (Aickin and Brading, 1984; see also Section 2.2.1), which in turn has been shown to affect the conductance of gap junctions (Turin and Warner, 1978; Spray *et al.*, 1981). This may contribute to the decrease in conductance observed in Cl^--deficient solutions in guinea pig taenia coli (Ohashi, 1970a) and terminal ileum (Bolton, 1973).

2.1.2c. Studies Measuring Cl⁻-Dependent Changes in Membrane Potential.

Probably the best method for obtaining a realistic estimate of P_{Cl} is from the instantaneous effect of changing the $[Cl^-]_o$ on membrane potential. The fact that in most cells, membrane potential is unaffected suggests a relatively low permeability. This can be quantified, relative to the K^+ permeability (P_K), by comparison of the changes caused by alteration of external $[Cl^-]$ and $[K^+]$ (Strickholm and Wallin, 1967) and a P_{Cl}/P_K ratio of 0.04 has been calculated for the guinea pig vas deferens (Aickin and Brading, 1983). This method does depend on the assumptions that the constant field equation is valid and that the applied changes in external ions have no immediate effect on any other parameter governing the membrane potential. Both assumptions can be questioned. An alternative approach is to fit the relationship between membrane potential and $[K^+]_o$ using the constant field equation and this has yielded a P_{Cl}/P_K ratio of 0.09 for guinea pig submucosal arterioles (Hirst and van Helden, 1982). This method obviously again assumes the validity of the constant field equation but has the more serious drawback that intracellular ion levels must be known. It is possible that the assumed $[Cl^-]_i$ (36 mM) was too low and this would have resulted in an overestimate of the ratio. Nevertheless, these calculations indicate that P_{Cl} is at least an order of magnitude smaller than P_K.

2.2. Mechanisms for Active Accumulation of Cl⁻

Since direct measurements of a_{Cl}^i have conclusively shown a far higher level than would be predicted by a passive distribution (see Section 2.1), a mechanism or mechanisms capable of transporting Cl^- into the cell against its electrochemical gradient must exist. The most convincing method for studying these mechanisms is by continuous measurement of the rise in a_{Cl}^i on readmission of Cl^- to Cl^--depleted cells using Cl_i^--sensitive microelectrodes. Simultaneous measurement of the membrane potential, implicit in this method, provides vital information about the electrochemical gradient for Cl^- and can indicate whether electrogenic mechanisms are involved. In addition, this method measures net movements of Cl^-. These are obviously clear advantages over the complexities of interpreting radioisotope flux data (e.g., see Section 2.1), but the technical difficulty of inserting ion-selective microelectrodes into smooth muscle cells has limited this approach to quiescent preparations.

2.2.1. Cl^-/HCO_3^- Exchange

The involvement of anion exchange in Cl^- accumulation by smooth muscle cells was initially investigated because this mechanism had previously been shown to be largely, if not exclusively, responsible for Cl^- accumulation in mammalian cardiac muscle (Vaughan-Jones, 1979b). As illustrated in Fig. 1, the rate of rise of a_{Cl}^i on readmission of external Cl^- to a Cl^--depleted cell of the guinea pig vas deferens is slowed by the nominal absence of CO_2 and HCO_3^- and greatly slowed by the presence of the anion exchange inhibitor DIDS. Very similar results have also been obtained in the smooth muscle of guinea pig ureter (Aickin and Vermuë, 1983). The half-time of the rise in a_{Cl}^i in guinea pig vas deferens increases from 3.0 ± 0.2 min ($n = 18$) in the presence of 3% CO_2, to 8.0 ± 0.9 min ($n = 17$) in the nominal absence of CO_2 and 33.1 ± 2.5 min ($n = 4$) in the presence of DIDS (Aickin and Brading, 1984). Comparable results were obtained by measurement of ^{36}Cl uptake in Cl^--depleted tissues, the half-time increasing from about 3 min in the presence of 3% CO_2, to 14 min in the nominal absence of CO_2 and 38 min in the presence of DIDS.

Figure 1 also illustrates the slowing of the decline in a_{Cl}^i on removal of Cl_o^- caused by the nominal absence of CO_2 and by the presence of DIDS described in Section 2.1.2a. The fact that the loss of Cl_i^- is slowed by the nominal absence of CO_2 and accelerated by its addition suggests that the loss is carrier-mediated and not via a conductance pathway. As already discussed, P_{Cl} in this preparation is very low (Section 2.1.2). In any event, loss of Cl^- through a significant Cl^- conductance would cause a depolarization, not the hyperpolarization observed in Fig. 1. In fact, the changes in membrane potential caused by alteration of $[Cl^-]_o$ vary, depending upon the duration of exposure to the previous solution. Removal of Cl_o^- after brief exposure to Cl^--containing solution (see Fig. 1) invariably causes a transient hyperpolarization, but after prolonged exposure to Cl^--containing solution causes a small depolarization (see Fig. 2). Readdition of Cl_o^- usually, but not invariably (see Fig. 2B), causes a small depolarization. These effects on membrane potential are not readily explained but their variability would argue against a direct link between the transmembrane movement of Cl^- and generation of current. Thus, the mechanisms for Cl^- transport are concluded to be electroneutral.

2.2.1a. How Does Cl^-/HCO_3^- Exchange Cause Active Cl^- Accumulation? These results seem to provide firm evidence for the involvement of Cl^-/HCO_3^-

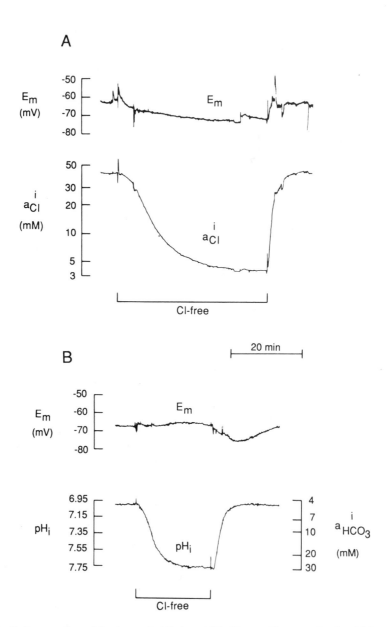

Figure 2. Pen recordings of the changes in (A) a^i_{Cl} and (B) pH_i caused by removal and readdition of Cl^-_o in the guinea pig vas deferens. Both recordings were made with double-barreled microelectrodes using RLIE to measure E_m and (A) the Corning Cl^- ion-exchanger to measure a^i_{Cl} or (B) the H^+-selective neutral carrier (supplied by Fluka; Ammann *et al.*, 1982) to measure pH_i. Both preparations were superfused with solutions equilibrated with 3% CO_2, 97% O_2 at 35°C. The pH_i trace (B) is also calibrated in terms of the intracellular HCO_3^- ion activity $a^i_{HCO_3}$. (Reproduced from Aickin and Brading, 1984.)

exchange in both the active accumulation of Cl^- and in its washout into Cl^--free solution. The best characterized example of this exchange is the band 3 protein of red blood cells. Although many of the elegant kinetic studies performed with erythrocytes would be exceptionally difficult to reproduce in smooth muscle, the results to date do not suggest that the anion exchanger of smooth muscle is fundamentally different from the band 3 protein. Both show the same affinity sequence (equivalent to Eisenman's series IV) and both appear to be asymmetric with respect to the internal and external sites (see Aickin and Brading, 1985b). Both also are electroneutral (see Aickin and Brading, 1984). Why then can the presence of the anion exchanger in smooth muscle result in active accumulation of Cl^- when it does not in the erythrocyte? Low permeability to Cl^- is not the answer, since erythrocytes also have a very low P_{Cl} (Knauf et al., 1977; Hunter, 1977). The answer lies simply in the fact that H^+ and HCO_3^- are not in equilibrium across the smooth muscle cell membrane, whereas they are in erythrocyte. Active regulation of pH_i provides the driving force to allow the accumulation of Cl^-. It is an interesting point that under normal conditions in the guinea pig vas deferens, anion exchange does not even reach equilibrium. E_{Cl} averages -24 mV (Aickin and Brading, 1982a) whereas the mean HCO_3^- equilibrium potential (E_{HCO_3}) is -17 mV (Aickin, 1984). The same is true in the smooth muscle of guinea pig ureter, where direct measurements of a_{Cl}^i and pH_i have given values for E_{Cl} of -18.6 mV (Aickin and Vermüe, 1983) and for E_{HCO_3} of -13.7 mV (C. C. Aickin, unpublished observations), and also apparently in rabbit aorta, where ^{36}Cl uptake would suggest an E_{Cl} of about -40 mV and the distribution of the weak acid DMO (5,5-dimethyloxazolidine-2,4-dione) would suggest an E_{HCO_3} of about -11.5 mV (Gerstheimer et al., 1987). Thus, there is more than enough energy stored in the HCO_3^- gradient to account for the high $[Cl^-]_i$ (see also Vaughan-Jones, 1986).

Figure 2 clearly shows the effects of operation of Cl^-/HCO_3^- exchange on pH_i in the guinea pig vas deferens and possibly provides the strongest evidence for the involvement of this mechanism in the active accumulation of Cl^-. In normal Krebs solution, pH_i is 7.06 ± 0.09 ($n = 52$; Aickin, 1984), i.e., about 0.8 pH unit more alkaline than predicted for a passive distribution of H^+ ions. Removal of Cl_o^-, in addition to causing a fall in a_{Cl}^i (Fig. 2A), causes a rise in pH_i (Fig. 2B). In other words, the fall in $[Cl^-]_i$ is paralleled by a rise in $[HCO_3^-]_i$ since, in the presence of a fixed concentration of CO_2, an increase in pH reflects an increase in $[HCO_3^-]$. This rise in $[HCO_3^-]_i$ occurs against the electrochemical gradient for HCO_3^- and is presumably driven by the very large outward Cl^- chemical gradient. pH_i stabilizes at a mean 7.79 ± 0.02 ($n = 37$; Aickin and Brading, 1984). Intracellular alkalinization on removal of Cl_o^- has also been observed in rabbit aorta (Gerstheimer et al., 1987), cultured rat aortic cells (Korbmacher et al., 1988), and guinea pig ureter (although in this preparation, the alkalosis is not maintained; C. C. Aickin, unpublished observations). When Cl^- is readmitted to the superfusate, pH_i rapidly falls to the level previously recorded in this solution (Fig. 2B) while a_{Cl}^i rises (Fig. 2A)—i.e., $[Cl^-]_i$ rises against the electrochemical gradient as $[HCO_3^-]_i$ falls. In both cases active inward transport of one anion is accompanied by an efflux of the other anion down the chemical gradient.Conclusive proof that these changes in pH_i are brought about by

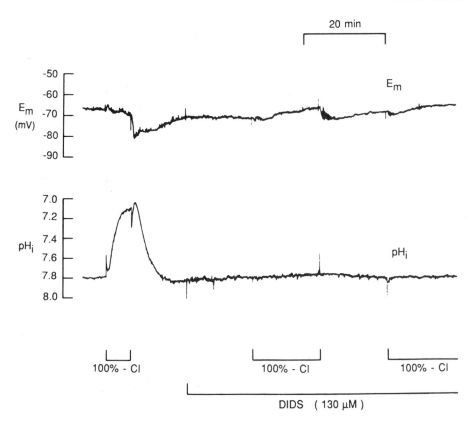

Figure 3. Pen recordings of an experiment illustrating that application of the anion exchange inhibitor DIDS abolishes the changes in pH_i normally observed in guinea pig vas deferens on alteration of $[Cl^-]_o$. The recording was made with a double-barreled microelectrode using RLIE to measure E_m and the H^+-selective neutral carrier (Fluka) to measure pH_i. The preparation was superfused with Cl^--free solution except for the intervals indicated and all solutions were equilibrated with 3% CO_2, 97% O_2 at 35°C. (Reproduced from Aickin and Brading, 1984.)

operation of Cl^-/HCO_3^- exchange is given by experiments like that illustrated in Fig. 3. Application of the anion exchange inhibitor DIDS completely prevents the acidification normally seen on readmission of Cl_o^- to Cl^--depleted cells and completely prevents the alkalinization normally seen on removal of Cl_o^-.

2.2.1b. Evidence of Cl^-/HCO_3^- Exchange in Taenia Coli and Vascular Smooth Muscle.

Taking all these results together, from direct measurements of a_{Cl}^i and pH_i and from ^{36}Cl fluxes, there can be no doubt that Cl^-/HCO_3^- exchange is involved in the active accumulation of Cl^- in the smooth muscle of both guinea pig vas deferens and ureter. It is also involved in the loss of Cl_i^- on exposure to Cl^--free solution and in the steady-state fluxes of Cl^- across the cell membrane. Although not as thoroughly investigated, Cl^-/HCO_3^- exchange also appears to be present in an example of spontaneously active smooth muscle, the guinea pig taenia coli (Brading, 1987), and in examples of vascular smooth muscle, namely rabbit aorta (Gerstheimer *et al.*, 1987) and cultured rat aortic cells (Korbmacher *et al.*, 1988). ^{36}Cl efflux is slowed by DIDS in both tissues. ^{36}Cl uptake is also slowed by DIDS in the rabbit aorta. In addition, efflux into Cl^--free solution (nonpermeant anion substitute) is slightly slowed by removal of CO_2 and HCO_3^- from the bathing solution in this preparation. All these observations are consistent with a role for Cl^-/HCO_3^- exchange in establishment of the high $[Cl^-]_i$ in these preparations, although of a lesser importance than in the vas deferens and ureter. The most striking effect of HCO_3^- withdrawal on ^{36}Cl fluxes in the rabbit aorta (Gerstheimer *et al.*, 1987) is, however, rather puzzling— namely a marked stimulation of the efflux into Cl^--containing solution (or Cl^--free solution with transportable anion substitutes) and a smaller stimulation of the steady-state uptake. These results perhaps suggest significant HCO_3^- self-exchange in the steady state in HCO_3^--containing solution or a slower translocation of the HCO_3^--loaded carrier.

2.2.2. Na^+, K^+, Cl^- Cotransport

Strong as the evidence is for the involvement of Cl^-/HCO_3^- exchange, this mechanism is not the complete answer to the high a_{Cl}^i, even though the HCO_3^- gradient is sufficiently large to drive Cl^- above the level observed under normal conditions. Most significantly, a_{Cl}^i still rises to its usual level on readmission of Cl_o^- in the presence of DIDS (see Fig. 1) even though Cl^-/HCO_3^- exchange is fully inhibited, as witnessed by the abolition of any HCO_3^- movement (see Fig. 3). In addition, the changes in a_{Cl}^i observed on removal and readdition of Cl_o^- occur more slowly than those in pH_i (see Fig. 2). The mean half-time for the rise in $[HCO_3^-]_i$ on removal of Cl_o^- was calculated to be 2.4 ± 0.1 min ($n = 8$) compared with that for the concomitant fall in a_{Cl}^i of 3.2 ± 0.2 min ($n = 16$), while that for the fall in $[HCO_3^-]_i$ on readdition of Cl_o^- was 1.0 ± 0.0 min ($n = 16$) compared with that for the accompanying rise in a_{Cl}^i of 3.0 ± 0.2 min ($n = 18$; Aickin and Brading, 1984). Thus, Cl^- continues to be accumulated after the HCO_3^- movement has apparently ceased. Finally, the net change of a_{Cl}^i observed when a cell is exposed to Cl^--free solution is significantly larger (mean 38.1 mM; Aickin and Brading, 1984) than that calculated for the concomitant net change of $a_{HCO_3}^i$ (mean 28.6 mM; Aickin and Brading, 1984).

2.2.2a. Historical Perspective. The possible involvement of cations in the active uptake of Cl^- by smooth muscle was in fact investigated before that of anion exchange. Casteels (1971) initially suggested that Cl^- uptake of the guinea pig taenia coli was linked to K^+ uptake by the Na^+ pump based on the observation that Cl^- became passively distributed after ouabain treatment (Casteels, 1966). In retrospect, this effect on the Cl^- distribution was far more likely to have been caused indirectly, through the dramatic reduction in the transmembrane Na^+ gradient observed in this preparation on inhibition of the Na^+ pump (e.g., Casteels, 1966; Axelsson and Holmberg, 1971; Brading, 1973), but at the time there was little indication for mechanisms of ion transport other than the Na^+ pump. A decrease in K^+ efflux on substitution of Cl_o^- with impermeant anions was also cited as evidence for the supposed linked uptake of K^+ and Cl^- by the Na^+ pump (Casteels, 1971).

Investigation of the connection between cations and Cl^- transport was continued by Widdicombe and Brading (1980), who provided evidence for coupled movements of Na^+ and Cl^-, also in the guinea pig taenia coli. They showed that uptake of either ion was reduced if the other ion was removed extracellularly, that ^{36}Cl efflux was increased on inhibition of the Na^+ pump, and that ^{24}Na efflux was reduced by removal of Cl_o^-. At about the same time, linked uptake of Na^+ and Cl^- was described in the squid giant axon (Russell, 1979), another preparation in which Cl^- is accumulated to a high level, and the first evidence for Na^+,K^+,Cl^- cotransport was reported in Ehrlich ascites cells (Geck et al., 1980). An absolute requirement for K^+ on the Na^+,Cl^- cotransport in squid axons axons was added later (Russell, 1983), interestingly confirming the original speculation of Keynes (1963), and the discovery of Na^+,K^+,Cl^- cotransport extended to many different cell types (for recent reviews see Chipperfield, 1986; O'Grady et al., 1987). Some indication for the presence of this triple cotransport mechanism in smooth muscle was in fact available in 1980, although the pieces of evidence had not been put together in any formal way. Removal of Cl_o^- in the guinea pig taenia coli caused a reduction in both K^+ efflux (Casteels, 1971) and Na^+ efflux (Widdicombe and Brading, 1980).

Further indication for the involvement of a cotransport mechanism in regulation of $[Cl^-]_i$ in smooth muscle was obtained from observations in the presence of the loop diuretic furosemide. Furosemide, often taken to be a diagnostic inhibitor of Na^+,K^+,Cl^- cotransport, decreased the $[Cl^-]_i$ of dog carotid artery (Villamil et al., 1979) and rabbit vascular smooth muscles (Kreye et al., 1981) and reduced the steady-state Cl^- efflux (Kreye et al., 1981). However, large doses were used and at these millimolar concentrations furosemide is known to inhibit Cl^-/HCO_3^- exchange in a variety of preparations (e.g., Brazy and Gunn, 1976; Ellory et al., 1982; Reuss et al., 1984; Warnock et al., 1984) including the smooth muscle of guinea pig vas deferens (Aickin and Brading, 1985a). Cl^-/HCO_3^- exchange is present in vascular smooth muscle (see Section 2.2.1) and its nonspecific inhibition could have contributed to the results.

2.2.2b. Direct Evidence of Na^+,K^+,Cl^- Cotransport in Smooth Muscle.

The first conclusive demonstration of the presence of a Na^+,K^+,Cl^- cotransport system in smooth muscle came from measurements of isotopic flux in cultured cells from embryonic rat aorta (Owen, 1985). The stimulus for these experiments was the report of a defective Na^+,K^+,Cl^- cotransport in erythrocytes of patients with essential hypertension (Garay et al., 1980). Clearly the presence of a cotransport system with a similar defect in vascular smooth muscle cells could be extremely relevant to the etiology of hypertension. In this investigation, ^{86}Rb uptake was used to assay Na^+,K^+,Cl^- cotransport: Rb^+ is known to substitute quantitatively for K^+ on this cotransporter in many tissues (e.g., Geck et al., 1980; Chipperfield, 1981; Hannafin et al., 1983). A large proportion of the flux was ouabain-sensitive (53%) but 33% was inhibited by 10 μM bumetanide, another loop diuretic, but considerably more potent (10–100 times) than furosemide. The ouabain-insensitive flux was decreased by the same amount in the absence of Na_o^+ or Cl_o^- as it was in the presence of bumetanide and the effects of ion substitution and application of bumetanide were not additive. Neither amiloride not SITS had any effect on the bumetanide-sensitive flux. Inhibition of ouabain-insensitive Rb^+ uptake by 1 mM furosemide was later reported in cultured cells from adult rat aorta (Little *et al.*, 1986).

2.2.2c. Net Cl^- Uptake by Na^+,K^+,Cl^- Cotransport.

More recently, clear proof of the involvement of Na^+,K^+,Cl^- cotransport in active accumulation of Cl^- by smooth muscle cells has been obtained by direct measurement of a_{Cl}^i in guinea pig vas deferens (Aickin, 1987b). In this study, the active accumulation of Cl^- that remains after inhibition of Cl^-/HCO_3^- exchange (see Section 2.2.1 and Fig. 1) was investigated. As illustrated in Fig. 4, the rise in a_{Cl}^i observed on readdition of Cl_o^- to Cl^--depleted cells in the presence of the anion exchange inhibitor DIDS is prevented by the absence of either Na_o^+ or K_o^+. This rise is also reversibly inhibited by the loop diuretics in the order: bumetanide > piretanide > furosemide, i.e., in the classical sequence for inhibition of Na^+,K^+,Cl^- cotransport (e.g., Palfrey *et al.*, 1980; Ellory and Stewart, 1982). Evidence for the involvement of the cotransporter under normal conditions, i.e., the presence of a functional anion exchange, is less convincing—but it is necessarily so. Under normal conditions, Cl^-/HCO_3^- exchange accounts for such a large proportion of Cl^- accumulation in this preparation that it had already been predicted that inhibition of the other mechanism would have little affect on the rate of

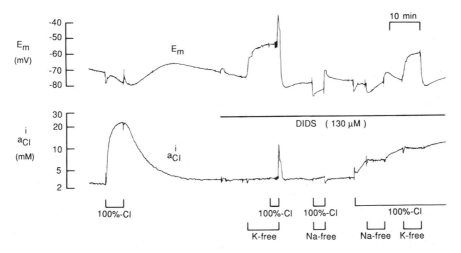

Figure 4. Pen recordings of an experiment to investigate the cation-dependence of the DIDS-insensitive accumulation of Cl_i^- in guinea pig vas deferens. A double-barreled microelectrode was used, filled with RLIE, to measure E_m, and the Corning Cl^--sensitive ion-exchanger, to measure a_{Cl}^i. The preparation was superfused with Cl^--free solution except where indicated and all solutions were equilibrated with 3% CO_2, 97% O_2 at 35°C. Prolonged exposure to K^+-free solution caused vigorous contractions (action potentials are only visible as a slight thickening of the E_m trace, due to the low-frequency response of the RLIE) and these partially dislodged the electrode. Recovery of the impalement was spontaneous on readmission of external K^+. (Reproduced from Aickin, 1987b.)

Cl$^-$ accumulation (Aickin and Brading, 1984). However, comparison of the rates of rise of a_{Cl}^i and concomitant fall of the calculated intracellular HCO$_3^-$ activity led to the conclusion that the second mechanism might be solely responsible for accumulation of about the last 6 mM. Recent results indeed show that under normal conditions, the rate of rise of a_{Cl}^i observed when Cl$^-$ is readmitted to Cl$^-$-depleted cells is not measurably altered by the absence of Na$_o^+$ or K$_o^+$ or by the presence of bumetanide at a concentration (10 μM) sufficient to halt the rise in a_{Cl}^i in the presence of DIDS (Aickin, 1987b). The final level of a_{Cl}^i reached is, however, reduced by 6–8 mM in the presence of bumetanide (10 μM) or piretanide (100 μM) and by about 15 mM in the absence of Na$_o^+$ (Aickin, 1987b), although it is unaffected by the nominal absence of K$_o^+$ (C. C. Aickin, unpublished observation). In other words, the results essentially confirm the predictions. The rather large reduction in the final level of a_{Cl}^i observed in the absence of Na$_o^+$ almost certainly results from an indirect effect of Na$^+$-dependent pH$_i$ changes on Cl$^-$/HCO$_3^-$ exchange, as well as from inhibition of the cotransport mechanism. Preliminary results show that readdition of Cl$_o^-$ and simultaneous removal of Na$_o^+$ causes a substantial intracellular acidification, over and above that observed in normal conditions (C. C. Aickin, unpublished observations). This pronounced acidification (see also Aickin, 1985) probably inhibits Cl$^-$ accumulation by Cl$^-$/HCO$_3^-$ exchange. On the other hand, the failure of removal of K$_o^+$ to affect the total accumulation of Cl$_i^-$ possibly results from a high affinity of the Na$^+$,K$^+$,Cl$^-$ cotransport for K$^+$ (for recent reviews see Chipperfield, 1986; O'Grady et al., 1987) and from the passive leak of K$_i^+$ maintaining a low [K$^+$] in the immediate pericellular space (see also Aickin et al., 1984; Aickin, 1987a).

2.2.2d. ^{36}Cl Flux via Na$^+$,K$^+$,Cl$^-$ Cotransport. Comparable results have also been obtained recently using radioisotopic flux in the guinea pig vas deferens (Brading, 1987). Uptake of ^{36}Cl by Cl$^-$-depleted tissues and efflux of ^{36}Cl into Cl$^-$-free solution is reduced by the absence of Na$_o^+$. This effect on at least the efflux can be attributed to inhibition of the cotransport mechanism, since it is still observed in the presence of DIDS and it is blocked by the presence of furosemide (2 mM) or piretanide (100 μM). Analysis of the total tissue Cl$^-$ content showed a slight reduction in Na-free solution and in the presence of furosemide but not in the presence of DIDS. These results are reminiscent of those found in rabbit vascular tissues (Kreye et al., 1981; Gerstheimer et al., 1987) and are consistent with the presence of two mechanisms responsible for Cl$^-$ accumulation. Inhibition of anion exchange has no effect on the steady-state [Cl$^-$]$_i$ (see also Aickin and Brading, 1984) since presumably any leak can be compensated by the cotransport mechanism. On the other hand, inhibition of cotransport causes a fall in [Cl$^-$]$_i$ since, for whatever reason (see Vaughan-Jones, 1986), anion exchange can only accumulate Cl$^-$ to some 6–8 mM below the normal level. It is an interesting point that in the sheep heart Purkinje fiber, where anion exchange is solely responsible for the high [Cl$^-$]$_i$ (Vaughan-Jones, 1986), a_{Cl}^i is about 20 mM (Vaughan-Jones, 1979a) lower than in smooth muscle under normal conditions but closer to the value observed in smooth muscle with cotransport inhibited.

2.2.2e. Different Smooth Muscles Have Different Relative Proportions of Cl$^-$/HCO$_3^-$ Exchange and Na$^+$,K$^+$,Cl$^-$ Cotransport. It does appear that the contribution of the two mechanisms, anion exchange and anion,cation cotransport, varies in different smooth muscles. Brading (1987) found that ^{36}Cl efflux was more potently inhibited by DIDS than by furosemide in the guinea pig vas deferens, but that the converse was true in the guinea pig taenia coli. Similarly, inhibition of ^{36}Cl efflux by removal of Na_o^+ was much more pronounced in taenia coli than in vas deferens. Thus, contransport seems to be better developed in taenia coli than in vas deferens and anion exchange more important in vas deferens than in taenia coli. Judging from the reduction of ^{36}Cl uptake in rabbit aorta (Gerstheimer *et al.*, 1987) caused by the presence of DIDS (40%) compared with that caused by the presence of furosemide (60%), it would seem that cotransport is the more important mechanism in this preparation. This could well underlie the relatively small effect of CO_2 removal on ^{36}Cl efflux into Cl^--free solution (nonpermeant substitute) observed in rabbit aorta (Gerstheimer *et al.*, 1987) compared with the pronounced effect seen in vas deferens (Aickin and Brading, 1984).

2.2.2f. Net Na^+ Flux Associated with Na^+,K^+,Cl^- Cotransport.

In the same way that measurement of changes in pH_i on alteration of external $[Cl^-]$ confirmed the involvement of Cl^-/HCO_3^- exchange in transmembrane Cl^- transport (Aickin and Brading, 1984), measurement of changes in a_{Na}^i could be expected to add to the evidence of the participation Na^+,K^+,Cl^- cotransport. As illustrated in Fig. 5, readmission of Cl_o^- to a Cl^--depleted cell of the guinea pig vas deferens causes a significant rise in a_{Na}^i concurrently with the rise in a_{Cl}^i. Although this result is consistent with coupled entry of Na^+ and Cl^- on the cotransport mechanism, it must be remembered that these intracellular changes are also associated with a considerable fall in pH_i (see Fig. 2). Intracellular acidosis results in the effective extrusion of acid equivalents by Na^+-dependent mechanisms (Aickin, 1985, 1988; Weissberg *et al.*, 1987; Korbmacher *et al.*, 1988). In very simple terms, protons are effectively extruded at the expense of Na^+ entry. Preliminary experiments indicate that part of the rise in a_{Na}^i is due to Na^+ entry through this route. Inhibition of the concomitant acidosis by application of DIDS reduced the rise in a_{Na}^i (C. C. Aickin, unpublished observations). Further experiments are required before the change in a_{Na}^i due solely to operation of Na^+,K^+,Cl^- cotransport can be quantified.

2.3. Involvement of Cl^- Transport in the Regulation of pH_i

The first description of a mechanism capable of effectively extruding acid equivalents against the electrochemical gradient, and thus of regulating pH_i, was of a Na^+-dependent, Cl^-/HCO_3^- exchange (Thomas, 1976, 1977). Although this mechanism appears to be solely responsible for pH_i regulation in some preparations (e.g., snail neurons, Thomas, 1977; barnacle muscle, Boron *et al.*, 1981; squid giant axon, Boron and Russell, 1983; crayfish stretch receptor neurons, Moser, 1985), it can be joined by other mechanisms (e.g., crayfish neurons, Moody, 1981; frog muscle, Abercrombie *et al.*, 1983; leech neurons, Schlue and Thomas, 1985; crayfish muscle, Galler and Moser, 1986; leech glial cells, Deitmer and Schlue, 1987) or even be totally absent (e.g., mouse soleus muscle, Aickin and Thomas, 1977b; sheep heart Purkinje fibers, Ellis and Thomas, 1976; Ellis and MacLeod, 1985; human neutrophils, Simchowitz

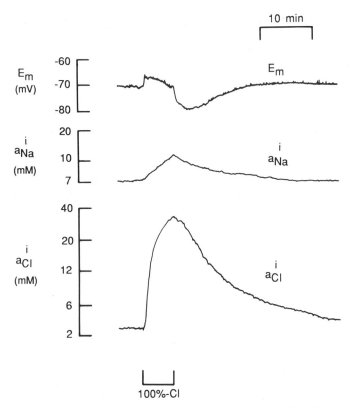

Figure 5. Pen recordings of the changes in membrane potential (E_m), Na_i^+ activity (a_{Na}^i) and Cl_i^- activity (a_{Cl}^i) observed on addition and removal of Cl_o^- in the guinea pig vas deferens. Measurements of a_{Na}^i and a_{Cl}^i were made in different cells under identical conditions, solutions equilibrated with 3% CO_2, 97% O_2 at 35°C. Double-barreled microelectrodes were used: both with RLIE to record E_m; one with the Na^+-sensitive neutral carrier (ETH 227 supplied by Fluka; Steiner *et al.*, 1979) to record a_{Na}^i, and the other with the Corning Cl^- ion-exchanger to record a_{Cl}^i. The preparations were superfused with Cl^--free solution except for the interval indicated.

and Roos, 1985). Unlike the omnipresent Na^+,K^+ pump for Na^+ extrusion, there is not a single mechanism for the effective extrusion of protons, but rather a variety of mechanisms (see *CIBA Found. Symp.* **139**).

Na^+-*dependent,* Cl^-/HCO_3^- exchange does not exist in the smooth muscle of guinea pig ureter (Aickin, 1988), nor probably in vascular smooth muscle (Korbmacher *et al.*, 1988), but a Na^+-*independent* anion exchange does (see Section 2.2.1) and its operation can have significant effects on pH_i (e.g., Figs. 2 and 3). Nevertheless, the normal transmembrane chemical gradients of Cl^- and HCO_3^- are such that Cl^- would tend to enter the cell and HCO_3^- leave, thus imposing an acid load on the cell. Hence, Cl^-/HCO_3^- exchange would not be expected to contribute to acid extrusion under normal conditions. This has been confirmed by the lack of effect of both Cl^--free conditions (Aickin, 1988) and the presence of DIDS (Aickin, 1988; Korbmacher *et al.*,

1988) on the rate of pH_i recovery from moderate acidosis. However, in the absence of Na_o^+, when other mechanisms for the effective extrusion of acid equivalents are inhibited, extreme intracellular acidosis results in some HCO_3^--dependent recovery of pH_i (Aickin, 1985, 1988). This is probably due to Cl^-/HCO_3^- exchange, activated because of the very low $[HCO_3^-]_i$. "Reversal" of the anion exchanger under these conditions has, in fact, been predicted from thermodynamic considerations (see Vanheel *et al.*, 1984; Vaughan-Jones, 1986).

Regulation of any parameter must comprise both a mechanism for its reduction, when it is in excess, and a mechanism for its increase, when it is too low. This is usually visualized as a "one-way" regulation with respect to intracellular ion levels, i.e., in simple terms, as a pump and a leak, or in the case of regulation of pH_i, a pump for the removal of excess acid equivalents and both a leak and metabolic production for the increase of intracellular acid equivalents. Adequate though this "one-way" regulation is for maintenance of $[Na^+]_i$ and apparently pH_i in many cell types, true "two-way" regulation of pH_i occurs in both cardiac (Vaughan-Jones, 1982) and smooth muscle (Aickin, 1988). In other words, transport mechanisms are involved in the recovery from both acidosis *and* alkalosis.

As illustrated in Fig. 6, recovery of pH_i from an alkaline load in the smooth muscle of guinea pig ureter is inhibited by the presence of the anion exchange inhibitor

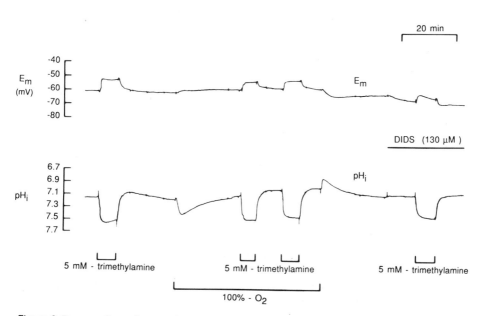

Figure 6. Pen recordings of an experiment to investigate the role of anion exchange in the recovery from intracellular alkalosis in the guinea pig ureter. The recordings were made with a double-barreled microelectrode using RLIE to measure E_m and the H^+-selective neutral carrier (Fluka) to measure pH_i. Intracellular alkalosis was produced by application of trimethylamine. The preparation was superfused with solutions equilibrated with 5% CO_2, 95% O_2 at pH 7.47 except for the interval indicated and was maintained at 35°C. Nominally CO_2-free solution was equilibrated with 100% O_2 and buffered to pH 7.47 with Na-HEPES. (Reproduced from Aickin, 1988.)

DIDS. In this experiment the weak base trimethylamine was used to alkaline load the cell. A slow acidification then occurs in the continued presence of trimethylamine and hence pH_i overshoots the previous level recorded in normal Krebs solution when the weak base is removed (see Thomas, 1984, for a full description of the process involved in alteration of pH_i on addition and removal of a weak base). After application of DIDS, neither the slow acidification nor the resultant overshoot in pH_i is observed. Not only is the recovery from alkalosis inhibited by the presence of DIDS, it is also inhibited by Cl^--free conditions (Aickin, 1988). These results are very reminiscent of the original observations that led to the concept of alkaline regulation in the sheep heart Purkinje fiber (Vaughan-Jones, 1982). As in the Purkinje fiber, they strongly suggest that Cl^-/HCO_3^- exchange mediates an effective *inward* transport of acid equivalents on intracellular alkalosis.

It is interesting to note that these conclusions from direct measurement of pH_i confirm the earlier suggestions of Ighoroje and Spurway (1985), made from observation of changes in vascular tone in the isolated rabbit ear. In their experiments, tone was interpreted to mirror pH_i—alkalosis (due to addition of NH_4^+-caused a relaxation and acidosis (due to removal of NH_4^+), a contraction. Just as pH_i recovers from both these perturbations (in smooth muscle, see Aickin, 1985, 1988) so too did the vascular tone. The presence of the stilbene disulfonate derivative SITS or absence of Cl^- slowed recovery of tone from the NH_4^+ induced relaxation but had no effect on the recovery from contraction observed on removal of NH_4^+. Thus, a major role for Cl^-/HCO_3^- exchange in the elimination of an alkaline load was suggested.

3. SKELETAL MUSCLE

3.1. Cl^- Distribution and Sarcolemmal Cl^- Permeability

Frog skeletal muscle was the first preparation in which the cell membrane was shown to be permeable to Cl^- (Boyle and Conway, 1941). Before this classical paper, biological membranes were generally believed to be impermeable to small anions (e.g., Fenn, 1936) and afterwards, the understanding of transmembrane ion distribution and permeabilities developed into the proposal of the constant field equation (Goldman, 1943; Hodgkin and Katz, 1949). Unlike the confused story in smooth muscle (see Section 2.1), the results of all the early studies in skeletal muscle were consistent with a large Cl^- permeability. The constancy of the Donnan equilibrium for K^+ and Cl^- described by Boyle and Conway (1941), the influence of $[Cl^-]_o$ on membrane potential recorded by Hodgkin and Horowicz (1959), and the large isotopic flux measured by Adrian (1961) together provide a very convincing body of evidence for a considerable conductive pathway. Subsequent studies have shown that a large Cl^- conductance (G_{Cl}) is a common feature of skeletal muscle throughout the animal kingdom (for a comprehensive recent review see Bretag, 1987; Blatz, this volume). Reported estimates of G_{Cl} in skeletal muscle have also always exceeded those for G_K. For example, G_{Cl} is 2 times G_K in frog (Hodgkin and Horowicz, 1959), 8–10 times G_K

in elasmobranch fish (Hagiwara and Takahashi, 1974), and 20 times G_K in rat extensor digitorum longus (DeCoursey *et al.*, 1978).

3.1.1. In Frog Skeletal Muscle, Active Cl⁻ Uptake May Be Shunted by a High P_{Cl}

Because of this very large conductance, Cl^- is generally believed to be passively distributed across the skeletal muscle fiber membrane: E_{Cl} is believed to be equal to E_m. Measurement of $[Cl^-]_i$ in frog skeletal muscle by ^{36}Cl efflux (Adrian, 1961) or by ion analysis (e.g., Harris, 1963) indeed confirmed the low level predicted from a passive distribution, around 3 mM. However, at the time there was no reason to suspect anything other than a passive distribution and it might almost be considered that the values obtained were a greater credit to the meticulous application of the methods than they were as a test of the transmembrane Cl^- gradient! The great disparity between intra- and extracellular levels and the large permeability to Cl^- make it very difficult to distinguish the intracellular fraction (see Adrian, 1961).

The first indication that $[Cl^-]_i$ might not be simply dictated by the membrane potential and $[Cl^-]_o$ under all conditions came from an investigation of the influence of pH on ionic permeability in frog muscle by Hutter and Warner (1967). They showed that changes in G_{Cl} were principally responsible for the pH sensitivity of the resting membrane conductance. But their striking observation was that when pH_o was first reduced for a period (low G_{Cl}) and then increased, there was a significant, transient depolarization. Their interpretation was that Cl^- was accumulated within the fiber during the period of reduced G_{Cl}. Such an accumulation requires the presence of an active uptake process. More recently, Hironaka and Morimoto (1980) have shown that increasing external pH causes a depolarization in frog muscle, at a $[K^+]_o$ of 2.5 mM, but a hyperpolarization soon after changing to 5 mM. This again led to the conclusion of active inward Cl^- transport, in this case with E_{Cl} significantly depolarizing to E_m under normal conditions. E_{Cl}, assumed to be equal to the membrane potential at which alteration of pH_o would have no effect, was calculated to be -88.5 mV while E_m was -94.8 mV.

3.1.2. In Some Mammalian Skeletal Muscle, Cl⁻ May Be Actively Accumulated

The Cl^- conductance of mammalian skeletal muscle does not show the same rapid changes on alteration of pH_o as observed in amphibian muscle. Fifteen to twenty minutes is required for the full effects to develop in rat diaphragm (Palade and Barchi, 1977a). Thus, the trick of suddenly changing pH_o to reveal a nonpassive Cl^- gradient cannot be applied (see Aickin *et al.*, 1988b). Nevertheless, in a careful study of the effect of $[Cl^-]_o$ on membrane potential, Dulhunty (1978) showed that at least some mammalian muscles differ from amphibian muscle in another respect. Her investigation followed the same lines as that of Hodgkin and Horowicz (1959) in frog muscle, but, unlike in the frog, she found that membrane potential of some rat and mouse muscles was affected by Cl^-_o in the steady state. The membrane potential hyper-

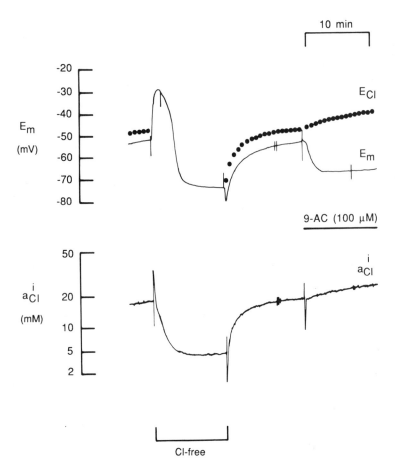

Figure 7. Measurement of the membrane potential (E_m) and Cl_i^- activity (a_{Cl}^i) in a rat lumbrical muscle during the removal and readdition of Cl_o^- and the inhibition of the Cl^- conductance by application of 9-anthracene carboxylic acid (9-AC). The Cl^- equilibrium potential (E_{Cl}), calculated from the recorded a_{Cl}^i, has been plotted as dots above the recording of E_m. The recording was made with a double-barreled microelectrode using RLIE to record E_m and the Corning Cl^- ion-exchanger to record a_{Cl}^i. The preparation was maintained in the nominal absence of CO_2 in solutions equilibrated with 100% O_2, buffered with PIPES-SO_4 to pH 7.4 at 20°C. Tetrodotoxin was present throughout to prevent spontaneous fibrillation.

polarized by as much as 20 mV in low-[Cl^-] solution (see Fig. 7 for an example of this behavior in rat lumbrical muscle). She therefore concluded that under normal conditions Cl^- was actively accumulated by these mammalian muscles to a level considerably higher than that predicted by a passive distribution. This accumulation would result in the transmembrane Cl^- gradient exerting a significant depolarizing influence on the resting membrane potential. Other explanations of the steady-state hyperpolarization in low-[Cl^-] solution would involve either a complex relationship between the P_{Na}/P_K ratio and both K_o^+ and Cl_o^- (see Dulhunty, 1978) or a significant contribution of an electrogenic mechanism to the resting potential, which would be greatly enhanced by the reduced membrane conductance.

Unfortunately, not all mammalian muscles show this dependence of the membrane potential on $[Cl^-]_o$. Dulhunty (1978) found the steady-state hyperpolarization in low-$[Cl^-]$ solution in mouse extensor digitorum longus and soleus and rat sternomastoid but not in rat diaphragm. Human intercostal muscle (Kwieciński *et al.*, 1984) is unaffected in the steady state and goat intercostal muscle shows a steady-state depolarization in low $[Cl^-]$ (Bryant and Morales-Aguilera, 1971). Nor do all researchers agree on the dependence of a given muscle. For example, both Vyskočil (1977) and Elsner and Westphal (1979) reported a steady-state hyperpolarization in the rat diaphragm while Palade and Barchi (1977a), like Dulhunty (1978), found no effect.

Another, less obvious line of investigation has also led to the conclusion that Cl^- is actively accumulated by mammalian skeletal muscle. In experiments using an extracellular vibrating probe, Caldwell and Betz (1984) demonstrated the presence of a steady, endogenous outward current in rat lumbrical muscles. Investigation of this current (Betz *et al.*, 1984) suggested that it was carried by Cl^-: it disappeared in Cl^--free solution, was blocked by application of the Cl^- channel inhibitor 9-anthracene carboxylic acid (9-AC; Palade and Barchi, 1977b) and by solutions of low pH. For such a Cl^- current to exist, $[Cl^-]_i$ would have to be higher than predicted for a passive distribution and the Cl^- conductance nonhomogeneous. There is, in fact, support for the presence of a Cl^--accumulating mechanism in the rat lumbrical muscle. First, it is one of the preparations to show a steady-state hyperpolarization in Cl^--free solution (see Fig. 7). Second, application of 9-AC causes a hyperpolarization of about 10–15 mV (see Fig. 7). On the assumption that 9-AC selectively inhibits the Cl^- conductance (Palade and Barchi, 1977b), this, like the steady-state hyperpolarization in Cl^--free solution, suggests that the normal transmembrane Cl^- distribution has a depolarizing influence on E_m. Interestingly, 9-AC does not cause a hyperpolarization in every muscle. It has no effect on E_m in the rat diaphragm (Palade and Barchi, 1977b; Blum and Westphal, 1981), but then, possibly, neither does low $[Cl^-]$ in the steady state in this preparation (Palade and Barchi, 1977a; Dulhunty, 1978). And third, application of furosemide (10 μM) or bumetanide (1 μM), well-established inhibitors of Cl^-,cation cotransport, also causes a 10- to 15-mV hyperpolarization (Betz *et al.*, 1984). Although furosemide is known to inhibit the Cl^- conductance (e.g., Bretag *et al.*, 1980), very much higher concentrations are required (2.5 mM). Thus, it is unlikely that the hyperpolarization was caused by a reduction in P_{Cl} in this instance. Other results described by Betz and his co-workers (Betz *et al.*, 1984; Harris and Betz, 1987; Aickin *et al.*, 1988a,b) have strongly suggested that a loop diuretic-sensitive, Na^+,K^+,Cl^- cotransport mechanism is responsible for active accumulation of Cl^- in this preparation (see Section 3.2).

There is some support for a high $[Cl^-]_i$ in mammalian muscle from indirect methods of determination. Lipicky and Bryant (1966) measured levels as much as four to seven times higher than predicted for a passive distribution in goat intercostal muscle, whereas Westphal and colleagues (Westphal and Limbourg, 1972; Renner and Westphal, 1975) concluded there was a rather more modest accumulation in rat diaphragm. However, the obvious problems associated with estimation of cellular $[Cl^-]$

from the total tissue content must throw doubt on the values obtained. In addition, these muscles give little (rat diaphragm) or no (goat intercostal) indication of a non-passive Cl^- distribution, as judged from the effect of Cl_o^- on the steady-state membrane potential (see above).

Direct measurements of a_{Cl}^i with Cl^--sensitive microelectrodes under normal conditions have hardly proved more convincing. Kernan and co-workers (Kernan *et al.*, 1974) made the first measurements in frog sartorius and concluded that $[Cl^-]_i$ was higher than predicted for a passive distribution. E_{Cl} was appreciably less negative than E_m. However, these measurements were made with a solid-state, Cl^--sensitive microelectrode (Neild and Thomas, 1974) which, at least in the snail neuron, suffers from an intracellular offset resulting in the recording of an artificially high a_{Cl}^i. In a far more exacting study, also in frog sartorius, Bolton and Vaughan-Jones (1977) recorded an E_{Cl} only slightly less negative than E_m under normal conditions. They used liquid ion-exchange, Cl^--sensitive microelectrodes, which (as described in Section 2.1 and see Chapter 1) are not perfectly selective for Cl^- and can therefore give an artificially high reading because of the presence of interfering anions. They concluded that under most conditions, the difference between E_{Cl} and E_m could be ascribed to experimental error. Several other reports of the measurement of a_{Cl}^i in amphibian (Armstrong *et al.*, 1977; Macchia and Baumgarten, 1979; Baumgarten and Fozzard, 1981) and mammalian skeletal muscle (McCaig and Leader, 1984; Donaldson and Leader, 1984; Leader *et al.*, 1984; Kernan and Westphal, 1984; Harris and Betz, 1987; Aickin *et al.*, 1988a,b) have also failed to give conclusive evidence for a nonpassive distribution under normal conditions. One such example in the rat lumbrical muscle is illustrated in Fig. 7. a_{Cl}^i in this preparation is apparently slightly higher than that predicted for a passive distribution by a mean 1.4 ± 1.0 mM ($n = 60$; Aickin *et al.*, 1988a). The calculated E_{Cl} is therefore slightly positive to E_m. However, in the absence of Cl_o^-, a_{Cl}^i stabilizes at a mean apparent 1.7 ± 1.0 mM ($n = 24$; Aickin *et al.*, 1988b), presumably reflecting the presence of other anions intracellularly which are sensed by the Cl^- electrode. Although the number of intracellular interfering anions is probably greater in Cl^--free, than Cl^--containing solution, this degree of interference could well account for the apparent nonpassive distribution.

Overall it seems that Cl^- is at least very close to being passively distributed across the sarcolemma under normal conditions. This, though, does not imply that Cl^- is not actively transported into the sarcoplasm. The very large Cl^- permeability would simply short-circuit the effect of such a transport, at least as far as causing a significant accumulation of Cl^-. It is, however, conceivable that the combination of a high P_{Cl} and inward Cl^- transport could exert a depolarizing influence on E_m without the Cl^- distribution being very much different from passive. In this case, the Cl^- distribution would dictate the membrane potential rather than vice versa.

3.1.3. Reducing P_{Cl} Uncovers Active Cl^- Accumulation in Both Frog and Mammalian Skeletal Muscle

Given the possibility that there is an inward active transport mechanism, its effect on the Cl^- distribution should be most obvious when the Cl^- permeability is low, as for example under normal conditions in both smooth and cardiac muscle (see Section

2.1). This approach was first used by Bolton and Vaughan-Jones (1977) in the frog sartorius. Their crucial experiment, inspired by the earlier conclusions of Hutter and Warner (1967), showed that reduction of pH_o caused a rise in a_{Cl}^i such that E_{Cl} became substantially less negative that E_m. This result, as carefully considered by the authors, is difficult to explain on an artifactual basis, and must be taken as concrete evidence for the existence of a Cl^- uptake mechanism.

Other methods for reduction of P_{Cl} have been used in mammalian skeletal muscle, both physiological (by denervation) and pharmacological (by application of 9-AC). An initial study of the effect of denervation on intracellular ion activities showed that although a_{Cl}^i rose in the rat extensor digitorum longus, it did not rise any more than was expected from the observed depolarization (Leader *et al.*, 1984). However, these measurements were made only up to 3 days after denervation when P_{Cl} remains unaffected (Camerino and Bryant, 1976; Lorkovic and Tomanek, 1977). One or two weeks after denervation, when P_{Cl} has fallen, a_{Cl}^i values about 5.5 mM higher than predicted for a passive distribution have been recorded in rat lumbrical muscle (Harris and Betz, 1987). Unfortunately, an unusually high residual apparent a_{Cl}^i was recorded in Cl^--free solution in these denervated fibers (8.1 ± 1.1 mM, $n = 8$), quite high enough to account for the apparent accumulation of Cl^-. Much more convincingly, as illustrated in Fig. 7, application of 9-AC causes an immediate rise in a_{Cl}^i in the face of membrane hyperpolarization (Harris and Betz, 1987; Aickin *et al.*, 1988a,b). E_{Cl} diverges from E_m and can become up to 50 mV less negative than E_m (Aickin *et al.*, 1988b). Significantly, application of 9-AC has no effect on either E_m or the apparent a_{Cl}^i in Cl^--free solution, suggesting that the changes observed in Cl^--containing solution do not arise from an artifact in the method (Aickin et al., 1988a,b).

3.1.4. Possible Role of Active Cl^- Uptake in Pathological Conditions That Reduce P_{Cl}

The demonstration of active accumulation of Cl^- clearly raises a new perspective on the consequences of a lowered Cl^- conductance and their contribution to the behavior of muscle under these conditions. It has long been known that the Cl^- conductance is greatly depressed in myotonic muscle (Lipicky and Bryant, 1966). It is now generally accepted that the depressed conductance is more than symptomatic, but rather is causative of the complaint. Pharmacological reduction of P_{Cl} induces clinical myotonia (Bryant and Morales-Aguilera, 1971) and the ability of a chemical to induce myotonia has even been used to screen for Cl^- channel blockers (e.g., Kwieciński, 1981; Palade and Barchi, 1977b). The large reduction in resting membrane conductance resulting from the decreased Cl^- conductance obviously will have a considerable effect on the electrical stability of the fiber. Mathematical modeling of the action potential has suggested that this reduction alone is sufficient to account for the repetitive firing characteristic of myotonic behavior (Adrian and Marshall, 1976). However, because of the presence of inward Cl^- transport, the decreased Cl^- conductance will allow a_{Cl}^i to rise. Depending on the ionic mechanism of the transport process, this will also affect other intracellular ions which may have a direct bearing on membrane excitability. It is perhaps significant that myotonia can be relieved clinically by administration of acetazolamide (Kwieciński, 1980), which also causes metabolic acidosis (e.g., Griggs *et al.*, 1978), and *in vitro* by elevation of CO_2 (Birnberger and Kelpzig, 1979). These results lend credence to the supposition that neither the reduced Cl^- conductance nor the elevated a_{Cl}^i is solely responsible for the myotonic condition. This is further supported by the observation that reduction of Cl^- conductance by a decrease in $[Cl^-]_o$ induces myotonic activity (Rüdel and Senges, 1972) but probably lowers a_{Cl}^i. Nevertheless, it is thought-provoking that denervation, which causes a fall in Cl^- conductance and almost certainly a rise in a_{Cl}^i (Harris and Betz, 1987), prevents the induction of myotonia (e.g., Iyer *et al.*, 1981).

3.2. Active Accumulation of Cl^- by Na^+, K^+, Cl^- Cotransport

The foregoing evidence of active accumulation of Cl^- in skeletal muscle under conditions when G_{Cl} is reduced simultaneously poses the question of the ionic mechanism of the uptake process and provides an ideal means for investigating the mechanism. Such investigation was first attempted in rat lumbrical muscle that had been denervated for 10 days or more (Harris and Betz, 1987). Steady-state measurements showed that the absence of external Na^+ or K^+, or the presence of furosemide (10 μM), reduced a_{Cl}^i to the level expected for a passive distribution. Significantly, each of these treatments also caused a hyperpolarization of the membrane potential. If it is argued that the high apparent a_{Cl}^i recorded in these fibers was caused by interference from other anions adding to the signal of the ion-exchanger, Cl^--sensitive microelectrodes, the relative effect of such interference should be larger at more negative potentials (lower a_{Cl}^i; e.g., see Deisz and Lux, 1982). Thus, these results also provide good evidence that the nonpassive Cl^- distribution recorded in these denervated fibers was real. Earlier work in both mammalian cardiac (Vaughan-Jones, 1977b) and smooth muscle (Aickin and Brading, 1984) showed that Cl^-/HCO_3^- exchange was largely responsible for active Cl^- accumulation. However, in rat skeletal muscle, application of the anion-exchange inhibitor SITS has no detectable effect on the Cl^- disequilibrium. (It should be noted that the experiments were performed in the nominal absence of CO_2, when any contribution from anion exchange would have been small.) In normal, innervated muscle, removal of external Na^+ or K^+ resulted in no significant effects on a_{Cl}^i. However, when 9-AC was applied and a significant difference between E_{Cl} and E_m recorded, application of furosemide (10 μM) reversibly reduced E_{Cl} and E_m.

This approach has recently been extended with continuous recording of a_{Cl}^i and E_m in rat lumbrical muscle fibers (Aickin *et al.*, 1988a,b). In the presence of 9-AC, Cl^--depleted fibers rapidly accumulate Cl^- on readmission of Cl_o^- such that E_{Cl} becomes significantly less negative than E_m. This is illustrated in Fig. 8. In the absence of K_o^+ (Fig. 8A) or Na_o^+ (Fig. 8B), the rise of a_{Cl}^i on readmission of Cl_o^- is slowed and restricted such that E_{Cl} does not become less negative than E_m. Removal of both cations is no more effective than removal of only one and readdition of the cations causes an immediate rise in a_{Cl}^i and divergence of E_{Cl} and E_m (Aickin *et al.*, 1988a,b). This confirms the earlier steady-state measurements (Harris and Betz, 1987) and indicates that the Na^+ and K^+ requirements are on a single mechanism. In these experi-

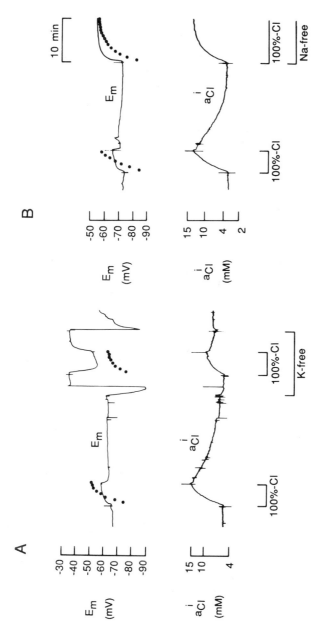

Figure 8. Pen recordings of experiments investigating the role of (A) K^+ and (B) Na^+ in the accumulation of Cl_i^- by rat lumbrical muscles. The Cl^- conductance was inhibited throughout by the presence of 100 μM 9-AC. Double-barreled microelectrodes were used with RLIE to measure E_m and the Corning Cl^- ion-exchanger to measure a_{Cl}^i. E_{Cl}, calculated from the recorded a_{Cl}^i, has been plotted as dots on the recording of E_m. Both preparations were superfused with Cl^--free solutions except where indicated otherwise and were maintained in the nominal absence of CO_2. Solutions were equilibrated with 100% O_2 and buffered to pH 7.4 with PIPES-SO_4 at 20°C. Tetrodotoxin was present throughout.

ments, furosemide (10 μM) decreased but did not eliminate active accumulation of Cl^-. But it did greatly reduce the stimulatory effect of readdition of external K^+ and/or Na^+ (Aickin *et al.*, 1988a,b). Incomplete inhibition of Na^+,K^+,Cl^- cotransport by furosemide at this concentration is observed in other preparations (see Chipperfield, 1986) but higher concentrations add significantly to the intracellularly recorded a^i_{Cl} (C. C. Aickin, unpublished observations; see also Chao and Armstrong, 1987). Again, experiments were conducted in the nominal absence of CO_2 and HCO_3^-, but changing to solutions buffered with 5% CO_2/24 mM HCO_3^- had no effect on the rate or extent of Cl^- accumulation—nor did the application of DIDS (Aickin *et al.*, 1988a,b). Thus, it seems clear that anion exchange plays no role in the active accumulation of Cl^- in skeletal muscle. Rather, this seems to be performed exclusively by a Na^+- and K^+-dependent mechanism probably equivalent to the Na^+,K^+,Cl^- cotransport system recently identified in mammalian smooth muscle (see Section 2.2.2) and cultured chick heart cells (Liu *et al.*, 1987).

Earlier reports in both amphibian and mammalian muscle are consistent with these findings. Although Bolton and Vaughan-Jones (1977) made no detailed investigation of the mechanism responsible for the accumulation of Cl^- in solutions of low pH in frog sartorius, they concluded that anion exchange was an unlikely candidate. Their best indications for active accumulation were obtained in the nominal absence of CO_2 and HCO_3^-. More recently, Macchia (1982) has presented some rather intriguing flux studies in toad hind limb muscles, which led him to conclude there was an electroneutral inward cotransport of Na^+ and Cl^- (although the results were not inconsistent with Na^+,K^+,Cl^- cotransport.) Results in mammalian skeletal muscle had given rather strong indications of the mechanism involved in the inward transport of Cl^-. Kernan (1968) reported that the rat extensor digitorum longus gained less Na^+ on inhibition of the Na^+ pump if Cl^-_o was removed and equally that Na^+-loaded muscles lost more Na^+ on reactivation of the Na^+ pump in the absence of Cl^-_o. More recently, he has shown directly that the rate of rise of a^i_{Na} on inhibition of the Na^+ pump is slowed by the presence of furosemide and that this rise in a^i_{Na} is accompanied by a rise in a^i_{Cl} which is also inhibited by the presence of furosemide (Kernan, 1986). Although these results were not presented as evidence for the mechanism of active transport of Cl^-, they clearly show the presence of a Cl^--dependent Na^+ entry. Betz and his colleagues (Betz *et al.*, 1984), on the other hand, had distinct reasons to believe that their results were evidence for the mechanism of active Cl^- uptake. Their investigation of an endogenous outward current led to the conclusion that the current was carried by Cl^-—a conclusion that requires the active accumulation of Cl^- (see Section 3.1). This current was greatly reduced or abolished by application of furosemide (10 μM) or bumetamide (1 μM) and by the removal of Na^+, and it was temporarily reversed following depolarization on removal of K^+_o before again being abolished. Thus, inward Cl^- transport via a Na^+- and K^+-dependent mechanism was inferred. Many researchers have been, perhaps forgivably, skeptical about the basis of these conclusions but, in the light of the recent investigations using direct measurement of a^i_{Cl} (Harris and Betz, 1987; Aickin *et al.*, 1988a,b), there can be little doubt that the results were correctly interpreted.

3.3. Involvement of Cl^- Transport in the Regulation of pH_i

Because pH_i in skeletal muscle is maintained at a level considerably more alkaline than that predicted for a passive distribution (e.g., Bolton and Vaughan-Jones, 1977; Aickin and Thomas, 1977a; Roos and Boron, 1978; Vigne et al., 1984), a simple equilibrating anion-exchange mechanism would tend, under normal conditions, to cause an efflux of HCO_3^- in exchange for an influx of Cl^-. This acidifying effect would be far more pronounced in skeletal muscle than smooth muscle (see Sections 2.2.1 and 2.3) since the disparity in the Cl^- and HCO_3^- gradients is much greater— E_{HCO_3} is about -20 mV whereas E_{Cl}, approximating E_m, is -80 to -90 mV (e.g., Bolton and Vaughan-Jones, 1977; Aickin and Thomas, 1977a). If $E_{HCO_3} = -90$ mV, pH_i would be around 6.0. In other words, an anion-exchange mechanism similar to the band 3 protein of erythrocytes would cause a considerable intracellular acid load. The input of some form of energy, for example by coupling turnover of the exchange either to an influx of Na^+, as in the well-characterized Na^+-dependent, Cl^-/HCO_3^- exchange mechanism (see Thomas, 1977; Boron et al., 1981; Boron and Russell, 1983), or to the hydrolysis of ATP, could reverse the direction of the exchanger and so convert it into a mechanism poised for extrusion of acid equivalents.

3.3.1. Energy-Dependent Cl^-/HCO_3^- Exchange in Mammalian Skeletal Muscle

As illustrated in Fig. 9, recovery of pH_i from acidosis in the mouse soleus is significantly slowed by the nominal absence of CO_2 and HCO_3^-. The rate constant of the recovery is decreased by a mean $35 \pm 10\%$ ($n = 7$; Aickin and Thomas, 1977b). Since the intracellular buffering power is also reduced by the nominal absence of CO_2 (from 58 meq H^+/pH unit per liter to 45 meq H^+/pH unit per liter; Aickin and Thomas, 1977a), the observed slowing of the pH_i recovery represents an even greater slowing of the effective extrusion of acid equivalents. This would suggest an involvement of HCO_3^-. Further indication of a role for anion exchange in acid extrusion is given by the inhibitory effect of the anion-exchange inhibitor SITS. In the presence of 5% CO_2, application of SITS decreases the rate of pH_i recovery from acidosis by about 30% (Aickin and Thomas, 1977b). Shortly after publication of these results, Roos and Boron (1978) painstakingly repeated some of the experiments in rat diaphragm, using the distribution of DMO to determine pH_i. Despite the limitations of this technique, they were able to demonstrate transient, if time-averaged, changes in pH_i caused by application and removal of acid loads. They concluded that the nominal absence of CO_2 and HCO_3^- reduced acid extrusion by about 20%, but it should be noted that this conclusion is entirely dependent on the accuracy of their choice of intracellular buffering power. They were unable to find any effect of SITS.

Dependence of acid extrusion on the presence of CO_2 and HCO_3^- and sensitivity to the stilbene disulfonate derivatives is insufficient information to characterize the mechanism involved. Na^+-dependent, Cl^-/HCO_3^- exchange, Na^+,HCO_3^- cotransport, and Na^+-independent Cl^-/HCO_3^- exchange are all dependent on the presence

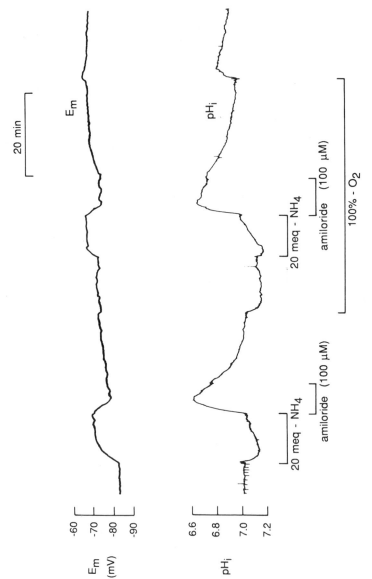

Figure 9. Pen recordings of an experiment illustrating the contribution of a CO_2- and HCO_3^--dependent mechanism to the recovery from intracellular acidosis in the mouse soleus. The recordings were made with a conventional voltage electrode, filled with 2.5 M KCl, and a separate, recessed-tip, pH-sensitive microelectrode (Thomas, 1974) inserted into the same fiber. Intracellular acidosis was produced by removal of NH_4^+ after a short application. The preparation was superfused with solutions equilibrated with 5% CO_2, 95% O_2 at pH 7.4 except for the interval indicated and maintained at 37°C. Nominally CO_2-free solutions were equilibrated with 100% O_2, buffered to pH 7.4 with Na-HEPES. The contribution of Na^+/H^+ exchange to the recovery was temporarily inhibited by application of amiloride. (Reproduced from Aickin and Thomas, 1977b.)

of CO_2 while these three plus a Na^+,monocarboxylate cotransport are sensitive to the stilbene disulfonate derivatives (see Knauf and Rothstein, 1971; Thomas, 1977; Boron *et al.*, 1981; Boron and Russell, 1983; Boron and Boulpaep, 1983; Boron *et al.*, 1988). As shown in Figs. 9 and 10, recovery from acidosis in the mouse soleus muscle is largely, but not completely inhibited by the absence of Na_o^+ and equally inhibited by the presence of amiloride, a now-accepted inhibitor of Na^+/H^+ exchange (e.g., see Aronson, 1985). Further, and most significantly, either the nominal absence of CO_2 or the presence of SITS adds to the inhibition caused by the absence of Na_o^+ or the presence of amiloride, as shown in Figs. 9 and 10. It could be argued that Li^+ was a poor choice of Na^+ substitute, although unfortunately the other substitutes that were tried caused contraction and dislodged the intracellular electrodes! Lithium was subsequently shown to be transported by the Na^+/H^+ exchange mechanism (e.g., Kinsella and Aronson, 1981; Moody, 1981; Piwnica-Worms *et al.*, 1985; Ellis and MacLeod, 1985). However, application of amiloride did not add to the inhibition caused by the absence of Na_o^+, when Li^+ was the substitute (Aickin and Thomas, 1977b). Lithium would have to be transported exclusively by a Na^+-dependent, Cl^-/HCO_3^- exchange mechanism or amiloride would have to be unable to block Li^+/H^+ exchange (but see Ives *et al.*, 1983; Gunther and Wright, 1983) for these results to be interpretable as anything other than evidence for a Na^+-*independent*, anion-exchange mechanism. Significantly, Roos and Boron (1978) showed no more than a 50% inhibition of acid extrusion over a 2-hr period in Na^+-free solution substituted by Mg^{2+} and sucrose, neither of which would seem likely to be transported in place of Na^+. Some indication that the suggested anion-exchange mechanism does have the required input of metabolic energy in mouse soleus muscle is given by the observation that the pH_i recovery from acidosis believed to be mediated by this mechanism alone (recovery in the presence of amiloride) has a Q_{10} of 6.9, within the temperature range of 28–37°C (Aickin and Thomas, 1977b).

One of the most important pieces in the jigsaw of evidence for the involvement of Cl^-/HCO_3^- exchange in acid extrusion in the mouse soleus is still missing. No investigation of the effect of Cl^- removal has been attempted in this preparation. However, the absence of Cl^- caused a 50% reduction in acid extrusion in the rat diaphragm (Roos and Boron, 1978). Strangely, this experiment was performed in the nominal absence of CO_2 and HCO_3^- when an anion-exchange mechanism would be expected to have been already largely inhibited. The pronounced inhibition, therefore, perhaps suggests a very high affinity of the exchanger for HCO_3^- and an adequate intracellular production of CO_2 and HCO_3^- from metabolism to ensure operation of the exchanger when CO_2 and HCO_3^- are omitted from the experimental solution (e.g., see Aickin and Brading, 1984). Such could explain the relatively small degree of inhibition observed in the absence of external CO_2 and much greater inhibition observed on removal of Cl_o^-. However, this result is difficult to reconcile with the apparent lack of effect of SITS in this preparation. Clearly this problem needs further investigation before we can be completely satisfied with the unique anion-exchange mechanism suggested by the results in mouse soleus.

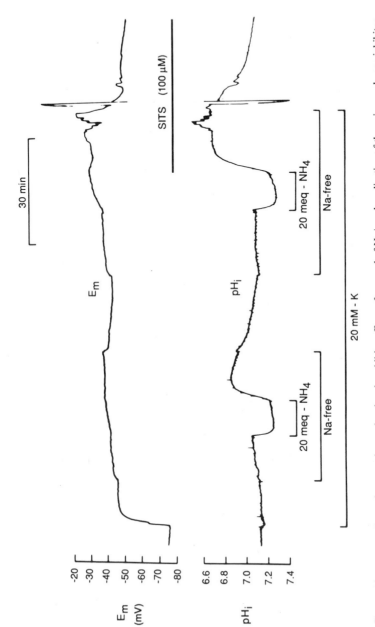

Figure 10. Pen recordings of an experiment showing the additive effects of removal of Na_o^+ and application of the anion-exchange inhibitor SITS on the recovery from intracellular acidosis in the mouse soleus. The recordings were made with a single-barreled conventional, 2.5 M KCl-filled, and a recessed-tip pH-sensitive microelectrode inserted into the same fiber. Intracellular acidosis was induced by removal of NH_4^+. The preparation was superfused with solutions equilibrated with 5% CO_2, 95% O_2 (pH 7.4) at 37°C throughout and the $[K^+]_o$ was raised to 20 mM to minimize contraction caused by the removal of Na_o^+. Equal and opposite changes in the E_m and pH_i traces recorded during the last 30 min shown were caused by instability in the voltage electrode impalement. [Reproduced with permission, from the *Annual Review of Physiology*, Vol. 48, 1986 by Annual Reviews Inc. (Aickin, 1986).]

3.3.2. Na^+-Dependent Cl^-/HCO_3^- Exchange in Frog Skeletal Muscle

Regulation of pH_i in amphibian skeletal muscle is not exactly the same as in mammalian skeletal muscle. First, the presence of a large, passive leak of HCO_3^- means that normally polarized fibers show no recovery from an acidosis induced by addition of CO_2 and HCO_3^- (Bolton and Vaughan-Jones, 1977; Abercrombie et al., 1983). Nevertheless, because pH_i is maintained at a considerably more alkaline level than predicted for a passive distribution, even in the face of the large HCO_3^- conductance, acid equivalents must be extruded from the fiber. The mechanisms involved in this extrusion can obviously be most easily studied when net passive movements of HCO_3^- are absent. Two approaches are possible: investigation of the recovery from acidosis either in the absence of CO_2 and HCO_3^- (see Putnam and Roos, 1985) or in the presence of CO_2 and HCO_3^- in depolarized fibers where membrane potential is close to the HCO_3^- equilibrium potential (Abercrombie et al., 1983). Clearly the latter is best suited to reveal a contribution from anion-exchange mechanisms. This approach has indeed shown that recovery from an acid load in the presence of 5% CO_2 is considerably faster if the fibers are depolarized at constant $[Cl^-]_o$ rather than at constant $K^+ \times Cl^-$ product, i.e., recovery is faster when $[Cl^-]_i$ is high. This suggests an involvement of Cl^- transport in the effective extrusion of acid equivalents, even if under rather unphysiological conditions! A 50% inhibition of the rate of acid extrusion under these conditions by the presence of SITS further supports the presence of an anion-exchange mechanism. Unlike in the mammalian skeletal muscle, this mechanism appears to be Na^+-dependent. Recovery from acidosis is virtually completely inhibited by the absence of Na_o^+ (substituted by N-methyl-D-glucamine) but only about 50% inhibited by the presence of amiloride. The effects of SITS and amiloride are at least partially additive but inhibition in the presence of both pharmacological agents is not as great as that caused by the absence of Na_o^+. Evidence for the operation of this Na^+-dependent, Cl^-/HCO_3^- exchange mechanism at normal $[Cl^-]_i$ is weak, but perhaps suggested by the slight inhibitory effect of SITS on the rate of acid extrusion in fibers depolarized at constant $K^+ \times Cl^-$ product, both by itself and in the presence of amiloride. It should, however, be pointed out that identification of transport processes from their pharmacological sensitivity can be unwise. The large number of processes inhibited by the stilbene disulfonate derivatives has already been mentioned but there are also instances where processes, apparently having all the characteristics of a mechanism sensitive to this class of drugs, fail to be inhibited by them (Aickin, 1988). In this respect it is worth noting that in hypertonic solutions, frog fibers depolarized with constant $[Cl^-]_o$ recovered faster from acidosis than those depolarized at constant $K^+ \times Cl^-$ product, but showed no sensitivity to SITS (Abercrombie and Roos, 1983).

3.3.3. Na^+-Independent Acid Extrusion in Cultured Chick Skeletal Muscle

The presence of Na^+/H^+ exchange in the extrusion of acid equivalents seems to be universal to vertebrate skeletal muscle and compelling evidence for the importance of this mechanism has been presented in cultured chick skeletal muscle (Vigne et al.,

1982, 1984). It is, however, notable that these experiments were performed in the nominal absence of CO_2 and HCO_3^- when any contribution from anion exchange would have been small. Nevertheless, it is interesting that inhibition of Na^+/H^+ exchange, either by potent derivatives of amiloride or by abolition of the transmembrane Na^+ gradient, resulted in a stable pH_i apparently significantly alkaline to that predicted by a passive distribution (Vigne *et al.*, 1984). This would suggest the presence of a Na^+-independent mechanism capable of effectively extruding acid equivalents.

3.3.4. Alkali Extrusion from Skeletal Muscle

The participation of transport mechanisms in the recovery of pH_i from alkalosis has not been investigated in skeletal muscle. As in most other preparations, passive permeation of H^+, OH^-, or HCO_3^- and metabolic production of acid has been assumed to be responsible (e.g., Aickin and Thomas, 1977a; Roos and Boron, 1978). Nevertheless, there is evidence to suggest that anion exchange is not involved, in contrast to both cardiac and smooth muscle (see Section 2.3). The slow acidification observed during continued exposure to an ammonium salt in mouse soleus muscle was not measurably affected by the nominal absence of CO_2 and HCO_3^- (see Fig. 9) or by the presence of SITS (Aickin, 1977).

4. CONCLUSIONS

Far from the early concept that the passage of Cl^- across the sarcolemma was simply one of passive diffusion via a very high sarcolemmal conductance, we now know that Cl^- is actively transported into the sarcoplasm, be it of smooth, skeletal, or cardiac muscle. The consequences of this transport are, of course, dependent on the electrical permeability of the sarcolemma to Cl^-. In both smooth and cardiac muscle, where the electrical permeability is now known to be low, $[Cl^-]_i$ is far higher than predicted for a passive distribution. In skeletal muscle, where the Cl^- permeability normally dominates the resting membrane conductance, $[Cl^-]_i$ is close to, if not the same as, that predicted by the Nernst relation between the membrane potential and $[Cl^-]_o$. However, it now seems a moot point whether the membrane potential dictates the $[Cl^-]_i$ or vice versa.

Surprisingly, inward transport of Cl^- is not mediated by a single mechanism, analogous to the universal Na^+,K^+-ATPase for extrusion of Na^+. Depending on the particular tissue, Na^+-dependent Cl^-/HCO_3^- exchange, Na^+-independent Cl^-/HCO_3^- exchange, and Na^+,K^+,Cl^- cotransport can be involved either alone or in combination. Nor is Cl^- transport restricted to the regulation of $[Cl^-]_i$. It is also involved in the regulation of pH_i. In smooth muscle, this seems to be limited to the action of Na^+-independent Cl^-/HCO_3^- exchange, most importantly in the recovery from intracellular alkalosis. But under conditions of extreme acidosis, the exchange appears to reverse and so assist recovery by effectively extruding acid equivalents. There is as yet no indication for the involvement of transport processes in the recovery from intracellular alkalosis in skeletal muscle but, on the other hand, Cl^- transport does appear to be involved in recovery from intracellular acidosis in this muscle type. Na^+-dependent, Cl^-/HCO_3^- exchange contributes to this recovery in amphibian skeletal muscle, at least to a clearly demonstrable extent when $[Cl^-]_i$ is elevated. This mechanism does not seem to be present in mammalian skeletal muscle, where there is evidence for a unique anion-exchange mechanism—unique in as much as it must be linked to a source of energy other than that of the net uptake of Na^+ to drive the uphill movement of HCO_3^- into the cell. In this case, the counter movement of Cl^- is only assumed and further experiments are required to characterize this mechanism beyond its HCO_3^--dependence and sensitivity to SITS.

The presence of inward Cl^- transport mechanisms in all types of muscle raises the question of why the muscle cell should require an elevated $[Cl^-]_i$. Clearly in both smooth and cardiac muscle, the considerable accumulation of Cl_i^- provides a significant depolarizing battery which could be used in the generation of electrical events. There has been considerable debate as to whether voltage-activated Cl^- currents are associated with the action potential, either contributing to its shape and duration or contributing to the pacemaker current, in both cardiac (e.g., Carmeliet, 1961; Millar and Vaughan-Williams, 1981; Noble, 1984) and smooth muscle (e.g., Ohashi, 1970b; Vassort, 1981). Although the present consensus of opinion assigns no role to Cl^- channels in intrinsic electrical activity of either muscle type (DiFrancesco and Noble, 1985; Tomita, 1981; Vassort, 1981), it would seem unwise to discount the possibility following the demonstration of voltage-activated Cl^- channels in preparations from the calf cardiac sarcolemma (Coronado and Latorre, 1982), rat carotid artery (Shoemaker *et al.*, 1985), and guinea pig uterus (Coleman and Parkington, 1987). In addition, there is good evidence that some transmitters activate Cl^- channels in a variety of smooth muscles—for example, histamine (Kirkpatrick, 1981) and catecholamines (Ohashi, 1971; Wahlström, 1973b; Marshall, 1977; Large, 1984; Byrne and Large, 1987). Accumulation of Cl^- in skeletal muscle is very much less, if indeed at all, detectable. Under normal conditions, for inward transport of Na^+, K^+, and Cl^- to occur simply for Cl^- to leak out of the fiber again seems a rather pointless exercise. Its end product, to all intents and purposes, is an increase in the Na^+ leak, which has to be counteracted with expenditure of energy by the Na^+ pump, and an influx of water. Perhaps there lies a clue—cotransport has of course been intimately associated with volume regulation since it was first described (e.g., Kregenow and Caryk, 1979; Geck *et al.*, 1980). It does, however, appear that even a very limited accumulation of Cl_i^- can provide the basis for an outward current, localized to the neuromuscular junction. Although the function of this current is not known, its precise location and persistence after denervation (Betz *et al.*, 1986) suggest that it may provide a target for the growing motor nerve.

ACKNOWLEDGMENT. I am grateful to the MRC for support through a senior research fellowship.

REFERENCES

Abercrombie, R. F., and Roos, A., 1983, The intracellular pH of frog skeletal muscle: Its regulation in hypertonic solutions, *J. Physiol. (London)* **345**:189–204.

Abercrombie, R. F., Putnam, R. W., and Roos, A., 1983, The intracellular pH of frog skeletal muscle: Its regulation in isotonic solutions, *J. Physiol. (London)* **345**:175–187.

Adrian, R. H., 1961, Internal chloride concentration and chloride efflux of frog muscle, *J. Physiol. (London)* **156**:623–632.

Adrian, R. H., and Marshall, M. W., 1976, Action potentials reconstructed in normal and myotonic muscle fibres, *J. Physiol. (London)* **258**:125–143.

Aickin, C. C., 1977, Intracellular pH of mammalian skeletal muscle, Ph.D. thesis, Bristol University.

Aickin, C. C., 1981, A double-barrelled micro-electrode suitable for measurement of intracellular chloride activity (a_{Cl}^i) in guinea-pig vas deferens, *J. Physiol. (London)* **320**:4P–5P.

Aickin, C. C., 1984, Direct measurement of intracellular pH and buffering power in smooth muscle cells of guinea-pig vas deferens, *J. Physiol. (London)* **349**:571–585.

Aickin, C. C., 1985, The effect of Na^+ and HCO_3^- ions on recovery from an acid load in the smooth muscle of guinea-pig ureter, *J. Physiol. (London)* **369**:80P.

Aickin, C. C., 1986, Intracellular pH regulation by vertebrate muscle, *Annu. Rev. Physiol.* **48**:349–361.

Aickin, C. C., 1987a, Investigation of factors affecting the intracellular sodium activity in the smooth muscle of guinea-pig ureter, *J. Physiol. (London)* **385**:483–505.

Aickin, C. C., 1987b, Na,K,Cl co-transport is involved in Cl accumulation in the smooth muscle of isolated guinea-pig vas deferens, *J. Physiol. (London)* **394**:87P.

Aickin, C. C., 1988, Movements of acid equivalents across the mammalian smooth muscle membrane, *CIBA Found. Symp.* **139**:3–22.

Aickin, C. C., and Brading, A. F., 1982a, Measurement of intracellular chloride in guinea-pig vas deferens by ion analysis, [36]chloride efflux and micro-electrodes, *J. Physiol. (London)* **326**:139–154.

Aickin, C. C., and Brading, A. F., 1982b, The effect of CO_2/HCO_3 and DIDS on Cl movements in guinea-pig vas deferens, *J. Physiol. (London)* **327**:74P–75P.

Aickin, C. C., and Brading, A. F., 1983, Towards an estimate of chloride permeability in the smooth muscle of guinea-pig vas deferens, *J. Physiol. (London)* **336**:179–197.

Aickin, C. C., and Brading, A. F., 1984, The role of chloride–bicarbonate exchange in the regulation of intracellular chloride in guinea-pig vas deferens, *J. Physiol. (London)* **349**:587–606.

Aickin, C. C., and Brading, A. F., 1985a, Advances in the understanding of transmembrane ionic gradients and permeabilities in smooth muscle obtained by using ion-selective micro-electrodes, *Experientia* **41**:879–887.

Aickin, C. C., and Brading, A. F., 1985b, The effects of bicarbonate and foreign anions on chloride transport in smooth muscle of the guinea-pig vas deferens, *J. Physiol. (London)* **366**:267–280.

Aickin, C. C., and Thomas, R. C., 1977a, Micro-electrode measurement of the intracellular pH and buffering power of mouse soleus muscle fibres, *J. Physiol. (London)* **267**:791–810.

Aickin, C. C., and Thomas, R. C., 1977b, An investigation of the ionic mechanism of intracellular pH regulation in mouse soleus muscle fibres, *J. Physiol. (London)* **273**:295–316.

Aickin, C. C., and Vermuë, N. A., 1983, Microelectrode measurement of intracellular chloride activity in smooth muscle cells of guinea-pig ureter, *Pfluegers Arch.* **397**:25–28.

Aickin, C. C., Brading, A. F., and Burdyga, Th. V., 1984, Evidence for sodium–calcium exchange in the guinea-pig ureter, *J. Physiol. (London)* **347**:411–430.

Aickin, C. C., Betz, W. J., and Harris, G. L., 1988a, Inhibition of resting Cl conductance (G_{Cl}) reveals active accumulation of Cl by Na,K,Cl co-transport in isolated rat skeletal muscle, *Pfluegers Arch.* **411** (Suppl. 1):R188.

Aickin, C. C., Betz, W. J., and Harris, G. L., 1988b, Intracellular chloride and the mechanism for its accumulation in rat lumbrical muscle, *J. Physiol. (London)* **411**:437–455.

Ammann, D., Lanter, F., Steiner, R. A., Schulthess, P., Shijo, Y., and Simon, W., 1981, Neutral carrier based hydrogen ion selective microelectrode for extra- and intracellular studies, *Anal. Chem.* **53**:2267–2269.

Armstrong, W. M., Wojtkowski, W., and Bixenman, W. R., 1977, A new solid-state microelectrode for measuring intracellular chloride activities, *Biochim. Biophys. Acta* **465:**165–170.

Aronson, P. S., 1985, Kinetic properties of the plasma membrane Na$^+$-H$^+$ exchanger, *Annu. Rev. Physiol.* **47:**545–560.

Ashley, C. C., Ellory, J. C., Lea, T. J., and Ramos, M., 1978, The effect of inhibitors on ^{36}Cl efflux from barnacle muscle fibres, *J. Physiol. (London)* **285:**52P–53P.

Axelsson, J., and Holmberg, B., 1971, The effects of K-free solution on tension development in the smooth muscle taenia coli from the guinea-pig, *Acta Physiol. Scand.* **82:**322–332.

Baumgarten, C. M., and Fozzard, H. A., 1981, Intracellular chloride activity in mammalian ventricular muscle, *Am. J. Physiol.* **241:**C121–C129.

Betz, W. J., Caldwell, G. H., and Kinnamon, S. C., 1984, Physiological basis of a steady electric current in rat skeletal muscle, *J. Gen. Physiol.* **83:**175–192.

Betz, W. J., Caldwell, G. H., and Harris, G. L., 1986, Effect of denervation on a steady electric current generated at the end plate region of rat skeletal muscle, *J. Gen. Physiol.* **373:**97–114.

Birnberger, K. L., and Kelpzig, M., 1979, Influence of extracellular potassium and intracellular pH on myotonia, *J. Neurol.* **222:**23–35.

Blum, R., and Westphal, W., 1981, On the actions of a chloride conductance blocking agent (anthracene-9-COOH) on the resting potential of a single mammalian skeletal muscle fibre, *Pfluegers Arch.* **389:**R45.

Bolton, T. B., 1973, Effects of electrogenic sodium pumping on the membrane potential of longitudinal smooth muscle from the terminal ileum of guinea-pig, *J. Physiol. (London)* **228:**693–712.

Bolton, T. B., and Vaughan-Jones, R. D., 1977, Continuous direct measurement of intracellular chloride and pH in frog skeletal muscle, *J. Physiol. (London)* **270:**801–833.

Boron, W. F., and Boulpaep, E. L., 1983, Intracellular pH regulation in the renal proximal tubule of the salamander. Basolateral HCO$_3^-$ transport, *J. Gen. Physiol.* **81:**53–94.

Boron, W. F., and Russell, J. M., 1983, Stoichiometry and ion dependencies of the intracellular-pH-regulating mechanism in squid giant axons, *J. Gen. Physiol.* **81:**373–399.

Boron, W. F., McCormick, W. C., and Roos, A., 1981, pH regulation in barnacle muscle fibers: Dependence on extracellular sodium and bicarbonate, *Am. J. Physiol.* **240:**C80–C89.

Boron, W. F., Siebens, A. W., and Nakhoul, N. L., 1988, The role of monocarboxylate transport in the regulation of intracellular pH of renal proximal tubule cells, *CIBA Found. Symp.* **139:**91–105.

Boyle, P. J., and Conway, E. J., 1941, Potassium accumulation in muscle and associated changes, *J. Physiol. (London)* **100:**1–63.

Brading, A. F., 1973, Ion distribution and ion movements in smooth muscle, *Philos. Trans. R. Soc. London Ser. B* **265:**35–46.

Brading, A. F., 1987, The effect of Na ions and loop diuretics on transmembrane ^{36}Cl fluxes in isolated guinea-pig smooth muscles, *J. Physiol. (London)* **394:**86P.

Brading, A. F., and Tomita, T., 1968, Effects of anions on the volume of a smooth muscle, *Nature* **218:**276–277.

Brading, A. F., and Widdicombe, J. H., 1977, The use of lanthanum to estimate the numbers of extracellular cation-exchanging sites in the guinea-pig's taenia coli, and its effect on transmembrane monovalent ion movements, *J. Physiol. (London)* **266:**255–273.

Brazy, P. C., and Gunn, R. B., 1976, Furosemide inhibition of chloride transport in human red blood cells, *J. Gen. Physiol.* **68:**583–599.

Bretag, A. H., 1987, Muscle chloride channels, *Physiol. Rev.* **67:**618–724.

Bretag, A. H., Dawe, S. R., Kerr, D. I. B., and Moskwa, A. G., 1980, Myotonia as a side effect of diuretic action, *Br. J. Pharmacol.* **71:**467–471.

Bryant, S. H., and Morales-Aguilera, A., 1971, Chloride conductance in normal and myotonic muscle fibres and the action of monocarboxylic aromatic acids, *J. Physiol. (London)* **219:**367–383.

Byrne, N. G., and Large, W. A., 1987, Action of noradrenaline on single smooth muscle cells freshly dispersed from the rat anococcygeus muscle, *J. Physiol. (London)* **389:**513–525.

Caldwell, J. H., and Betz, W. J., 1984, Properties of an endogenous steady current in rat muscle, *J. Gen. Physiol.* **83:**157–173.

Camerino, D., and Bryant, S. H., 1976, Effects of denervation and colchicine treatment on the chloride conductance of rat skeletal muscle fibers, *J. Neurobiol.* **7:**221–228.

Carmeliet, E. E., 1961, Chloride ions and the membrane potential of Purkinje fibres, *J. Physiol. (London)* **156:**375–388.

Casteels, R., 1966, The action of ouabain on the smooth muscle cells of the guinea-pig's taenia coli, *J. Physiol. (London)* **186:**131–142.

Casteels, R., 1969a, Calculation of the membrane potential in smooth muscle cells of the guinea-pig's taenia coli by the Goldman equation, *J. Physiol. (London)* **205:**193–208.

Casteels, R., 1969b, Ion content and ion fluxes in the smooth muscle cells of the longitudinal layer of the guinea-pig's vas deferens, *Pfluegers Arch.* **313:**95–105.

Casteels, R., 1971, The distribution of chloride ions in the smooth muscle cells of the guinea-pig's taenia coli, *J. Physiol. (London)* **214:**225–243.

Casteels, R., 1981, Membrane potential in smooth cells, in: *Smooth Muscle: An Assessment of Current Knowledge* (E. Bülbring, A. F. Brading, A. W. Jones, and T. Tomita, eds.), Arnold, London, pp. 105–126.

Casteels, R., and Kuriyama, H., 1966, Membrane potential and ion content in the smooth muscle of the guinea-pig's taenia coli at different external potassium concentrations, *J. Physiol. (London)* **184:**120–130.

Chao, A. C., and Armstrong, W. M., 1987, Cl^--selective microelectrodes: Sensitivity to anionic Cl^- transport inhibitors, *Am. J. Physiol.* **253:**C343–C347.

Chipperfield, A. R., 1981, Chloride dependence of frusemide- and phloretin-sensitive passive sodium and potassium fluxes in human red cells, *J. Physiol. (London)* **312:**435–444.

Chipperfield, A. R., 1986, The $(Na^+-K^+-Cl^-)$ co-transport system, *Clin. Sci.* **71:**465–476.

Coleman, H. A., and Parkington, H. C., 1987, Single channel Cl^- and K^+ currents from cells of uterus not treated with enzymes, *Pfluegers Arch.* **410:**560–562.

Coronado, R., and Latorre, R., 1982, Detection of K^+ and Cl^- channels from calf cardiac sarcolemma in planar lipid bilayer membranes, *Nature* **298:**849–852.

DeCoursey, T. E., Younkin, S. G., and Bryant, S. H., 1978, Neural control of chloride conductance in rat extensor digitorum longus muscle, *Exp. Neurol.* **61:**705–709.

Deisz, R. A., and Lux, H. D., 1982, The role of intracellular chloride in hyperpolarizing post-synaptic inhibition of crayfish stretch receptor neurones, *J. Physiol. (London)* **326:**123–138.

Deitmer, J. W., and Schlue, W.-R., 1987, The regulation of intracellular pH by identified glial cells and neurones in the central nervous system of the leech, *J. Physiol. (London)* **388:**261–283.

DiFrancesco, D., and Noble, D., 1985, A model of cardiac electrical activity incorporating ionic pumps and concentration changes, *Philos. Trans. R. Soc. London Ser. B* **307:**353–398.

Donaldson, P. J., and Leader, J. P., 1984, Intracellular ionic activities in the EDL of the mouse, *Pfluegers Arch.* **400:**166–170.

Dulhuntry, A. F., 1978, The dependence of membrane potential on extracellular chloride concentration in mammalian skeletal muscle, *J. Physiol. (London)* **276:**67–82.

Ellis, D., and MacLeod, K. T., 1985, Sodium-dependent control of intracellular pH in Purkinje fibres of sheep heart, *J. Physiol. (London)* **359:**81–105.

Ellis, D., and Thomas, R. C., 1976, Direct measurement of the intracellular pH of mammalian cardiac muscle, *J. Physiol. (London)* **262:**755–771.

Ellory, J. C., and Stewart, G. W., 1982, The human Cl-dependent Na-K cotransport system as a possible model for studying the action of loop diuretics, *Br. J. Pharmacol.* **75:**183–188.

Ellory, J. C., Dunham, P. B., Logue, P. J., and Stewart, G. W., 1982, Anion-dependent cation transport in erythrocytes, *Philos. Trans. R. Soc. London Ser. B* **299:**483–495.

Elsner, P., and Westphal, W., 1979, External chloride concentration and membrane resting potential of mammalian skeletal muscle fibres, *Pfluegers Arch.* **382:**R24.

Fenn, W. O., 1936, Electrolytes in muscle, *Physiol. Rev.* **16:**450–487.

Galler, S., and Moser, H., 1986, The ionic mechanism of intracellular pH regulation in crayfish muscle fibres, *J. Physiol. (London)* **374:**137–151.

Garay, R. P., Dagher, G., Pernollet, M.-G., Devynck, M.-A., and Meyer, P., 1980, Inherited defect in a Na^+,K^+-cotransport system in erythrocytes from essential hypertensive patients, *Nature* **284:**281–283.

Geck, P., Pietrzyk, C., Burckhardt, B. C., Pfeiffer, B., and Heinz, E., 1980, Electrically silent cotransport of Na, K and Cl in Ehrlich cells, *Biochim. Biophys. Acta* **600**:432–447.

Gerstheimer, F. P., Mühleisen, M., Nehring, D., and Kreye, V. A. W., 1987, A chloride–bicarbonate exchanging anion carrier in vascular smooth muscle of the rabbit, *Pfluegers Arch.* **409**:60–66.

Goldman, D. E., 1943, Potential, impedance and rectification in membranes, *J. Gen. Physiol.* **27**:37–60.

Griggs, R. C., Moxley, R. T., Riggs, J. E., and Engel, W. K., 1978, Effects of acetazolamide on myotonia, *Ann. Neurol.* **3**:531–538.

Gunther, R. D., and Wright, E. M., 1983, Na$^+$, Li$^+$ and Cl$^-$ transport by brush border membranes from rabbit jejunum, *J. Membr. Biol.* **74**:85–94.

Hagiwara, S., and Takahashi, K., 1974, Mechanism of anion permeation through the muscle fibre membrane of an elasmobranch fish, *Taeniura lymma, J. Physiol. (London)* **238**:109–127.

Hamon, G., Papadimitriou, A., and Worcel, M., 1976, Ionix fluxes in rat uterine smooth muscle, *J. Physiol. (London)* **254**:229–243.

Hannafin, J., Kinne-Saffran, E., Frieman, D., and Kinne, R., 1983, Presence of a sodium–potassium chloride cotransport system in the rectal gland of *Squalus acanthias, J. Membr. Biol.* **75**:73–83.

Harris, E. J., 1963, Distribution and movement of muscle chloride, *J. Physiol. (London)* **166**:87–109.

Harris, G. L., and Betz, W. J., 1987, Evidence for active chloride accumulation in normal and denervated rat lumbrical muscle, *J. Gen. Physiol.* **90**:127–144.

Hironaka, T., and Morimoto, S., 1980, Intracellular chloride concentration and evidence for the existence of a chloride pump in frog skeletal muscle, *Jpn. J. Physiol.* **30**:357–363.

Hirst, G. D. S., and van Helden, D., 1982, Ionic basis of the resting potential of submucosal arterioles in the ileum of the guinea-pig, *J. Physiol. (London)* **333**:53–67.

Hodgkin, A. L., and Katz, B., 1949, The effect of sodium on the electrical activity of the giant axon of the squid, *J. Physiol. (London)* **108**:37–77.

Hodgkin, A. L., and Horowicz, P., 1959, The influence of potassium and chloride ions on the membrane potential of single muscle fibres, *J. Physiol. (London)* **148**:127–160.

Holman, M. E., 1958, Membrane potentials recorded with high-resistance microelectrodes and the effects of changes in ionic environment on the electrical and mechanical activity of the smooth muscle of the taenia coli of guinea-pig, *J. Physiol. (London)* **141**:464–488.

Hunter, M. J., 1977, Human erythrocyte anion permeabilities measured under conditions of net charge transfer, *J. Physiol. (London)* **268**:35–49.

Hutter, O. F., and Warner, A. E., 1967, The pH sensitivity of the chloride conductance of frog skeletal muscle, *J. Physiol. (London)* **189**:403–425.

Ighoroje, A. D., and Spurway, N. C., 1985, How does vascular smooth muscle in the isolated rabbit ear artery adapt its tone after alkaline or acid loads? *J. Physiol. (London)* **367**:46P.

Inoue, I., 1985, Voltage-dependent chloride conductance of the squid axon membrane and its blockade by some disulfonic stilbene derivatives, *J. Gen. Physiol.* **85**:519–537.

Ives, H. E., Yee, V. J., and Warnock, D. G., 1983, Mixed type inhibition of the renal Na$^+$/H$^+$ antiporter by Li$^+$ and amiloride, *J. Biol. Chem.* **258**:9710–9716.

Iyer, V. G., Ranish, N. A., and Fenichel, G. M., 1981, Ionic conductance and experimentally induced myotonia, *J. Neurol. Sci.* **49**:159–164.

Jones, A. W., 1974, Altered ion transport in large and small arteries from spontaneously hypertensive rats and the influence of calcium, *Circ. Res.* **34/35**(Suppl. 1):117–121.

Jones, A. W., and Karreman, G., 1969, Ion exchange properties of the canine carotid artery, *Biophys. J.* **9**:884–909.

Jones, A. W., Somlyo, A. P., and Somlyo, A. V., 1973, Potassium accumulation in smooth muscle and associated ultrastructural changes, *J. Physiol. (London)* **232**:247–273.

Kernan, R. P., 1968, Membrane potential and chemical transmitter in active transport of ions by rat skeletal muscle, *J. Gen. Physiol.* **51**:204S–210S.

Kernan, R. P., 1986, Chloride-dependent sodium influx into rat skeletal muscle fibres measured with ion-selective micro-electrodes, *J. Physiol. (London)* **371**:146P.

Kernan, R. P., and Westphal, W., 1984, The ionic basis of membrane hyperpolarization in rat diaphragm on exposure to chloride deficient physiological saline, *Irish J. Med. Sci.* **153**:198.

Kernan, R. P., MacDermott, M., and Westphal, W., 1974, Measurement of chloride activity within frog sartorius muscle fibres by means of chloride-sensitive micro-electrodes, *J. Physiol. (London)* **241**:60P–61P.

Keynes, R. D., 1963, Chloride in the squid giant axon, *J. Physiol. (London)* **169**:690–705.

Kinsella, J. L., and Aronson, P. S., 1981, Amiloride inhibition of the Na^+-H^+ exchanger in renal microvillus membrane vesicles, *Am. J. Physiol.* **241**:F374–F379.

Kirkpatrick, C. T., 1981, Tracheobronchial smooth muscle, in: *Smooth Muscle: An Assessment of Current Knowledge* (E. Bülbring, A. F. Brading, A. W. Jones, and T. Tomita, eds.), Arnold London, pp. 385–395.

Knauf, P. A., and Rothstein, A., 1971, Chemical modification of membranes. 1. Effect of sulfhydryl and amino reactive reagents on anion and cation permeability of the human red blood cells, *J. Gen. Physiol.* **58**:190–210.

Knauf, P. A., Fuhrmann, G. F., Rothstein, S., and Rothstein, A., 1977, The relationship between anion exchange and net anion flow across the human red blood cell membrane, *J. Gen. Physiol.* **69**:363–386.

Korbmacher, C., Helbig, H., Stahl, F., and Wiederholt, M., 1988, Evidence for Na/H exchange and Cl/HCO_3 exchange in AIO vascular smooth muscle cells, *Pfluegers Arch.* **412**:29–36.

Kregenow, F. M., and Caryk, T., 1979, Co-transport of cations and Cl during the volume regulatory responses of duck erythrocytes, *Physiologist* **22**:73.

Kreye, V. A. W., Bauer, P. K., and Villhauer, I., 1981, Evidence for furosemide-sensitive active chloride transport in vascular smooth muscle, *Eur. J. Pharmacol.* **73**:91–95.

Kuriyama, H., 1963, The influence of potassium, sodium and chloride on the membrane potential of smooth muscle of taenia coli, *J. Physiol. (London)* **166**:15–28.

Kwieciński, H., 1980, Treatment of myotonic dystrophy with acetazolamide, *J. Neurol.* **222**:261–263.

Kwieciński, H., 1981, Myotonia induced by chemical agents, *Crit. Rev. Toxicol.* **8**:279–310.

Kwieciński, H., Lehmann-Horn, F., and Rüdel, R., 1984, The resting membrane parameters of human intercostal muscle at low, normal and high extracellular potassium, *Muscle Nerve* **7**:60–65.

Large, W. A., 1984, The effect of chloride removal on the responses of the isolated rat anococcygeus muscle to α_1-adrenoceptor stimulation, *J. Physiol. (London)* **352**:17–29.

Leader, J. P., Bray, J. J., MacKnight, A. D. C., Mason, D. R., McCaig, D., and Mills, R. G., 1984, Cellular ions in intact and denervated muscles of the rat, *J. Membr. Biol.* **81**:19–27.

Lee, C. O., and Fozzard, H. A., 1975, Activities of potassium and sodium ions in rabbit heart muscle, *J. Gen. Physiol.* **65**:695–708.

Lev, A. A., 1964, Determination of activity and activity coefficients of K and Na ions in frog muscle fibres, *Nature* **201**:1132–1134.

Lipicky, R. J., and Bryant, S. H., 1966, Sodium, potassium and chloride fluxes in intercostal muscle from normal goats and goats with hereditary myotonia, *J. Gen. Physiol.* **50**:90–111.

Little, P. J., Cragoe, E. J., and Bobik, A., 1986, Na–H exchange is a major pathway for Na influx in rat vascular smooth muscle, *Am. J. Physiol.* **251**:C707–C712.

Liu, S., Jacob, R., Piwnica-Worms, S. D., and Lieberman, M., 1987, (Na + K + 2Cl) cotransport in cultured embryonic chick heart cells, *Am. J. Physiol.* **253**:C721–C730.

Lorkovic, H., and Tomanek, R. J., 1977, Potassium and chloride conductances in normal and denervated rat muscles, *Am. J. Physiol.* **232**:C109–C114.

McCaig, D., and Leader, J. P., 1984, Intracellular chloride activity in the extensor digitorum longus (EDL) of the rat, *J. Membr. Biol.* **81**:9–17.

Macchia, D., 1982, Chloride self-exchange in toad skeletal muscle in vivo and in vitro, *Am. J. Physiol.* **242**:C207–C217.

Macchia, D., and Baumgarten, C. M., 1979, Is chloride passively distributed in skeletal muscle in vivo? *Pfluegers Arch.* **382**:193–195.

Marshall, J. M., 1977, Modulation of smooth muscle activity by catecholamines, *Fed. Proc.* **36**:2450–2455.

Millar, J. S., and Vaughan-Williams, E. M., 1981, Pacemaker selectivity: Influence on rabbit atria of ionic environment and of alinidine, a possible anion antagonist, *Cardiovasc. Res.* **15**:335–350.

Montrose, M., Randles, J., and Kimmich, G. A., 1987, SITS-sensitive Cl⁻ conductance pathway in chick intestinal cells, *Am. J. Physiol.* **253:**C693–C699.

Moody, W. J., 1981, The ionic mechanism of intracellular pH regulation in crayfish neurones, *J. Physiol. (London)* **316:**293–308.

Moser, H., 1985, Intracellular pH regulation in the sensory neurone of the stretch receptor of the crayfish (Astacus fluviatilis), *J. Physiol. (London)* **362:**23–38.

Neild, T. O., and Thomas, R. C., 1974, Intracellular chloride activity and the effects of acetylcholine in snail neurones, *J. Physiol. (London)* **242:**453–470.

Nelson, D. J., Tang, J. M., and Palmer, L. G., 1984, Single channel recordings of apical membrane chloride conductance in A6 epithelial cells, *J. Membr. Biol.* **80:**81–89.

Noble, D., 1984, The surprising heart: A review of recent progress in cardiac electrophysiology, *J. Physiol. (London)* **353:**1–50.

O'Grady, S. M., Palfrey, H. C., and Field, M., 1987, Characteristics and functions of Na-K-Cl cotransport in epithelial tissues, *Am. J. Physiol.* **253:**C177–C192.

Ohashi, H., 1970a, An estimate of the proportion of the resting membrane conductance of the smooth muscle of the guinea-pig taenia coli attributable to chloride, *J. Physiol. (London)* **210:**405–419.

Ohashi, H., 1970b, Effects of changes in ionic environment on the negative after-potential of the spike in rat uterine muscle, *J. Physiol. (London)* **210:**785–797.

Ohashi, H., 1971, The relative contribution of K and Cl to the total increase of membrane conductance produced by adrenaline on the smooth muscle of guinea-pig taenia coli, J. Physiol. (London) **212:**561–575.

Owen, N. E., 1985, Regulation of Na/K/Cl cotransport in vascular smooth muscle cells, *Biochem. Biophys. Res. Commun.* **125:**500–508.

Palade, P. T., and Barchi, R. L., 1977a, Characteristics of chloride conductance in muscle fibres of the rat diaphragm, *J. Gen. Physiol.* **69:**325–342.

Palade, P. T., and Barchi, R. L., 1977b, On the inhibition of muscle membrane chloride conductance by aromatic carboxylic acids, *J. Gen. Physiol.* **69:**879–896.

Palfrey, H. C., Feit, P. W., and Greenwald, P., 1980, cAMP-stimulated cation transport in avian erythrocytes: Inhibition by "loop" diuretics, *Am. J. Physiol.* **238:**C139–C148.

Piwnica-Worms, D., Jacob, R., Horres, C. R., and Lieberman, M., 1985, Na/H exchange in cultured chick heart cells. pH$_i$ regulation, *J. Gen. Physiol.* **85:**43–64.

Prosser, C. L., 1974, Smooth muscle, *Annu. Rev. Physiol.* **36:**503–535.

Putnam, R. W., and Roos, A., 1985, Aspects of pH$_i$ regulation in frog skeletal muscle, in: *Current Topics in Membranes and Transport: Na⁺–H⁺ Exchange, Intracellular pH and Cell Function* (P. S. Aronson and W. F. Boron, eds.), Academic Press, New York, pp. 35–36.

Renner, H., and Westphal, W., 1975, Potassium and chloride: A reassessment in mammalian skeletal muscle, *Pfluegers Arch.* **359:**R68.

Reuss, L., Lewis, S. A., Wills, N. K., Helman, S. I., Cox, T. C., Boron, W. F., Siebens, A. W., Guggino, W. B., Giebisch, G., and Schultz, S. G., 1984, Ion transport processes in basolateral membranes of epithelia, *Fed. Proc.* **43:**2488–2502.

Roos, A., and Boron, W. F., 1978, Intracellular pH transients in rat diaphragm muscle measured with DMO, *Am. J. Physiol.* **235:**C49–C54.

Rüdel, R., and Senges, J., 1972, Mammalian skeletal muscle: Reduced chloride conductance in drug-induced myotonia and induction of myotonia by low-chloride solution, *Naunyn-Schmiedebergs Arch. Pharmacol.* **274:**337–347.

Russell, J. M., 1979, Chloride and sodium influx: A coupled mechanism in the squid axon, *J. Gen. Physiol.* **73:**801–818.

Russell, J. M., 1983, Cation-coupled influx in squid axon. Role of potassium and stoichiometry of the transport process, *J. Gen. Physiol.* **81:**909–925.

Russell, J. M., and Brodwick, M. S., 1976, Chloride fluxes in the dialyzed barnacle muscle fiber and the effect of SITS, *Biophys. J.* **16:**156a.

Russell, J. M., and Brodwick, M. S., 1979, Properties of chloride transport in barnacle muscle fibers, *J. Gen. Physiol.* **73:**343–368.

Schlue, W.-R., and Thomas, R. C., 1985, A dual mechanism for intracellular pH regulation by leech neurones, *J. Physiol. (London)* **364:**327–338.

Shearin, N. L., and Walker, J. L., 1983, Direct intracellular measurement of chloride ionic activity and the relationship to membrane potential in colonic smooth muscle, *Fed. Proc.* **43:**10042.

Shoemaker, R., Naftel, J., and Farley, J., 1985, Measurement of K^+ and Cl^- channels in rat cultured vascular smooth muscle cells, *Biophys. J.* **47:**465a.

Simchowitz, L., and Roos, A., 1985, Regulation of intracellular pH in human neutrophils, *J. Gen. Physiol.* **85:**443–470.

Somlyo, A. P., Somlyo, A. V., and Schuman, H., 1979, Electron probe analysis of vascular smooth muscle. Composition of mitochondria, nuclei and cytoplasm, *J. Cell Biol.* **81:**316–335.

Spray, D. C., Harris, A. L., and Bennett, M. V. L., 1981, Gap junctional conductance is a simple and sensitive function of intracellular pH, *Science* **211:**712–715.

Steiner, R. A., Oehme, M., Ammann, D., and Simon, W., 1979, Neutral carrier sodium ion-selective microelectrode for intracellular studies, *Anal. Chem.* **51:**351–353.

Strickholm, A., and Wallin, B. G., 1967, Relative ion permeabilities in the crayfish giant axon determined from rapid external ion changes, *J. Gen. Physiol.* **50:**1929–1953.

Thomas, R. C., 1974, Intracellular pH of snail neurones measured with a new pH-sensitive glass microelectrode, *J. Physiol. (London)* **238:**159–180.

Thomas, R. C., 1976, Ionic mechanism of the H^+ pump in a snail neurone, *Nature* **262:**54–55.

Thomas, R. C., 1977, The role of bicarbonate, chloride and sodium ions in the regulation of intracellular pH in snail neurones, *J. Physiol. (London)* **273:**317–338.

Thomas, R. C., 1984, Experimental displacement of intracellular pH and the mechanism of its subsequent recovery, *J. Physiol. (London)* **354:**3P–22P.

Thomas, R. C., and Cohen, C. J., 1981, A liquid ion-exchanger alternative to KC1 for filling intracellular reference microelectrodes, *Pfluegers Arch.* **390:**96–98.

Tomita, T., 1981, Electrical activity (spikes and slow waves) in gastrointestinal smooth muscles, in: *Smooth Muscle: An Assessment of Current Knowledge* (E. Bülbring, A. F. Brading, A. W. Jones, and T. Tomita, eds.), Arnold, London, pp. 127–156.

Turin, L., and Warner, A. E., 1978, Carbon dioxide reversibly abolishes ionic communication between cells of early amphibian embryo, *Nature* **270:**56–57.

Vanheel, B., De Hemptinne, A., and Leusen, I., 1984, Analysis of $Cl^- -HCO_3^-$ exchange during recovery from intracellular acidosis in cardiac Purkinje strands, *Am. J. Physiol.* **246:**C391–C400.

Vassort, G., 1981, Ionic currents in longitudinal muscle of the uterus, in: *Smooth Muscle: An Assessment of Current Knowledge* (E. Bülbring, A. F. Brading, A. W. Jones, and T. Tomita, eds.), Arnold, London, pp. 353–366.

Vaughan, P., and Fong, C. N., 1978, Effects of SITS on chloride permeability in *Xenopus* muscle, *Can. J. Physiol. Pharmacol.* **56:**1051–1054.

Vaughan-Jones, R. D., 1979a, Non-passive chloride distribution in mammalian heart muscle: Micro-electrode measurement of the intracellular chloride activity, *J. Physiol. (London)* **295:**83–109.

Vaughan-Jones, R. D., 1979b, Regulation of chloride in quiescent sheep-heart Purkinje fibres studied using intracellular chloride and pH-sensitive micro-electrodes, *J. Physiol. (London)* **295:**111–137.

Vaughan-Jones, R. D., 1982, Chloride activity and its control in skeletal and cardiac muscle, *Philos. Trans. R. Soc. London Ser B* **299:**537–548.

Vaughan-Jones, R. D., 1986, An investigation of chloride–bicarbonate exchange in the sheep cardiac Purkinje fibre, *J. Physiol. (London)* **379:**377–406.

Vigne, P., Frelin, C., and Lazdunski, M., 1982, The amiloride-sensitive Na^+/H^+ exchange system in skeletal muscle cells in culture, *J. Biol. Chem.* **257:**9394–9400.

Vigne, P., Frelin, C., and Lazdunski, M., 1984, The Na^+-dependent regulation of the internal pH in chick skeletal muscle cells. The role of the Na^+/H^+ exchange system and its dependence on internal pH, *EMBO J.* **3:**1865–1870.

Villamil, M. F., Ponce, J., Amorena, C., and Müller, A., 1979, Effect of furosemide on the ionic composition of the arterial wall, *TIT J. Life Sci.* **9:**9–14.

Vyskočil, F., 1977, Diazepam blockade of repetitive action potentials in skeletal muscle fibers. A model of its membrane action, *Brain Res.* **133:**315–328.

Wahlström, B. A., 1973a, Ionic fluxes in the rat portal vein and the applicability of the Goldman equation in predicting the membrane potential from flux data. *Acta Physiol. Scand.* **89**:436–448.

Wahlström, B. A., 1973b, A study on the action of nonadrenaline on ionic content and sodium, potassium and chloride effluxes in rat portal vein, *Acta Physiol. Scand.* **89**:522–530.

Warnock, D. G., Greger, R., Dunham, P. B., Benjamin, M. A., Frizzell, R. A., Field, M., Spring, K. R., Ives, H. E., Aronson, P. S., and Sieffer, J., 1984, Ionic transport processes in apical membranes of epithelia, *Fed. Proc.* **43**:2478–2487.

Weissberg, P. L., Little, P. J., Cragoe, E. L., and Bobik, A., 1987, Na–H antiport in cultured rat aortic smooth muscle: Its role in cytoplasmic pH regulation, *Am. J. Physiol.* **253**:C193–C198.

Westphal, W., and Limbourg, P., 1972, Chloride movements in resting and stimulated mammalian skeletal muscle, *Pfluegers Arch.* **332**:R72.

White, M. M., and Miller, C., 1979, A voltage-gated anion channel from the electric organ of *Torpedo californica, J. Biol. Chem.* **254**:10161–10166.

Widdicombe, J. H., and Brading, A. F., 1980, A possible role of linked Na and Cl movement in active Cl uptake in smooth muscle, *Pfluegers Arch.* **386**:35–37.

Xie, X., Stone, D. K., and Racker, E., 1983, Determinants of clathrin coated vesicle acidification, *J. Biol. Chem.* **258**:14834–14838.

II

DIFFERENT TYPES OF Cl⁻ CHANNELS

A

Transmitter-Activated Anion Channels

Biophysical Aspects of GABA- and Glycine-Gated Cl⁻ Channels in Mouse Cultured Spinal Neurons

Joachim Bormann

1. INTRODUCTION

γ-Aminobutyric acid (GABA) and glycine are the main inhibitory transmitter substances in the mammalian CNS (Krnjević and Schwartz, 1967; Werman et al., 1967). Following their release from presynaptic nerve terminals, GABA and glycine molecules bind to postsynaptic receptors and cause an increase in membrane conductance to chloride ions (Barker and Ransom, 1978). The molecular components involved in the Cl⁻ conductance increase, the GABA receptor (GABAR) and glycine receptor (GlyR) channels, have been studied electrophysiologically by using patch-clamp current-recording techniques (Sakmann et al., 1983; Hamill et al., 1983; Bormann et al., 1987). The results obtained from mouse cultured spinal neurons are presented below in order to illustrate the conductance and gating properties of the two receptor-regulated membrane channels.

2. PERMEABILITY MEASUREMENTS

GABA- and glycine-activated whole-cell currents were investigated using 145 mM Cl⁻ on both faces of the cell membrane (Fig. 1A). At a membrane voltage of −70 mV, transmitter-induced inward current reflects the net outward movement of Cl⁻ ions (Fig. 1B). Under these conditions, the current reverses polarity at about 0 mV membrane potential (Fig. 1C), indicating Cl⁻-specific membrane channels. The permeability of GABAR and GlyR channels to other monovalent anions was determined in reversal potential measurements where the test anion replaced Cl⁻ in the patch pipette. The calculated permeabilities of organic anions (relative to Cl⁻) decreased with hydrated size. The largest anion to pass GABAR channels was propionate, indicating a pore diameter of 5–6 Å (Bormann et al., 1987).

Joachim Bormann • Max-Planck-Institut for biophysikalische Chemie, D-3400 Göttingen, Federal Republic of Germany; *present address:* Merz & Co., D-6000 Frankfurt 1, Federal Republic of Germany.

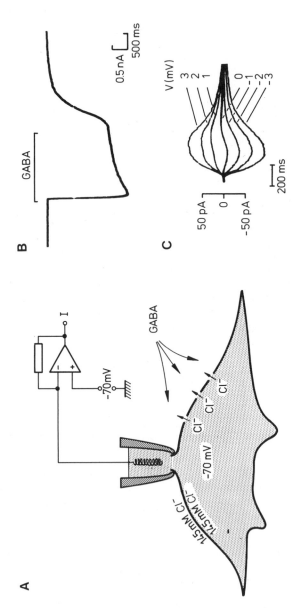

Figure 1. Whole-cell recording of GABA-activated membrane currents. (A) Schematic representation of the experimental arrangement. The cell is dialyzed against the high [Cl⁻] pipette solution while recording from GABAR channels in symmetrical [Cl⁻] solutions (145 mM). GABA is applied to the cell via the bath solution. (B) Whole-cell recording of the membrane current activated at −70 mV during the application of 10 μM GABA. The GABA-induced outward movement of Cl⁻ ions at negative membrane potentials is measured as an inward current (downward deflection of trace). (C) Reversal of GABA-activated currents close to 0 mV membrane holding potential. (Modified from Bormann *et al.*, 1987.)

3. CONDUCTANCE MEASUREMENTS

Single-channel currents were recorded from excised, outside-out, membrane patches in symmetrical Cl⁻ solutions of 145 mM (Fig. 2). At a fixed membrane voltage, glycine-activated currents have larger amplitudes (Fig. 2A) than GABA-activated currents (Fig. 2B). The single-channel current–voltage relations are linear for both receptor channels, crossing the voltage axis at 0 mV and indicating slope conductances of 45 pS for GlyR channels and 30 pS for GABAR channels (Fig. 2C). In

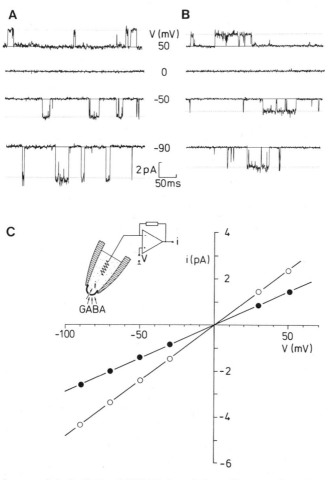

Figure 2. Conductance of single GlyR and GABAR channels in outside-out patches with equal (145 mM) [Cl⁻] on both membrane faces. Channels were activated by adding agonists at 10 μM to the bath (see inset of C). Currents, *i*, through GlyR channels (A) and GABAR channels (B) were recorded at different membrane holding potentials, V, as indicated on each trace. The current–voltage relations (C) are linear indicating slope conductances of 45 pS (GlyR, open symbols) and 30 pS (GABAR, filled symbols). (Modified from Bormann *et al.*, 1987.)

addition to these main conductance states, several other, less frequently occurring, states were observed. Both channels displayed at least four conductances, including for GlyR, 45, 30, 20, and 12 pS and for GABAR, 44, 30, 19, and 12 pS (Hamill *et al.*, 1983; Bormann *et al.*, 1987). Single-channel current–voltage relations were also obtained for various other inorganic anions. The conductance sequence for the anions tested was in the order $Cl^- > Br^- > I^- > SCN^- > F^-$ for both receptor channels, which compares to the sequence of relative permeabilities, $SCN^- > I^- > Br^- > Cl^- > F^-$. This nearly inverse relationship between the two sequences indicated the presence of binding sites that attract anions upon entering the channels (Bormann *et al.*, 1987).

4. GATING OF CHANNELS

Elementary currents through GABAR and GlyR channels have a burstlike appearance (see Fig. 2A,B), i.e., they are interrupted short closing gaps. The time course of channel gating could be described with a reaction scheme for agonist-activated channels (Del Castillo and Katz, 1957) where two agonist molecules, A, bind to the receptor R in a stepwise manner to form the complex A_2R, which then isomerizes to A_2R^*, the open channel:

$$R + A \underset{k_{-1}}{\overset{2k_1}{\rightleftarrows}} AR + A \underset{2k_{-1}}{\overset{k_1}{\rightleftarrows}} A_2R \underset{\alpha}{\overset{\beta}{\rightleftarrows}} A_2R^*$$

$$\text{closed} \qquad\quad \text{closed} \qquad\quad \text{closed} \qquad\quad \text{open}$$

The average burst duration is on the order of 50 msec, with two to three interruptions per burst. From the fine structure of current pulses, some of the rate constants in the above scheme were estimated (Sakmann *et al.*, 1983).

5. CONCLUSIONS

GABA- and glycine-activated membrane channels have been studied in cultured neurons in order to compare the mechanisms of the two major inhibitory transmitter substances in the mammalian CNS. Both channels exclude cations but allow the passage of anions as large as propionate, indicating a minimal pore diameter of 5–6 Å. The rate of ion transport through the channels is governed by binding sites that interact with anions during their passage from one side of the membrane to the other. The similarity in the biophysical properties, including the multi-conductance state behavior supports the view that both receptor channels permit Cl^- to permeate by a mechanism common to both.

Attempts to clone the GABAR channel (Schofield *et al.*, 1987) and the GlyR channel (Grenningloh *et al.*, 1987) have in fact revealed a considerable degree of homology in the amino acid sequences of the subunits forming the channels. Moreover, comparison of the subunit primary structures to that of the nicotinic acetylcholine receptor channel (Noda *et al.*, 1983) raises the possibility that all these chemically gated ion channels could be members of a common class of membrane channel.

REFERENCES

Barker, J. L., and Ransom, B. R., 1978, Amino acid pharmacology of mammalian central neurones grown in tissue culture, *J. Physiol. (London)* **280:**331–334.

Bormann, J., Hamill, O. P., and Sakmann, B., 1987, Mechanism of anion permeation through channels gated by glycine and γ-aminobutyric acid in mouse cultured spinal neurones, *J. Physiol. (London)* **385;** 243–286.

Del Castillo, J., and Katz, B., 1957, Interaction at end-plate receptors between different choline derivatives, *Proc. R. Soc. London Ser. B* **146:**369–381.

Grenningloh, G., Rienitz, A., Schmitt, B., Methfessel, C., Zensen, M., Beyreuther, K., Gundelfinger, E. D., and Betz, H., 1987, The strychnine-binding subunit of the glycine receptor shows homology with nicotinic acetylcholine receptors, *Nature* **328:**215–220.

Hamill, O. P., Bormann, J., and Sakmann, B., 1983, Activation of multiple-conductance state chloride channels in spinal neurones by glycine and GABA, *Nature* **305:**805–808.

Krnjević, K., and Schwartz, S., 1967, The action of aminobutyric acid on cortical neurones, *Exp. Brain Res.* **3:**320–336.

Noda, M., Takahashi, H., Tanabe, T., Toyosato, M., Kikyotani, S., Furutani, Y., Hirose, T., Takashima, H., Inayama, S., Miyata, T., and Numa, S., 1983, Structural homology of Torpedo californica acetylcholine receptor subunits, *Nature* **302:**528–532.

Sakmann, B., Hamill, O. P., and Bormann, J., 1983, Patch clamp measurements of elementary chloride currents activated by the putative inhibitory transmitters GABA and glycine in mammalian spinal neurones. *J. Neural. Transm. Suppl.* **18:**83–95.

Schofield, P. R., Darlison, M. G., Fujita, N., Burt, D. R., Stevenson, F. A., Rodriguez, H., Rhee, L. M., Ramachandran, J., Reale, V., Glencorse, T. A., Seeburg, P. H., and Barnard, E. A., 1987, Sequence and functional expression of GABA$_A$ receptor shows a ligand-gated receptor super-family, *Nature* **328:** 221–227.

Werman, R., Davidoff, R. A., and Aprison, M. H., 1967, Evidence for glycine as the principal transmitter mediating postsynaptic inhibition in the spinal cord of the cat, *J. Gen. Physiol.* **50:**1093–1094.

GABA-Gated Cl⁻ Currents and Their Regulation by Intracellular Free Ca²⁺

Norio Akaike

1. INTRODUCTION

Intracellular Ca^{2+} is well known to play a central role in the transduction of extra-cellular signals (Nishizuka, 1984) and in the control and modulation of fundamental cellular functions. In addition to well-established functions such as Ca^{2+}-mediated transmitter release from nerve terminals (Katz and Miledi, 1967), the liberation of hormones from neuroendocrine cells (Thorn *et al.*, 1978), and muscular contraction, several types of voltage-dependent ionic channels, including Ca^{2+} channels them-selves, have been found to be highly sensitive to changes in intracellular free Ca^{2+} concentration ($[Ca^{2+}]_i$) (Colquhoun *et al.*, 1981; Eckert *et al.*, 1981; Bader *et al.*, 1982; Maruyama and Petersen, 1982; Yellen, 1982; Ashcroft and Stanfield, 1984; Marty *et al.*, 1984; Owen *et al.*, 1984; Bregestovski *et al.*, 1986; see also Mayer *et al.*, this volume). Other prominent examples are the Ca^{2+}-activated K^+ conductance in molluscan neurons (Meech and Standen, 1975; Thompson, 1977; Aldrich *et al.*, 1979) and the Ca^{2+}-activated anion conductance in mouse spinal cord neurons (Owen *et al.*, 1984) and rat sensory neurons (Mayer, 1985). However, evidence increasingly sug-gests that not only electrically operated but also chemically gated neuronal membrane conductances are affected by such transient changes in $[Ca^{2+}]_i$ (Morita *et al.*, 1979; Miledi, 1980; Chemeris *et al.*, 1982). An increase in $[Ca^{2+}]_i$ not only facilitates the onset of desensitization to the nicotinic cation response of acetylcholine in the muscle endplate (Miledi, 1980) but also inhibits the nicotinic Cl^- response in *Aplysia* neurons (Chemeris *et al.*, 1982) as well as the nicotinic cation response in frog sympathetic ganglion cells (Morita *et al.*, 1979). $[Ca^{2+}]_i$ also behaves as a second messenger for many biologically active substances (Rasmussen and Barrett, 1984). In this chapter we report a powerful suppressive action of $Ca^{2+}]_i$ on the γ-aminobutyric acid (GABA)-gated Cl^- current in isolated bullfrog sensory neurons. We also report evidence for a saturable intracellular binding site for Ca^{2+} and other divalent cations which mediates this suppressive effect. We recorded macroscopic and single-channel membrane cur-rents from internally perfused sensory neurons of bullfrog dorsal root ganglia using

Norio Akaike • Department of Neurophysiology, Tohoku University School of Medicine, Sendai 980, Japan.

"concentration-clamp" (Akaike *et al.*, 1986, 1987) and patch-clamp (Hamill *et al.*, 1981) techniques.

2. METHODS

Dorsal root ganglia were dissected from decapitated American bullfrogs (*Rana catesbeiana*). The ganglion masses were digested in 10 ml normal Ringer solution containing 0.3% collagenase and 0.05% trypsin at pH 7.4 for 15 min at 37°C. Thereafter, single cells were isolated from the ganglion mass with finely polished pins under binocular observation, and left overnight at room temperature (about 22°C) in a culture medium consisting of equal parts of Ringer solution and an isotonic Eagle's minimum essential medium (NISSUI, Japan).

The ionic compositions of the standard solutions were as follows (in mM): internal, CsCl 95, Cs-aspartate 10, TEA-Cl 25, EGTA-Ca^{2+} buffer ($[Ca^{2+}] = 10^{-8}$ M); external, Tris-Cl 89, CsCl 2, $CaCl_2$ 2, TEA-Cl 25, glucose 5. The pH of all solutions was adjusted to 7.4 with Tris base and HEPES.

The membrane potential was measured through an Ag-AgCl wire in a Ringer–agar plug mounted on a suction pipette holder. The reference electrode was also an Ag-AgCl wire in a Ringer–agar plug. The resistance between the suction pipette filled with standard internal solution and the reference electrode was 200 to 300 kΩ. Both electrodes led to a voltage-clamp circuit, and membrane potential was controlled by a single-electrode voltage-clamp system switching at a frequency of 10 kHz and passing current for 36% of the cycle (Ishizuka *et al.*, 1984). Clamp currents were measured as the voltage drop across a 10-MΩ resistor in the feedback path of a headstage amplifier. In this system, the suction electrode could carry time-averaged currents exceeding 100 nA at a switching frequency of 10 kHz, without showing signs of polarization or other artifacts. Ca^{2+}-inward currents (I_{Ca}) elicited by depolarizing voltage pulses were corrected by subtracting the current response to a hyperpolarizing voltage pulse of identical amplitude and duration by means of a digital averaging oscilloscope (Nihon Kohden, Japan). Total I_{Ca} was measured as the area under the current waveform. Both current and voltage were monitored on a digital storage oscilloscope (National, VP-5730A), and were simultaneously recorded on an ink writing recorder (Rikadenki, R-22) and stored on a magnetic data recorder for later analysis (TEAC, MR-30).

A suction pipette technique was used for voltage-clamp and internal perfusion: neurons were sucked into the pipette tip and the aspirated membrane was ruptured by large square wave pulses of depolarizing current. Adequacy of internal perfusion was evaluated by ensuring that the reversal potential for GABA-induced Cl⁻ responses (E_{GABA}) was very close to the E_{Cl} (+4 mV) as calculated from the Nernst equation based on the Cl⁻ concentrations in the internal and external solutions. The "concentration-clamp" method was used for extremely rapid application of external solution within 2 msec, with or without an agonist (Akaike *et al.*, 1986). Results were expressed as mean ± S.E.M. and Students *t* test was applied for statistical significance of differences. All experiments were performed at room temperature.

3. RESULTS

3.1. Inhibition of GABA-Induced Depolarization by a Preceding Ca^{2+} Spike

Under current-clamp conditions, rapid application of GABA (10^{-5} M) in Ca^{2+}-free external solution evoked a large depolarization, and a relatively stable plateau was reached within 2 sec (Fig. 1A). This response was mediated through an increase in Cl^- conductance across the soma membrane (Akaike *et al.*, 1986; Inoue *et al.*, 1986). The GABA-induced I_{Cl} was not affected by 50 preceding Na^+ spikes (Fig. 1B). However, a single preceding Ca^{2+} spike dramatically reduced the subsequent response to GABA (Fig. 1C). This inhibitory effect was not simply due to the presence of Ca^{2+} in the external perfusate, since it could be reversed by adding Co^{2+}, a well-known blocker of voltage-sensitive Ca^{2+} entry (Ishizuka *et al.*, 1984).

3.2. Suppression of the GABA-Induced I_{Cl} by I_{Ca}

The suppression of the GABA-gated I_{Cl} by Ca^{2+} currents (I_{Ca}) was quantified in neurons immersed in a Na^+- and K^+-free Ringer solution containing 2 mM Ca^{2+}. Cells perfused with internal solution containing 0.5 mM EGTA-Ca buffer ($[Ca^{2+}]_i = 10^{-8}$ M) were voltage-clamped at a holding potential (V_H) of -50 mV, and I_{Ca} was evoked by 50-mV depolarizing command pulses of varying durations. Figure 2A illustrates the typical suppression of GABA-activated I_{Cl} by I_{Ca}, in which the GABA responses were progressively inhibited with increasing Ca^{2+} influx. A typical relationship between the total amount of I_{Ca} ($\int I_{Ca}$) and the GABA-induced peak response is shown in Fig. 2B. I_{Cl} was maximally inhibited when the Ca^{2+} influx exceeded 80

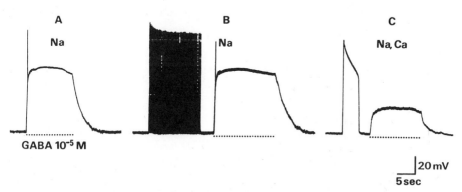

Figure 1. Effects of Ca^{2+} spike and Na^+ spikes on the GABA-induced depolarization. (A) 10^{-5} M GABA depolarized the membrane by 75 mV in Na^+ solution. The membrane potential was held at -80 mV. (B) Fifty Na^+ spikes elicited by depolarizing pulses at a frequency of 3.3 Hz had little influence on the 10^{-5} M GABA response (74 mV) in an external solution containing Na^+. (C) Marked suppression of 10^{-5} M GABA response after a Ca^{2+} spike (4.3-sec duration) in an external solution containing Na^+ and Ca^{2+}; depolarization of 32 mV. (Modified from Inoue *et al.*, 1986, 1987.)

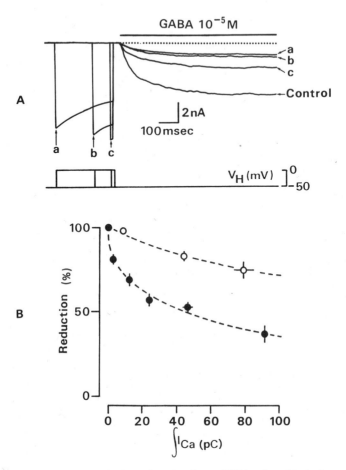

Figure 2. (A) Suppressive action of voltage-dependent I_{Ca} on GABA-activated I_{Cl} in frog sensory neurons perfused with internal solution containing 0.5 mM EGTA-Ca buffer. Recordings were obtained from the same cell. The symbols *a* to *c* in I_{Ca} and GABA-induced I_{Cl} refer to three separate, continuous same traces. The control is I_{Cl} without a preceding I_{Ca}. 0.5 mM EGTA-Ca buffer ($[Ca^{2+}]_i = 10^{-8}$ M). 10^{-5} M GABA-activated I_{Cl}, with and without a preceding I_{Ca}, elicited by the depolarizing voltage pulse from V_H −50 mV (the holding potential) to 0 mV for various durations, i.e., 300 (a), 100 (b), 20 (c) msec. The linear leakage and capacitative currents of I_{Ca} were subtracted by adding the currents produced by each of a pair of depolarizing and hyperpolarizing pulses of 50 mV by a signal averager. (B) Relationship between reduction of GABA-activated I_{Cl} and amount of Ca^{2+} influx; open and closed circles represent the data from cells perfused with 2.5 mM EGTA and 0.5 mM EGTA-Ca buffer ($[Ca^{2+}]_i = 10^{-8}$ M), respectively. V_H was −50 mV. I_{Cl} was activated by 10^{-5} M GABA and I_{Ca} was evoked by 50-mV depolarizing command pulses of various durations ranging from 10 to 100 msec. The amount of I_{Ca} was expressed in picocoulombs after correcting for capacitative and leakage currents. Horizontal and vertical bars give ± S.E.M. (Modified from Inoue *et al.*, 1987.)

pC. The degree of this I_{Ca}-mediated suppression of I_{Cl} depended on the chelating potency of the internal solution used. In cells perfused with 0.5 mM EGTA-Ca buffer, I_{Cl} was reduced to 50% of the control by an I_{Ca} evoked by a command pulse of 100-msec duration, while I_{Cl} was reduced by over 25% in cells perfused with 2.5 mM EGTA. The difference in the degree of I_{Cl} suppression by I_{Ca} in the two experimental conditions was statistically significant ($p < 0.01$).

3.3. Voltage-Dependence of Inhibition by I_{Ca}

[Ca²⁺]ᵢ stimulates not only a K⁺ conductance in molluscan neurons (Meech, 1972; Gorman and Thomas, 1980; Akaike *et al.*, 1983) but also a Cl⁻ conductance in spinal cord cells in culture (Owen *et al.*, 1984). The Ca²⁺-dependent Cl⁻ or K⁺ conductance is considerably voltage-dependent (Miledi and Parker, 1984; Meech and Standen, 1975). However, the inhibition of GABA-induced I_{Cl} by the preceding I_{Ca} was voltage-independent over a membrane potential range between −100 and −40

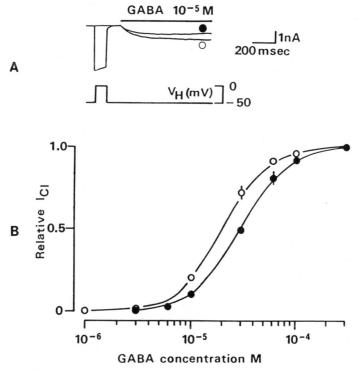

Figure 3. Effect of voltage-dependent I_{Ca} on the concentration–response curve of GABA-induced I_{Cl}. (A) Actual records of I_{Cl} elicited by 10⁻⁵ M GABA with (filled circle) and without (open circle) a preceding I_{Ca} evoked by a depolarizing command pulse of 100 msec from −50 to 0 mV. V_H was −50 mV. (B) GABA dose–response curve, with and without a preceding I_{Ca}. Recording conditions are the same as for A. Note that I_{Ca} shifted the GABA dose–response curve to the right in a parallel manner. Each point and vertical bar indicate the mean ± S.E.M. of six experiments. (Modified from Inoue *et al.*, 1987.)

mV. The voltage-independence suggests that the suppression of the GABA response is related to the decrease of "apparent" affinity of the GABA receptor for the agonist rather than to changes in Cl^- channel gating kinetics.

Furthermore, as shown in Fig. 3, I_{Ca} shifted the GABA concentration–response curve to the right without changing the maximum response indicating that the GABA-gated I_{Cl} was inhibited in a competitive manner. The result also indicates that an increase in $[Ca^{2+}]_i$ probably decreases the affinity of the GABA receptor for GABA.

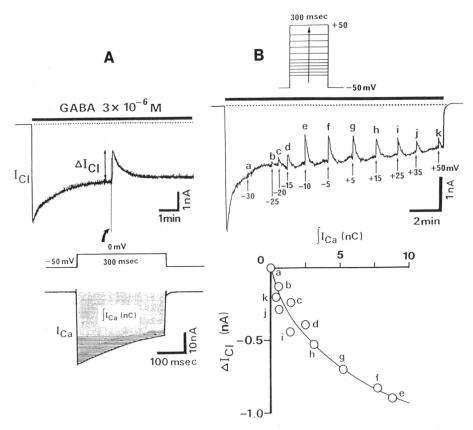

Figure 4. Effect of eliciting I_{Ca} during steady-state I_{Cl} responses to 3×10^{-6} M GABA. Internal solution contained 0.5 mM EGTA-Ca ($[Ca^{2+}]_i = 10^{-8}$ M). Holding potential was -50 mV. (A) The inhibition of steady-state I_{Cl} response induced by GABA. The inhibition is visible as an apparent current relaxation in the outward direction. I_{Ca} was elicited by a 50-mV command pulse having 300-msec duration. Note different time scales in top and bottom panels. Curved arrow indicates time when I_{Ca} was turned on by the 300-msec pulse. (B) Stimulation of I_{Ca} with varying voltage pulse amplitudes (arrows a–k) (See insets above for voltage pulse protocol) at times indicated by arrows. Numbers below arrows indicate the value of the voltage to which the 300-msec step was made. Lower panel is a plot of the total amount of Ca^{2+} influx ($\int I_{Ca}$) versus the change in GABA-gated I_{Cl} amplitude (ΔI_{Cl}).

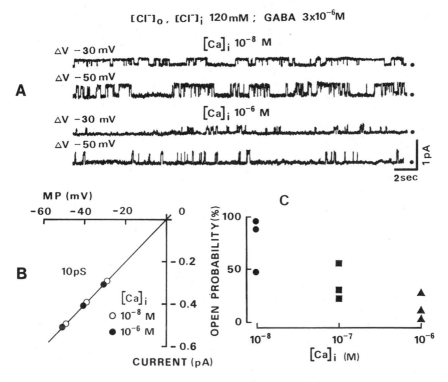

Figure 5. (A) Single GABA-activated Cl⁻-channel currents recorded in the "inside-out" configuration. The patch pipette contained 3×10^{-6} M GABA. The ionized Ca^{2+} concentration at the intracellular side of the cell membrane was 10^{-8} M $[Ca^{2+}]_i$ (upper two traces) or 10^{-6} M $[Ca^{2+}]_i$ (lower two traces). Closed state is indicated by black dots at right-hand end of each trace. Upward direction is inward current. Note the reduction in the duration and frequency of open-state intervals upon exposure to 10^{-6} M $[Ca^{2+}]_i$. Driving forces for Cl⁻ ($\triangle V_H$) were 30 and 50 mV. (B) The relationship of single-channel current amplitude and the membrane potential: control ($[Ca^{2+}]_i = 10^{-8}$ M) and in the presence of elevated $[Ca^{2+}]_i$ (10^{-6} M). Both experimental data points from both experiments were fit by a straight line corresponding to a unitary conductance of 10 pS. (C) Open probability decreased upon increasing $[Ca^{2+}]_i$ (three neurons). A and B were obtained from the same patch.

3.4. Mechanism of [Ca²⁺]ᵢ-Mediated Suppression of GABA-Gated I_{Cl}

Experiments were performed on neurons perfused externally and internally with Na^+- and K^+-free solutions. Thus, current flows across the plasma membrane were restricted to those carried by Cl⁻ and Ca^{2+}. When a steady-state GABA response was attained, a 50-mV depolarizing voltage command pulse of 300-msec duration was applied to the cells, which elicited an inward I_{Ca}. The I_{Ca} caused a transient relaxation in the GABA-gated I_{Cl}, which rapidly reached a minimum and slowly returned to the steady-state value over several tens of seconds (Fig. 4A). Figure 4B shows the inhibition of the GABA-induced steady-state response by I_{Ca} induced by various voltage

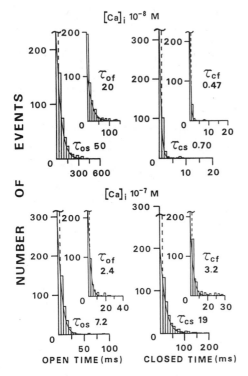

Figure 6. Open and closed time histograms for fast (τ_{of}, τ_{cf}) and slow (τ_{os}, τ_{cs}) opening and closing events at two different $[Ca^{2+}]_i$ respectively. Upon changing $[Ca^{2+}]_i$ from 10^{-8} M to 10^{-7} M, a prominent decrease in open-interval duration as well as an increase in closed-interval duration is observed. Open probability decreased from 98% to 57% while the number of events increased by 18%. Data were obtained from the same neuron.

activation threshold for voltage-dependent Ca^{2+} channels (-25 mV, arrow b), I_{Ca} was elicited and subsequently a transient relaxation in the GABA-gated I_{Cl} occurred. The I_{Ca}-mediated suppression of I_{Cl} was enhanced with increasing voltage pulse intensities (arrows b–e), but the suppressive effect gradually diminished at command pulses more positive than -5 mV (arrows f–k). This observation becomes important in the light of the fact that, while maximal I_{Ca} amplitude continues to increase with command potentials positive to -10 mV, inactivation of I_{Ca} is increasingly accelerated with more positive command pulses. This leads to a decrease in total current, i.e., a decrease in the amount of net Ca^{2+} entry into the cell. These findings are consistent with the hypothesis that the actual number of Ca^{2+} ions entering the cell rather than the peak current amplitude determines the effectiveness of I_{Cl} suppression. The graph below the command pulses having constant duration. When the command potential reached the recording shows the quantitative analysis of the inhibited I_{Cl} amplitude (ΔI_{Cl}) plotted as a function of total amount of I_{Ca} ($\int I_{Ca}$). This figure demonstrates that I_{Cl} was suppressed by increasing $\int I_{Ca}$ and that the relationship was hyperbolic indicating the inhibition of I_{Cl} by I_{Ca} was saturable. Thus, it may be assumed that the depressive action of I_{Ca} on the GABA-gated I_{Cl} is mediated by an increase in $[Ca^{2+}]_i$ which may act at an intracellular binding site.

3.5. Effect of $[Ca^{2+}]_i$ on GABA-Gated Unitary Cl^--Channel Currents

To verify the findings obtained in the whole-cell patch recording configuration, single GABA-activated Cl^--channel currents were recorded using the "inside-out" patch recording configuration, in which the intracellular face of the membrane was exposed to various $[Ca^{2+}]_1$. Figure 5A shows representative traces recorded at two different $[Ca^{2+}]_i$. In the presence of 10^{-6} M Ca^{2+} the duration of open intervals was greatly shortened compared to a 10^{-8} M Ca^{2+}, though single-channel conductance was unaffected (Fig. 5B). The change in open probability provoked by a successive increase in $[Ca^{2+}]_i$ in three different patches is illustrated in Fig. 5C.

Open and closed time histograms (Fig. 6) from the experiment shown in Fig. 5A show a clear decrease in the duration of open intervals with increased $[Ca^{2+}]_i$ as well as a prominent increase in the duration of closed intervals. The open probability decreased from 98 to 57% in this patch. Interestingly, this decrease in open probability was accompanied by an 18% increase in the number of opening and closing events. Furthermore, increases in $[Ca^{2+}]_i$ tended to produce flickering-like gating behavior of the Cl^- channel in an intermediate state just prior to the strong inhibition which occurred at 10^{-6} M $[Ca^{2+}]_i$.

4. DISCUSSION

In various preparations, both external and internal Ca^{2+} control membrane permeability to cations and/or anions (Meech, 1972; Colquhoun *et al.*, 1981; Miledi and Parker, 1984). Thus, there is the possibility that $[Ca^{2+}]_i$ may directly interfere with Cl^- channels activated by GABA. However, the present data do not support this supposition for the following reasons. (1) There was no voltage-dependence in the suppression of I_{Cl} by I_{Ca}. (2) I_{Ca} produced a parallel shift of the GABA dose–response curve to the right without changing the maximal current or the Hill coefficient ($n = 2$). If $[Ca^{2+}]_i$ inhibits the GABA response by transforming the GABA receptor–ionophore complexes to the inactive type, as noted in the case of the ACh-induced I_{Cl} by I_{Ca} in snail neurons (Chemeris *et al.*, 1982), then the mode of the inhibition would be noncompetitive. (3) GABA-induced I_{Cl} amplitude was inhibited in a hyperbolic manner with increasing $[Ca^{2+}]_i$, indicating saturating behavior or Ca^{2+}-mediated I_{Cl} inhibition. All these results taken together suggest that the increase in $[Ca^{2+}]_i$ acts by decreasing the affinity of GABA receptors on the soma membrane rather than blocking GABA-gated Cl^- channels.

The present experiments provide conclusive evidence that increases in the amount of Ca^{2+} current suppress the GABA-gated I_{Cl} in frog sensory neurons and that the suppressive modulation appears to be affected via an intracellular binding site for Ca^{2+}. As to the intracellular locus of the Ca^{2+} binding, two lines of evidence suggest that it is located within the membrane or at least very close to it. (1) The suppressive effect persists in the inside-out patch-clamp configuration in which the membrane is excised from the cell volume. (2) Intracellular perfusion with up to 10 mM of the

Ca^{2+}-EGTA buffer does not block this effect, but does greatly enhance the recovery time course. The latter finding is in agreement with a study on chromaffin cells, where Ca^{2+} buffers were found to be surprisingly ineffective in blocking the Ca^{2+}-activated K^+ current (Marty and Neher, 1985). Based on those results, Neher (1986) provided theoretical estimates of the $[Ca^{2+}]_i$ profile in the presence of chelators such as EGTA based on diffusional properties of both the chelator and Ca^{2+}. Under these conditions he concluded that $[Ca^{2+}]_i$ changes during Ca^{2+} influx are restricted to a narrow shell under the membrane, where chelating agents are subject to severe kinetic limitations. The results presented here, together with Neher's theoretical assessment suggest that the Ca^{2+} binding site mediating the Ca^{2+}-dependent depression of the GABA-gated I_{Cl} is a membrane-bound structure. Recently, Schofield *et al.* (1987) reported that the functional $GABA_A$ receptor–ionophore complex consists of α and β subunits and that GABA binding to the β subunits would induce a conformational change, resulting in Cl^- flux. In addition, a cAMP-dependent phosphorylation site on the interior is only associated with the β subunit. However, it is unknown how the phosphorylation site is related to the intracellular control of GABA response. It is interesting to speculate that $[Ca^{2+}]_i$ may bind to a Ca^{2+}-dependent phosphatase, which dephosphorylates a cAMP-sensitive phosphorylated site, and consequently may suppress the GABA response.

The modulatory effect of $[Ca^{2+}]_i$ on the GABA-gated I_{Cl} could be of great importance for the psysiological regulation as well as for the pathophysiology of neuronal excitability, should it prove to be a more general feature of both ganglionic and central neurons. In the *in vitro* hippocampus, development of a form of synaptic plasticity that is induced by tetanic stimulation of synaptic pathways [long-term potentiation (LTP)] has been shown to require activation of the NMDA subtype of the glutamate receptor. This is associated with a progressive depression of responses to iontophoretically applied GABA and of GABA-mediated inhibitory postsynaptic potentials (Stelzer *et al.*, 1987). Several studies have shown that both activation of the NMDA-receptor subtype and tetanic stimulation result in an increase in $[Ca^{2+}]_i$ (Krnjević *et al.*, 1986; MacDermott *et al.*, 1986). This Ca^{2+}-mediated process which is similar to that presented here might be explained as a loss of GABA-mediated inhibitory influence in hippocampal pyramidal cells. Such a mechanism might account for the development of LTP, an important model for synaptic plasticity in the CNS.

5. CONCLUSIONS

1. Evidence for a novel mechanism for the modulation of neuronal excitability has been obtained in internally perfused, isolated frog sensory neurons.

2. The increase of $[Ca^{2+}]_i$ suppresses in a hyperbolic manner the Cl^- response induced by the activation of the $GABA_A$ receptor.

3. Results suggest the possible existence of a saturable intracellular binding site for Ca^{2+}, which relates to GABA suppression.

4. From Ca^{2+}-buffering and excised patch-clamp studies the divalent cation binding site seems to be a membrane-bound structure.

5. With respect to the relatively slow time course of the $[Ca^{2+}]_i$ effect on GABA response, a catalytic intermediate (a Ca^{2+}-dependent enzyme, e.g., Ca^{2+}-dependent phosphatase) may be required to effect suppression of GABA response.

6. This Ca^{2+}-mediated modulation of an inhibitory neurotransmitter receptor might provide a model for molecular mechanisms that regulate changes in synaptic plasticity.

REFERENCES

Akaike, N., Brown, A. M., Dahl, G., Higashi, H., Isenberg, G., Tsuda, Y., and Yatani, A., 1983, Voltage-dependent activation of potassium current in *Helix* neurones by endogenous cellular calcium, *J. Physiol. (London)* **334:**309–324.

Akaike, N., Inoue, M., and Krishtal, O. A., 1986, "Concentration clamp" study of γ-aminobutyric acid-induced chloride current kinetics in frog sensory neurones, *J. Physiol. (London)* **379:**171–185.

Akaike, N., Yakushiji, T., Tokutomi, N., and Carpenter, D. O., 1987, Multiple mechanisms of antagonism of γ-aminobutyric acid (GABA) responses, *Cell. Mol. Neurobiol.* **7:**97–103.

Aldrich, R. W., Jr., Getting, P. A., and Thompson, S. H., 1979, Inactivation of delayed outward current in molluscan neurone somata, *J. Physiol. (London)* **291:**507–530.

Ashcroft, F. M., and Stanfield, P. R., 1984, Calcium dependence of the inactivation of calcium currents in skeletal muscle fibers of an insect, *Science* **213:**224–226.

Bader, C. R., Bertrand, D., and Schwartz, E. A., 1982, Voltage-activated and calcium-activated currents studied in solitary rod inner segments from the salamander retina, *J. Physiol. (London)* **331:**253–284.

Bregestovski, P., Redkozubov, A., and Alexev, A., 1986, Elevation of intracellular calcium reduces voltage-dependent potassium conductance in human T cells, *Nature* **319:**776–778.

Chemeris, N. K., Kazachenko, V. N., and Kurchikov, A. L., 1982, Inhibition of acetylcholine responses by intracellular calcium in *Lymnaea stagnalis* neurones, *J. Physiol. (London)* **323:**1–19.

Colquhoun, D., Neher, E., Reuter, H., and Stevens, C. F., 1981, Inward current channels activated by intracellular Ca in cultured cardiac cells, *Nature* **294:**752–754.

Eckert, R., Tillotson, D. L., and Brehm, P., 1981, Calcium-mediated control of Ca and K currents, *Fed. Proc.* **40:**2226–2232.

Gorman, A. L. F., and Thomas, M. V., 1980, Potassium conductance and internal calcium accumulation in a molluscan neurone, *J. Physiol. (London)* **308:**287–313.

Hamill, O. P., Marty, A., Neher, E., Sakmann, B., and Sigworth, F. J., 1981, Improved patch-clamp techniques for high resolution current recording from cells and cell-free membrane patches, *Pfluegers Arch.* **391:**85–100.

Inoue, M., Oomura, Y., Yakushiji, T., and Akaike, N., 1986, Intracellular calcium ions decrease the affinity of the GABA receptor, *Nature* **324:**156–158.

Inoue, M., Tokutomi, N., and Akaike, N., 1987, Modulation of the γ-aminobutyric acid-gated chloride current by intracellular calcium in frog sensory neurons, *Jpn. J. Physiol.* **37:**379–391.

Ishizuka, S., Hattori, K., and Akaike, N., 1984, Separation of ionic currents in the somatic membrane of frog sensory neurons, *J. Membr. Biol.* **78:**19–28.

Katz, B., and Miledi, R., 1967, A study of synaptic transmission in the absence of nerve impulses, *J. Physiol. (London)* **192:**407–436.

Krnjević, K., Morris, M. E., and Ropert, N., 1986, Changes in free calcium ion concentration recorded inside hippocampal pyramidal cells in situ, *Brain Res.* **374:**1–11.

MacDermott, A. B., Mayer, M. L., Westbrook, G. L., Smith, S. J., and Barker, J. L., 1986, NMDA-receptor activation increases cytoplasmic calcium concentration in cultured spinal cord neurones, *Nature* **321:**519.

Marty, A., and Neher, E., 1985, Potassium channels in cultured bovine adrenal chromaffin cells, *J. Physiol. (London)* **367:**117–141.

Marty, A., Tan, Y. P., and Trautman, A. J., 1984, Three types of calcium-dependent channels in rat lacrimal glands, *J. Physiol. (London)* **357**:293–325.

Maruyama, Y., and Petersen, O. H., 1982, Single-channel currents in isolated patches of plasma membrane from basal surface of pancreatic acini, *Nature* **299**:159–161.

Mayer, M., 1985, A calcium-activated chloride current generates the afterdepolarization of rat sensory neurones in culture, *J. Physiol. (London)* **364**:217–239.

Meech, R. W., 1972, Intracellular calcium injection causes increased potassium conductance in *Aplysia* nerve cells, *Comp. Biochem. Physiol.* **42A**:493–499.

Meech, R. W., and Standen, N. B., 1975, Potassium activation in *Helix aspersa* neurones under voltage clamp: A component mediated by calcium influx, *J. Physiol. (London)* **249**:211–239.

Miledi, R., 1980, Intracellular calcium and desensitization of acetylcholine receptors, *Proc. R. Soc. London Ser. B* **209**:447–452.

Miledi, R., and Parker, I., 1984, Chloride current induced by injection of calcium into *Xenopus* oocytes, *J. Physiol. (London)* **357**:173–183.

Morita, K., Kato, E., and Kuba, K., 1979, A possible role of intracellular Ca^{2+} in the regulation of the ACh receptor–ion channel complex of the sympathetic ganglion cell, *Kurume Med. J.* **26**:371–376.

Neher, E., 1986, Concentration profiles of intracellular calcium in the presence of a diffusible chelator, *Exp. Brain Res. Suppl.* **14**:80–96.

Nishizuka, Y., 1984, The role of protein kinase C in cell surface signal transduction and tumor promotion, *Nature* **308**:693–798.

Owen, D. G., Segal, M., and Barker, J. L., 1984, A Ca-dependent Cl⁻ conductance in cultured mouse spinal neurones, *Nature* **311**:567–570.

Rasmussen, H., and Barrett, P. Q., 1984, Calcium messenger system: An integrated view, *Physiol. Rev.* **64**: 938–984.

Schofield, P. R., Darlison, M. G., Fujita, N., Burt, D. R., Stephenson, F. A., Rodriguez, H., Rhee, L. M., Ramachandran, J., Reale, V., Glencorse, T. A., Seeburg, P. H., and Barnard, E. A., 1987, Sequence and functional expression of the GABA_A receptor shows a ligand-gated receptor super-family, *Nature* **328**:221–227.

Stelzer, A., Slater, N. T., and ten Bruggencate, G., 1987, Activation of NMDA receptors blocks GABAergic inhibition in an *in vitro* model of epilepsy, *Nature* **326**:698–701.

Thompson, S. H., 1977, Three pharmacologically distinct potassium channels in molluscan neurones, *J. Physiol. (London)* **265**:465–488.

Thorn, N. A., Russell, J. T., Torp-Pedersen, C., and Treiman, M., 1978, Calcium and neurosecretion, *Ann. N.Y. Acad. Sci.* **307**:618–639.

Yellen, G., 1982, Single Ca^{2+}-activated nonselective cation channels in neuroblastoma, *Nature* **296**:357–359.

Pharmacology and Physiology of Cl⁻ Conductances Activated by GABA in Cultured Mammalian Central Neurons

Jeffrey L. Barker, N. L. Harrison, and David G. Owen

1. INTRODUCTION

Cl^- permeability mechanisms regulated by neurotransmitters acting via receptor proteins are expressed in a wide variety of excitable tissues throughout all evolutionary forms studied. *In vivo* the permeability can be regulated physiologically by neurotransmitters like γ-aminobutyric acid (GABA) or glycine acting on receptors at subsynaptic membranes. Synaptically activated Cl^- flux can be found in the adult at all levels of spinal and supraspinal regions of the mammalian CNS where it usually functions to depress cellular excitability in a transient manner. A brief pause in ambient electrical activity lasting milliseconds is typically recorded during this physiologically elaborated synaptic signal. This in turn usually suppresses secretory activity derived indirectly from Na^+ action potential invasion of presynaptic terminals. Thus, synaptic signals involving Cl^- conductances effectively uncouple central neurons from one another and in so doing shape patterns of cellular excitation and neuronal circuit activity.

These synaptic signals have been studied for the last 50 years *in vivo, in situ,* and *in vitro* using experimental strategies and preparations of increasing simplicity. Recent studies employing biochemical and molecular genetic techniques have led to the hypothesis that GABA receptor/Cl^- channel complexes are composed of a variable number of discrete subunits enclosing a transmembrane aqueous domain (Scholfield *et al.*, 1987; Levitan *et al.*, 1988; Pritchett *et al.*, 1988). Interpretation of electrophysiological results on functional GABA receptors developing *in vitro* has led to the notion that the movement of Cl^- through the aqueous pore expressed in native membranes is probably gated by the coordinate actions of GABA molecules acting on two of the homologous subunits (Barker and Ransom, 1978a; Sakmann *et al.*, 1983).

Jeffrey L. Barker, N. L. Harrison, and David G. Owen • Laboratory of Neurophysiology, National Institute of Neurological Disorders and Stroke, National Institutes of Health, Bethesda, Maryland 20892. *Present address of N.L.H.:* Department of Anesthesia and Critical Care, University of Chicago, Chicago, Illinois. *Present address of D.G.O.:* Electrophysiology Section, Wyeth Research UK Ltd, Taplow Near Slough, Bucks, United Kingdom.

We and many others have used electrophysiological methods to examine some of the details involved in Cl^- permeability regulated by GABA and other amino acid neurotransmitters in embryonic CNS cells maintained *in vitro*. We have used electrophysiological recording strategies to study this phenomenon in central neurons differentiating in dissociated cultures and, more recently, voltage-sensitive indicator-dye staining of cells acutely suspended in the stream of a flow cytometer. Using the latter strategy we have found that both resting and amino acid-regulated Cl^- permeability mechanisms are expressed well before birth at the level of the cell body by the majority of embryonic spinal and supraspinal CNS cells (Mandler *et al.*, 1988; Fiszman *et al.*, 1988). The precise relationships between the expression of GABA-activated Cl^- permeability detected in embryonic cells and the differentiation of functional synapses utilizing GABA receptors have not been elucidated. However, Cl^--dependent responses to pharmacological applications of GABA and inhibitory synaptic signals involving GABA-mediated Cl^- permeability readily develop *in vitro* where they can be studied in some detail.

In this brief and necessarily eclectic review we will describe some of the details derived from pharmacological and physiological studies of GABA's actions on Cl^- conductances in cultured CNS neurons. These results may help to explain one of the actions of a widespread neurotransmitter that regulates Cl^- flux across many types of CNS neurons.

2. MONOLAYER CULTURE

The strategy we have used to examine GABA receptor-coupled Cl^- conductance mechanisms differentiating *in vitro* involves culturing the embryonic mammalian CNS as a virtual monolayer where the cells become readily accessible to quantitative analysis. The monolayer character permits application of contemporary electrophysiological recording techniques involving either sharp, high-resistance microelectrodes or blunt, low-resistance patch pipettes. Although the culture conditions represent a clear departure from the normal organization and development of CNS tissue, neurons differentiate with many properties characteristic of cells studied *in vivo*. Initially, recordings of GABA-activated Cl^- conductance mechanisms were made from embryonic mouse spinal cord neurons cultured for several weeks using high-resistance microelectrodes filled with molar concentrations of K^+ electrolytes. With these techniques it has become possible to record electrical responses to GABA applied discretely to different regions of the surface membrane of individual neurons.

3. GABA ACTIVATES Cl⁻ CONDUCTANCES IN CULTURED CNS NEURONS

The results of studies of pharmacologic applications of GABA to cultured cells show that spinal and supraspinal neurons exhibit a decidedly nonuniform distribution in their response to GABA (Barker and Ransom, 1978a; Fig. 1). Applications of this amino acid at or about the cell body of spinal cord neurons usually produce hyper-

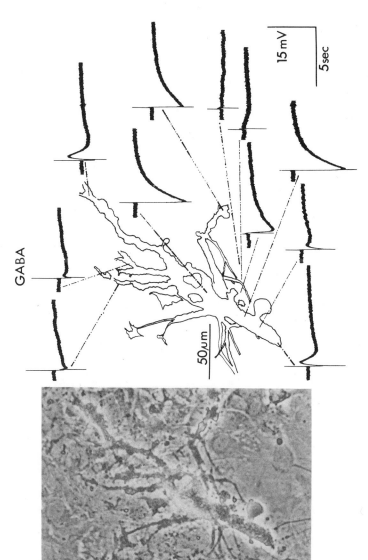

Figure 1. GABA activates simple and complex electrical responses in the same cultured mouse spinal neuron. The cell was recorded with a K^+-acetate-filled high-resistance microelectrode and GABA was iontophoresed by cationic current (application period marked by upward-and-downward transients on the voltage traces). Application of GABA (100-msec × 5-nA pulses) at different sites on the cell surface evokes different transmembrane potential responses. Applications at tips of two neuronal processes (at 2 and 7 o'clock) trigger complex responses consisting of transient depolarizing components superimposed on hyperpolarizing events. Applications in other regions evoke predominantly hyperpolarizing potentials with the most intense responses being at the level of the cell body. Applications to the bottom of the culture plate do not trigger a response unless the pipette is located within 20 μm of the cell surface in which case the response is markedly delayed and attenuated. (Modified from Barker and Ransom, 1978a.)

polarizing responses, while applications near the tips of processes induce complex effects with both depolarizing and hyperpolarizing components. The depolarizing component extrapolates to about 0 mV while the hyperpolarizing response reverses polarity at about −60 mV. Not all cultured central neurons express such complex responses to GABA. For example, embryonic rat hippocampal neurons cultured for several weeks consistently display a uniform hyperpolarizing response to GABA (Segal and Barker, 1984a).

The hyperpolarizing response to GABA in both spinal and hippocampal neurons and its reversal potential, E_{GABA}, are dependent on the Cl⁻ electrochemical gradient across the neuronal membrane. When cells kept in a low-Cl⁻ medium are recorded with K⁺ acetate-containing microelectrodes, E_{GABA} is equal to, or slightly more positive than the resting membrane potential. In this ionic environment, GABA induces small increases in membrane conductance (Barker and Ransom, 1978a; Fig. 2A). Acute application of Cl⁻ in the vicinity of such a recorded cell invariably leads to a significant hyperpolarizing shift in E_{GABA} (Fig. 2B) along with a substantial increase in membrane conductance. Loading cells with Cl⁻ by using microelectrodes filled

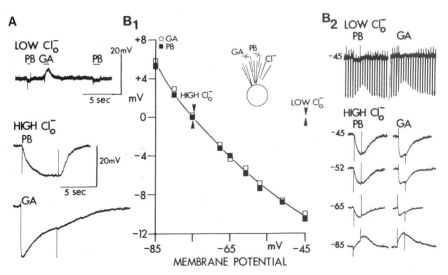

Figure 2. Cl⁻-dependency of electrical responses to GABA and pentobarbital (PB) in cultured mouse spinal cord neurons. GABA and PB were iontophoresed at the cell body surface of a neuron whose transmembrane potential was recorded with a K⁺-acetate-filled microelectrode. Initially the extracellular medium contained 9 mM Cl⁻. Culture medium containing 159 mM Cl⁻ was applied by pressure from another pipette located near the cell (schematized in the inset of panel B1). There was little transmembrane voltage response to PB or GABA (panel A, top traces, and B2, top traces) in the 9 mM Cl⁻-containing extracellular medium ("low Cl⁻ₒ") although there was a modest increase in input conductance (panel B2), indicating that the apparent lack of voltage response occurs because the resting potential (−45 mV) was close to the equilibrium potential for the responses, as shown in the plot of agonist responses at various membrane potentials (panel B1). Application of the medium containing 159 mM Cl⁻ ("high Cl⁻ₒ") shifted the reversal potential some 30 mV in the hyperpolarizing direction so that hyperpolarizing responses to GABA and PB now occurred at the resting membrane potential. (From Barker and Ransom, 1978b.)

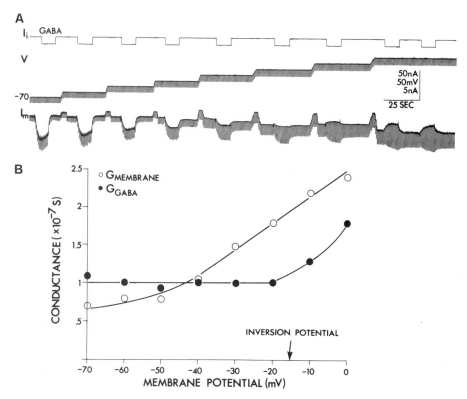

Figure 3. Reversal of GABA-induced currents in Cl⁻-loaded cells. The cultured mouse spinal neuron was recorded with two microelectrodes filled with KCl using the voltage-clamp technique. GABA was iontophoresed onto the cell body during the periods indicated in the top trace marked I_i in panel A. The membrane potential was clamped at −70 mV and then stepped in 10-mV increments to 0 mV. Current responses to pulses of GABA applied at each potential invert polarity at about −15 mV (panel A). Each response is associated with an increase in conductance. (B) Membrane conductance ($G_{membrane}$) and conductance responses to GABA (G_{GABA}) are plotted as a function of membrane potential. $G_{membrane}$ begins to increase at about −50 mV, while G_{GABA} is relatively constant until more depolarized potentials at which point it increases. (J. L. Barker and B. R. Ransom, unpublished observations.)

with KCl always shifts E_{GABA} in a depolarizing direction (Barker and Ransom, 1978a; Fig. 3).

From these and a number of other experiments involving ion substitution, we have concluded that transmembrane Cl⁻ movements account for most, if not all, of the hyperpolarizing conductance response to GABA in these cells. Other observations have shown that the macroscopic conductance change evoked by identical applications of GABA to spinal and hippocampal neurons at depolarized membrane potentials is relatively higher than at hyperpolarized potentials (Fig. 3; Segal and Barker, 1984a). Thus, GABA seems to be more effective at activating Cl⁻ conductance in cells that are depolarized. Some clinically important drugs like pentobarbital are also capable of activating hyperpolarizing, Cl⁻ conductance mechanisms at anesthetic concentrations

(Barker and Ransom, 1978b; Fig. 2). This effect probably results from direct activation of the GABA receptor-coupled Cl⁻ channel complex (Jackson *et al.*, 1982). Like GABA, pentobarbital's effects intensify on depolarized cells. Since GABA receptor-coupled Cl⁻ channels are virtually ubiquitous, the latter activity may account for some of the CNS depressant actions of the drug.

Without altering the physiological Cl⁻ electrochemical gradient, E_{GABA} can easily be manipulated by changing the transmembrane electrical gradient (Fig. 4). For example, sustained (minutes-long) conditioning of the cell at potentials de- or hyperpolarized to the resting potential using current-clamp recording techniques causes the reequilibration of the Cl⁻ gradient, as reflected in the time course of changes in the amplitudes of GABA-activated voltage responses and associated E_{GABA} values, within about 10–20 sec (Fig. 4C). Similar results have been obtained with glycine, which also activates Cl⁻ conductances in mammalian spinal neurons (Barker and Ransom, 1978a; Fig. 4). Three results indicate that Cl⁻ is not passively distributed across the plasma membrane in most cultured mammalian central neurons under these experimental conditions. Rather, E_{GABA}, which we presume to reflect the reversal potential for Cl⁻,

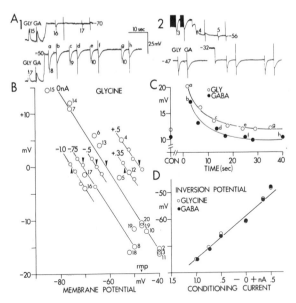

Figure 4. Inversion potentials of GABA- and glycine-evoked responses can be altered by changing the electrical gradient. The cultured mouse spinal neuron has been recorded with a K⁺-acetate-filled microelectrode and GABA and glycine applied by brief iontophoretic pulses. (A) Voltage responses to the brief pulses are illustrated at various membrane potentials. (A1, top trace) The cell has been current-clamped at a potential hyperpolarized to −70 mV and responses to glycine (marked by "15") and GABA elicited. Then the cell has been rapidly polarized to −70 mV and two pairs of agonist pulses applied (glycine-induced responses are marked by "16" and "17"). (Bottom trace) Paired pulses of glycine and GABA have been applied before and after the cell has been polarized to −50 mV. The lettered responses are plotted in panel C. (B) The inversion potentials (arrowheads) of voltage responses to glycine are plotted as a function of the DC current used (−1.0, −0.75, −0.5, 0, +0.35, +0.5 nA) to condition cells at various potentials hyper- and depolarized to rest (about −47 mV). The inversion potentials of both GABA and glycine are identical and related in linear manner to the conditioning current over the range used (−1.0 to +0.5 nA) (panel D). (C) Time course of changes in the amplitudes of GABA- and glycine-evoked responses at the resting potential following a conditioning period of hyperpolarization. Equilibration of Cl⁻ ions with the relaxation to the resting potential requires about 10 sec. Letters refer to responses illustrated in panel A1. (T. L. Barber and B. R. Ransom, unpublished observations.)

is more negative than the resting membrane potential under physiological conditions. This implies that Cl^- must be actively extruded, making E_{Cl} more negative than the resting membrane potential. Hence, the hyperpolarizing actions of GABA, by activating Cl^- conductance mechanisms sufficiently to transiently dominate membrane properties, bring the cell membrane potential close to E_{Cl}. The mechanisms responsible for active extrusion of Cl^- across the neuronal plasma membrane have not been elucidated (see Alvarez-Leefmans, this volume). Nevertheless, it is clear that long-lasting changes in the electrical gradient can readily and reversibly perturb E_{Cl} and thus E_{GABA}.

E_{GABA} is found to have a more positive value than the resting membrane potential in many types of mammalian peripheral ganglion neurons as well as in frog motoneurons and dorsal root ganglion cells (Bührle and Sonnhof, 1985; see Alvarez-Leefmans, this volume). In each cell type, GABA can activate the Cl^- conductance sufficiently to drive the transmembrane potential close to E_{Cl}. The peripheral ganglion cells that respond to GABA are not invested with GABAergic synapses and therefore a functional role for GABA here is presumably extrasynaptic. The role of GABA-activated Cl^- conductance in mammalian central neurons is functionally inhibitory to depolarizing forms of cation (e.g., Na^+, K^+, and Ca^{2+}) excitability. However, rapid applications of GABA to Cl^--loaded cells in which E_{Cl} lies about 20 mV depolarized to threshold for action potentials can also trigger action potentials (Barker, 1983). Therefore, conditions might arise in CNS neurons in which GABA could be functionally excitatory via activation of Cl^- conductance mechanisms.

4. GABA-ACTIVATED Cl⁻ CONDUCTANCE HAS UNIQUE PROPERTIES

We have used two-electrode voltage-clamp techniques in conjunction with Cl^- loading to study some of the details underlying the Cl^- conductance activated by GABA and other amino acids like glycine and β-alanine. By loading cells with Cl^- and then clamping them at relatively hyperpolarized potentials, it is possible to obtain an adequate driving force (the difference between the "holding" potential and E_{Cl}) and signal-to-noise ratio for the Cl^- current. Under these conditions, E_{Cl} is more positive than E_m and brief applications of GABA or glycine from pipettes placed within micrometers of the cell surface produce inward currents due to an efflux of Cl^- over the range of potentials more negative than the reversal potential. Cl^- currents triggered by brief pulses of GABA are typically slower to rise and fall than those generated by glycine independent of the current magnitude and the position of the iontophoresis pipette (Fig. 5). Similar differences between the time courses of voltage responses evoked by the amino acids can be seen in Fig. 4. Glycine-activated electrical events are invariably faster to decay.

We have studied the possible mechanisms accounting for the observed differences in the time course of the pharmacologically activated conductances by using fluctuation analysis of sustained responses. Fluctuation analysis involves statistical treatment of the complex noisy-looking current signal that is evoked during long-lasting agonist applications (Fig. 6A). The "noise" or rapid fluctuations in the current signal can be

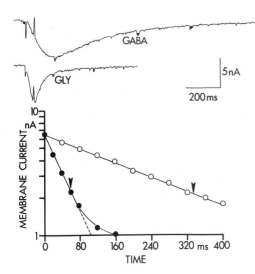

Figure 5. Cl⁻ currents activated by GABA relax more slowly than those activated by glycine. The cultured mouse spinal neuron was recorded under voltage clamp at −70 mV with two KCl-filled microelectrodes. GABA and glycine were iontophoresed onto the cell body with current pulses lasting 50 msec. Both agonists evoke currents of about 6.5 nA but the current activated by glycine relaxes about five times faster than that triggered by GABA. (J. L. Barker and B. R. Ransom, unpublished observations.)

amplified, filtered, acquired by computer and digitized, then later analyzed spectrally. The resulting spectrum (see Fig. 7B for examples) can often be fitted by a single Lorentzian term. From this and the measurement of membrane current variance, which is directly proportional to the amplitude of the current response, it is possible to *estimate* elementary Cl⁻ conductance properties (see Fig. 6B).

Analyses of many spectra calculated from sustained amino acid-induced Cl⁻ currents evoked in mouse spinal neurons show that GABA and glycine trigger Cl⁻ channel activity whose electrical properties appear to be different for each amino acid (McBurney and Barker, 1978; Barker and McBurney, 1979a). Average values of estimated elementary conductance and duration for channels activated by GABA, glycine and β-alanine are schematized in Fig. 6B. The values shown are averaged from cells in which at least two or all three of the amino acids had been applied. These estimates of elementary ion channel properties are derived from the activities of several hundred to several thousand ion channels. They show that glycine activates channels whose average conductance is about 150% greater and whose duration is about 20–25% as long as that triggered by GABA (Barker and McBurney, 1979a; Barker et al., 1982). The Cl⁻ charge transferred during GABA-triggered events is about twice that evoked by glycine (Fig. 6B). Thus, the differences in the time courses of the macroscopic Cl⁻ currents recorded in the whole cell in response to the amino acids correlate well with the estimated effective duration of channels activated by the amino acids.

Bicuculline, a relatively specific, competitive antagonist of GABA binding to receptors on membranes derived from rat CNS tissue (Wong and Iversen, 1985), readily blocks Cl⁻ conductance and associated current variance activated by GABA (Barker et al., 1983; Fig. 7). The membrane current variance induced during GABA-evoked responses is blocked by bicuculline in direct proportion to the decrease in amplitude of the response (Fig. 7A2). Spectral analysis of the current response de-

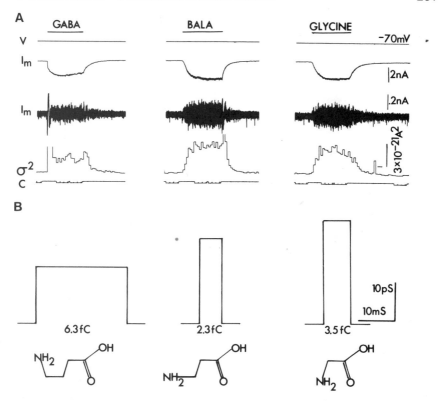

Figure 6. Fluctuation analysis of membrane current responses to GABA, glycine, and β-alanine (BALA) yields estimates of different Cl⁻ channel properties. (A) The spinal cord neuron was voltage-clamped at −70 mM (V) using two microelectrodes filled with KCl. Cl⁻ inward currents of about the same amplitude (upper traces marked I_m) were evoked by pressure applications to the cell body of 20 to 50 μM of each of the three amino acids. Each response is associated with an increase both in current fluctuations (shown in the AC-coupled, 10×-amplified lower traces marked I_m) and in current variance (trace labeled σ^2). Disproportionately more variance is observed during the glycine- and BALA-evoked responses. Traces marked C indicate periods when data were sampled for fluctuation analysis. (B) Schematic representations of the estimated average electrical properties of the Cl⁻ channels underlying the amino acid-elicited responses. The charge estimated for each average-sized microsopic event is indicated in femtocoulombs (fC). The channel opened by GABA was longer-lasting but lower in conductance than that activated by BALA or glycine. The total charge triggered by GABA per unitary event was considerably more than that activated by glycine or BALA. The chemical structures of each amino acid are shown at the bottom. The results reveal that for these particular responses, the estimated electrical properties of the ion channels are unique to each amino acid transmitter. (Adapted from Barker *et al.*, 1982.)

creased by bicuculline indicates little if any change in the spectrum. Taken together, the results strongly suggest that bicuculline blocks a proportion of the GABA-activated Cl⁻ ion channel activity in a roughly all-or-none manner, presumably by displacing GABA from its receptors. The Cl⁻ channels still operating in the presence of bicuculline are similar if not identical to those participating in the unblocked response.

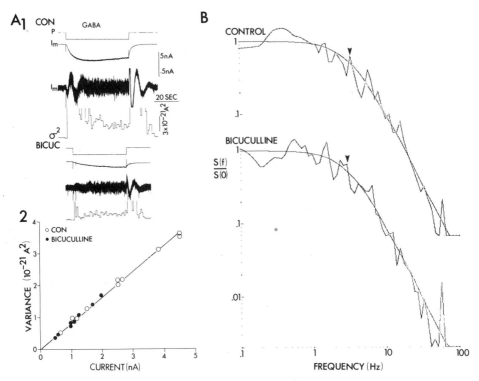

Figure 7. Fluctuation analysis of bicuculline-induced depression of GABA-activated Cl⁻ currents. The recording was made with two KCl-filled microelectrodes and the cell membrane potential was held at −70 mV by voltage clamp. (A1) GABA was applied by identical pressure pulses (P) during the indicated time under control conditions (CON) and in the presence of bicuculline (BICUC) in such a way as to generate a series of Cl⁻ currents (upper I_m). The fluctuations apparent on the lower I_m trace during the control response and the corresponding variance (σ^2) are depressed by bicuculline. (A2) Plot of variance as a function of the current response amplitude for a range of responses evoked under control conditions (○) and in the presence of bicuculline (●) showing that the linear relationship between variance and current amplitude is not changed by the drug. The slope of the mean-to-variance plot together with the driving force acting on the conductance give an estimated elementary Cl⁻ channel conductance of 15 pS. (B) Normalized power spectra reflecting the distribution of power (log) as a function of frequency (log) for spectra calculated from fluctuations generated under control conditions and in the presence of bicuculline. The jagged line represents the experimental data, while the smooth curve shows the best-fitting Lorentzian equation that accounts for the data. In each spectrum the arrowhead indicates the corner frequency (f_c) at which power falls by 50%. This is unchanged for bicuculline, indicating no apparent difference in the estimated average open-time for channels activated under control and experimental conditions. Estimated mean open-time was 39.8 msec in control and 37.9 msec in bicuculline. (Modified from Barker *et al.*, 1983.)

5. Cl⁻ CHANNELS ACTIVATED BY AGONISTS

Tight (gigaohm)-seal, on-cell, and excised-patch-clamp analyses of Cl⁻ ion channel activities in micrometer patches of cell body membranes have been carried out by an increasing number of laboratories on vertebrate central and peripheral neurons and on several types of nonneuronal cells (for details see Bormann, this volume; Akaike, this volume). The results from these studies indicate that many patches of cell body membrane have Cl⁻ channels whose activity is coupled to GABA receptors. However, the Cl⁻ channels activated by $0.5-10.0$ μM GABA in isolated membrane patches behave consistently in a more complex manner than predicted from fluctuation analysis of population responses in whose cells (see Mathers, 1985a), but the latter correspond to results obtained with tight-seal patch-clamp recordings of whole-cell Cl⁻ currents evoked in cultured rat hippocampal neurons (Ozawa and Yuzaki, 1984). For example, there are several distinct moments in the opening and closing kinetics of the Cl⁻ channels, which often assume a main conductance state and a variable number of subconductance states. The longer-lived ion channel events are often interrupted by brief closings, leading to the notion of "burst duration" rather than "channel duration." More importantly, the average values of mainstate conductance and longer-lived ion channel kinetics, which together account for most (80–90%) of the Cl⁻ flux through the patch (Jackson et al., 1982; Hamill et al., 1983; Bormann et al., 1987), closely correspond to the electrical properties of the channel predicted from fluctuation analysis of responses in spinal neurons (McBurney and Barker, 1978; Barker and McBurney, 1979a,b; Study and Barker, 1981; Barker et al., 1982). This correspondence between microscopic measurements and properties inferred from macroscopic activity also appears to hold for Cl⁻ channels activated by glycine, muscimol, and (−)-pentobarbital (Mathers, 1985b). Individual Cl⁻ channel activity recorded directly in a micrometer patch may be a particular example of the population of channel activities observed macroscopically in the whole cell. Although single- and population-channel studies *on the same cell* have not been carried out in a systematic manner, the present set of results suggest that *there may be a relatively uniform set of Cl⁻ channel properties activated pharmacologically on each neuron instead of ion channels whose properties vary with synaptic and extrasynaptic locations.*

Although GABA and glycine activate Cl⁻ channels each of which, in general, possesses unique electrical properties, nevertheless, there are several lines of evidence suggesting that the channels activated by these two agonists may not be so different. When applied discretely to different sites on multipolar, cultured spinal cord neurons, there is a significant correlation between the responses evoked by the two transmitters (R. E. Study and J. L. Barker, unpublished observations). Conductance responses to the agonists occlude one another as well as those responses evoked by β-alanine, which also activates Cl⁻ conductance (Barker and McBurney, 1979a). Many, but not all, patches of cell body membrane respond to both agonists (Bormann et al., 1987). Once opened, the channels require the same energy to close (Mathers and Barker, 1981a). The ion channels activated by each transmitter have identical subconductance states and anion permselectivities (see Bormann, this volume). All of these observations have gradually led to the notion that GABA and glycine might trigger Cl⁻ flux through similar if not physically identical channel mechanisms in the membrane. The rate of Cl⁻ flux and the effective duration would then be determined by the transmitter.

Somewhat surprisingly, glycine receptors in the spinal cord exist in direct postsynaptic apposition to GABAergic terminals (Triller et al., 1987). What are the properties of signals generated at these synapses? Are they like those activated by

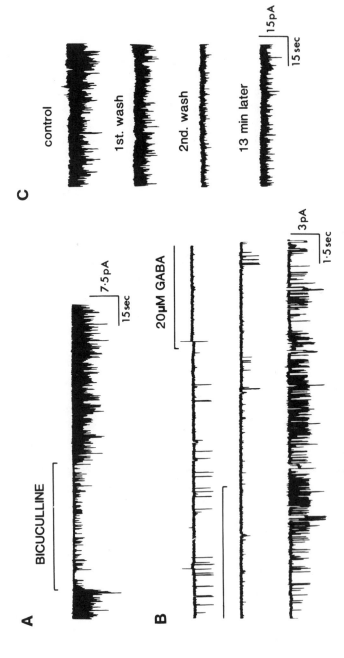

Figure 8. Spontaneous bicuculline-sensitive Cl⁻ channels are present in mouse nerve-cell body membrane patches. An "outside-out" patch was held at the tip of a pipette above the monolayer. The physiological monovalent cations were replaced by TRIS⁺ so as to isolate elementary Cl⁻ currents. (A) This and many other cell-body patches exhibit spontaneous Cl⁻ current activity at a membrane potential similar to that usually found under *in situ* "resting conditions" (−60 mV). Bicuculline, but not strychnine, applied to the receptor-bearing surface eliminates most of the Cl⁻ currents in a reversible manner. (B) Relatively high concentrations of GABA transiently suppress all of this activity, which recovers at a greater frequency of openings. (C) Superfusion of the monolayer with saline decreases the intensity of ambient activity in a reversible manner. (Modified from MacDonald *et al.*, 1985.)

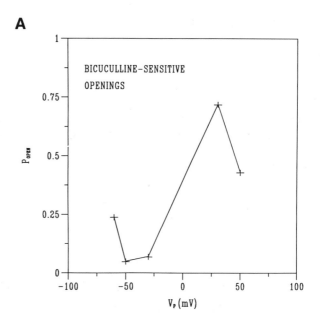

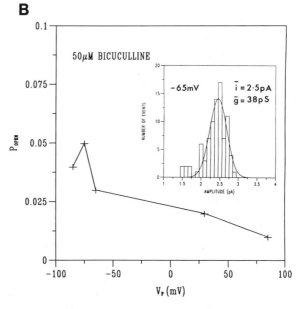

Figure 9. Bicuculline-sensitive Cl⁻ channels exhibit voltage-sensitive behavior different from bicuculline-insensitive channels. Data obtained from an experiment similar to that shown in Fig. 8. (A) The probability of bicuculline-sensitive Cl⁻ channels being open increases dramatically at depolarized potentials. (B) Conversely, the probability of bicuculline-insensitive channels being open decreases and becomes very small at depolarized potentials. (Modified from MacDonald *et al.*, 1985.)

GABA, by glycine, or neither? Recent molecular biological analysis of the two receptor structures has led to the conclusion that there are extensive homologies between GABA and glycine receptors and also between GABA receptors and those activated by acetylcholine at nicotinic sites which activate cation channels (Grenningloh *et al.*, 1987; Scholfield *et al.*, 1987). Probably, in the near future it will be possible to provide an explanation, at the molecular level, for the different mainstate conductances and kinetics of the type of channel (or channels) activated by the two amino acids as well as for the specificity of modulatory substances on each of the receptor-coupled properties.

In the course of recording ion channel activity in membrane patches excised from the cell bodies of cultured neurons under experimental conditions designed to isolated Cl^- ion currents, it became clear that many patches exhibited spontaneous levels of activity independent of added transmitter. For example, patches of membrane excised and everted in such a way as to expose the receptor-bearing surface to the extracellular bathing solution ("outside-out" configuration) often express spontaneous Cl^- channel activity in the absence of transmitter added to the media. Much of this channel activity can be eliminated by high (50 μM) concentrations of the competitive GABA receptor-antagonist bicuculline, by desensitizing (20 μM) concentrations of GABA, and by continuous superfusion of the culture chamber with saline (Fig. 8). The properties of these "spontaneous" bicuculline-sensitive Cl^- channels are similar to those of channels activated by pharmacological applications of GABA in terms of mainstate/substate conductances and biexponential kinetics (MacDonald *et al.*, 1985; Barker *et al.*, 1987b). The probability of finding one of these bicuculline-sensitive channels open is increased at depolarized potentials (Fig. 9A). These Cl^- channel activities might result from ambient levels of GABA released from the preparation into the recording saline. There is some Cl^- channel activity remaining in the presence of high concentrations of bicuculline sufficient to block GABA-mediated activity. These bicuculline-resistant Cl^- channels open less at depolarized potentials and exhibit a unitary conductance level rather than multiple states (Fig. 9B). These voltage-sensitive Cl^- channel activities are likely to occur through mechanisms independent of those activated by GABA. Other Cl^- channels in somal patches are quite sensitive to Ca^{2+} ions at the cytoplasmic membrane face (see Mayer *et al.*, this volume) and some GABA-operated Cl^- channels are themselves subject to dramatic modulation by intracellular Ca^{2+} (see Akaike, this volume; Taleb *et al.*, 1987). All of these results indicate that Cl^- fluxes across central mammalian neuronal membranes can occur via several porelike avenues, the regulation of which may involve overlapping transmitter mechanisms, as well as membrane potential and Ca^{2+} ions.

6. GABAmimetic Cl⁻ CHANNEL DURATION CORRELATES WITH GABAmimetic BINDING

Fluctuation analysis of GABAmimetic activation of Cl^- channels in cultured mouse spinal cord neurons showed that structural analogues of GABA activated ion channels having similar conductance, but different lifetimes than those opened by

GABA (Barker and Mathers, 1981). For instance, the lifetimes of channels activated by muscimol, a GABAmimetic that is about twice as potent as GABA (on a molar basis) at generating Cl^- conductance, were about twice as long as those channels activated by GABA (Barker and Mathers, 1981; Mathers and Barker, 1981b). Thus, at these Cl^- channels, estimated duration, not elementary conductance, may be a molecular correlate of molar potency. Hence, relative potency of agonists on a molar basis can be correlated with their ability to stabilize Cl^- channels in a conducting state for varying periods of time. A more detailed patch-clamp analysis of the relationship between structure of GABAmimetics and Cl^- channel kinetics needs to be carried out to establish the molecular basis for these observations. Somewhat surprisingly, recent studies of GABA-activated Cl^- channels in patches of cultured mammalian central neurons have revealed a concentration-dependency to the open-state kinetics over the range of 0.05–5.0 μM (MacDonald et al., 1989). Perhaps the briefer openings reflect mono- rather than biliganded receptors (Jackson et al., 1982).

We have compared the estimated Cl^- channel durations and radioligand binding of GABAmimetics. The binding of GABAmimetics to nerve cell membranes obtained from normally developed adult membranes was determined by their ability to competitively displace radiolabeled GABA (Olsen and Snowman, 1983; Wong and Iversen, 1985). There is a significant correlation between estimated channel durations (lifetimes) and agonist concentration required to displace GABA under equilibrium binding conditions (Fig. 10A1). Channel lifetime also correlates significantly with the two rates of unbinding from equilibrium (Fig. 10A2,3) as well as with their ability to displace radiolabeled bicuculline (Fig. 10B). The more potent an agonist is in competitively interacting with receptor proteins, the more effective it is in stabilizing channels in the open, conducting state. Although these biophysical and biochemical data are in excellent qualitative agreement, implying that the two assays are measuring a common property of the GABA receptor/Cl^- channel differentiated both *in vivo* and *in vitro*, they are not in quantitative agreement even when carried out at the same temperature (Yang and Olsen, 1987).

GABA and muscimol activate ion channels whose predominant properties are virtually identical to those inferred from fluctuation analysis when many hundreds to thousands of channels are operating simultaneously. But these micromolar concentrations of agonist are higher than many of the dissociation constants measured in equilibrium binding studies. The different methods of assay of membrane receptor binding and function and their corresponding levels of sensitivity and resolution may help to explain the discrepancy. Superficially, the binding constant data appear to be in close agreement with the bicuculline- and picrotoxin-sensitive pharmacological effects of GABA and muscimol on acutely isolated, intact early embryonic mammalian and avian CNS cells measured using voltage-sensitive dyes and a flow cytometer where the effects of nanomolar concentrations of agonist can readily be detected (Fiszman et al., 1988; Mandler et al., 1988). As embryogenesis proceeds and synapses emerge, the potency of these agonists in flow physiological assays *decreases* (Fiszman et al., 1988; A. Prasad et al., unpublished observations), approaching micromolar values typically reported for embryonic cells cultured for several weeks. The developmental transfor-

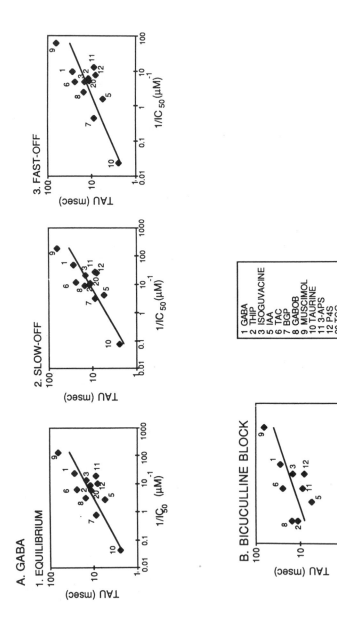

Figure 10. $T_{GABAmimetic}$ correlates with GABA receptor binding properties. $T_{GABAmimetic}$ values [TAU (msec)] are derived from fluctuation analyses of Cl⁻ currents activated by various GABAmimetics (Barker and Mathers, 1981 and unpublished observations; see box for legend). Half-maximal displacement (IC₅₀) values are from Olsen and Snowman (1983) who studied competitive displacement of GABA under equilibrium conditions (A1), during unbinding (A2,3) and competitive displacement of bicuculline (B) at equilibrium. Log–log plots of the pairs of $T_{GABAmimetic}$ and IC₅₀ values show that there is a linear relationship between the two parameters. In each case the relationship is statistically significant.

mation of high to low potency for activation of putative Cl^- conductance mechanisms occurs in rat hippocampus and chick spinal cord and appears to parallel the differentiation of GABAergic synapses.

7. GABAergic INHIBITORY POSTSYNAPTIC CURRENTS IN CULTURED HIPPOCAMPAL NEURONS

Many neurons maintained in culture for at least several days make functional contacts with each other, generating synaptic signals that either excite or inhibit postsynaptic target neurons. Some neurons cultured from the embryonic rat hippocam-

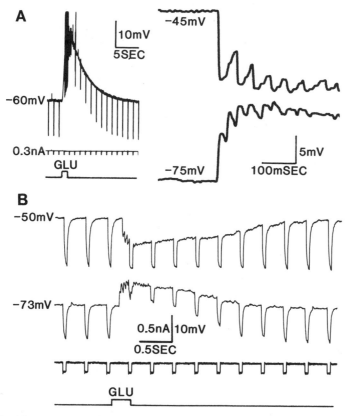

Figure 11. Rapidly summating inhibitory synaptic transmission between cultured hippocampal neurons. The cells were recorded with K^+-acetate-containing microelectrodes. (A) The excitatory amino acid glutamate was applied to a presynaptic cell so as to evoke a summating barrage of postsynaptic potentials (PSPs) in the postsynaptic neuron. These are illustrated with the cell held at −45 and −75 mV. The PSPs effectively clamp the neuron at about −60 mV. (B) Constant-current stimuli applied to the postsynaptic neuron reveal that the input resistance is reduced by about 50% at the peak of the summating PSP burst. The conductance increases manifested by the decrease in voltage responses during the summation evoked by the constant-current stimuli occur whether the cells are held at −50 or −73 mV. (Modified from Segal and Barker, 1984a.)

pus can be depolarized to generate action potentials which, in turn, trigger inhibitory postsynaptic potentials (IPSPs) in target cells (Segal and Barker, 1984b; Fig. 11A). A burst of these IPSPs effectively clamps the target cell at potentials decidedly hyperpolarized relative to the threshold for generating all-or-none action potentials. With K^+-acetate or K^+-gluconate recording electrodes, the potential at which IPSPs summate is invariably hyperpolarized relative to the cell's resting potential. IPSP summation is always associated with an increase in membrane conductance (Fig. 11B). The polarity of the IPSP and its reversal potential, E_{IPSP}, are both dependent on the Cl^- electrochemical gradient across the neuronal membrane (Segal and Barker, 1984b). Furthermore, these signals are blocked by the $GABA_A$ receptor antagonists bicuculline and picrotoxin (Segal and Barker, 1984b). Additional evidence for the GABAergic nature of these synaptic pathways found between cultured rat hippocampal neurons is the fact that the presynaptic neurons can be stained immunocytochemically for the presence of glutamic acid decarboxylase, the enzyme that metabolizes glutamate to GABA, or for GABA itself. All these results indicate that GABAergic neurons routinely survive in cultures of dissociated hippocampal cells and can generate synaptic signals with relatively stable properties for the duration of the recording period, which sometimes lasts for hours.

Inhibitory postsynaptic currents (IPSCs) underlying synaptic potentials have been studied using two-electrode voltage-clamp techniques with Cl^--loaded cells (Segal and Barker, 1984b) and more recently with tight-seal, whole-cell recording techniques in non-Cl^--loaded cells (Harrison *et al.*, 1987a; Fig. 12). The synaptically evoked signals appear to have similar properties whether recorded with two-electrode or single patch electrode voltage-clamp techniques. Under two-electrode voltage-clamp conditions the IPSCs recorded in Cl^--loaded cells reverse polarity in the range -20 to 0 mV, while in whole-cell recording (non-Cl^--loaded neurons) IPSCs reverse at about -60 mV (Fig. 12a). Both the IPSPs and the IPSCs reverse polarity at E_{GABA} (Fig. 12b).

The IPSCs peak within milliseconds and decay with a variable rate usually following a single-exponential time course (Segal and Barker, 1984b; Harrison *et al.*, 1987a; Barker and Harrison, 1988). The range of exponential time constant (T_{IPSC}) values overlaps that inferred for the mean open-time of GABA-activated Cl^- channels (T_{GABA}) recorded in cultured hippocampal neurons (Segal and Barker, 1984a,b). IPSCs also decay exponentially in hippocampal slice neurons but appear to be shorter than many of those recorded in culture even at the same potential and temperature (Collingridge *et al.*, 1984). In some cultured neurons, T_{IPSC} and T_{GABA} increase at depolarized potentials (Segal and Barker, 1984a,b; Barker and Harrison, 1988). Similar effects of voltage on the time constant of decay of GABA-mediated, Cl^--dependent synaptic signals recorded in rat hippocampal neurons (Collingridge *et al.*, 1984) and on glycine-mediated Cl^- signals in Mauthner cells of the goldfish have also been reported (Faber and Korn, 1987).

The peak conductance associated with IPSCs and with GABA-induced responses usually increases at depolarized potentials. Comparison of experimental values with predictions from the Goldman–Hodgkin–Katz equation has led us to conclude that the outward rectification observed at the macroscopic level in cultured hippocampal neu-

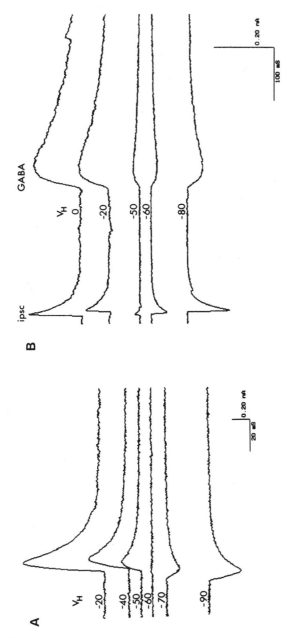

Figure 12. E_{IPSC} and E_{GABA} correspond to each other. The cultured rat hippocampal neurons were recorded using the tight-seal, patch-clamp recording technique in the whole-cell configuration with a K^+-gluconate-containing pipette. A series of inhibitory postsynaptic currents (IPSCs) was triggered by successively exciting the presynaptic neuron while clamping the postsynaptic cell at different potentials. (a) The IPSCs reverse polarity close to -60 mF. (b) IPSCs and Cl^--dependent current responses to brief pulses of GABA reverse polarity between -50 and -60 mV. (Modified from Barker and Harrison, 1988.)

rons is due primarily to the Cl^- chemical gradient across the cell membrane. On the other hand, the outward rectification of GABA-induced Cl^- currents recorded in hippocampal CAl pyramidal neurons in the slice appears to be greater than that which can be accounted for by the constant field equation (Ashwood *et al.*, 1987). Whatever the mechanism responsible for the outward rectification, the fact is that GABAergic synaptic signals in rat hippocampal neurons, whether in culture or slices, are naturally more effective when the postsynaptic cell is depolarized. It seems that not only the Cl^- gradient across the membrane but also the voltage-sensitivity of channel kinetics contribute to the observed rectification. In cultured chick cerebral neurons, macroscopic Cl^- current responds to an increase of applied GABA at depolarized potentials (Weiss *et al.*, 1988), but this has been attributed to a depolarization-induced increase in the probability that channels would open rather than a change in mean open-time (Weiss, 1988). On the other hand, patches excised from adult guinea pig hippocampal pyramidal and dentate granule cells express GABA-activated channels that increase both open-state probability and elementary conductance at depolarized potentials (Gray and Johnston, 1985). Furthermore, outward rectification of single GABA-activated channels recorded from patches of hippocampal neurons, occurs even under conditions of symmetrical Cl^- concentrations. From these results it would appear that more than one mechanism may be responsible for the outward rectification of the IPSC which underlies the greater efficacy of GABA-mediated Cl^--dependent synaptic signals recorded in depolarized postsynaptic cells.

T_{IPSC} at -40 mV ranges from about 10 msec to about 40 msc (Barker and Harrison, 1988; Fig. 13). The rate of decay of the IPSC is *not* correlated with the peak conductance of the IPSC. IPSCs of quite similar conductance can be both short- and long-lasting. Under optimal recording conditions it is possible to measure IPSC rise-

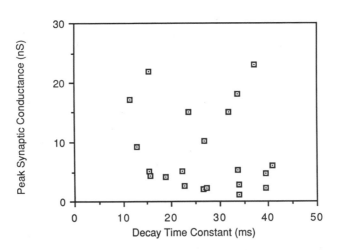

Figure 13. Peak synaptic conductance is unrelated to T_{IPSC}. The peak synaptic conductance evoked in 21 hippocampal neurons at -40 mV was calculated following measurement of E_{IPSC} and then plotted as a function of T_{IPSC}. There is no correlation between the peak conductance and the time constant of IPSC decay, T_{IPSC}. (Modified from Barker and Harrison, 1988.)

time with sufficient accuracy. In those cells where rise-time and T_{IPSC} can be accurately measured, the former correlates significantly with the latter. The faster rising IPSCs are correspondingly quicker to decay. Theoretically, the time course of the IPSC is determined by the rate of GABA release from presynaptic terminals, diffusion in the synaptic cleft to postsynaptic receptors, binding to and subsequent activation of postsynaptic receptor-coupled Cl⁻ channels, followed by diffusion away from the receptors. Other important aspects to be considered in this scheme include the kinetics inherent in the complex receptor-binding/channel-activation process and the relative position of the synaptic boutons with respect to the cell body, where the IPSC is actually recorded. The exponential term describing the mean open-time (or burst duration) of hundreds of GABA receptor-coupled Cl⁻ channels activated simultaneously by GABA that has been estimated from fluctuation analyses on whole-cell recordings aptly describes both the time constant of the long openings of individual Cl⁻ channels activated by GABA and the T_{IPSC}. Thus, the variable kinetic properties of the receptor/Cl⁻ channel complex could help to account for the fourfold range in T_{IPSC} values. Different cells likely express different receptor-activated Cl⁻-channel kinetics which, on any one cell, may be similar at synaptic and extrasynaptic sites. Location of synaptic boutons at membrane sites distant from the cell body and the electrotonic properties of the cell could lead to "filtering" of the synaptic signal, which is recorded at the cell body, more so than if the terminals invested the cell body.

These possibilities can be tested by examining the spatial distribution of investing boutons ending on postsynaptic neurons and correlating this distribution with short- and long-lasting IPSCs. This could be done using conventional histochemical techniques (to reveal the distribution of innervating boutons) and inferring T_{GABA} from fluctuation analysis of GABA-induced Cl⁻ currents or measuring directly GABA-activated ion channels in membrane patches in the same cell as the synaptic signal is recorded. This combined methodology will permit answering questions such as: is there any apparent correlation between the T_{IPSC} value and the relative distribution of GABAergic innervation; i.e., are fast signals associated with predominantly somal and circumsomal innervation whereas do slow signals correspond to terminals arranged on processes? Does T_{GABA} closely approximate T_{IPSC} in the same cell at the same potential?

GABAergic IPSCs are sensitive to a variety of clinically important drugs. By comparing the effects of the drugs on pharmacologically evoked Cl⁻ currents with their actions on IPSCs, it appears that Cl⁻ channel kinetics, not conductance, is the receptor property most commonly affected by these drugs (Barker and McBurney, 1979b; Segal and Barker, 1984b). Some drugs that depress the electrical excitability of CNS circuits either globally (like barbiturates) or discretely (like benzodiazepines) increase hippocampal IPSCs. For example, (−)-pentobarbital, a general anesthetic, increases T_{IPSC} with little if any change in IPSC amplitude both in cultured hippocampal neurons (Segal and Barker, 1984b) and in hippocampal CA1 pyramidal cells recorded in slices (Collingridge et al., 1984). From fluctuation analysis of macroscopic currents, (−)-pentobarbital increases T_{GABA} (Study and Barker, 1981). At the microscopic level, the racemic mixture of (±)-pentobarbital induces GABA-activated channels that exhibit bursts of activity characterized by periods of rapidly occurring open-

and-closed states (Mathers, 1985b). The latter are grouped together in such a way that the envelope of activity enclosing an individual burst of openings closely resembles a single opening of long duration, especially if recorded with low band-pass filtering. It is thus possible that the effects of (−)-pentobarbital on T_{GABA} can be better explained in terms of the results obtained with the higher resolution afforded by the patch-clamp technique and should be reinterpreted from these direct observations of individual channel behavior in the presence of the drug. Alternatively, the racemic mixture of pentobarbital isomers may have pharmacological effects on Cl⁻ channel behavior different from the (−) isomer.

Steroids, also used clinically to induce general anesthesia, appear to act in a manner quite similar to (−)-pentobarbital in that they prolong T_{GABA} and T_{IPSC}, but they are considerably more potent than pentobarbital on a molar basis (Barker et al., 1987a; Harrison et al., 1987a). Naturally occurring steroid metabolites of certain hormones like progesterone behave like the synthetic steroid used clinically (Majewska et al., 1986; Harrison et al., 1987b). If these metabolites could gain access to GABA receptor/Cl⁻-channel complexes, then patterns of steroid hormone synthesis and release might be important in tuning GABAergic transmission.

Diazepam, a benzodiazepine with anxiolytic action in vivo, as well as other benzodiazepines modestly increase IPSC amplitude and/or T_{IPSC} (Segal and Barker, 1984b; Vicini et al., 1986). Diazepam increases the rate at which GABA opens Cl⁻ channels and modestly increases T_{GABA} (Study and Barker, 1981; Redmann and Barker, 1984; Vicini et al., 1987). The enhancement of IPSC amplitude likely reflects the increased rate of channel opening, while the changes in T_{IPSC} may result from more persistently opened channels that retain an exponential closing rate.

All of these results with clinically important drugs, together with those on the voltage-sensitive kinetics of T_{GABA} and T_{IPSC}, strongly suggest that the Cl⁻ channel's opening and closing kinetics (which shape the synaptic signal) are subject to intrinsic (e.g., voltage) and extrinsic (e.g., modulator/drug) regulation and that T_{IPSC} is a natural approximation both of the exponential distribution of long-duration channel openings recorded across the somatic membrane and of T_{GABA}. The peak of the IPSC corresponds to the near-synchronous activation of tens to hundreds of channels, while T_{IPSC} reflects the exponentially decreasing number of channels remaining open for longer and longer periods.

8. CONCLUSION

Cl⁻ flux mechanisms are preserved in most if not all central neurons maintained in monolayer culture. The precise contribution of Cl⁻ to resting potentials in mammalian central neurons has not been clearly elucidated. If anything, resting potentials appear modestly depolarized when cultured spinal or hippocampal cells are loaded with Cl⁻, and lowering [Cl⁻]$_o$ transiently depolarizes cultured mouse spinal cord cells (Barker and Ransom, 1978a). The existing Cl⁻ electrochemical gradient can be tapped in the majority of mammalian neurons studied thus far via amino acid receptor-coupled activation of Cl⁻ channels. The receptors become functional on embryonic cells well before birth and become distributed in a nonuniform manner along the surface of complex multipolar neurons differentiating in monolayer culture. The distribution may well be related to the investment of GABAergic terminals. Pharmacological studies of the structural requirements for Cl^- channel activation indicate that the channels differentiated *in vitro* have properties common to those assayed biochemically in normally developed tissue and that extrasynaptic and synaptic receptor properties may be similar if not identical. Each amino acid triggers a unique set of ion-channel properties, although there are molecular aspects common to all amino acid-activated Cl^- channels. Recordings from pairs of cultured hippocampal neurons show that functional GABAergic transmission involving Cl^- ion permeation mechanisms differentiates *in vitro* where it serves to inhibit excitatory activity as it usually does *in vivo*. The time course of the synaptic signal recorded is likely a reflection of both the intrinsic properties of the Cl^- channels activated by GABA as well as the location of innervating terminals along the cell surface. Clinically important drugs and naturally occurring steroids modulate the intensity and duration of the synaptic signal primarily by altering the kinetics rather than the conductance of the Cl^- channels activated by GABA. Thus, Cl^- fluxes activated by GABA develop before birth apparently prior to synaptogenesis and functional synaptic activity. In the well-differentiated CNS they are utilized to uncouple neurons from one another in a transient manner, thereby shaping the complex patterns of electrical activity ambient throughout the vertebrate CNS.

REFERENCES

Ashwood, T. J., Collingridge, G. L., Herron, C. E., and Wheal, H. V., 1987, Voltage-clamp analysis of somatic gamma-aminobutyric acid responses in adult rat hippocampal CAl neurons *in vitro, J. Physiol. (London)* **384:**27–37.

Barker, J. L., 1983, Chemical excitability in vertebrate central neurons, in: *The Clinical Neurosciences,* Volume 5 (W. D. Wills, ed.), Churchill–Livingston, Edinburgh, pp. 121–141.

Barker, J. L., and Harrison, N. L., 1988, Outward rectification of inhibitory postsynaptic currents in cultured rat hippocampal neurons, *J. Physiol. (London)* **403:**41–55.

Barker, J. L., and McBurney, R. N., 1979a, GABA and glycine may share the same conductance channel on cultured mammalian neurones, *Nature* **277:**234–236.

Barker, J. L., and McBurney, R. N., 1979b, Phenobarbitone modulation of postsynaptic GABA receptor function on cultured mammalian neurons, *Proc. R. Soc. London Ser. B* **206:**318–326.

Barker, J. L., and Mathers, D. A., 1981, GABA analogues activate channels of different duration in cultured mouse spinal neurons, *Science* **212:**358–361.

Barker, J. L., and Ransom, B. R., 1978a, Amino acid pharmacology of mammalian central neurones grown in tissue culture, *J. Physiol. (London)* **280:**331–354.

Barker, J. L., and Ransom, B. R., 1978b, Pentobarbitone pharmacology of mammalian central neurones grown in tissue culture, *J. Physiol. (London)* **280:**355–372.

Barker, J. L., McBurney, R. N., and MacDonald, J. F., 1982, Fluctuation analysis of neutral amino acid responses in cultured mouse spinal neurons, *J. Physiol. (London)* **322:**365–387.

Barker, J. L., McBurney, R. N., and Mathers, D. A., 1983, Convulsant-induced depression of amino acid responses in cultured mouse spinal neurons studied under voltage clamp, *Br. J. Pharmacol.* **80:**619–629.

Barker, J. L., Harrison, N. L., Lange, G. D., and Owen, D. G., 1987a, Potentiation of gamma-aminobutyric acid-activated chloride conductance by a steroid anesthetic in cultured rat spinal neurones, *J. Physiol. (London)* **386:**485–501.

Barker, J. L., Dufy, B., Harrington, J. W., Harrison, N. L., MacDermott, A. B., MacDonald, J. F., Owen, D. G., and Vicini, S., 1987b, Signals transduced by gamma-aminobutyric acid in cultured central nervous system neurons and thyrotropin releasing hormone in clonal pituitary cells, *Ann. N.Y. Acad. Sci.* **494:**1–38.

Bormann, J., Hamill, O. P., and Sakmann, B., 1987, Mechanism of anion permeation through channels gated by glycine and gama-aminobutyric acid in mouse cultured spinal neurones, *J. Physiol. (London)* **385:**243–286.

Bührle, C. P., and Sonnhof, U., 1985, The ionic mechanism of postsynaptic inhibition in motoneurones of the frog spinal cord, *Neuroscience* **14:**581–592.

Collingridge, G. L., Gage, P. W., and Robertson, B., 1984, Inhibitory postsynaptic currents in rat hippocampal CA1 neurones, *J. Physiol. (London)* **356:**551–564.

Faber, D. S., and Korn, H., 1987, Voltage-dependence of glycine-actived Cl⁻ channels: A potentiometer for inhibition, *J. Neurosci.* **7:**807–811.

Fiszman, M., Novotny, E. A., Lange, G. D., and Barker, J. L., 1988, Functional GABA$_A$ receptors are expressed in the early embryonic rat hippocampus and striatum, *FASEB J.* **2:**A1734.

Gray, R., and Johnston, D., 1985, Rectification of single GABA-gated chloride channels in adult hippocampal neurons, *J. Neurophysiol.* **54:**134–142.

Grenningloh, G., Rienitz, A., Schmitt, B., Methfessel, C., Zenson, M., Bayreuther, K., Gundelfinger, E. D., and Betz, H., 1987, The strychnine-binding subunit of the glycine receptor shows homology with nicotinic acetylcholine receptors, *Nature* **328:**215–220.

Hamill, O. P., Bormann, J., and Sakmann, B., 1983, Activation of multiple-conductance state chloride channels in spinal neurones by glycine and GABA, *Nature* **305:**805–808.

Harrison, N. L., Vicini, S., and Barker, J. L., 1987a, A steroid anesthetic prolongs inhibitory postsynaptic currents in cultured rat hippocampal neurons, *J. Neurosci.* **7:**604–609.

Harrison, N. L., Majewska, M. D., Harrington, J. W., and Barker, J. L., 1987b, Structure–activity relationships for steroid interaction with the gamma-aminobutyric acid$_A$ receptor complex, *J. Pharmacol. Exp. Ther.* **241:**346–353.

Jackson, M. B., Lecar, H., Mathers, D. A., and Barker, J. L., 1982, Single channel currents activated by GABA, muscimol, and (−) pentobarbital in cultured mouse spinal neurons, *J. Neurosci.* **2:**889–894.

Levitan, E. S., Blair, A. C. L., Dionne, V. E., and Barnard, E. A., 1988, Biophysical and pharmacological properties of cloned GABA$_A$ receptor subunits expressed in Xenopus oocytes, *Neuron* **1:**773–781.

McBurney, R. N., and Barker, J. L., 1978, GABA-induced conductance fluctuations in spinal neurons, *Nature* **274:**596–597.

MacDonald, J. F., Owen, D. G., and Barker, J. L., 1985, Voltage-sensitive spontaneously-occurring chloride channels in cultured spinal neurons, *J. Neurosci. Abstr.* **11:**200.

MacDonald, R. L., Rogers, C. J., and Twyman, R. E., 1989, Kinetic properties of the GABA$_A$ receptor main conductance state of mouse spinal cord neurons in culture, *J. Physiol. (London)* **410:**479–499.

Majewska, M. D., Harrison, N. L., Schwartz, R. D., Barker, J. L., and Paul, S. M., 1986, Steroid hormone metabolites are barbiturate-like modulators of the GABA receptor, *Science* **323:**1004–1007.

Mandler, R. N., Schaffner, A. E., Novotny, E. A., Lange, G. D., and Barker, J. L., 1988, Functional GABA and glycine receptors precede glutamate receptors during the development of the rat spinal cord, *Soc. Neurosci. Abstr.* **13:**1226.

Mathers, D. A., 1985a, Spontaneous and GABA-induced single-channel currents in cultured murine spinal neurons, *Can. J. Physiol. Pharmacol.* **63:**1228–1233.

Mathers, D. A., 1985b, Pentobarbital promotes bursts of gamma-aminobutyric acid-activated single channel currents in cultured mouse central neurons, *Neurosci. Lett.* **60:**121–126.

Mathers, D. A., and Barker, J. L., 1981a, GABA- and glycine-activated channels in cultured mouse spinal neurons require the same energy to close, *Brain Res.* **224:**441–445.

Mathers, D. A., and Barker, J. L., 1981b, GABA and muscimol open channels of different lifetimes on cultured mouse spinal neurons, *Brain Res.* **204:**242–247.

Olsen, R. W., and Snowman, A., 1983, [³H] Bicuculline methochloride binding to low-affinity gamma-aminobutyric acid receptor sites, *J. Neurochem.* **41:**1653–1663.

Ozawa, S., and Yuzaki, M., 1984, Patch-clamp studies of Cl⁻ channels activated by gamma-aminobutyric acid in cultured hippocampal neurones of the rat, *Neurosci. Res.* **1:**275–293.

Pritchett, D. B., Sontheimer, H., Gorman, C. M., Kettenmann, H., Seeburg, P. H., and Scholfield, P. R., 1988, Transient expression shows ligand gating and allosteric potentiation of GABA$_A$ receptor subunits, *Science* **242:**1306–1308.

Redmann, G. A., and Barker, J. L., 1984, Diazepam and voltage increase GABA-activated Cl⁻ ion channel opening kinetics in cultured mouse spinal neurons, *Soc. Neurosci. Abstr.* **10:**642.

Sakmann, B., Hamill, O. P., and Bormann, J., 1983, Patch-clamp measurements of elementary chloride currents activated by the putative inhibitory transmitters GABA and glycine in mammalian spinal neurons, *J. Neural Transm.* (Suppl.) **1:**83–95.

Scholfield, P. R., Darlison, M. G., Fujita, N., Burt, D. R., Stephenson, F. A., Rodriguez, H., Rhee, L. M., Ramachandram, J., Reale, J., Glencorse, T. A., Seeburg, P. H., and Barnard, E. A., 1987, Sequence and functional expression of the GABA$_A$ receptor shows a ligand-gated receptor super-family, *Nature* **328:**221–223.

Segal, M., and Barker, J. L., 1984a, Rat hippocampal neurons in culture: Properties of GABA-activated Cl⁻ ion conductance, *J. Neurophysiol.* **52:**500–515.

Segal, M., and Barker, J. L., 1984b, Rat hippocampal neurons in culture: Voltage clamp analysis of inhibitory connections, *J. Neurophysiol.* **52:**469–487.

Study, R. E., and Barker, J. L., 1981, Diazepam and (−) pentobarbital: Fluctuation analysis reveals different mechanisms for potentiation of GABA responses in cultured central neurons, *Proc. Natl. Acad. Sci. USA* **78:**7180–7184.

Taleb, A., Trouslard, J., Demeneix, B. A., Feltz, P., Bossu, J.-L., Dupont, J.-L., and Feltz, A., 1987, Spontaneous and GABA-evoked chloride channels on pituitary intermediate lobe cells and their internal Ca requirements, *Pfluegers Arch.* **409:**620–631.

Triller, A., Cluzeaud, R., and Korn, H., 1987, Gamma-aminobutyric acid-containing terminals can be apposed to glycine receptors at central synapses, *J. Cell Biol.* **104:**947–956.

Vicini, S., Alho, H., Costa, E., Mienville, J.-M., Santi, M. R., and Vaccarino, F. M., 1986, Modulation of gamma-aminobutyric acid-mediated inhibitory synaptic currents in dissociated cortical cell cultures, *Proc. Natl. Acad. Sci. USA* **83:**9269–9273.

Vicini, S., Mienville, J.-M., and Costa, E., 1987, Actions of benzodiazepine and beta-carboline derivatives on gamma-aminobutyric acid-activated Cl⁻ channels recorded from membrane patches of neonatal rat cortical neurons in culture, *J. Pharmacol. Exp. Ther.* **243:**1195–1201.

Weiss, D. S., 1988, Membrane potential modulates the activation of GABA-gated channels, *J. Neurophysiol.* **59:**514–527.

Weiss, D. S., Barnes, E. M., and Hablitz, J. J., 1988, Whole-cell and single-channel recordings of GABA-gated currents in cultured chick cerebral neurons, *J. Neurophysiol.* **59:**495–513.

Wong, E. H. F., and Iversen, L. L., 1985, Modulation of [³H] diazepam binding in rat cortical membranes by GABA$_A$ agonists, *J. Neurochem.* **44:**1162–1167.

Yang, J. S.-J., and Olsen, R. W., 1987, Gamma-aminobutyric acid receptor binding in fresh mouse brain membranes at 22° C ligand-induced changes in affinity, *Mol. Pharmacol.* **32:**266–277.

Acetylcholine-Activated Cl⁻ Channels in Molluscan Nerve Cells

Vladimir N. Kazachenko

1. INTRODUCTION

Although the acetylcholine receptor (AChR) coupled to the cationic channel has received the most attention (see, e.g., Adams, 1981; Popot and Changeux, 1984; Hucho, 1986; Skok *et al.*, 1987), it is nevertheless true that in many animals there exist AChR coupled to Cl⁻ channels, which seem to play a significant role in both synaptic transduction of signals and humoral regulation. Nicotinic AChR coupled to Cl⁻ channels are abundant in neuronal membranes of various molluscs, e.g., *Helix* (Kerkurt and Thomas, 1964), *Cryptophallus* (Chiarandini and Gerschenfeld, 1967; Chiarandini *et al.*, 1967), *Onchidium* (Sawada, 1969), *Aplysia* (Frank and Tauc, 1964; Kehoe, 1967, 1972; Sato *et al.*, 1968; Blankenship *et al.*, 1971), *Navanax* (Levitan *et al.*, 1970; Levitan and Tauc, 1972), *Lymnaea stagnalis* (Kislov, 1974; Kislov and Kazachenko, 1974), *Planorbarius corneus* (Ger *et al.*, 1980), and others (see Gerschenfeld, 1973).

Activation of Cl⁻ channels by AChR leads to the hyperpolarization of some neurons (H neurons) (e.g., Tauc and Gerschenfeld, 1962; Kerkurt and Thomas, 1964; Chiarandini and Gerschenfeld, 1967) and to the depolarization of others (D neurons) (e.g., Frank and Tauc, 1964; Chiarandini *et al.*, 1967; Kislov, 1974; Kislov and Kazachenko, 1974; Ger *et al.*, 1980). It should be noted that H and D neurons differ not only with respect to intracellular Cl⁻ concentration ($[Cl^-]_i$) but also with respect to the AChR themselves. Thus, the AChR in D neurons are inhibited by hexamethonium whereas those of H neurons are not (Tauc and Gerschenfeld, 1962; Zeimal and Vulfius, 1968).

A wide range of studies have been carried out on AChR of neurons of the freshwater mollusc *Lymnae stagnalis*. The main results can be summarized as follows:

1. The receptors are of the nicotinic type (Zeimal and Vulfius, 1968) and are coupled to Cl⁻ channels (Kislov, 1974; Kislov and Kazachenko, 1974).
2. Sequences of effectiveness of mono- and bisquaternary cholinomimetics and cholinolytics have been established (Zeimal and Vulfius, 1968; Kazachenko, 1979; Kurchikov *et al.*, 1985).

Vladimir N. Kazachenko • Institute of Biological Physics, USSR Academy of Sciences, Pushchino, Moscow Region, 142292, USSR.

3. The receptors are selectively inhibited by *d*-tubocurarine, hexamethonium (Zeimal and Vulfius, 1968), and the "Ca-antagonist" D-600 (Bregestovski and Iljin, 1980).

4. The AChR possesses COO^- groups and disulfide bonds which probably play an important role in its operation (Iljin *et al.*, 1976; Bregestovski *et al.*, 1977; Vulfius *et al.*, 1979; Vulfius and Iljin, 1980).

5. Conformational states are different for resting, active, and desensitized forms of the receptor (Bregestovski *et al.*, 1977; Vulfius *et al.*, 1979). The desensitization is a two-component process (Andreev *et al.*, 1984).

6. Activity of the AChR is inhibited by intracellular cAMP (dialyzed neuron) as well as via activation of dopamine or serotonin receptors (Akopyan *et al.*, 1980).

7. The substrates of the tricarboxylic acid cycle modulate the AChR activity (Andreev *et al.*, 1986).

8. The AChR are inhibited in the range of depolarizing membrane potentials (Kazachenko and Kislov, 1973, 1974, 1977; Kazachenko *et al.*, 1979) due to Ca^{2+} entering the cell through voltage-sensitive Ca^{2+} channels (Kazachenko, 1979; Kazachenko *et al.*, 1981a,b; Chemeris *et al.*, 1982; Chemeris and Iljin, 1985; Ivanova *et al.*, 1987).

9. Some properties of single Cl^- channels (e.g., conductance and its sublevels, open-time) have been revealed (Bregestovski and Redkozubov, 1986; Ivanova *et al.*, 1986, 1987).

Besides the ACh-activated Cl^- channels, the somatic membrane of *Lymnaea* neurons has another type of Cl^- conductance, one that is activated by both extracellular alkali metal cations ($Rb^+ > K^+ > Cs^+$) and Ca_i^{2+} (Kislov, 1974; Kislov and Kazachenko, 1975, 1977; Geletyuk and Kazachenko, 1985). One may suggest that alkali metal cations and ACh activate the same Cl^- channels (see Hughes *et al.*, 1987). However, no competition has been revealed between K^+ and ACh in activation of the Cl^- conductance (Kislov, 1974).

In this chapter, special attention will be paid to two phenomena: (1) effect of membrane potential on interaction of ACh with its receptors, and (2) inactivation of the receptors induced by elevation of $[Ca^{2+}]_i$. These phenomena are responsible for the appearance of "on"- and "off"-relaxations of the ACh-induced currents as the membrane potential is shifted in a stepwise manner in either a hyperpolarizing or a depolarizing direction.

The data to be discussed in the following sections were obtained on completely isolated (Kostenko *et al.*, 1974) nondialyzed and/or dialyzed neurons using the voltage-clamp technique (Chemeris *et al.*, 1982; Kurchikov and Kazachenko, 1984a).

2. Cl⁻-SELECTIVITY OF ION CHANNELS OPENED BY ACh

Application of ACh to the somata of *Lymnaea* neurons increases the membrane conductance (Chemeris *et al.*, 1982). Under voltage-clamp conditions and when microelectrodes were filled with a 2.5 M solution of either KCl or CsCl, inward currents

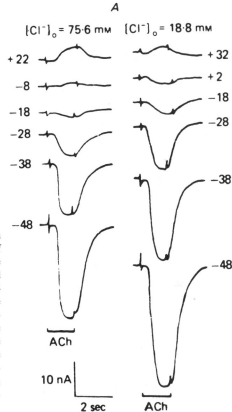

Figure 1. ACh-induced currents (I_{ACh}) in a *Lymnaea stagnalis* neuron at different levels of membrane potential (indicated at the sides of the records) under conditions of high (75.6 mM) and low (18.8 mM) [Cl⁻]$_o$. ACh (2 μM) was applied 4 sec after setting the membrane potential. To reduce the outward K⁺ current flowing through the voltage-dependent channels, CsCl microelectrodes were used. Between the ACh applications the membrane potential was held at −48 mV. Here and elsewhere the ionic composition (mM) of the standard physiological solution was: NaCl, 56; KCl, 1.6; CaCl₂, 4; MgCl₂, 4; pH adjusted to 7.5 using Tris-HCl. (Modified from Chemeris *et al.*, 1982.)

(I_{ACh}) appeared (Fig. 1) and had a reversal potential, V_r, of −16 ± 6.3 mV (S.D., n = 20). With 2.5 M K⁺-propionate-filled microelectrodes, V_r was −55.4 ± 7.6 mV (n = 23). However, if the microelectrodes were filled with 0.1 M KCl, CsCl, K⁺-propionate, or K₂SO₄, the reversal potential was always the same, −37 ± 5 mV (n = 17). These data suggest that under physiological conditions, V_r is more positive than the resting potential, and the neurons investigated belong to the D type.

To study further the ionic nature of the ACh-induced currents, the influence of the ionic composition of the bathing solution on V_r was tested. It was found that only variations in [Cl⁻]$_o$ caused appreciable changes of V_r. This is illustrated in Fig. 1 where V_r shifted by about 30 mV when the [Cl⁻]$_o$ was decreased from 75.6 to 18.8 mM.

The measurement of reversal potential in the depolarized potential range is made difficult by the development of a large K⁺ conductance activated by depolarization. In the experiment of Fig. 1 this was obviated by using a CsCl-filled electrode; the intracellular accumulation of Cs⁺ produced a partial blockage of the voltage-dependent K⁺ channels. Thus, we conclude that the ACh-activated current is mediated by Cl⁻ channels.

3. INTERACTION OF ACh WITH THE AChR AND
SOME PROPERTIES OF THE CHANNELS

It is now well known that the reaction between ACh and its receptors can be influenced by the membrane potential. Moreover, it is this phenomenon that permitted the determination of the reaction rate constants using the voltage-jump relaxation method. The influence of the membrane potential on the channel open-time was first observed on vertebrate muscle and *Electrophorus* electroplax while studying the decay of endplate currents (Gage and Armstrong, 1968; Gage and McBurney, 1975; Kordas, 1969, 1972a,b; Magleby and Stevens, 1972a,b), ACh-induced current fluctuations (Katz and Miledi, 1972; Anderson and Stevens, 1973), voltage-jump-induced relaxations (Adams, 1975; Neher and Sakmann, 1975; Sheridan and Lester, 1975, 1977), and single-channel recording (Neher and Sakmann, 1976). All these methods of analysis revealed that the channel open-time increases with hyperpolarization.

In *L. stagnalis* neurons there exist two mechanisms of membrane potential action on the ACh-induced responses: (1) an increase of the channel open-time caused by hyperpolarization (Kurchikov and Kazachenko, 1979, 1984a,b; Bregestovski and Iljin, 1980; Kurchikov *et al.*, 1985; Bregestovski and Redkozubov, 1986), and (2) inhibition of G_{ACh} by depolarizing pulses due to Ca^{2+} entering the cells through voltage-dependent Ca^{2+} channels (Kazachenko and Kislov, 1973, 1974, 1977; Kazachenko, 1979; Kazachenko *et al.*, 1979, 1981a,b; Chemeris *et al.*, 1982; Chemeris and Iljin, 1985). In this section I intend to describe the first phenomenon and some related data.

3.1. Effect of Membrane Hyperpolarization on G_{ACh}

Instantaneous current-voltage, $I-V$, relations of the ACh-sensitive membrane are linear over a wide range of membrane potentials (V_m), at different [ACh] (Fig. 2). The slope of the instantaneous $I-V$ relation is determined only by the value of [ACh]. The linearity of the $I-V$ relation probably indicates that (1) under the experimental conditions used the ionic channels have no rectification and (2) the properties of both the AChR and the channel remain unchanged immediately after V_m shifted to a new level.

Due to a voltage jump, I_{ACh} relaxes from an instantaneous value, I_0, toward a new steady-state level, I, in a few tens of milliseconds (Fig. 2A). Corresponding steady-state $I-V$ relations are essentially nonlinear at [ACh] < 0.5 M and at low temperatures (Fig. 2B). Steady-state chord conductance, G_{ACh}, increases exponentially with membrane hyperpolarization depending on V_m ([ACh] < 0.5 M):

$$G_{ACh} = G_{ACh}(O) \, e - V_m/V_h \tag{1}$$

where $G_{ACh}(O)$ is the G_{ACh} value at $V_m = 0$ and V_h is a temperature-dependent constant: $V_h \simeq 200$ mV at 20°C and $V_h \simeq 80$ mV at 2–5°C (Kurchikov and Kazachenko, 1984a).

Similar dependencies of G_{ACh} on V_m were found while studying ACh-activated cationic channels (Kordas, 1972a,b; Dionne and Stevens, 1975; Lester *et al.*, 1975; Rang, 1974; see also Skok *et al.*, 1987).

At high [ACh], G_{ACh} becomes potential-independent ($V_m < -50$ mV) (Fig. 3). A

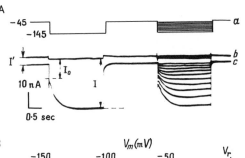

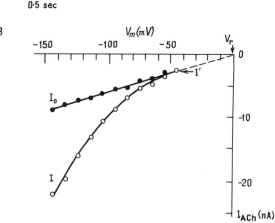

Figure 2. Effects of varying membrane potential on the ACh-induced currents. (A) Relaxations of the ACh-induced currents arising from membrane potential jumps from the holding level (−45 mV) to more negative values; a, membrane potential (in mV); b, currents in the absence of ACh; currents in the presence of ACh (1 M). (B) Instantaneous (○) and steady-state (●) *I–V* relations of the ACh-sensitive membrane at 5°C. (From Kurchikov and Kazachenko, 1984a.)

sharp decrease of G_{ACh} at $V_m > -50$ mV is due to the inhibitory action of Ca_i^{2+} on the AChR (see Section 4).

3.2. Dose–Response Relationships

Typical relations between G_{ACh} and [ACh] for two different values of membrane potential are shown in Fig. 4. In this case the apparent dissociation constant (K_D) is

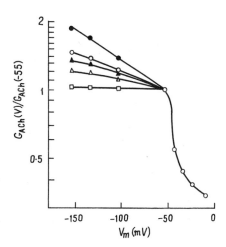

Figure 3. Dependence of steady-state relative conductance [$G_{ACh}(V_m)/G_{ACh}(-55)$] on V_m at various [ACh] (in μM) : 0.5 (●), 1 (○), 2 (▲), 3 (△), and 5 (□). $G_{ACh}(-55)$ is the conductance at $V_m = -55$ mV. (From Kurchikov and Kazachenko, 1984a.)

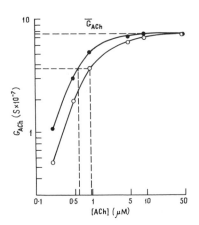

Figure 4. Dependence of steady-state G_{ACh} on [ACh] at two values of membrane potential (in mV): −65 (●) and −165 (CI). Temperature 22°C. (From Kurchikov and Kazachenko, 1984a.)

equal to 1 µM at −65 mV and 0.6 µM at −165 mV. Thus, at the more negative potential, the K_D value decreases. At physiological transmembrane potential values (−50 mV), the average K_D is 2.2 ± 1.3 µM (Kazachenko, 1979). The K_D for ACh in molluscan neurons is close to that in eel electroplax (Weber and Changeux, 1974; Lester *et al.*, 1978) and frog muscle endplate (Dreyer and Peper, 1975).

The Hill coefficient for the ACh dose–response (G_{ACh}) relation was approximately 2.3 (Kurchikov and Kazachenko, 1984a). This suggests that activation of one Cl⁻ channel (or AChR) may require the binding of two or three ACh molecules to the receptor.

3.3. Relaxations of the ACh-Induced Current from Hyperpolarizing Voltage Jumps

After step changes in membrane potential, the ACh-induced current relaxes from its instantaneous value, I_0, to a new steady-state one (Fig. 5). A corresponding change in G_{ACh} is monoexponential (Kurchikov and Kazachenko, 1984b):

$$G_{ACh}(t) = G_\infty - (G_\infty - G_O) \, e^{-t/\tau} \qquad (2)$$

where G_O is the instantaneous value of G_{ACh} at $t = 0$, G_∞ is the steady-state G_{ACh} value at a new level of V_m, and τ is the relaxation time constant.

τ was found to be determined by the value of V_m but not V_h (holding potential), i.e., τ has no "memory." This differs from the current relaxations induced during membrane depolarizations (> -40 mV) (see Section 4 and MacDermott *et al.*, 1980).

At low [ACh] (< 0.2 µM), τ depends exponentially on V_m (Kurchikov and Kazachenko, 1984b):

$$\tau(V_m) = \tau(O) \, e^{-V_m/V_h} \qquad (3)$$

where V_h is a constant equal to 182 ± 19 mV (S.D., $n = 22$) at 18–24°C, and 85 ± 22 mV ($n = 14$) at 2–5°C. Thus, G_{ACh} and τ both depend on V_m in a similar manner (Eq. 1).

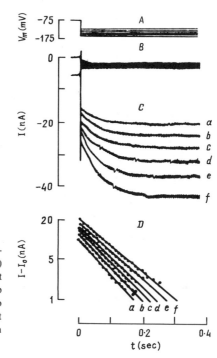

Figure 5. Relaxations of the ACh-induced current at hyperpolarizing voltage jumps. (A) Membrane potential. (B) Current relaxations in the absence of ACh. (C) Current relaxations in the presence of ACh (0.1 M) at V_m jump from −75 mV to various more negative values: −125 to −175 mV (a–f). (D) Relaxations of ACh-induced current ($I - I_O$) in semilogarithmic coordinates (see Fig. 2). (From Kurchikov and Kazachenko, 1984b.)

Due to voltage jumps, both on- and off-relaxations of the ACh-induced currents can be seen. The properties of the off-relaxations are similar to those of the on-relaxations.

The relative amplitude of the current relaxation decreases with increasing [ACH]. The relaxations disappear almost completely at saturating [ACh] (5–10 μM).

The rate constant of relaxations (τ^{-1}) increases with increasing [ACh] (Fig. 6) in the following manner:

$$\tau^{-1} = \beta[ACh] + \alpha \tag{4}$$

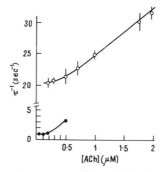

Figure 6. Dependence of relaxation time constant, τ^{-1}, on [ACh]. Membrane potential is shifted from −60 to −160 mV. ○, 22°C; ●, 3°C. (From Kurchikov and Kazachenko, 1984b.)

where β may depend on [ACh]. However, when [ACh] > 0.3 μM, the dependence becomes linear and β and α no longer depend on [ACh]. In the range of [ACh] between 0.05 and 0.2 μM, τ^{-1} remains practically unchanged, a result that is in agreement with β[ACh] $\ll \alpha$. Hence, for this range of [ACh] values, $\tau = \alpha^{-1}$, where α^{-1} is believed to be the mean open-time of an ACh-activated channel (Anderson and Stevens, 1973; Neher and Sakmann, 1975, 1976; Sheridan and Lester, 1977; Ascher *et al.*, 1978; Selyanko *et al.*, 1979; MacDermott *et al.*, 1980).

The membrane potential influences α^{-1} but not β. At [ACh] < 0.2 μM and 20°C $\beta = 6.8 \pm 4.1$ M^{-1}·sec^{-1} (S.D., $n = 21$). At the same temperature, α^{-1} is 31.8 $\pm$ 3.1 msec ($n = 12$) and 59.7 $\pm$ 4 msec ($n = 11$) at -50 and -160 mV, respectively. Changes in temperature affect α^{-1} more than β. The temperature coefficient Q_{10} equals -3 and 1.2 for α and β, respectively ([ACh] < 0.2 μM) (Kurchikov and Kazachenko, 1984b).

In the experiments described (Kurchikov and Kazachenko, 1984a,b), the time course of the relaxations is monoexponential. Similar relaxations have been reported for other cells (Anderson and Stevens, 1973; Neher and Sakmann, 1975; Sheridan and Lester, 1977; Ascher *et al.*, 1978; MacDermott *et al.*, 1980). Double-exponential relaxations have been observed in neurons of rabbit (Derkach, 1986), rat submandibular ganglion (Rang, 1981), and *Helix aspersa* (Finkel, 1983; see also Skok *et al.*, 1987).

The value of α^{-1} (32 msec) at -50 mV and 20°C (Kurchikov and Kazachenko, 1984b) is much higher than those for the AChR at the frog endplate (1 msec) (Anderson and Stevens, 1973; Neher and Sakmann, 1975, 1976) as well as those in the neurons of *Aplysia* (Ascher *et al.*, 1978), frog (MacDermott *et al.*, 1980), and rabbit (Selyanko *et al.*, 1979), where α^{-1} ranges between 5 and 6 msec. Nevertheless, the α^{-1} value found in *Lymnaea* neurons is close to those in rat neurons (Ascher *et al.*, 1979), chromaffin cells (Hamill *et al.*, 1981), and the slow components of α^{-1} in the double-exponential relaxations found in rabbit (Derkach, 1986) and rat (Rang, 1981) neurons.

An increase of α^{-1} with membrane hyperpolarization has been observed in the majority of studies in the ACh-activated cationic channels (Anderson and Stevens, 1973; Sheridan and Lester, 1977; Ascher *et al.*, 1979; Selyanko *et al.*, 1979; MacDermott *et al.*, 1980). However, the data on the Cl⁻ channel in molluscan neurons are contradictory. For instance, in *Aplysia* neurons α^{-1} increased with hyperpolarization in some cases (Adams *et al.*, 1976) but remained unchanged in others (Gardner, 1980; Simonneau *et al.*, 1980). On the other hand, in *H. aspersa* neurons α^{-1} was found to decrease with hyperpolarization (Finkel, 1983). Bregestovski and Iljin (1980) also observed the increase of α^{-1} in *L. stagnalis* neurons.

The constant β, as a rule, does not depend on the membrane potential (Neher and Sakmann, 1975; Sheridan and Lester, 1977; Ascher *et al.*, 1978) and increases nonlinearly with [ACh] (Sheridan and Lester, 1977).

The effect of membrane potential on α^{-1} is probably mediated through its action on a limiting step of the interaction between ACh and its receptor. In accordance with some earlier data, the limiting step of the reaction is a conformational change of the AChR macromolecule during the channel transition from closed to open state (Magleby and Stevens, 1972a,b; Anderson and Stevens, 1973). In this particular case, the membrane potential could influence the positions of some electrical charges and/or dipoles of the receptor in the membrane.

As postulated by another hypothesis, the rate-limiting step of the reaction is the dissociation of ACh molecules from the AChR (Kordas, 1969, 1972a,b; Adams and Sakmann, 1978; Kurchikov et al., 1985). According to this hypothesis, binding of the agonist molecule to the receptor leads to immersion of the complex into the membrane and thus gives rise to the voltage sensitivity. There are some data that are consistent with both the hypothesis regarding α^{-1} and its voltage sensitivity (Colquhoun and Sakmann, 1983).

3.4. Noise Analysis of the ACh-Induced Current

The technique of noise analysis gives information not only on kinetic parameters, but also on the conductance of single ionic channels, γ (Katz and Miledi, 1972). In *Lymnae* ganglia, noise analysis was carried out using the voltage-clamp method on nondialyzed (Bregestovski, 1980; Bregestovski and Redkozubov, 1986) and dialyzed (Ivanova et al., 1986, 1987) neurons. In the first case, the noise power spectrum could be approximated by a single Lorentzian, α^{-1} being 15.6 msec (-40 mV) and 20 msec (-100 mV), and γ being 3.5 pS ($[Cl^-]_i/[Cl^-]_o = 20/80$ mM). In this case the values of α^{-1} and the degree of its voltage-dependence are ~1.5-fold less than those obtained from the current relaxations (Kurchikov and Kazachenko, 1984a,b). The experiments on dialyzed neurons (Ivanova et al., 1986, 1987) gave $\alpha^{-1} = 31.4 \pm 5$ msec and $\gamma = 8.4 \pm 2$ pS ($[Cl^-]_i/[Cl^-]_o = 100$ mM). In this case the α^{-1} is very close to 31.8 msec obtained by the relaxation measurement method (Kurchikov and Kazachenko, 1984b). In 25% of the experiments of Ivanova et al., (1986), approximation of the noise power spectrum required more than a single Lorentzian function. No special studies were performed to address this latter issue.

Noise measurements in neurons of other molluscs showed that α^{-1} could be a one- (Simonneau et al., 1980) or two-component (Gardner and Stevens, 1980; Ascher and Erulkar, 1983; Finkel, 1983) parameter. The fast component of α^{-1} was in the range of 3 to 5 msec, the slow one in the range of 20 to 140 msec. In the same experiment, γ varied from 0.58–2 pS to 16 pS.

3.5. Single-Channel Measurements

Activity of single-channel ACh-activated Cl⁻ channels in "outside-out" patches was studied by Bregestovski and Redkozubov (1986) (Fig. 7). It was shown that duration of the channel openings had two components, fast (4 msec) and slow (20 msec). The single-channel openings were grouped usually into bursts. The burst behavior of the ACh-activated channels was also observed by other workers (Ascher and Erulkar, 1983; Colquhoun and Sakmann, 1983).

The single-channel conductance, γ, varied from 3 to 12 pS (Bregestovski and Redkozubov, 1986). Moreover, a number of multiple conductance substates of a single channel were detected. The nearest substates differed from each other by about 2 pS. In this respect the ACh-activated Cl⁻ channels of *Lymnaea* neurons are quite similar to

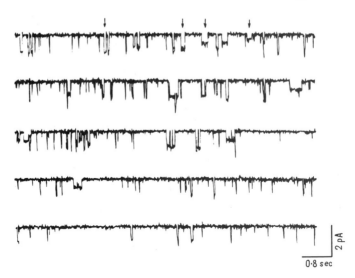

Figure 7. Recording of currents from a single ACh-activated channel. $V_h = -90$ mV. The arrows indicate the current sublevels. "Outside-out" membrane patch. The solution inside the pipette contained (in mM): CsCl 75, CaCl₂ 2.5, EGTA 5.5 (pCa = 7), HEPES 5 (pH 7.2). The composition of the bath solution was (in mM): Tris-HCl 54, CaCl₂ 4, MgCl₂ 4, glucose 7 (pH 7.4). (From Bregestovski and Redkozubov, 1986, with permission.)

the anionic (Bormann *et al.*, 1987; Hughes *et al.*, 1987) and cationic (Cull-Candy and Usowicz, 1987; Jachr and Stevens, 1987) channels activated by amino acid transmitters in mammalian neurons.

A comparison of the data obtained by current relaxation measurements, noise power spectrum, or single-channel activity demonstrates their essential differences. It is quite possible that the first two methods give some averaged (effective) values of the channel parameters.

3.6. Interactions of ACh Analogues with the AChR

The effects of ACh analogues such as tetramethylammonium (TMA), carbachol (CCh), and suberyldicholine (SubCh) on the AChR of *Lymnaea* neurons have been compared. The results are listed in Table 1.

Evidently the apparent dissociation constant, K_D, for monoquaternary compounds exceeds that for SubCh and some other bisquaternary ammonium compounds (Kurchikov *et al.*, 1985). However, the maximal conductance, $\bar{G}_X$ (corresponding to saturating drug concentrations), for SubCh is ~ 2 times less than those for the other compounds. Nevertheless, γ for all compounds tested is approximately the same. Next, for the sequence of monoquaternary compounds, α^{-1} falls with G_{ACh} and K_D. For SubCh α^{-1} is twofold less than that for ACh; the opposite situation was observed at the frog neuromuscular junction (Katz and Miledi, 1972; Colquhoun *et al.*, 1975; Neher and

Table 1. Characteristics of the ACh Receptors in Lymnaea stagnalis Neurons

v_m (mV):	K_D (mM)	$\bar{G}_x/\bar{G}_{ACh}{}^a$		Q (pS)		α^{-1} (msec)	
	-50	-50	-55	-50 to -100	-100	-50	-100
ACh	2.2	1.0	1.0	3.5	8.4	31.8	31.4
TMA	—	—	1.1	3.5	8.4	—	9.4
CCh	21	1.2	0.9	3.5	8.4	—	6.4
SubCh	0.87	0.65	0.56	3.5	8.4	15	14.7
Reference[b]	(1)	(1)	(2)	(3)	(4)	(2)	(4)

[a]$\bar{G}_{ACh}$, $\bar{G}_x$, averaged values of saturated conductance (Fig. 4) activated by ACh or its analogues; $\bar{G}_{ACh} = 1.1 \times 10^{-6}$ s.
[b]References: (1) Kazachenko (1979); (2) Kurchikov et al. (1985); (3) Bregestovski and Redkozubov (1986); (4) Ivanova et al. (1986).

Sakmann, 1976). Comparison of the effects of ACh and SubCh leads to the conclusion that the efficacy of ACh is higher than that of SubCh, but the affinity of ACh is less.

4. INACTIVATION OF THE ACh RESPONSES

4.1. Effect of Membrane Depolarization on G_{ACh}

It can be seen from Fig. 1 that when the cell is depolarized, I_{ACh} decreases more than expected from a simple ohmic behavior. Moreover, G_{ACh} drastically falls at V_m more positive than -50 mV (Fig. 3). This phenomenon, which was first observed in *Helix* and *Aplysia* neurons by Frank and Tauc (1964), was termed the depolarization-induced "inactivation" (Kazachenko and Kislov, 1973, 1974, 1977; Kazachenko *et al.*, 1979, 1981a,b; Chemeris *et al.*, 1982).

A similar phenomenon was reported by Gardner (1980), who, however, attributed it to inadequate voltage-clamp. Earlier, some objections to this latter interpretation were discussed and the conclusion was reached that the inactivation induced by depolarization arises from an actual decrease of the ACh-induced conductance (Kazachenko *et al.*, 1979; Chemeris *et al.*, 1982).

The inactivation is further illustrated in Fig. 8, in which G_{ACh} has been analyzed at various membrane potentials for two values of $[Cl^-]_o$. The change of G_{ACh} observed with depolarizations more positive than -50 to -40 mV is rather abrupt, but it is interesting to observe that it occurs in the same potential range in the two solutions.

4.2. Voltage-Jump-Induced Relaxations

It is rather difficult to study the ACh response inactivation by comparing the responses to successive ACh applications at different levels of membrane potential. The difficulties are mostly due to the fact that the potential-induced, outward K^+ currents are often larger than the ACh-induced currents. One way to get over the difficulties is to study I_{ACh}, after having blocked as much as possible the outward K^+

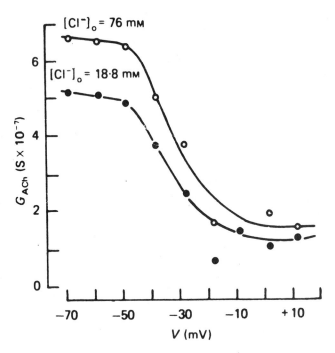

Figure 8. Potential dependence of the chord G_{ACh} for two values of $[Cl^-]_o$. The data were obtained from the experiment shown in Fig. 1. Between the ACh applications (2 μM), the membrane potential was held at −48 mV. (Modified from Chemeris *et al.*, 1982.)

currents. The method of internal dialysis facilitates such a blockage, since the K⁺ current can be strongly reduced by substitution of internal K⁺ by Cs⁺. An example of such a voltage-jump experiment performed on a neuron dialyzed with CsCl is illustrated in Fig. 9. The membrane potential was stepped from a holding value, V_h (−60 mV), to a test potential, V_m (−30 or 0 mV), first in the absence, then during perfusion of ACh. The value of the ACh-induced current, I_{ACh}, was determined by subtracting the current recorded before the application of ACh from the current recorded in the presence of ACh. Immediately after the potential jump to V_m, the ACh-induced current takes an instantaneous value, $I_O(V_m)$, then decreases toward a steady-state value, $I_\infty(V_m)$. Conversely, when at the end of the depolarizing pulse, the membrane potential is brought back to the holding potential, V_h, the ACh-induced current takes an instantaneous value, $I_O(V_h)$, then relaxes toward a steady-state value, $I_\infty(V_h)$—a small correction being required to correct for desensitization.

The transitions from $I_O(V_m)$ to $I_\infty(V_m)$ and from $I_O(V_h)$ to $I_\infty(V_h)$ are described by two relaxation processes, the "on"- and "off"-relaxations. From Fig. 9C it appears that the on-relaxation is sometimes slightly sigmoid. As for the off-relaxation, it is satisfactorily described by an exponential in Fig. 9A, and, more generally, for all experiments in which the depolarizing pulse lasted for more than 0.5 sec. However, if the depolarizing pulse was shorter than 0.5 sec, the off-relaxations usually showed two

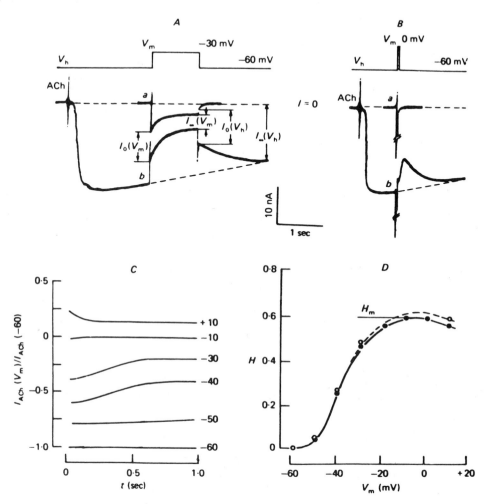

Figure 9. A relaxation experiment on a dialyzed neuron. (A) "On-relaxation" and "off-relaxation" of I_{ACh} produced by prolonged membrane depolarization (1 sec, 30 mV). (B) Biphasic "off-relaxation" of I_{ACh} arising after a short depolarizing pulse (50 msec, 60 mV). In A and B: V_h, holding potential; V_m, potential during the pulse; a, current in the absence of ACh; b, current in the presence of ACh. The artifacts indicate the beginning of ACh application. (C) "On-relaxations" of I_{ACh} at various levels of V_m (indicated to the right of the curves). The relaxations were obtained by subtraction of the inward current in the absence of ACh from the total current in the presence of ACh in an experiment similar to that in A. The origin of the t axis is the moment of peak value of inward currents in the absence of ACh. (D) Potential dependence of the degree of inactivation (H) obtained from either the "on-relaxations" (○) or the "off-relaxations" (●). The data were taken from the experiment shown in A, B, and C. In all the experiments illustrated, 2 μM ACh was applied; $V_h = -60$ mV. CsCl was used in the internal medium. (From Chemeris *et al.*, 1982.)

components: a fast outward component corresponding to a decrease of I_{ACh} preceded the inward component corresponding to the phenomenon observed with long depolarizing pulses (Fig. 9B).

Both the on-relaxation observed during a depolarizing pulse and the slow off-relaxation observed after repolarization are qualitatively similar to what could be expected if the system under study resembled the frog neuromuscular junction excited by ACh in having mean channel open-time decreased by depolarization (see Section 3). Nevertheless, the relaxations illustrated in Fig. 9 do not appear to be linked with the voltage-dependence of the channel open-time. The mean open-time of the channel is in the 10 msec range, and the corresponding relaxations, which may indeed be present, would be undetectable at the slow speed used for recording the currents illustrated in Fig. 9. Furthermore, the fast relaxations linked with the voltage-dependence of the channel open-time decrease sharply when the [ACh] is increased and, if present in the experiments of Fig. 9 (and the following ones), are probably quite small as the [ACh] was above 1 μM.

For all practical purposes, therefore, the relaxations illustrated in Fig. 9 do not appear to be contaminated to a serious extent by the relaxations linked with the voltage-dependence of the channel open-time.

4.3. Steady-State Inactivation

For a quantitative description of the inactivation of the ACh response, the degree of inactivation could be defined by comparing the instantaneous and steady-state values of the ACh-induced current. The comparison could be done by comparing the values either at V_m or at V_h.

If one uses the values measured during the depolarization, the degree of inactivation can be defined by

$$H = 1 - I_{ACh}(t)/I_O(V_m) \tag{5}$$

where $I_{ACh}(t)$ is I_{ACh} at time t after shifting the membrane potential from the holding potential V_h to test potential V_m. H tends toward a steady-state level, H_∞, when $I_{ACh}(t)$ approaches the steady-state value, $I_\infty(V_m)$. Figure 9D shows the potential dependence of H calculated by applying Eq. (5) to the experiment illustrated in Fig. 9C.

The evaluation of the steady-state inactivation by this method is rather fast but has some limitations, e.g., (1) the necessity of reducing the K^+ currents, (2) the low accuracy of the evaluation of I_{ACh} even after K^+ current reduction, and finally, (3) the danger of a contamination by the "fast" relaxations discussed above.

One can also evaluate the inactivation by analyzing the current after the end of the depolarizing pulse. In this case the value of H is given by

$$H = 1 - I_O(V_h)/I_\infty(V_h) \tag{6}$$

where $I_O(V_h)$ is the instantaneous value of I_{ACh} after returning to V_h, and $I_\infty(V_h)$ is the steady-state value of I_{ACh} at V_h, corrected for desensitization. The fact that the slow

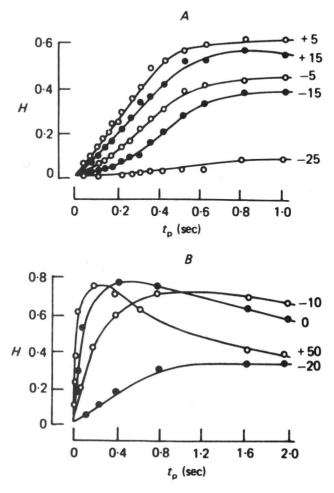

Figure 10. Influence of the duration and of the magnitude of the depolarizing pulse on the inactivation of the ACh response. Isolated, nondialyzed neurons. The membrane potential was depolarized by pulses of various duration from −53 mV (A) or −50 mV (B) to the different levels indicated to the right of the curves. The amount of inactivation, H, was calculated using the off-relaxations. Note the different time scales in A and B. [ACh] = 2 μM. (From Chemeris *et al.*, 1982.)

off-relaxation is exponential allows one to calculate $I_O(V_h)$ with precision (by back-extrapolation of the exponential), thus eliminating the possible contamination of the early part of the decay by the "fast" relaxations. This was done in Fig. 9D. The good agreement between the values of H obtained by Eq. (5) and those obtained by Eq. (6) indicates that the two methods of evaluation of H are, for most purposes, equivalent.

Figure 9D also shows that H increases with depolarization until it reaches a maximum value, H_m, at about 0 mV. Sometimes H decreases at more positive values of V_m (see, e.g., Figs. 11 and 12).

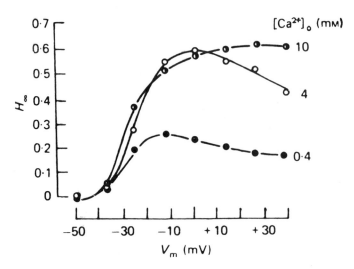

Figure 11. Effect of $[Ca^{2+}]_o$ on the potential dependence of the inactivation. The numbers to the right of the curves indicate the values of $[Ca^{2+}]_o$. Isolated, nondialyzed neuron. H was calculated using the off-relaxations. $[ACh] = 2\ \mu M$, $V_h = -50\ mV$, $t_p = 2\ sec$. (From Chemeris *et al.*, 1982.)

Figure 10 illustrates the influence of the duration (t_p) and of the height of the depolarizing pulse on the amount of inactivation. The inactivation was evaluated using Eq. (6), i.e., by analyzing the off-relaxation. The curves illustrate three properties of the inactivation: (1) at moderate values of V_m the relation between H and the pulse duration (t_p) is usually sigmoid; (2) when t_p increases, H reaches a steady-state value, H_∞, which depends on the value of V_m; (3) inactivation decreases, sometimes quite markedly, when t_p exceeds 1 sec.

When the "off-relaxations" of I_{ACh} were studied at different levels of desensitization, desensitization was found to influence neither the amplitude nor the time constants of relaxations. This is illustrated in Table 2 in Chemeris *et al.*, (1982).

4.4. Ca_i^{2+} Initiates the AChR Inactivation

Earlier, a variety of hypotheses about the mechanism of the inactivation were considered and rejected (Kazachenko *et al.*, 1979). A possible role of Ca^{2+} entering the cell in the course of membrane depolarization remained uncertain. This section is devoted to the evidence that the AChR inactivation is caused by Ca^{2+}. The fact that the potential range for AChR inactivation is roughly similar to the range of activation of the Ca^{2+} channels (Naruschevichus *et al.*, 1979; Kostyuk, 1980) may serve as the first argument in favor of the Ca^{2+} hypothesis about inactivation. The following data obtained on nondialyzed and dialyzed neurons support this hypothesis.

4.4.1. Changes in $[Ca^{2+}]_o$

If the receptor inactivation is caused by a rise in $[Ca^{2+}]_i$ due to Ca^{2+} entry, an influence of $[Ca^{2+}]_o$ on H can be expected.

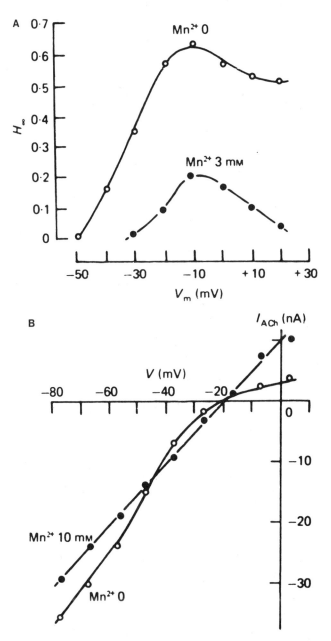

Figure 12. Decrease of inactivation by Mn^{2+}. Isolated nondialyzed neuron. The inactivation was estimated from the analysis of the off-relaxations. (A) Effect of Mn^{2+} (3 mM) on H. [ACh] = 2 μM, V_h = −49 mV, t_p = 2 sec. (B) Effect of Mn^{2+} (10 mM) on the I–V relation of the ACh-sensitive membrane. ACh (2 μM) was applied 4 sec after rectangular shift of the membrane from −50 mV to various levels. CsCl-filled microelectrodes (see Fig. 1). ○, currents induced by ACh before application of Mn^{2+}; ●, currents measured after addition of Mn^{2+}. As seen, Mn^{2+} decreases I_{ACh} in the range −50 to −80 mV. This phenomenon probably can be ascribed to a shortening of the channel open-time similar to that produced by another "Ca-antagonist," D-600 (Bregestovski and Iljin, 1980). (From Chemeris *et al.*, 1982.)

Three different values of $[Ca^{2+}]_o$ were tested: 0.4, 4.0, and 10 mM. As expected, H_m was decreased (more than 3-fold) when $[Ca^{2+}]_o$ was reduced from 4.0 to 0.4 mM (Fig. 11). Increasing $[Ca^{2+}]_o$ from 4.0 to 10 mM had no appreciable effect on H_m, an effect that may result from the saturation of the Ca^{2+} channel conductance occurring at high $[Ca^{2+}]_o$ (Naruschevichus *et al.*, 1979; Brown *et al.*, 1980; Kostyuk, 1980). The bell-like shape of the relation between H and V_m (Figs. 9, 11) can also be explained by a "Ca^{2+} hypothesis," since the Ca^{2+} currents (I_{Ca}) decrease at positive values of the membrane potential as the driving force for Ca^{2+} decreases.

Substitution of Ca_o^{2+} for equimolar amounts of Sr^{2+}, Ba^{2+}, or Mg^{2+} showed that Sr^{2+} or Ba^{2+} entering the cells through the voltage-sensitive Ca^{2+} channels also inactivates the AChR. Mg^{2+} is ineffective. With respect to their ability to inhibit the ACh-induced responses, the divalent cations follow the sequence: $Ca^{2+} > Sr^{2+} > Ba^{2+}$ (Chemeris *et al.*, 1989), which is opposite to that of their permeability through Ca^{2+} channels (Naruschevichus *et al.*, 1979).

4.4.2. Blockage of the Ca²⁺ Channels

Manganese ions are known to block the voltage-dependent Ca^{2+} channels (Hagiwara and Nakajima, 1966). As seen in Fig. 12, Mn^{2+} at 3 mM partially blocked the AChR inactivation at all values of V_m. When used at 10 mM, Mn^{2+} eliminated the inactivation completely. The linearity of the $I–V_m$ relation of the cholinoreceptive membrane indicates absence of the inactivation (Fig. 12B). Thus, the blockage of I_{Ca} prevents the ACh-receptor inactivation.

4.4.3. Increase in [Ca²⁺]ᵢ

The next type of experiment was performed on dialyzed neurons in which it is believed that $[Ca^{2+}]_i$ was the same in the vicinity of the inner membrane surface and in the internal dialysis solution.

The protocol of the experiments was the following. The membrane potential was fixed at -60 mV. In control experiments ACh was applied during intracellular perfusion of a Ca^{2+}-free solution (Fig. 13A,a). Several minutes after washing away the ACh, the internal Ca^{2+}-free solution was replaced by a Ca^{2+}-containing one. Intracellular perfusion with this Ca^{2+} solution elicited additional currents, probably the Ca^{2+}-activated K^+ (Meech, 1974) and/or Cl^- ones (Geletyuk and Kazachenko, 1985). When these currents had reached a stable value (after several minutes), ACh was applied again. In this case, ACh responses were inhibited (Fig. 13A,b). The reduction of I_{ACh} was reversible and the ACh responses increased again when the internal perfusion was switched back to Ca^{2+}-free solution (Fig. 13A,c).

For the experiments with Ca_i^{2+}, the degree of the inactivation was determined according to

$$H' = 1 - I_2/I_1 \qquad (7)$$

where I_1 is the ACh-induced current in the absence of Ca_i^{2+} and I_2 is the current after introduction of Ca_i^{2+}.

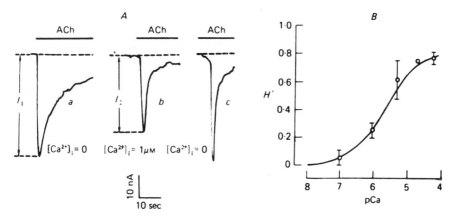

Figure 13. Inhibitory effects of Ca_i^{2+} on the ACh-induced current in dialyzed neurons. (A) Example of an inhibition of I_{ACh} by intracellular perfusion of 1 μM Ca^{2+}. Interrupted lines indicate zero levels of I_{ACh}. An outward current caused by Ca_i^{2+}-enriched solution (Meech, 1974) is not shown in the figure. (B) Relation between H' and $[Ca^{2+}]_i$. The data were taken from four cells. The solid line was drawn according to Eq. (8) with $K_D = 2.1$ μM. Error bars give ± S.D. [ACh] = 1 μM, $V_h = -60$ mV. (From Chemeris *et al.*, 1982.)

Figure 13B summarizes the data on the relation between $[Ca^{2+}]_i$ and the inactivation of the ACh-induced current. The maximal inhibition of the ACh responses is achieved at a $[Ca^{2+}]_i$ of about 0.1 mM (pCa = 4). The continuous line was drawn in accordance with the empiric equation:

$$H' = 0.8 \, [Ca^{2+}]_i/([Ca^{2+}]_i + K_D) \tag{8}$$

where K_D is the apparent dissociation constant. The K_D for the data shown in Fig. 13B was 2.1 μM.

4.4.4. Off-Relaxations of the ACh-Induced Current

The relaxations of the ACh-induced current observed at the end of a depolarizing pulse were analyzed from the point of view of the "Ca^{2+} hypothesis" of inactivation.

As already mentioned in the description of Fig. 9, short depolarizing pulses produced biphasic "off-relaxations" in which a fast outward component precedes a slow inward component resembling the one observed with long depolarizing pulses. These biphasic relaxations were observed after pulses varying in length from 0.05 to 0.5 sec. The fast outward component usually lasted for about 200–300 msec. Its origin is still unclear. It could indicate a delay in the development of the inactivation, if the limiting step in this phenomenon was a process slower than Ca^{2+} accumulation in the cell (or near the membrane surface).

The slow inward "off-relaxations," observed after both short and long pulses, usually decrease exponentially. The time constant of the decay, τ, increases with V_m and with the pulse duration t_p (Fig. 14) and for values of t_p larger than 0.5–1.0 sec τ reaches a maximum value, τ_∞. The minimum value of τ, τ_O, does not appear to depend

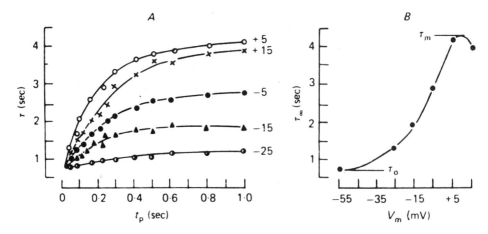

Figure 14. Effect of the membrane potential during the pulse (V_m) on τ, the time constant of the off-relaxation. (A) Dependence of τ on the duration (t_p) of the pulse. The values of V_m are shown near the curves. (B) Dependence of τ on V_m. The values of τ are the values of τ in A at $t_p = 1$ sec. Isolated, nondialyzed neurons, [ACh] = 2 μM, $V_h = -55$ mV. (From Chemeris *et al.*, 1982.)

on V_m. On the other hand, τ_∞ depends on V_m. Figure 14B illustrates that the relation between τ_∞ and V_m is sigmoid and that τ has a maximum value, τ_m, when V_m is about 0 mV.

In the context of the Ca^{2+} hypothesis, it can be assumed that the slow off-relaxation reflects the progressive decrease in the $[Ca^{2+}]_i$ near the inner membrane surface. This assumption is supported by experiments in which EGTA was introduced in the cell interior (Fig. 15). These experiments were performed on dialyzed neurons without Ca_i^{2+}. It was found that 1 mM EGTA decreased both the amplitude of the off-relaxations and τ, by about two- and three-fold, respectively (Fig. 15). At 5 mM, EGTA eliminated the inactivation and the off-relaxations completely. Also, EGTA decreased the amplitude of the on-relaxations.

The experiments with EGTA suggest that the rate-limiting step in the return of the receptors from the inactivated state to the resting state is the disappearance of Ca^{2+} in the submembrane layer.

4.5. Further Observations on the AChR Inactivation

4.5.1. Investigation of the AChR Inactivation Using Current Noise Analysis

Ivanova *et al.* (1987) studied inactivation of the AChR in dialyzed neurons using current noise analysis. First, they revealed the existence of two groups of neurons with different sensitivities of the AChR to Ca_i^{2+}. In most cells (~80%) an increase in $[Ca^{2+}]_i$ from 20 nM to 5 μM resulted in inhibition of ACh-induced currents by an average of 76%, whereas in the remainder (~20%) the effect was only to reduce ACh-induced current by ~25%.

Further, in the control experiments (20 mM $[Ca^{2+}]_i$) the conductance of a single

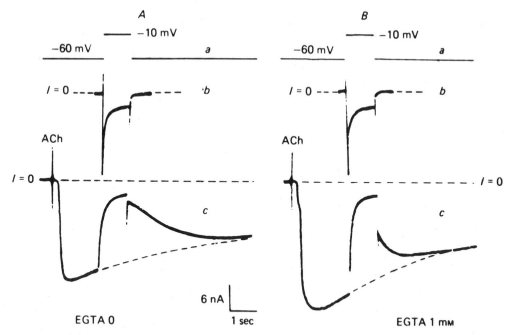

Figure 15. Influence of internal EGTA on the relaxation of I_{ACh}. (A) Currents in the absence of EGTA. (B) Currents in the presence of 1 mM EGTA. In A and B: a, membrane potential; b, currents recorded before and during the depolarizing pulse in the absence of ACh; c, currents recorded in the presence of ACh. The artifacts indicate the beginning of ACh application. Dialyzed neuron. Internal CsCl. In both cases the internal solution was nominally Ca^{2+}-free. [ACh] = 2 μM, $V_h = -60$ mV. The two records of the currents (before and after ACh application) are separated to improve clarity (cf. Fig. 9A and B). (From Chemeris *et al.*, 1982.)

Cl⁻ channel, γ, was 8.8 pS, and the open-time, α^{-1}, 29.9 msec (-100 mV). At $[Ca^{2+}]_i = 5$ μM these parameters were 13.3 pS and 22.3 msec, respectively. That is, γ was increased by 1.5-fold, probably due to an increase in probability of occurrence of the channel subconductance upper levels (see Geletyuk and Kazachenko, 1987) and α^{-1} somehow decreased. However, the total charge transferred in the course of a single-channel opening did not change.

In the cells with a low sensitivity to Ca_i^{2+}, the parameters of the Cl⁻ channels were: $\gamma - 9.1$ pS and $\alpha^{-1} = 31.6$ msec. Increase of $[Ca^{2+}]_i$ had practically no effect.

The data described show that the AChR inactivation is accompanied by a decrease in probability of the channel open state due to increase in duration of periods between channel openings.

4.5.2. Dose–Response Relationships

It was found that the apparent dissociation constant (K_D) characterizing the interactions between ACh and AChR in dialyzed neurons shifts to higher values of [ACh] with an increase of $[Ca^{2+}]_i$; $\bar{G}_{ACh}$ remains unchanged (Chemeris *et al.*, 1989). The shift exactly corresponds to the degree of inhibition (or inactivation) of the ACh-induced current. The data suggest that Ca_i^{2+} decreases the affinity of the AChR to ACh. In other words, when Ca^{2+} is present at the inner membrane side, the ACh molecules at the outer membrane side either cannot bind to the AChR, or being bound, cannot open the channel.

On the other hand, as shown from the off-relaxations in dialyzed neurons, the degree of the inactivation (H) corresponding to a given value of $[Ca^{2+}]_i$ decreases with elevation of [ACh] at the outer membrane side (Chemeris *et al.*, 1989). The inactivation caused by the increase of $[Ca^{2+}]_i$ up to 2.5 μM disappears almost completely at 5–10 μM ACh. It is seen from Fig. 4 that this range of [ACh] corresponds to saturation of the dose–response relation (G_{ACh} versus [ACh]). The time constant of the off-relaxations does not depend on [ACh]. The inactivation appears again when $[Ca^{2+}]_i$ increases further. These data allow one to conclude that if an ACh molecule is bound to the external site of the AChR, Ca_i^{2+} either is not capable of binding to the inner site of the AChR, or having bound, cannot inhibit it.

The data are qualitatively similar to those reported by Chang and Neumann (1976) on purified AChR and compatible with the idea that binding of ACh to the AChR requires one bound Ca^{2+} to be released.

In general, these data may suggest that the binding of ACh or Ca^{2+} to the AChR leads to transmission of conformational changes from one binding site to another site on the opposite side of the membrane.

4.5.3. Temperature

The AChR inactivation processes have low temperature coefficients (Q_{10}) (Chemeris *et al.*, 1982). The Q_{10}'s for H_∞ and H_m range from 1.5 to 1.8. The time constant τ is also slightly dependent on temperature. The Q_{10} for τ_∞ ranges from 1.1 to 1.3. The Q_{10} for τ_O and τ_m are also not large (1.1–1.5) (cf. Q_{10}'s for the time constants of the ACh-induced current relaxations in the hyperpolarizing range of V_m).

The weak temperature dependence of τ may indicate that this time constant is linked to some binding process (e.g., absorption of Ca^{2+} by cytoplasmic proteins) rather than to an active transport system.

4.6. Some Implications of the Data

4.6.1. Comparative Analysis of the Two Types of ACh-Induced Current Relaxations

The slow relaxations observed during and after a depolarizing pulse (more positive than ~ -40 to -50 mV) appear to constitute a phenomenon quite different from that revealed by relaxations in the hyperpolarizing range (more negative than ~ -40 to -50 mV). In particular, the two types of relaxations present four types of differences. The relaxations related to the channel open-time have the following features (Section 3). (1) They are fast, the time constants varying from tens to hundreds of milliseconds. (2) Their time constant decreases with [ACh], and at an [ACh] of ~5 μM it disappears completely. (3) Their time constant depends on V_m but not on the amplitude or the duration of a preceding pulse. (4) The temperature-dependence of these relaxations is rather strong, the Q_{10} being about 3. The relaxations reflecting the "depolarizing inactivation" process have different features. (1) Their time constant is slow, up to several seconds. (2) Their time constant does not depend on [ACh]. (3) Their time constant depends on the amplitude and duration of the depolarizing pulse. (4) The temperature-dependence of the time constant is negligible (1.1–1.5).

On the other hand, both types of relaxations are similar in two points. (1) Their relative amplitude decreases with [ACh]. The relaxations disappear almost completely at 5–10 μM, i.e., at the concentrations corresponding to saturation of the dose–response relations (G_{ACh} versus [ACh]). (2) Their apparent dissociation constant for interaction of ACh with the AChR increases when the membrane potential is shifted toward positive values. In other words, both membrane hyperpolarization and lowering [Ca²⁺]ᵢ increase the affinity of the AChR.

In spite of the similarities mentioned, the differences between the two types of relaxations indicate the existence of two distinct phenomena. It remains possible, however, that in some experiments the "fast" relaxations may complicate the analysis of the "slow" ones, and the evaluation of the degrees of inactivation, H.

4.6.2. Estimates of Submembrane [Ca²⁺]ᵢ

In the experiment on nondialized neurons, the maximum degree of inactivation was 0.74 ± 0.14 (S.D., $n = 16$). This value is close to the maximum value (0.8) of inactivation obtained on dialyzed neurons (Fig. 13B, Eq. 8). These data suggest that the degree of inactivation (0.74) observed after membrane depolarization corresponds to an increase in [Ca²⁺]ᵢ near the inner membrane surface of about 20 to 30 μM.

Various estimates of the changes in [Ca²⁺]ᵢ for whole cells have been reported. Using the metallochromic dye Arsenazo III, Gorman and Thomas (1978) calculated an increase of $3-6 \times 10^{-8}$ M in [Ca²⁺]ᵢ for the depolarizing phase of the pacemaker cycle in *Aplysia* neurons. According to the data of Ahmed and Connor (1979), the changes in [Ca²⁺]ᵢ are even larger, about 2×10^{-7} M. These data do not allow an evaluation of the instantaneous "submembrane" changes of [Ca²⁺]ᵢ (see below).

A large increase in submembrane [Ca²⁺] would reduce the Ca²⁺ equilibrium potential (E_{Ca}) significantly. By measuring changes of E_{Ca}, Krasts (1978) showed that, in the inner submembrane layer of a molluscan neuron, the concentration of free Ca²⁺ during the overshoot of the action potential may reach 70 μM. Gorman and Thomas (1980) using the same idea and the data of their own experiments with Arsenazo III, calculated a maximal change in [Ca²⁺]ᵢ near the inner membrane of *Aplysia* neurons and found it to be 7 μM for 300-msec depolarizing pulses. These authors noted, however, that the real changes in the perimembrane [Ca²⁺] could be higher due to membrane invaginations.

Some theoretical calculations have been made to estimate submembrane changes of [Ca²⁺]. Smith and Zucker (1980) attempted such an evaluation using a mathematical model of intracellular calcium movements. The model predicts changes in submembrane $[Ca^{2+}]_i$ from a resting level of 0.1–0.3 μM (DiPolo *et al.*, 1976; Alvarez-Leefmans *et al.*, 1981) to a peak of several micromolar during pulses to +15 mV. More recently, Simon and Llinás (1985) have estimated the calcium concentrations within hundreds of angstroms of the membrane, resolving the three-dimensional diffusion equation of calcium entering through a single open Ca^{2+} channel. As can be seen from their Fig. 3 (Simon and Llinás, 1985), the steady-state calcium concentration in the vicinity of the Ca^{2+} channel may reach ~200 μM.

Thus, both experimental and theoretical data suggest that a significant increase in the free submembrane $[Ca^{2+}]$ can take place after activation of Ca^{2+} channels. This increase appears to be sufficient to explain the inactivation of the ACh conductance observed in the course of neuronal depolarization.

4.6.3. Disappearance of Ca²⁺

The data presented in this chapter suggest that the slow "off-relaxations" of the ACh-induced current reflect a decrease in $[Ca^{2+}]_i$ in the submembrane region rather than the time course of a transition of the receptors from an inactivated state to the noninactivated one. Some data from the literature support this conclusion. Stinnakre and Tauc (1973) presented data from *Aplysia* neurons in which one can see that the half-time of decay of aequorin luminiscence increases from about 100 to 250 msec when the duration of the depolarizing pulse is prolonged from 75 to 240 msec (Fig. 3A of Stinnakre and Tauc, 1973). Similarly, the half-time increases from about 1 to 4 sec when the amplitude of the pulse (2.5 sec) is increased from about 62 to 150 mV (Fig. 3B of Stinnakre and Tauc, 1973). A similar phenomenon was observed on *Anisodoris nobilis* neurons (Eckert *et al.*, 1977).

In the experiments with aequorin, the relaxations observed after short pulses were faster than those seen in our experiments. As shown elsewhere, the fast disappearance of the aequorin luminescence can be ascribed to the facilitation of the aequorin response that is observed at low calcium concentrations (Ahmed and Connor, 1979; Smith and Zucker, 1980). When the free $[Ca^{2+}]_i$ was studied using Arsenazo III, which gives an absorbance signal more linearly dependent on $[Ca^{2+}]$, the relaxations were slower. For example, Smith and Zucker (1980) measured a half-time of disappearance of the signal of 1 sec or more after a single depolarizing prepulse (+15 mV, 0.3 sec) or after a train of pulses (Fig. 9 of Smith and Zucker, 1980). According to Gorman and Thomas (1978), the half-time of the absorbance signal decay, after a pacemaker cycle, was about 5 sec (Fig 3A of Gorman and Thomas, 1978). Similarly, a decay of several seconds was reported for the absorbance signal of Ca^{2+}-activated K^+-tail current (Fig. 7 of Thompson, 1977; Fig. 10 of Ahmed and Connor, 1979). Thus, the kinetics of the "off-relaxations" connected with the ACh-response inactivation are similar to those of the disappearance of free Ca^{2+}.

Smith and Zucker (1980) believed that the major part of the free Ca^{2+} entering the cell is rapidly bound (in a few milliseconds) by the cytoplasm, the rest then being either slowly sequestered by intracellular organelles (e.g., mitochondria) or gradually extruded by a membrane pump. This conclusion may be inconsistent with the present observations concerning the slight temperature dependence of τ, which suggests that the "off-relaxations" arise mainly from sequestration of Ca^{2+} by cytoplasmic proteins.

As indicated by Chemeris *et al.* (1982), τ is proportional to $H/(I - H)$. As H varies in direct proportion to $[Ca^{2+}]_i/([Ca^{2+}]_i + K_D)$ (Eq. 8), one may conclude that τ linearly increases with $[Ca^{2+}]_i$. The origin of this dependence is unclear. It may arise either from the multicomponent character or from the pre-steady-state regime of Ca^{2+} sequestration.

4.6.4. Possible Mechanism of the ACh-Response Inactivation by Ca^{2+}

The data discussed strongly suggest that the AChR inactivation is induced by Ca_i^{2+}. Ca^{2+} may bind directly to internal ACh receptor sites and, as a result, inactivate the receptor. The possibility of Ca^{2+} binding to an inner site of the AChR was postulated by Nastuk and Parsons (1970) and Ca^{2+} binding was demonstrated on purified AChR (Rübmassen *et al.*, 1978). Ca^{2+} could affect the receptor through its phospholipid environment (Magazanik and Vyskočil, 1970) or through some intracellular substances whose concentrations depend on $[Ca^{2+}]_i$ (Kretsinger and Nelson, 1976; Rasmussen and Goodman, 1977). Sufficient information on these possibilities is lacking, but it has been found that intracellular cAMP is not involved (Chemeris and Iljin, 1985). Nevertheless, the simple Langmuir-like form of H'-dependence on $[Ca^{2+}]_i$ (Eq. 8) and the slight temperature-dependence of the inactivation may indicate that the phenomenon investigated is brought about by binding of Ca^{2+} to the AChR sites.

The dependence of H' on $[Ca^{2+}]_i$ (Fig. 13B) is compatible with a scheme in which inactivation of one AChR requires the binding of only one Ca^{2+} (Chemeris *et al.*, 1982):

$$R + Ca_i \rightleftarrows R* \cdot Ca_i \qquad (9)$$

where $R* \cdot Ca_i$ is the inactive complex of Ca^{2+} with the AChR. The fact that a coefficient of 0.8 has to be introduced in describing the relation between H' and $[Ca^{2+}]_i$ (Eq. 8) may suggest the existence of at least two populations of AChR, one of which has little sensitivity to Ca^{2+}. The existence of two populations of neurons with different sensitivities of their AChR to Ca_i^{2+} was shown by Ivanova *et al.* (1987). On the other hand, the coefficient of 0.8 may indicate the presence of an additional step, e.g., an isomerization of the $R \cdot Ca_i$ complex, in reaction (9):

$$R + Ca_i \underset{K_{-1}}{\overset{K_1}{\rightleftharpoons}} R \cdot Ca_i \underset{K_{-2}}{\overset{K_2}{\rightleftharpoons}} R* \cdot Ca_i \qquad (10)$$

Here, $R \cdot Ca_i$ is a complex of Ca^{2+} with nonactivated receptor. In this case, H_m is determined in the following way:

$$H_m = K_2/(K_2 + K_{-2}) \qquad (11)$$

In such a scheme, the incomplete inhibition of the ACh sensitivity by high $[Ca^{2+}]_i$ is ascribed to the ratio of the isomerization rate constants.

In general terms, a mechanism for AChR inactivation may be postulated as follows (Chemeris *et al.*, 1989). A Ca^{2+} ion binds to the inner site of the AChR and inactivates it when the latter is only in the resting state. Bound Ca^{2+} causes some conformational changes in the receptor corresponding to the inactivation state. These changes are transmitted to the outer receptor site and prevent either binding of the ACh molecules to the receptor site or the channel opening, even when a complex of ACh with the receptor has been formed. When Ca^{2+} leaves the binding site, the receptor relaxes to its initial resting state. On the other hand, an ACh molecule being bound to the outer receptor site induces some conformational changes leading to the channel opening and subsequent receptor desensitization. The conformational changes are transmitted to the inner receptor site. Only when it is in the opened state can the AChR be inactivated by Ca_i^{2+}.

Thus, there is some inconsistency between the properties of the AChR states as identified by being opened (and/or desensitized) by ACh or inactivated by Ca^{2+} ions. More detailed data possibly can be obtained by use of the single-channel recording technique.

5. CONCLUSION

Comparison of the data shows that the nicotinic AChR coupled to either anionic or cationic channels have many similarities in terms of kinetic and pharmacological properties. The structure (primary amino acid sequence, at least) of the AChR coupled to the Cl^- channel is unknown, but it does not seem unreasonable to assume that it is analogous to the structure of the nicotinic AChR coupled to cationic channels.

The Cl^- channels activated by ACh in the molluscan nerve cells exhibit burstlike activity and multiple subconductance states (Bregestovski and Redkozubov, 1986). All substates occur with a high probability. This is similar to the behavior of the conductance substates in mammalian neurons activated by amino acids (Bormann *et al.*, 1987; Cull-Candy and Usowicz, 1987; Hughes *et al.*, 1987; Jachr and Stevens, 1987); however, the opposite situation is observed at the neuromuscular junction where the only conductance substate appears extremely rarely (see Colquhoun and Sakmann, 1985). Both the burstlike activity and the subconductance states excessively complicate the kinetic description of the channels investigated. Hence, one may think that the relaxation and noise measurements of the channel activity give no information on all the stages of the receptor reaction with ACh. The methods probably give some average (effective) values of the rate constants and single-channel conductance, γ.

The off-relaxations related to the Ca^{2+}-induced AChR inactivation reflect most likely the disappearance of submembrane Ca_i^{2+} following cell repolarization. As in the case of interaction of ACh with the receptors, inactivation mechanisms are due to the existence of the Cl^- channel conductance substates.

The channel multiple conductance substates are believed to be a result of an oligomeric structure of the channels consisting of a number of identical elementary channel-subunits (protochannels) (e.g., Geletyuk and Kazachenko, 1983, 1985; Kazachenko and Geletyuk, 1984; Fox, 1987). If this is the case, additional questions may arise in connection with the interaction between the receptors and ACh (or Ca^{2+}). For example, how can a single molecule of ACh or Ca^{2+} activate or inactivate several protochannels simultaneously? This and other related problems call for resolution.

ACKNOWLEDGMENTS. I am grateful to my colleagues Drs. N. K. Chemeris, A. N. Kislov, and A. L. Kurchikov for allowing use of our joint materials described here, Drs. P. D. Bregestovski and A. E. Redkozubov for permission to reproduce Fig. 7 (Fig. 4 in Bregestovski and Redkozubov, 1986), and Mrs. Olga N. Shvirst for assistance in English and preparing the manuscript.

REFERENCES

Adams, D. J., Gage, P. W., and Hamill, O. P., 1976, Voltage sensitivity of inhibitory postsynaptic currents in *Aplysia* buccal ganglia, *Brain Res.* **115**:506–511.

Adams, P. R., 1975, Kinetics of agonist conductance changes during hyperpolarization at frog end-plates, *Br. J. Pharmacol.* **53**:308–310.

Adams, P. R., 1981, Acetylcholine receptor kinetics, *J. Membr. Biol.* **58**:161–174.

Adams, P. R., and Sakmann, B., 1978, A comparison of current–voltage relations for full and partial agonists, *J. Physiol. (London)* **283**:621–644.

Ahmed, Z., and Connor, J. A., 1979, Measurements of calcium influx under voltage clamp in molluscan neurones using the metallochrome dye Arsenazo III, *J. Physiol. (London)* **286**:61–82.

Akopyan, A. R., Chemeris, N. K., Iljin, V. I., and Veprintsev, B. N., 1980, Serotonin, dopamine and intracellular cyclic AMP inhibit the responses of nicotinic cholinergic membrane in snail neurons, *Brain Res.* **201**:480–484.

Alvarez-Leefmans, F. J., Rink, T. J., and Tsien, R. Y., 1981, Free calcium ions in neurones of Helix aspersa measured with ion-selective microelectrodes, *J. Physiol. (London)* **315**:531–548.

Anderson, C. R., and Stevens, C. F., 1973, Voltage clamp analysis of acetylcholine produced endplate current fluctuation at frog neuromuscular junction, *J. Physiol. (London)* **235**:655–691.

Andreev, A. A., Veprintsev, B. N., and Vulfius, C. A., 1984, Two component desensitization of nicotinic receptor induced by acetylcholine agonist in *Lymnaea stagnalis* neurones, *J. Physiol. (London)* **353**:375–391.

Andreev, A. A., Vulfius, C. A., Budantsev, A. Y., Kondrashova, M. N., and Grishina, E. V., 1986, Depression of neuron responses to acetylcholine by combined application of norepinephrine and substrates of the tricarboxylic acid cycle, *Cell. Mol. Neurobiol.* **6**:407–420.

Ascher, P., and Erulkar, S., 1983, Cholinergic chloride channels in snail neurons, in: *Single-Channel Recording* (E. Neher and B. Sakmann, eds.), Plenum Press, New York, pp. 401–407.

Ascher, P., Marty, A., and Neild, T. O., 1978, Life time and elementary conductance of the channels mediating the excitatory effects of acetylcholine in *Aplysia* neurones, *J. Physiol. (London)* **278**:177–206.

Ascher, P., Large, W. A., and Rang, H., 1979, Studies on the mechanism of action of acetylcholine antagonists on rat parasympathetic ganglion cells, *J. Physiol. (London)* **295**:139–170.

Blankenship, J. E., Watchel, H., and Kandel, E. R., 1971, Ionic mechanisms of excitatory, inhibitory and dual synaptic actions mediated by an identified interneuron in the abdominal ganglion of *Aplysia, J. Neurophysiol.* **34**:76–92.

Bormann, J., Hamill, O. P., and Sakmann, B., 1987, Mechanism of anion permeation through channels activated by glycine and γ-aminobutyric acid in mouse cultured spinal neurones, *J. Physiol. (London)* **385**:243–286.

Bregestovski, P. D., 1980, Noise analysis in *Lymnaea* nerve cells, in: *Physiology and Biochemistry of Transmitter Processes,* Nauka, Moscow, p. 30 (in Russian).

Bregestovski, P. D., and Iljin, V. I., 1980, Effect of calcium antagonist D-600 on the postsynaptic membrane, *J. Physiol. (Paris)* **76:**515–522.

Bregestovski, P. D., and Redkozubov, A. E., 1986, Acetylcholine activated single chloride channels in *Lymnaea stagnalis* neurons, *Biol. Membr.* **3:**960–968 (in Russian).

Bregestovski, P. D., Iljin, V. I., Jurchenko, O. P., Veprintsev, B. N., and Vulfius, C. A., 1977, Acetylcholine receptor conformational transition on excitation masks disulphide bonds against reduction, *Nature* **270:**71–73.

Brown, A. M., Akike, N., Tsuda, Y., and Morimoto, K., 1980, Ion migration and inactivation in calcium channel, *J. Physiol. (Paris)* **76:**395–402.

Chang, H. W., and Neumann, E., 1976, Dynamic properties of isolated acetylcholine receptor proteins: Release of calcium ions caused by acetylcholine binding, *Proc. Natl. Acad. Sci. USA* **73:**3364–3368.

Chemeris, N. K., and Iljin, V. I., 1985, Intracellular regulation of ionic currents through chemoexcitable neurone membrane, *Proceedings of the 16th FEBS Congress, Part B,* VNU Science Press, pp. 373–378.

Chemeris, N. K., Kazachenko, V. N., Kislov, A. N., and Kurchikov, A. L., 1982, Inhibition of acetylcholine responses by intracellular calcium in *Lymnaea stagnalis* neurones, *J. Physiol. (London)* **323:**1–19.

Chemeris, N. K., Iljin, V. I., and Kazachenko, V. N., 1989, The dependences of the Ca^{2+}-induced ACh-receptor inactivation on the concentrations of ACh and Ca^{2+}, *Biofizika* in press.

Chiarandini, D. J., and Gerschenfeld, H. M., 1967, Ionic mechanism of cholinergic inhibition in molluscan neurons, *Science* **156:**1595–1596.

Chiarandini, D. J., Stefani, E., and Gerschenfeld, H. M., 1967, Ionic mechanism of cholinergic excitation in molluscan neurons, *Science* **156:**1597–1599.

Colquhoun, D., and Sakmann, B., 1983, Bursts of openings in transmitter-activated ion channels, in: *Single-Channel Recording* (E. Neher and B. Sakmann, eds.), Plenum Press, New York, pp. 345–364.

Colquhoun, D., and Sakmann, B., 1985, Fast events in single channel currents activated by acetylcholine and its analogues at the frog muscle end-plate, *J. Physiol. (London)* **369:**501–557.

Colquhoun, D., Dionne, V. E., Steinbach, J. H., and Stevens, C. F., 1975, Conductance of channels opened by acetylcholine-like drugs in muscle end-plate, *Nature* **253:**204–206.

Cull-Candy, S. G., and Usowicz, M. M., 1987, Multiple conductance channels activated by excitatory amino acids in cerebellar neurones, *Nature* **325:**525–528.

Derkach, V. A., 1986, Relaxations of acetylcholine-induced current in the neurons of sympathetic ganglion, *Dokl. Akad. Nauk Ukr. SSR Ser. B* **4:**60–62 (in Russian).

Dionne, V. E., and Stevens, C. F., 1975, Voltage dependence of agonist effectiveness of the frog neuromuscular junction: Resolution of a paradox, *J. Physiol. (London)* **251:**245–270.

DiPolo, R., Requena, J., Brinley, F. J., Jr., Mullins, L. J., Scarpa, A., and Tiffert, T., 1976, Ionized calcium concentrations in squid axons, *J. Gen. Physiol.* **67:**433–467.

Dreyer, F., and Peper, K., 1975, Density and dose–response curve of acetylcholine receptors in frog neuromuscular junction, *Nature (London)* **253:**641–643.

Eckert, R., Tillotson, D., and Ridgway, E. B., 1977, Voltage dependent facilitation of Ca^{2+} entry in voltage-clamp aequorin injected molluscan neurons, *Proc. National. Acad. Sci. USA* **74:**1748–1752.

Finkel, A. S., 1983, A cholinergic chloride conductance in neurones of *Helix aspersa, J. Physiol. (London)* **344:**119–135.

Fox, J. A., 1987, Ion channel subconductance states, *J. Membr. Biol.* **97:**1–8.

Frank, K., and Tauc, L., 1964, Voltage clamp studies on molluscan neurone membrane properties, in: *Cell Functions of Membrane Transport* (J. E. Hoffman, ed.), Prentice–Hall, Englewood Cliffs, N.J., pp. 113–135.

Gage, P. W., and Armstrong, C. M., 1968, Miniature end-plate currents in voltage-clamped muscle fibers, *Nature* **218:**363–365.

Gage, P. W., and McBurney, R. N., 1975, Effects of membrane potential, temperature and neostigmine on the conductance changes caused by a quantum of acetylcholine at the toad neuromuscular junction, *J. Physiol. (London)* **244:**385–407.

Gardner, D., 1980, Membrane-potential effects on an inhibitory postsynaptic conductance in *Aplysia* buccal ganglia, *J. Physiol. (London)* **304:**165–180.

Gardner, D., and Stevens, C. F., 1980, Rate-limiting step of inhibitory postsynaptic current decay in *Aplysia* buccal ganglia, *J. Physiol. (London)* **304:**145–164.

Geletyuk, V. I., and Kazachenko, V. N., 1983, Single potassium-dependent Cl⁻ channel in molluscan neurons: Multiplicity of the conductance substates, *Dokl Akad. Nauk SSSR* **268:**1245–1247 (in Russian).

Geletyuk, V. I., and Kazachenko, V. N., 1985, Single Cl⁻ channels in molluscan neurons: Multiplicity of the conductance states, *J. Membr. Biol.* **86:**9–15.

Geletyuk, V. I., and Kazachenko, V. N., 1987, Synchronization of potassium channel activity of mollusc neurons induced by ferricyanide and barium, *Biofizika* **32:**73–77 (in Russian).

Ger, B. A., Zeimal, E. V., and Kachman, A. H., 1980, Ionic mechanisms of fast (nicotinic) phase of acetylcholine-induced response in identified *Planobarius corneus* neurons, *Neurofiziologia* **12:**533–540 (in Russian).

Gerschenfeld, H. M., 1973, Chemical transmission in invertebrate central nervous systems and neuromuscular junctions, *Physiol. Rev.* **53:**1–119.

Gorman, A. L. F., and Thomas, M. V., 1978, Changes in intracellular concentration of free calcium ions in a pace-maker neurone measured with the metallochromic indicator dye Arsenazo III, *J. Physiol. (London)* **275:**357–376.

Gorman, A. L. F., and Thomas, M. V., 1980, Intracellular calcium accumulation during depolarization in molluscan neurone, *J. Physiol. (London)* **308:**259–285.

Hagiwara, S., and Nakajima, S., 1966, Differences in Na^+ and Ca^{2+} spikes as examined by application of tetrodotoxin, procain and manganese ions, *J. Gen. Physiol.* **49:**793–806.

Hamill, O. P., Marty, A., Neher, E., Sakmann, B., and Sigworth, F. J., 1981, Improved patch-clamp techniques for high-resolution current recording from cells and cell-free membrane patches, *Pfluegers Arch.* **391:**85–100.

Hucho, F., 1986, The nicotinic acetylcholine receptor and its ion channel, *Eur. J. Biochem.* **158:**211–226.

Hughes, D., McBurney, R. N., Smith, S. M., and Zorec, R., 1987, Caesium ions activate chloride channels in rat cultured spinal cord neurones, *J. Physiol. (London)* **392:**231–251.

Iljin, V. I., Bregestovski, P. D., and Vulfius, E. A., 1976, Influence of pH on the properties of cholinoreceptive membrane in *Lymnaea* neurones, *Neurofiziologia* **8:**640–643 (in Russian).

Ivanova, T. T., Iljin, V. I., Iljasov, F. E., and Veprintsev, B. N., 1986, Average characteristics of neuronal membrane chloride channels activated by different n-cholinomimetics, *Dokl. Akad. Nauk SSSR* **290:**1264–1267 (in Russian).

Ivanova, T. T., Iljin, V. I., Iljasov, F. E., Chemeris, N. K., and Veprintsev, B. N., 1987, Microscopic characteristics of modulation of cholinoreceptive membrane functions by intracellular Ca^{2+} in molluscan neurons, *Biofizika* **32:**295–299 (in Russian).

Jachr, C. F., and Stevens, C. F., 1987, Glutamate activates multiple single channel conductances in hippocampal neurones, *Nature* **325:**522–525.

Katz, B., and Miledi, R., 1972, The statistical nature of the acetylcholine potential and its molecular components, *J. Physiol. (London)* **224:**665–699.

Kazachenko, V. N., 1979, Inactivation of cholinoreceptors caused by intracellular calcium, *Postgraduate paper*, Pushchino (in Russian).

Kazachenko, V. N., and Geletyuk, V. I., 1984, The potential-dependent K^+ channel in molluscan neurons is organized in a cluster of elementary channels, *Biochim. Biophys. Acta* **733:**132–142.

Kazachenko, V. N., and Kislov, A. N., 1973, Voltage–current relations of the cholinoreceptive membrane of *Lymnaea stagnalis* neurons, VINITI 5421 (in Russian).

Kazachenko, V. N., and Kislov, A. N., 1974, Interrelation between electroexcitable and chemosensitive membranes, in: *Biophysics of Living Cell,* Volume 4, Part 2 (G. M. Frank, ed.), Pushchino, pp. 45–49 (in Russian).

Kazachenko, V. N., and Kislov, A. N., 1977, Influence of membrane potential on operation of cholinoreceptors, in: *Biophysics of Complex Systems and Radiation Violations* (G. M. Frank, ed.), Nauka, Moscow, pp. 15–16 (in Russian).

Kazachenko, V. N., Kislov, A. N., and Veprintsev, B. N., 1979, Cholinoreceptive membrane inactivation caused by depolarization of *Lymnaea stagnalis* neurons, *Comp. Biochem. Physiol.* **63C**:61–66.

Kazachenko, V. N., Kislov, A. N., Kurchikov, A. L., and Chemeris, N. K., 1981a, Inactivation of cholinoreceptors caused by intracellular calcium in *Lymnaea stagnalis* neurons, *Dokl. Akad. Nauk SSSR* **257**:1255–1257 (in Russian).

Kazachenko, V. N., Kislov, A. N., Kurchikov, A. L., and Chemeris, N. K., 1981b, Intracellular calcium initiates the receptor inactivation, *Biofizika* **26**:1052–1056 (in Russian).

Kehoe, J. S., 1967, Pharmacological characteristics and ionic bases of a two component post-synaptic inhibition, *Nature* **215**:1503–1505.

Kehoe, J. S., 1972, Ionic mechanisms of a two component cholinergic inhibition in *Aplysia* neurons, *J. Physiol. (London)* **225**:85–114.

Kerkurt, G. A., and Thomas, R. C., 1964, The effect of anion injection and changes in the external and internal potassium chloride concentration on the reversal potentials of IPSP and acetylcholine, *Comp. Biochem. Physiol.* **11**:199–213.

Kislov, A. N., 1974, Studying of activation of the membrane chloride conductance by alkali metal ions in isolated giant molluscan neurons, Postgraduate paper, Pushchino (in Russian).

Kislov, A. N., and Kazachenko, V. N., 1974, Ion currents through activated chemosensitive membrane, in: *Biophysics of Living Cell*, Volume 4 Part 2 (G. M. Frank, ed.), Pushchino, pp. 39–44 (in Russian).

Kislov, A. N., and Kazachenko, V. N., 1975, Potassium activation of chloride conductance in the isolated snail neurons, *Stud. Biophys.* **48**:151–153.

Kislov, A. N., and Kazachenko, V. N., 1977, Chemosensitive somatic membrane of neuron activated by alkali metal ions, in: *Biophysics of Complex Systems and Radiation Violations* (G. M. Frank, ed.), Nauka, Moscow, pp. 14–15 (in Russian).

Kordas, M., 1969, The effect of membrane polarization on the time course of the end-plate current in frog sartorius muscle, *J. Physiol. (London)* **204**:493–502.

Kordas, M., 1972a, An attempt at an analysis of the factors determining the time course of the endplate current. I. The effect of prostigmine and the ratio of Mg^{2+} to Ca^{2+}, *J. Physiol. (London)* **224**:317–332.

Kordas, M., 1972b, An attempt at an analysis of the factors determining the time course of the endplate current. II. Temperature, *J. Physiol. (London)* **224**:333–348.

Kostenko, M. A., Geletyuk, V. I., and Veprintsev, B. N., 1974, Completely isolated neurones in the mollusc *Lymnaea stagnalis*. A new objective for nerve cell biology investigation, *Comp. Biochem. Physiol.* **49A**:89–100.

Kostyuk, P. G., 1980, Calcium ionic channels in electrically excitable membranes, *Neurosciences* **5**:945–959.

Krasts, I. V., 1978, The amplitude of the action potential and calcium ion gradient on the membrane of mollusc neurone, *Comp. Biochem. Biophys.* **60A**:195–197.

Kretsinger, R. H., and Nelson, D. J., 1976, Calcium in biological systems, *Coord. Chem. Rev.* **18**:29–124.

Kurchikov, A. L., and Kazachenko, V. N., 1979, Two components of the current relaxations in chemoreceptive membrane of *Lymnaea stagnalis* neurons, VINITI 429–79 (in Russian).

Kurchikov, A. L., and Kazachenko, V. N., 1984a, Influence of membrane hyperpolarization on cholinoreceptive membrane conductance, *Biol. Membr.* **1**:289–293 (in Russian).

Kurchikov, A. L., and Kazachenko, V. N., 1984b, Kinetics of interaction of acetylcholine with the cholinoreceptors in molluscan neurons, *Biol. Membr.* **1**:384–388 (in Russian).

Kurchikov, A. L., Kazachenko, V. N., and Veprintsev, B. N., 1985, Interaction of mollusc cholinoreceptors with suberyldicholine and other agonists, *Biol. Membr.* **2**:525–533 (in Russian).

Lester, H. A., Changeux, J.-P., and Sheridan, R. E., 1975, Conductance increases produced by bath application of cholinergic agonists to *Electrophorus electricus* electroplaques, *J. Gen. Physiol.* **675**:797–816.

Lester, H. A., Koblin, D. D., and Sheridan, R. E., 1978, Role of voltage-sensitive receptors to nicotinic transmission, *Biophys. J.* **21**:181–194.

Levitan, H., and Tauc, L., 1972, Acetylcholine receptor: Topographic distribution and pharmacological properties of two receptor types on a single molluscan neurone, *J. Physiol. (London)* **222**:537–558.

Levitan, H., Tauc, L., and Segundo, J. P., 1970, Electrical transmission among neurones in the buccal ganglion of a mollusc, *Navanax inermis, J. Gen. Physiol.* **55:**484–496.

MacDermott, A. B., Connor, E. A., Dionne, V. E., and Parsons, R. L., 1980, Voltage clamp study of fast excitatory synaptic currents in bullfrog sympathetic ganglion cells, *J. Gen. Physiol.* **75:**39–60.

Magazanik, L. G., and Vyskočil, F., 1970, Dependence of acetylcholine desensitization on the membrane potential of frog muscle fibre and on the ionic changes in the medium, *J. Physiol. (London)* **210:**507–518.

Magleby, K. L., and Stevens, C. F., 1972a, The effect of voltage on the time course of end-plate currents, *J. Physiol. (London)* **223:**151–171.

Magleby, K. L., and Stevens, C. F., 1972b, A quantitative description of end-plate currents, *J. Physiol. (London)* **223:**173–197.

Meech, R. W., 1974, The sensitivity of *Helix aspersa* neurones to injected clacium ions, *J. Physiol. (London)* **237:**259–277.

Naruschevichus, E. V., Chemeris, N. K., Ponomarjov, V. N., and Akopjan, A. R., 1979, Study of the dependences of inward current on extracellular concentrations of calcium and strontium ions in isolated *Lymnaea stagnalis* neurons, *Neurophyziologia* **79:**362–366 (in Russian).

Nastuk, W. L., and Parsons, R. L., 1970, Factors in the inactivation of postjunctional membrane receptors of frog skeletal muscle, *J. Gen. Physiol.* **56:**218–249.

Neher, E., and Sakmann, B., 1975, Voltage dependence of drug-induced conductance in frog neuromuscular junction, *Proc. Natl. Acad. Sci. USA* **72:**2140–2144.

Neher, E., and Sakmann, B., 1976, Single-channel currents recorded from membrane at denervated frog muscle fibres, *Nature* **260:**799–802.

Popot, J.-L., and Changeux, J.-P., 1984, Nicotinic receptor of acetylcholine: Structure of an oligomeric integral membrane protein, *Physiol. Rev.* **64:**1162–1239.

Rang, H. P., 1974, Acetylcholine receptor, *Q. Rev. Biophys.* **7:**283–399.

Rang, H. P., 1981, The characteristics of synaptic currents and responses to acetylcholine of rat submandibular ganglion cells, *J. Physiol. (London)* **311:**23–55.

Rasmussen, H., and Goodman, D. B. P., 1977, Relationships between calcium and cyclic nucleotides in cell activation, *Physiol. Rev.* **57:**421–508.

Rübmassen, H., Eldefrawi, A. T., Eldefrawi, M. E., and Hess, G., 1978, Characterization of the calcium-binding sites of the purified acetylcholine receptor and identification of the calcium-binding subunits, *Biochemistry* **17:**3818–3825.

Sato, M. G., Austin, H., Yai, H., and Marashi, J., 1968, The nicotinic permeability changes during acetylcholine-induced responses in *Aplysia* ganglion cells, *J. Gen. Physiol.* **51:**312–345.

Sawada, M., 1969, Ionic mechanisms of the activated subsynaptic membrane in *Onchidium* neurons, *J. Physiol. Soc. Jpn.* **31:**491–504.

Selyanko, A. A., Skok, V. I., and Derkach, V. A., 1979, Potential dependence of excitatory postsynaptic current in the neurons of mammalian sympathetic ganglion, *Dokl. Akad. Nauk SSSR* **247:**1007–1009 (in Russian).

Sheridan, R. E., and Lester, H. A., 1975, Relaxation measurements of the acetylcholine receptor, *Proc. Natl. Acad. Sci. USA* **72:**3496–3500.

Sheridan, R. E., and Lester, H. A., 1977, Rates and equilibria at the acetylcholine receptor of *Electrophorus* electroplaques. A study of neurally evoked postsynaptic currents and of voltage-jump relaxations, *J. Gen. Physiol.* **70:**187–219.

Simon, S. M., and Llinás, R. P., 1985, Compartmentalization of the submembrane calcium activity during calcium influx and its significance in transmitter release, *Biophys. J.* **48:**485–498.

Simonneau, M., Tauc, L., and Baux, G., 1980, Quantal release of construction examined by current fluctuation analysis at an identified neuronal synapse of *Aplysia, Proc. Natl. Acad. Sci. USA* **77:**1661–1665.

Skok, V. I., Selyanko, A. A., and Derkach, V. A., 1987, *Neuronal Cholinoreceptors,* Nauka, Moscow (in Russian).

Smith, S. J., and Zucker, R. S., 1980, Aequorin response facilitation and intracellular calcium concentration in molluscan neurones, *J. Physiol. (London)* **300:**167–196.

Stinnakre, J., and Tauc, L., 1973, Calcium influx in active neurones detected by injected aequorine, *Nature* **242**:113–115.

Tauc, L., and Gerschenfeld, H. M., 1962, A cholinergic mechanism of inhibitory synaptic transmission in a molluscan nervous system, *J. Neurophysiol.* **25**:236–262.

Thompson, S. H., 1977, Three pharmacologically distinct potassium channels in molluscan neurones, *J. Physiol. (London)* **265**:465–488.

Vulfius, C. A., and Iljin, V. I., 1980, Investigation of acetylcholine receptors by the method of chemical modification, *Gen. Pharmacol.* **11**:19–25.

Vulfius, C. A., Yurchenko, O. P., Iljin, V. I., Bregestovski, P. D., and Veprintsev, B. N., 1979, Acetylcholine receptor of *Lymnaea stagnalis* neurones, in: *The Cholinergic Synapse* (S. Tucek, ed.,), Elsevier, Amsterdam, pp. 293–302.

Weber, M., Changeux, J.-P., 1974, Binding of *Naja nigricollis* [³H] α-toxin to membrane fragments from *Electrophorus electricus* and *Torpedo* electric organs. II. Effect of cholinergic agonsists and antagonists on the binding of the tritiated α-neurotoxin, *Mol. Pharmacol.* **10**:15–34.

Zeimal, E. V., and Vulfius, E. A., 1968, The action of cholino-mimetics and cholinolytics on the gastropod neurons, in: *Neurobiology of Invertebrates* (J. Salanki, ed.), Academiai Kiado, Budapest, pp. 255–265.

GABA-Activated Bicarbonate Conductance

Influence on E_{GABA} and on Postsynaptic pH Regulation

K. Kaila and J. Voipio

1. INTRODUCTION

γ-Aminobutyric acid (GABA) is a transmitter compound with a wide distribution and an exclusively inhibitory role in both vertebrate and invertebrate nervous systems (Gerschenfeld, 1973; Krnjević, 1974). Apart from its action on vertebrate GABA$_B$-type receptors (see Dutar and Nicoll, 1988), the inhibitory effect of GABA is based on the opening of postsynaptic Cl^- channels (Boistel and Fatt, 1958; Siggins and Gruol, 1986). An increase in postsynaptic Cl^- conductance is also characteristic of glycine-mediated inhibition in vertebrates (Siggins and Gruol, 1986) and of acetylcholine-mediated inhibition in some invertebrate synapses (e.g., Kerkut and Thomas, 1964). The inhibitory effect of an increase in postsynaptic Cl^- conductance is due to the fact that in most excitable cells, the equilibrium potential of chloride is at a level more negative than the threshold for action-potential generation.

Due to their easy accessibility, the GABA-activated channels of crustacean neuromuscular synapses have provided a fruitful model system in the study of the ionic mechanisms of synaptic inhibition (e.g., Boistel and Fatt, 1958; Takeuchi and Takeuchi, 1965, 1967, 1969; Dudel, 1977; Dudel et al., 1980). The selectivity of crustacean GABA-sensitive anion channels (Takeuchi and Takeuchi, 1967; Motokizawa et al., 1969; Mason et al., 1990) is very similar to that of vertebrates (Ito et al., 1962; Edwards, 1982; Bormann et al., 1987).

We have recently shown that in crayfish muscle fibers, in the presence of CO_2/HCO_3^-, GABA brings about a fall in postsynaptic pH_i and a depolarization that cannot be attributed to a current carried by Cl^- (Kaila and Voipio, 1987). The mechanism underlying these effects of GABA is activation of a bicarbonate current mediated by the inhibitory postsynaptic channels. A role for bicarbonate in the ionic basis of

K. Kaila and J. Voipio • Department of Zoology, Division of Physiology, University of Helsinki, SF-00100 Helsinki, Finland.

synaptic inhibition is of much consequence, since, as will be shown below, it provides a novel connection between the regulation of nervous inhibition and of intraneuronal pH.

At the moment, the experimental data that suggest a physiologically significant role for a postsynaptic bicarbonate conductance come from work on crayfish muscle fibers (Kaila and Voipio, 1987; Kaila *et al.*, 1989b, 1990) and on the crayfish stretch-receptor neuron (Voipio *et al.*, 1988). It is obvious, however, that postsynaptic effects of the kind observed in crayfish muscle fibers and neurons upon exposure to GABA are likely to be induced by any transmitter that gates a channel with a significant permeability to HCO_3^-.

1.1. Selectivity of Transmitter-Sensitive Anion Channels

A strict dependence on Cl^- movements of the actions of an inhibitory transmitter *in vivo* requires that, with respect to all ions present both extracellularly and intracellularly, the inhibitory anion channels are absolutely selective for Cl^-. In a large number of both vertebrate and invertebrate preparations, it has indeed been possible to exclude a significant contribution of cationic currents to GABA-mediated inhibition (e.g., Boistel and Fatt, 1958; Lux, 1971; Deisz and Lux, 1982; Bormann *et al.*, 1987). The first systematic investigation of the anion selectivity of inhibitory postsynaptic channels was performed by Coombs *et al.*, (1955). In these experiments it was found that injection not only of Cl^- but also of bromide, nitrate, and thiocyanate changed the glycine-mediated hyperpolarizing IPSP of cat motoneurons into a depolarizing response. In contrast, injection of anions such as phosphate, acetate, and glutamate had no effect on the IPSP. Similar results were obtained in experiments on anion channels activated by ACh in snail neurons (Kerkut and Thomas, 1964). A comparison of the effects of a large number of injected anions on the IPSP reversal potential (E_{IPSP}) in cat motoneurons suggested that anions with a hydrated size up to 1.2–1.3 times that of Cl^- are able to permeate the inhibitory channel but that larger anions are impermeant (Araki *et al.*, 1961; Ito *et al.*, 1962). This implied that the selectivity filter of glycine-activated Cl^- channels in mammalian neurons has a diameter of about 0.29–0.33 nm.

Most anions, existing at reasonable concentrations in the intracellular and extracellular milieu *in situ* are fairly large, organic molecules. The only exceptions to this are chloride and bicarbonate, which has a Stokes diameter (Robinson and Stokes, 1959) of 0.41 nm, about 1.7 times larger than that of Cl^-. In most of the early work cited above, injection of bicarbonate did not have a measurable effect on E_{IPSP}. This, however, does not necessarily mean that the transmitter-sensitive anion channels studied by the injection method were impermeable to bicarbonate. As shown by Thomas (1976), in the nominal absence of extracellular Co_2, intraneuronal injection of bicarbonate does not lead to an appreciable increase in the cytoplasmic concentration of this anion. This is due to the fact that the injected HCO_3^- anions will combine with H^+ ions, thereby forming carbonic acid, which readily leaves the cell as CO_2. The main effect of an intracellular injection of bicarbonate is therefore a rise in pH_i (Thomas, 1976). Thus, the failure of the injection method to detect a significant bicarbonate permeability of the inhibitory postsynaptic channels may reflect an inherent limitation of the technique itself rather than a zero bicarbonate permeability of the channels examined.

However, in experiments on cat cortical neurons, injection of bicarbonate was reported to produce a depolarizing shift in E_{IPSP} and, on the basis of the effects of a variety of anions, it was suggested that the equivalent diameter of the GABA-activated inhibitory channels in cortical neurons is 0.5–0.8 nm, much larger than the pore diameter of the glycine-activated channels in spinal neurons (Kelly *et al.*, 1969).

Recent work, carried out using patch-clamp techniques, has indicated that both GABA- and glycine-activated channels in cultured mouse spinal neurons show, in fact, a significant permeability to bicarbonate (Bormann *et al.*, 1987). The permeabilities of large anions such as formate, bicarbonate, acetate, and, in the case of the GABA-activated channel, propionate were strictly correlated to apparent ionic size as given by the Stokes diameter, and they yielded an estimate of 0.56 nm for the pore of the GABA-activated channel and 0.52 nm for the glycine-activated channel. Measurements of the reversal potential of the transmitter-activated current indicated that the GABA- and glycine-activated channels of cultured spinal neurons of the mouse have a HCO_3^-/Cl^- permeability ratio of about 0.2 and 0.1, respectively.

1.2. Bicarbonate as a Carrier of Transmitter-Activated Current

As will be shown in more detail below, bicarbonate has some interesting properties as a carrier of current in biological membranes. First, if a postsynaptic anion channel shows a significant permeability to HCO_3^-, the reversal potential of the channel-mediated current is, of course, determined by the distribution of both Cl^- and HCO_3^-. Due to active regulation of pH_i in nerve cells (e.g., Roos and Boron, 1981; Thomas, 1984), the HCO_3^- equilibrium potential (E_{HCO_3}) is more positive than the resting membrane potential (E_m). Therefore, activation of a bicarbonate conductance at the level of resting E_m leads to an inwardly directed (depolarizing) current. Another important consequence of activation of a transmitter-induced HCO_3^- conductance is that the concomitant electrogenic efflux of HCO_3^- brings about a fall in postsynaptic pH_i and an increase in the pH_o close to the postsynaptic membrane, i.e., an alkalosis in the synaptic cleft.

2. PREDICTIONS BASED ON ELEMENTARY THERMODYNAMICS

On the basis of the extensive work carried out on the selectivity properties of inhibitory postsynaptic anion channels (Coombs *et al.*, 1955; Araki *et al.*, 1961; Ito *et al.*, 1962), Eccles (1964) concluded that "chloride is the only permeable anion that normally exists in a concentration sufficient to contribute appreciably to the inhibitory current." If this is true, then, *in situ*, and in experiments under physiological conditions, the reversal potential of the current evoked by activity of an inhibitory terminal or by exogenously applied inhibitory transmitter must be identical to E_{Cl}. This conclusion has gained wide acceptance among neurophysiologists and, in fact, a commonly used experimental approach is to postulate the identity $E_{Cl} = E_{GABA}$ (where E_{GABA} is the reversal potential of the GABA-induced current) in order to obtain an estimate of

$[Cl^-]_i$. As shown below, such an approach is dangerous since a permeability to HCO_3^- of a transmitter-sensitive anion channel will give rise to a significant, pH-sensitive difference between the reversal potential and E_{Cl}.

2.1. Nonequilibrium Distributions of Anions across Biological Membranes

The special features in the causes and effects of conductive, transmembrane movements of bicarbonate are based on its being an anion of a weak acid (carbonic acid, H_2CO_3). If channels with suitable selectivity properties are located in a membrane, the electrochemical HCO_3^- driving force will determine the direction of conductive net movements of bicarbonate. However, a weak acid is also able to permeate the lipid matrix of a biological membrane in its protonated, neutral form (or, in the case of bicarbonate, as the anhydride, CO_2). A net flux of the neutral species, followed by its dissociation into the anionic form plus a proton, will have a major influence on the distribution of the anion (Roos and Boron, 1981). Clearly, a mechanism of the latter kind does not apply to anions of strong acids, such as Cl^-.

2.1.1. Anion of a Strong Acid: Cl⁻

At thermodynamic equilibrium, a monovalent anion (A^-), such as Cl^-, is distributed across a biological membrane as predicted by the familiar Nernst equation:

$$E_m = E_A = (RT/F) \ln ([A^-]_i/[A^-]_o)$$

where E_m is the membrane potential, E_A is the equilibrium potential of the anion, $[A^-]_i$ is its intracellular and $[A^-]_o$ its extracellular concentration, and the constants R, T, and F have their usual meaning. It is important to point out that a passive, Nernstian distribution does not necessarily mean that there is no active membrane transport of the anion. A sufficient condition for the attainment of a passive distribution is that there is a conductive pathway for the anion that swamps the net movements of A^- mediated by active membrane transport. Such a situation prevails, for instance, in crayfish muscle fibers, where the Cl^- conductance of the resting membrane is sufficiently high to maintain E_{Cl} very close to resting E_m even during substantial net movements of chloride on Cl^-/HCO_3^- exchange.

In order to generate a driving force, $E_m - E_A$, for a strong-acid anion, the membrane has to be equipped with an active transporter for A^- that is capable of maintaining the anion distribution in a nonequilibrium state despite a possible conductive "leak." For instance, an intraneuronal Cl^- concentration lower than predicted from a passive distribution can be due to Cl^- extrusion on a K^+,Cl^- cotransport mechanism (Deisz and Lux, 1982; Aickin et al., 1982), while an intracellular accumulation of Cl^- may result from the activity of a Na^+,K^+,Cl^- cotransport mechanism (Russell, 1983; Ballanyi and Grafe, 1985; Alvarez-Leefmans et al., 1988).

It should be noted that, in the case of a strong electrolyte, the anion *itself* has to be transported. As shown below, this situation differs from that related to weak-acid anions, since an electrochemical driving force due to a weak-acid anion depends on the transmembrane pH gradient (see Roos and Boron, 1981).

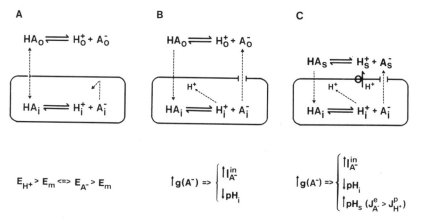

Figure 1. Influence of a conductance for a weak-acid anion (A^-), such as HCO_3^-, on membrane current, intracellular pH, and extracellular surface pH. (A) Due to the high membrane permeability of the nondissociated species (HA), its extracellular and intracellular concentrations are equal. If pH_i is higher than predicted from a passive distribution of H^+, the equilibrium potential of the anion $(E_A = E_H)$ is more positive than the membrane potential (E_m). (B) An electrogenic efflux of A^-, mediated by a conductance pathway $[g(A^-)]$, leads to an inward current (I_A^{in}). The shift to the right in the equilibrium of the intracellular dissociation reaction promotes a net influx of the neutral form, which gives rise to a progressive fall in pH_i. (C) In the vicinity of the extracellular surface, the net transmembrane flux of acid equivalents produces an alkalosis (this effect requires a small-enough H^+-buffering power of the medium to allow development of an extracellular pH gradient; see Gutknecht and Tosteson, 1973). The change in pH_s depends on the total net fluxes of acid equivalents across the membrane. The alkalosis caused by channel-mediated efflux of A^- is abolished as soon as activation of acid extrusion (in response to the fall in pH_i) leads to an efflux of acid equivalents (J_H^p) that is equal to the conductive efflux (J_A^e) of A^-. Subscripts "i," "o," and "s" refer to intracellular, extracellular, and surface, respectively.

2.1.2. Anion of a Weak Acid: HCO_3^-

The nondissociated species (HA) of a weak acid such as H_2CO_3 (or its anhydride, CO_2) is usually much more permeant across biological membranes than the corresponding anion. When a cell is exposed to a weak acid, the high membrane permeability to the neutral HA form will lead to its equilibration so that $[HA]_o = [HA]_i$. A plausible assumption is that the dissociation constant of HA has the same value in the extracellular and intracellular medium:

$$([H^+]_o \times [A^-]_o)/[HA]_o = ([H^+]_i \times [A^-]_i)/[HA]_i$$

Following equilibration of HA this is equivalent to

$$[A^-]_i/[A^-]_o = [H^+]_o/[H^+]_i$$

and, therefore,

$$E_A = E_H$$

Consequently, if a neuron is in a CO_2/HCO_3^--containing solution (*in situ* or *in vitro*),

$$E_{HCO_3} = E_H$$

As is evident from the Nernst equation, E_H is given (at 20°C) by

$$E_H = (pH_i - pH_o) \times 58 \text{ mV}$$

To summarize: if the nondissociated species of a permeant weak acid is at thermodynamic equilibrium across a cell membrane, the equilibrium potential of its anion is identical to the equilibrium potential of H^+. Consequently, it is obvious that the driving force ($E_m - E_A$) of the anion is equivalent to the H^+ driving force ($E_m - E_H$). This is very important since all animal cells, including excitable ones, typically maintain their pH_i at a level about one unit higher than predicted from a passive distribution of H^+ (Roos and Boron, 1981; Thomas, 1984) and therefore $E_m - E_H$ is often about -60 mV. This means that in a neuron with a membrane potential of -70 mV, E_H would be about -10 mV.

The difference in the mechanisms underlying the distribution of weak- and strong-acid anions has often been overlooked. For instance, a positive shift in E_{GABA}, induced by formate (a weak-acid anion capable of passing through GABA-sensitive channels), was reported in experiments on lobster muscle fibers (Motokizawa et al., 1969). The authors attributed the depolarizing shift in E_{GABA} to an intracellular accumulation of formate anion. However, the nature of this accumulation, which most likely reflected the equilibration of the neutral form (see also Sharp and Thomas, 1981), remained unexplained.

An important feature of an electrochemical driving force due to a weak-acid anion is that from a thermodynamic point of view, it is immaterial whether $E_m - E_H$ (and thereby $E_m - E_A$) is maintained by active transport of the anion itself (or of any other weak-acid anion), or by transport of protons. For instance, in the presence of CO_2, the HCO_3^- driving force is increased in exactly the same way by active extrusion of H^+ (e.g., on Na^+/H^+ exchange) or by inward transport of HCO_3^- (e.g., on Cl^-/HCO_3^- exchange).* It is important to point out here that a change in E_H can also be induced by generation or consumption of intracellular acid equivalents, or by a change in pH_o.

Finally, it may be of interest to note that channels permeable to weak-acid anions are not the only conductive pathways capable of mediating movements of acid equivalents across cell membranes. It is evident that, on the basis of its reversal potential ($E_H = E_A$), a current caused by channel-mediated permeation of a weak-acid anion is indistinguishable from a current due to H^+ channels (Thomas and Meech, 1982; Byerly et al., 1984; Meech and Thomas, 1987) or from a current due to a weak-acid protonophore (see Kaila et al., 1989a; McLaughlin and Dilger, 1980).

2.2. Physiological Consequences of Activation of a Bicarbonate Conductance

Let us assume that pH_i in a nerve cell is 7.2, pH_o is 7.3, and the resting membrane potential is -70 mV. In the presence of CO_2/HCO_3^-, $E_{HCO_3} = E_H = -6$ mV which gives an HCO_3^- driving force of -64 mV. As shown schematically in Fig. 1 (A and B),

---

*In the present paper we assume that the hydration of CO_2 is not a rate-limiting step in the pH_i and pH_s changes discussed. The role of carbonic anhydrase activitiy will be dealt with elsewhere (Kaila et al., 1990).

activation of a membrane conductance to bicarbonate [$g(A^-)$] will lead to a depolarizing current (inward current, I_A^{in}) that is paralleled by an increase in H_i^+, i.e., by a fall in pH_i.

The predicted influence on pH_i of activation of a bicarbonate conductance is examined more closely in Fig. 1C. Here, the cell membrane is equipped with an extruder of acid equivalents (not necessarily an H^+ extruder) which, of course, is a prerequisite for the initial condition assumed, i.e., $E_H > E_m$. The fall in pH_i caused by a bicarbonate conductance will proceed until the net efflux of acid equivalents on the extruder (J_H^p; superscript "p" for pump) is equal to the net efflux of HCO_3^- mediated by the anion channel (J_A^e). In all cells studied, activation of acid extrusion is steeply dependent on pH_i (Boron and Russell, 1983; Boron *et al.*, 1988; Aronson, 1985; Mahnensmith and Aronson, 1985). This general property of acid-extruding mechanisms will facilitate the attainment of a new steady-state level of pH_i during maintained activation of a bicarbonate conductance.

Provided that the pH-buffering power of the aqueous medium is not prohibitively high, a transmembrane flux of acid equivalents will give rise to a pH gradient in the vicinity of the membrane (Gutknecht and Tosteson, 1973). As shown in Fig. 1C, activation of a bicarbonate conductance is expected to produce an increase in extracellular surface pH as long as the active extrusion of acid does not match the conductive efflux of bicarbonate. This means that there is an increase in surface pH as long as $J_H^p < J_{HCO_3}^e$, i.e., as long as pH_i is decreasing.

In view of the very general assumptions made, the above predictions should apply to the activation of a membrane conductance to any weak-acid anion. Recent experiments on crayfish muscle fibers have shown, indeed, that GABA-induced effects of the kind depicted in Fig. 1 on membrane current, pH_i, and pH_s are observed not only in the presence of HCO_3^-, but also following application of solutions containing formate and acetate (see Kaila, 1988; Mason *et al.*, 1990). Formate has been widely used in experiments on the selectivity of inhibitory anion channels and it shows a high permeability in all preparations examined (e.g., Ito *et al.*, 1962; Kerkut and Thomas, 1964; Takeuchi and Takeuchi, 1967; Kelly *et al.*, 1969; Motokizawa *et al.*, 1969; Bormann *et al.*, 1987).

3. INFLUENCE OF GABA-ACTIVATED BICARBONATE CONDUCTANCE ON MEMBRANE POTENTIAL AND ON E_{GABA}

3.1. GABA-Activated Anion Movements in Excitable Cells

If IPSPs are due to channels that, under physiological conditions, allow the passage of Cl^- ions only, then E_{IPSP} must be identical to E_{Cl}. Accordingly, the postsynaptic effect of an inhibitory transmitter may manifest itself as a hyperpolarization, a depolarization, or no change in membrane potential at all, depending on whether intracellular chloride is actively extruded, accumulated, or passively distributed across the plasmalemma of the postsynaptic neuron (see Siggins and Gruol, 1986).

Until lately, ionic mechanisms of depolarizing effects of GABA have largely been deduced on the basis of indirect evidence (see Barker and Ransom, 1978; Alger and

Nicoll, 1979; Gallagher *et al.*, 1983; Siggins and Gruol, 1986). However, recent work, employing Cl^--selective microelectrodes, has shown that in rat sympathetic neurons (Ballanyi and Grafe, 1985) and in frog sensory neurons (Alvarez-Leefmans *et al.*, 1988), the depolarizing effect of GABA can be explained on the basis of an outwardly directed electrochemical gradient of Cl^-, maintained by active uptake of chloride via the Na^+,K^+,Cl^- cotransporter. As shown below, however, a depolarizing effect of GABA is not necessarily indicative of a situation where $E_{Cl} > E_m$. A significant bicarbonate permeability of inhibitory postsynaptic channels may also lead to a transmitter-induced depolarization.

3.2. Effects of GABA on Membrane Potential in Crayfish Muscle Fibers

In crayfish muscle fibers, GABA has little effect on membrane potential if the extracellular medium is nominally CO_2-free. This is due to the fact that the resting Cl^- conductance of these muscle fibers is high, and therefore, Cl^- ions are in equilibrium across the cell membrane (Dudel and Rüdel, 1969; Dudel, 1977; Kaila *et al.*, 1989b). If, however, GABA is applied in the presence of CO_2/HCO_3^-, a significant depolarization takes place (see Fig. 2). This depolarizing effect of GABA is sensitive to picrotoxin, but it cannot be explained on the basis of a Cl^- current, since ion-sensitive electrode measurements of the Cl_i^- activity (a_{Cl}^i) showed that a corresponding shift in E_{Cl} does not occur. In fact, the depolarization induced by GABA in the presence of

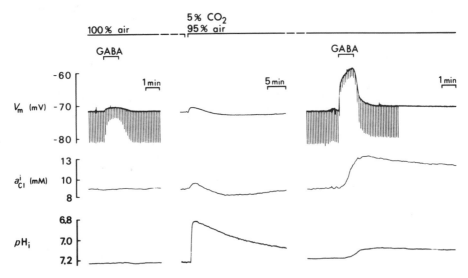

Figure 2. In crayfish muscle fibers, GABA gives rise to a CO_2/HCO_3^--dependent depolarization which is paralleled by an increase in a_{Cl}^i and by a fall in pH_i. This experiment shows the effects of bath-applied GABA (5×10^{-4} M) on membrane potential, on input resistance (as measured by hyperpolarizing current pulses of 100 nA, 0.3 sec; two-microelectrode current clamp), on a_{Cl}^i, and on pH_i. The effects of GABA were first measured in a solution equilibrated with air (left). After recovery of pH_i from exposure to 5% CO_2/95% air (center), GABA was reapplied. In the presence of CO_2/HCO_3^- (pH_o 7.4), GABA produced a marked depolarization paralleled by an influx of Cl^- as indicated by the increase in a_{Cl}^i (right). (From Kaila and Voipio, 1987, by permission of *Nature*.)

HCO_3^- was paralleled by an increase in the a_{Cl}^i, indicative of an electrogenic net influx of Cl^- which tends to oppose the depolarization. Consistent with this, the depolarizing effect of GABA was accentuated in a Cl^--free, HCO_3^--containing solution (Kaila *et al.*, 1989b). In the presence of CO_2, GABA also produced a fall in pH_i (see also Fig. 5). All these effects are consistent with the view that the depolarizing action of GABA is due to activation of a postsynaptic bicarbonate conductance.

3.3. Effect of Bicarbonate Permeability on the Reversal Potential of Postsynaptic Anion Channels

3.3.1. Influence of CO_2/HCO_3^- on E_{GABA} in Crayfish Muscle Fibers

In accordance with the effects of GABA on membrane potential in crayfish muscle fibers, measurements of E_{GABA} by means of a three-microelectrode voltage clamp showed (see Fig. 3) that, in a CO_2-free solution, E_{GABA} equals E_{Cl}, which (due to the high background Cl^- permeability; see above) is virtually identical to the resting membrane potential. However, in the presence of CO_2/HCO_3^-, there is a positive shift of 10–15 mV in E_{GABA}, although (under steady-state conditions) neither E_m nor E_{Cl} is affected.

Simultaneous measurements of E_{GABA} and pH_i in the presence of HCO_3^- yielded an estimated permeability ratio P_{HCO_3}/P_{Cl} of about 0.3 for the GABA-activated channels in crayfish muscle (Kaila and Voipio, 1987; Kaila *et al.*, 1989b). This estimate for the relative bicarbonate permeability is somewhat larger than that obtained for mammalian GABA-activated channels by Bormann *et al.* (1987). However, we will show below that, under physiological conditions, a permeability ratio of 0.2 is sufficiently high to result in a significant deviation of the reversal potential of inhibitory postsynaptic current from E_{Cl}.

Further evidence supporting the conclusion that, in crayfish muscle fibers, the CO_2/HCO_3^--dependent depolarizing shift in E_{GABA} is attributable to a bicarbonate current came from voltage-clamp experiments that showed that, in the absence of Cl^-, it is possible to identify a GABA-activated, picrotoxin-sensitive current carried by bicarbonate (Kaila and Voipio, 1987). In crayfish muscle fibers, picrotoxin is a noncompetitive antagonist of the GABA-induced Cl^- conductance (Takeuchi and Takeuchi, 1969). Simultaneous measurements of E_{GABA} and pH_i showed that, as expected, when the muscle fibers are bathed in a Cl^--free, HCO_3^--containing solution, E_{GABA} is close to the estimated value of E_{HCO_3} (Kaila *et al.*, 1989b).

3.3.2. Dependence on pH_i and pH_o of the Reversal Potential of a Bicarbonate-Permeable Anion Channel

If an anion channel is permeable to both Cl^- and HCO_3^-, the reversal potential of the current mediated by the channel (E_{rev}) must be sensitive to changes in the extracellular and intracellular concentrations of these two anions. This means that at a constant partial pressure of CO_2, E_{rev} should be sensitive (in a predictable manner) to changes in intracellular and extracellular pH.

Figure 4 illustrates the predicted dependence of E_{rev} on pH_i and on pH_o as

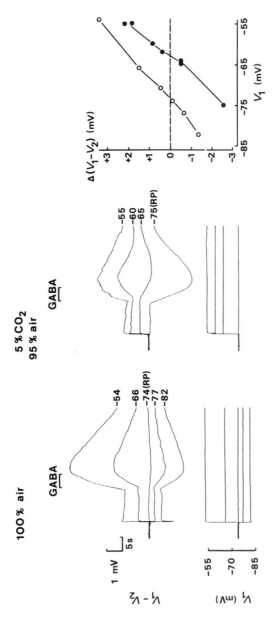

Figure 3. In crayfish muscle fibers, HCO_3^- ions induce a shift in E_{GABA} to a voltage level much more positive than E_{Cl}. In this experiment, the influence of CO_2/HCO_3^- on E_{GABA} in a crayfish muscle fiber was examined using a three-microelectrode voltage clamp (voltage control was imposed on the V_1 electrode located in the center of the fiber; the V_2 electrode was midway between center and end; and the current-passing electrode was close to V_1). pH_i (7.15) was simultaneously measured (now shown) in order to obtain an estimate of $[HCO_3^-]_i$. GABA (5×10^{-4} M) was applied in pulses of 5 sec. The peak changes in $V_1 - V_2$, measured in the absence (○) and presence (●) of 5% CO_2/HCO_3^- (pH₀ 7.4) have been plotted as a function of holding potential (right). In the CO_2-free solution, E_{GABA} is very close to the resting membrane potential (RP). In the presence of CO_2/HCO_3^-, there is a positive shift in E_{GABA}–RP of 12 mV. In both solutions, E_{Cl} is identical to the resting potential. (From Kaila and Voipio, 1987, by permission of *Nature.*)

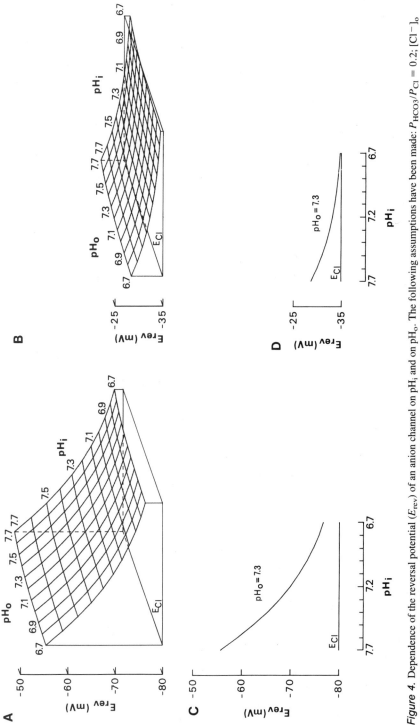

Figure 4. Dependence of the reversal potential (E_{rev}) of an anion channel on pH_i and on pH_o. The following assumptions have been made: $P_{HCO3}/P_{Cl} = 0.2$; $[Cl^-]_o = 130$ mM; $P_{CO2} = 40$ mm Hg; $T = 37°C$. In A, $[Cl]_i = 6.52$ mM, yielding an E_{Cl} of -80 mV. In this situation, E_{rev}, as calculated using the GHK voltage equation, shows a steep dependence on pH, but is only slightly affected by changes in pH_o. In B, $[Cl]_i = 35.1$ mM and $E_{Cl} = -35$ mV. In this case, E_{rev} shows a rather shallow dependence on both pH_i and pH_o. In C and D, the dependence of E_{rev} on pH_i is shown more accurately at a physiologically normal level of pH_o, 7.3.

calculated on the basis of the Goldman–Hodgkin–Katz voltage equation (Goldman, 1943; Hodgkin and Katz, 1949):

$$E_{rev} = \frac{RT}{F} \ln \frac{[Cl^-]_i + b[HCO_3^-]_i}{[Cl^-]_o + b[HCO_3^-]_o}$$

where b is the relative permeability of HCO_3^- with respect to Cl^- (P_{HCO_3}/P_{Cl}) and R, T, and F have their usual meaning.*

The parameters for constructing the graphs in Fig. 4 have values approximating conditions that prevail in the mammalian nervous system: the permeability ratio P_{HCO_3}/P_{Cl} of the anion channel is 0.2 (see Bormann et al., 1987); $[Cl^-]_o$ is 130 mM (e.g., Hansen and Zeuthen, 1981) and $[HCO_3^-]_o$ is 20 mM at pH_o 7.3 (Kraig et al., 1983; Mutch and Hansen, 1984), which corresponds to a CO_2 partial pressure of about 40 mm Hg at 37°C (see Pontén and Siesjö, 1966; Siggaard-Andersen, 1974; Wichser and Kazemi, 1975).

Two situations will be examined. In the first one (Fig. 4A,C), the inhibitory anion channel is located in the postsynaptic membrane of a cell that has a low a_{Cl}^i, and, consequently, a relatively negative E_{Cl} (−80 mV). In the other case to be examined (Fig. 4B,D), the anion channel is in the postsynaptic membrane of a neuron that actively accumulates Cl^- and therefore has a more positive E_{Cl} (−35 mV; see Ballanyi and Grafe, 1985; Alvarez-Leefmans et al., 1988). A neuron of this kind would produce a depolarizing IPSP. In both cases, E_{Cl} is assumed to be constant despite variation in pH_i and pH_o. This is justified in the present context aimed at elucidating the *immediate* effect on E_{rev} of a change in extracellular and intracellular pH.

There are several features in Fig. 4 that merit comment. First, it should be noted that at physiological levels of pH_i and pH_o (7.2 and 7.3, respectively), a HCO_3^-/Cl^- permeability ratio of 0.2 is sufficient to produce a deviation by 10 mV of E_{rev} from E_{Cl} in a cell with a low a_{Cl}^i (Fig. 4A,C). This means that, if the selectivity properties of GABA-activated anion channels in vertebrate neurons in general are similar to those of the cultured mouse neurons studied by Bormann et al. (1987), it would be clearly erroneous to assume that E_{GABA} is identical to E_{Cl}. It is also evident from the data shown that a HCO_3^-/Cl^- permeability ratio of 0.2 is sufficient to produce a depolarizing IPSP in a neuron with an E_{Cl} that is identical to, or only slightly more negative than, the resting membrane potential.

Another interesting consequence of a bicarbonate permeability is that, in a neuron with a low $[Cl]_i$, it will make E_{rev} very steeply dependent on pH_i (Fig. 4A). The dependence on pH_o of E_{rev} is much smaller. As is evident from Fig. 4C, if the membrane potential of a neuron with a low a_{Cl}^i were at −68 mV, a change in pH_i from 7.0 to 7.3 would be sufficient to abolish the driving force (about 6 mV) of the inhibitory current and a further rise in pH_i would turn an originally hyperpolarizing IPSP into a depolarizing one.

---

*In the above equation, it is more orthodox to use ion activities instead of concentrations. However, at the level of ionic strength relevant in the present context, the activity coefficients of the two anions are almost identical (Meier et al., 1980; Ammann, 1986), and, therefore, using concentrations instead of activities has a negligible effect on the calculated reversal potentials.

The electrical consequences of a bicarbonate permeability are much smaller in a cell that actively accumulates Cl_i^- (Fig. 4B,D). At a pH_i of 7.2 and a pH_o of 7.3, E_{rev} − E_{Cl} is only 1.5 mV.

Finally, it may be worthwhile emphasizing that the above quantitative predictions are based on the assumption of a constant partial pressure of CO_2. Therefore, Fig. 4 cannot be used to predict the effects on E_{rev} and on synaptic transmission of changes in pH_i that take place during a respiratory acidosis or alkalosis (see, e.g., Balestrino and Somjen, 1988).

4. INFLUENCE OF GABA-ACTIVATED BICARBONATE CONDUCTANCE ON POSTSYNAPTIC pH REGULATION

4.1. Effect of GABA on pH_i in Muscle Fibers and Neurons of the Crayfish

As explained above (Section 2.2), activation of a bicarbonate conductance is expected to produce a fall in pH_i. Figure 5 shows the results of a typical experiment on a crayfish muscle fiber, where pH_i, pH_s, and membrane potential were simultaneously recorded. In the absence of CO_2, application of GABA produced a depolarization of less than 3 mV which was associated with a slight fall (0.05 unit) in pH_i and with an increase in pH_s of 0.025 unit. In contrast to this, exposure of the fiber to GABA in the presence of 30 mM HCO_3^- (pH 7.4) brings about a pronounced depolarization (19 mV) paralleled by a fall of 0.4 unit in pH_i and by a transient increase of almost 0.1 unit in pH_s.

All the effects of GABA shown in Fig. 5 can be blocked by application of picrotoxin (see Kaila and Voipio, 1987), which supports the conclusion that the change in pH_i is secondary to opening of GABA-gated anion channels and not to some other action of the transmitter. Further support for this view came from measurements (Kaila et al., 1990) that indicated that the instantaneous efflux of HCO_3^- ($J^e_{HCO_3}$) shows a dependence on GABA concentration that is similar to that of the transmitter-activated Cl^- conductance. In these experiments, $J^e_{HCO_3}$ was calculated from

$$J^e_{HCO_3} = -\beta \times dpH_i/dt$$

(where β is the H_i^+-buffering power; see Roos and Boron, 1981; Kaila et al., 1990), and the GABA-induced Cl^- conductance was measured by means of a three-microelectrode voltage clamp (Kaila et al., 1990).

In addition to the muscle fibers of crayfish, another classical preparation in the study of GABA-mediated inhibition is the stretch-receptor neuron (e.g., Kuffler and Eyzaguirre, 1955; Iwasaki and Florey, 1969; Deisz and Lux, 1982). We have recently found that effects on pH_i and membrane potential similar to those produced by GABA in crayfish muscle are also observed in the stretch-receptor neuron (Voipio et al., 1988).

In muscle fibers and nerve cells of the crayfish, the effects of GABA on E_m, pH_i, and pH_s, observed in the presence of CO_2/HCO_3^-, are fully attributable to the bicarbonate permeability of the GABA-activated anion channels. However, it is not immedi-

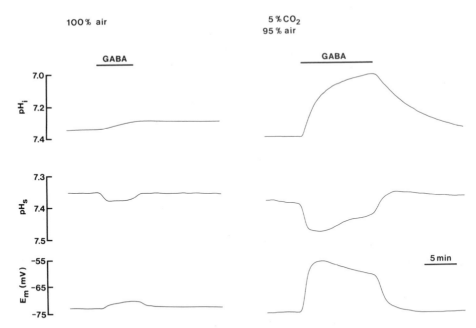

Figure 5. Effect of GABA (2×10^{-4} M) on pH_i, pH_s, and membrane potential in a crayfish muscle fiber. In a solution equilibrated with 5% CO_2 (pH 7.4), GABA induces a marked fall in pH_i which is associated with a depolarization and a transient increase in pH_s (note higher gain in the latter trace). Both solutions contained 10 mM HEPES. (K. Kaila and J. Voipio, unpublished experiment.)

ately clear what the small but consistent effects seen in a nominally CO_2-free solution are due to (e.g., Fig. 5). There are several possibilities that should be considered. First, it is possible that CO_2, generated by cellular energy metabolism, gives rise to the effects of GABA on pH_i and pH_s in a CO_2-free solution. There is, in fact, preliminary evidence indicating that such a mechanism may be of significance in the stretch-receptor neuron (Voipio *et al.,* 1988). Another possibility is that GABA-activated channels are permeable to H^+ and/or OH^- ions. (In this context it may be worthwhile pointing out that the equilibrium potential of OH^- is identical to that of H^+: $E_H = E_{OH}$.) To account for the fall in pH_i observed in the absence of CO_2 via H^+ or OH^- fluxes would require that the permeability of GABA-activated channels to H^+ or OH^- is several orders of magnitude higher than their permeability to Cl^- since their concentrations are several orders of magnitude smaller.

4.2. Effect of GABA on Extracellular pH_s

4.2.1. GABA-Induced Changes in pH_s in Crayfish Muscle Fibers

As predicted by the scheme in Fig. 1C, the increase in pH_s, observed following GABA application, should be at its largest when the rate of fall of pH_i is at its maximum. In the recording in Fig. 5, pH_s was measured using an H^+-selective microelectrode, with a shape similar to that of a patch pipette, gently pressed on the surface of the muscle fiber. During the initial, steep fall in pH_i, pH_s peaks at a value 0.08 unit more alkaline than before GABA exposure. When the intracellular acidosis due to GABA levels off, the channel-mediated efflux of HCO_3^- must equal the extrusion of acid on the sodium-dependent Cl^-/HCO_3^- exchanger (Galler and Moser, 1986), and therefore pH_s relaxes toward its pre-GABA baseline (see Section 2.2).

In contrast to the absolute values of pH_i recorded by aid of an intracellular H^+-selective microelectrode, the measured values of pH_s do not have a similar quantitative nature. They are best regarded as qualitative measurements, which are helpful in determining whether a given change in pH_i depends on a transmembrane movement of acid equivalents or whether pH_i is altered by intracellular generation or consumption of H^+. The absolute value of a change in pH_s depends, among other factors, on the buffering power of the extracellular medium (Gutknecht and Tosteson, 1973; Vanheel *et al.*, 1986; Thomas, 1988). It is therefore obvious that the measured values of GABA-induced changes of pH_s in crayfish muscle fibers are likely to be much smaller than those present close to the postsynaptic membrane, i.e., in the synaptic cleft. However, an upper limit for the change in $[H^+]_o$ experienced, for instance, by ionizable groups of channel proteins in the postsynaptic membrane, can be given.

The upper limit of the GABA-induced pH change at the extracellular surface can be estimated as follows. Let us assume that at the postsynaptic membrane, the GABA-induced HCO_3^- conductance completely abolishes the H^+ electrochemical gradient. The H^+ electrochemical gradient across the plasma membrane of crayfish muscle fibers is about 60 mV. If GABA is applied in the presence of CO_2/HCO_3^-, the depolarization of the membrane will dissipate part of the gradient. Assuming that the depolarization at the postsynaptic membrane is 15 mV, a complete dissipation of the H^+ electrochemical gradient requires an increase in pH_s of $(60 - 15)/58$ pH units, i.e., a surface alkalosis of about 0.8 unit. It should be noted that, as shown in Fig. 5, the increase in pH_s reaches its maximum at the start of GABA exposure, in advance of a significant fall in the value of pH_i. However, the above estimate does not take into account that a fall in pH_i close to the postsynaptic membrane (a fall in intracellular pH_s; see Meech and Thomas, 1987) may contribute to the dissipation of the proton gradient and, therefore, the theoretical upper limit for the alkalosis at the extracellular surface is likely to be too high. In view of these considerations, it can be concluded that the extracellular pH electrode senses at least 10–15% (see Fig. 5) of the pH change that takes place in the synaptic cleft. This is a rather large fraction when considering that the inhibitory synapses are located in sparsely distributed spots in deep invaginations of the plasma membrane (Takeuchi and Takeuchi, 1965; Atwood, 1976).

4.2.2. Activity-Induced Changes in pH_o in Vertebrate Nervous Tissue

A large number of reports indicate that activity in vertebrate nervous tissue gives rise to changes in pH_o (Urbanics *et al.*, 1978; Kraig *et al.*, 1983; Endres *et al.*, 1986; Krishtal *et al.*, 1987; Rice and Nicholson, 1988; Chesler and Chan, 1988; Syková, 1988). In some cases, an increase in the production of metabolic acid seems to account for prolonged, activity-induced extracellular acidoses (Kraig *et al.*, 1983; Endres *et al.*, 1986; Spuler *et al.*, 1987). A different type of extracellular acidosis, related to synaptic activity, has been detected in hippocampal slices of the rat, where single electrical stimuli of presynaptic pathways were found to induce a rapid (duration about 10 msec) fall in pH_o (Krishtal *et al.*, 1987). This acid transient was suggested to be due to the release of the acid contents of synaptic vesicles (Melnik *et al.*, 1985) into the synaptic cleft, or to the incorporation of active vesicular proton pumps (Stadler and Tsukita, 1984) into the presynaptic membrane.

As discussed above, activation of inhibitory postsynaptic anion channels with a significant HCO_3^- permeability is expected to lead to an increase in pH_o. To our knowledge, the only work on the effects of an inhibitory transmitter on pH_o in vertebrate nervous tissue is that by Ballanyi and Grafe (1985), who showed that in rat sympathetic ganglia *in vitro*, application of GABA produces an acidosis of the extracellular space. However, the results of the above authors show that, in the sympathetic ganglion, an extracellular acidosis is brought about also by application of carbachol and by synaptic stimulation. A noteworthy feature of all these experimental procedures is that they give rise to a simultaneous increase in $[K^+]_o$. It is, therefore, possible that the changes in pH_o were mainly due to the concomitant changes in $[K^+]_o$. In fact, the data of Ballanyi and Grafe (1985; their Fig 7) reveal a strong correlation between pH_o and $[K^+]_o$. This is not unexpected in view of the large number of reports that indicate a close coupling between regulation of pH_o and K_o^+ in nervous tissue (Kraig *et al.*, 1983; Chesler and Kraig, 1987; Chesler and Chan, 1988; Rice and Nicholson, 1988; Syková, 1988), and the existing data suggest a role for both neurons and glia.

In experiments on the rat cerebellar cortex, Kraig *et al.* (1983) found that local surface stimulation produced an initial alkaline shift in pH_o, followed by a long-lasting acidification. The initial extracellular alkalosis was selectively inhibited by Mn^{2+}, which may suggest an involvement of synaptic transmission in its generation (see Katz, 1969; but also see Mutch and Hansen, 1984, who report an absence of a similar effect of Mn^{2+} in rat cerebral cortex). Recent work has demonstrated that stimulation evokes an extracellular alkalosis in a number of preparations (e.g., Krishtal *et al.*, 1987; Rice and Nicholson, 1988; Syková, 1988). It is possible that, at least in some cases, the underlying mechanism is a channel-mediated flux of H^+ (Chesler and Chan, 1988).

An initial extracellular alkalosis in the rat cerebellar cortex, followed by an acid shift, is also observed during spreading depression (Kraig *et al.*, 1983). Electrode measurements of the concentration of various extracellular "probe" anions have shown that during spreading depression, there is a marked increase in the anion permeability of cells in the cerebellar cortex (Phillips and Nicholson, 1979). On the basis of the size of the largest permeant ion (hexafluoroarsenate) and smallest impermeant ion (α-naphthalene sulfonate) included in the survey, a lower limit of 0.6 nm and an upper limit of 1.2 nm was obtained for the equivalent diameter of the channels activated. This range is compatible with the estimate of 0.5–0.8 nm postulated by Kelly *et al.* (1969) for GABA-sensitive anion channels in cortical neurons. It is therefore possible that at least part of the anion channels that are activated during spreading depression are postsynaptic inhibitory channels (Phillips and Nicholson, 1979; Phillis and Ochs, 1971). It will be interesting to see in future work whether some of the pH_o changes described above are attributable to bicarbonate fluxes mediated by inhibitory postsynaptic channels.

5. OTHER PHYSIOLOGICAL IMPLICATIONS OF A TRANSMITTER-INDUCED BICARBONATE CONDUCTANCE

Above, we have examined the immediate effects of a transmitter-activated bicarbonate conductance on membrane current, pH_i, and pH_s. An interesting question that emerges now is whether these effects might play a modulatory role in synaptic inhibition.

5.1. Dependence of E_{IPSP} on pH_i, and the Effectiveness of Synaptic Inhibition

It is generally agreed that the effectiveness of postsynaptic inhibition is primarily based on short-circuiting of excitatory input, caused by the transmitter-induced increase in Cl^- conductance. However, both theoretical considerations (Jack *et al.*, 1975) as well as experimental results (Lux *et al.*, 1970; Lux, 1971; Llinás *et al.*, 1974; Raabe and Gumnit, 1975; Iles and Jack, 1980) clearly show that a reduction or removal of the hyperpolarizing component of postsynaptic inhibition alone can give rise to a significant decrease in the effectiveness of inhibition.

As shown in Fig. 4, if inhibitory postsynaptic channels are permeable to bicarbonate, then, in a neuron that does not actively accumulate chloride, the reversal potential of the transmitter-activated current will show a rather steep dependence on pH_i. This provides a possible mechanism whereby changes in pH_i might have a direct effect on the inhibitory postsynaptic potential and, consequently, on neuronal excitability. Such a mechanism would give rise to an increase in neuronal excitability following an increase in postsynaptic pH_i (see Fig. 4).

One of the most common agents used to induce a displacement of pH_i in experiments *in vitro* is ammonia (NH_3/NH_4^+; see Roos and Boron, 1981). In the present context, it is of interest that application of ammonium salts leads to an increase in nervous excitability and to epileptiform seizures (for references, see Lux, 1971). In experiments on cat spinal motoneurons, Lux *et al.* (1970) observed that NH_3/NH_4^+ acts predominantly on postsynaptic inhibition, and reversibly decreases the driving force of the hyperpolarizing IPSP. Subsequent studies showed that NH_3/NH_4^+ has a similar effect in a variety of preparations (Llinás *et al.*, 1974; Raabe and Gumnit, 1975; Iles and Jack, 1980; Deisz and Lux, 1982; Aickin *et al.*, 1982).

Qualitatively, the effects of NH_3/NH_4^+ observed in most of the work cited above are consistent with a scheme wherein application of ammonium salts produces an increase in intraneuronal pH, and thereby an increase in $[HCO_3^-]_i$ leading to a positive shift in E_{IPSP} (see Fig. 4). However, at least in some preparations studied (see Iles and Jack, 1980), the concentration required to produce an effect on E_{IPSP} is rather low to be compatible with a mechanism involving an effect on pH_i. An alternative possibility to account for the influence of ammonium salts on E_{IPSP} is that the charged species (NH_4^+) inhibits Cl^- extrusion on K^+,Cl^- cotransport which leads to an increase in Cl_i^- and to a positive shift in E_{Cl} (Lux, 1971; Deisz and Lux, 1982; Aickin *et al.*, 1982). The possibility of such a "cationic" mechanism (Aickin *et al.*, 1982) is due to the chemical similarity between NH_4^+ and alkali cations (especially K^+), and it complicates the interpretation of the effects of NH_3/NH_4^+ on E_{IPSP}. In future work, it will be interesting to examine the effects on E_{IPSP} and on nervous excitability of weak bases which, in their protonated form, do not have potassium-like effects on membrane transporters and channels.

5.2. Sensitivity of Ion Channels to Changes in Intracellular and Extracellular pH

It has been known for a long time that the activity of several enzymes shows a steep dependence on pH, and this phenomenon is likely to play a central role in the regulation of various metabolic processes (e.g., Busa, 1986). Therefore, it is not surprising that the properties of various ion channels are strongly affected by changes in both intracellular (Brodwick and Eaton, 1978; Moody, 1980, 1984) and extracellular pH. For instance, an intracellular acidosis inhibits delayed K^+ currents in squid axons (Wanke et al., 1979) and snail neurons (Meech, 1979). A fall in pH_i also enhances Na^+ inactivation and depresses peak Na^+ currents in squid axons (Wanke et al., 1980; Carbone et al., 1981). The possibility that a transmitter-induced bicarbonate conductance might have an effect on excitability, mediated by the pH_i sensitivity of ion channels, ought to be investigated in future work.

Perhaps even more dramatic effects are seen in some preparations following a change in pH_o. In various types of neurons, a rapid fall in pH_o gives rise to a Na^+ current (e.g., Krishtal and Pidoplichko, 1980; Gruol et al., 1980) which is due to a proton-induced, reversible transformation of voltage-dependent Ca^{2+} channels to H^+-activated, Na^+-permeable channels (Konnerth et al., 1987). In experiments on chick dorsal root ganglion cells, it was found that the H^+-induced Na^+ conductance is inactivated by small decrements in pH_o, and (at a physiological $[Ca^{2+}]_o$) half-maximum inactivation occurs at pH 7.3 (Konnerth et al., 1987). This indicates that the transformation process is sensitive to small fluctuations of pH_o around its physiological level.

A particularly interesting target for a modulating influence of a bicarbonate conductance is provided by certain GABA-activated channels which themselves show a sensitivity to changes in pH_o. For instance, in crayfish muscle fibers, an increase in pH_o produces a decrease in the GABA-activated anion conductance (Takeuchi and Takeuchi, 1967). In the presence of bicarbonate, such a pH sensitivity of the postsynaptic conductance might provide a basis for negative feedback, where a progressive increase in postsynaptic anion conductance would damp itself due to a simultaneous rise in pH_o.

5.3. Influence of a Postsynaptic Acid Load on Intracellular Ion Activities

As explained in detail above, activation of a transmitter-gated bicarbonate conductance imposes a postsynaptic, intracellular acid load. Even if the postsynaptic neuron is equipped with a powerful (or, perhaps more accurately, with a steeply pH_i-dependent) acid-extrusion mechanism capable of maintaining pH_i close to its steady-state level even during the activation of a HCO_3^- conductance, the acid-extrusion process itself may bring about significant changes in the activities of intracellular ions, such as Na^+ and Cl^-.

If extrusion of acid equivalents in the postsynaptic cell is mediated by Cl^-/HCO_3^- exchange, a postsynaptic acid load can lead to a decrease in a_{Cl}^i and, consequently, to a negative shift in E_{Cl}. If, on the other hand, acid extrusion is mediated by a Na^+-dependent mechanism (such as Na^+/H^+ exchange or Na^+-dependent Cl^-/HCO_3^- exchange), a postsynaptic acid load will give rise to an increase in a_{Na}^i. Again, an increase in Na_i^+ might induce a negative shift in E_{Cl} by acting on a Na^+,K^+,Cl^- cotransport system.

It is obvious that, at the present time, the above problems have not been addressed experimentally. However, in future work activation of a bicarbonate conductance may prove to be a widely spread property of inhibitory transmission. Therefore, we call attention to the fact that HCO_3^- can act in three different roles in the postsynaptic neuron: as a carrier of current, as a carrier of conductive fluxes of acid equivalents, and as a substrate (real or apparent) of acid-extrusion mechanisms.

ACKNOWLEDGMENT. The original research work of the authors has been supported by grants from the Academy of Finland.

REFERENCES

Aickin, C. C., Deisz, R. A., and Lux, H. D., 1982, Ammonium action on post-synaptic inhibition in crayfish neurones: Implications for the mechanism of chloride extrusion, *J. Physiol. (London)* **329:** 319–339.

Alger, E., and Nicoll, R. A., 1979, GABA-mediated biphasic inhibitory responses in hippocampus, *Nature* **281:**315–317.

Alvarez-Leefmans, F. J., Gamiño, S. M., Giraldez, F., and Noguerón, I., 1988, Intracellular chloride regulation in amphibian dorsal root ganglion neurones studied with ion-selective microelectrodes, *J. Physiol. (London)* **406:**225–246.

Ammann, D., 1986, *Ion-Selective Microelectrodes,* Springer-Verlag, Berlin.

Araki, T., Ito, M., and Oscarsson, O., 1961, Anion permeability of the synaptic and non-synaptic motoneurone membrane, *J. Physiol. (London)* **159:**410–435.

Aronson, P. S., 1985, Kinetic properties of the plasma membrane Na^+-H^+ exchanger, *Annu. Rev. Physiol.* **47:**545–560.

Atwood, H. L., 1976, Organization and synaptic physiology of crustacean neuromuscular systems, *Prog. Neurobiol.* **7:**291–391.

Balestrino, M., and Somjen, G. G., 1988, Concentration of carbon dioxide, interstitial pH and synaptic transmission in hippocampal formation of the rat, *J. Physiol. (London)* **396:**247–266.

Ballanyi, K., and Grafe, P., 1985, An intracellular analysis of γ-aminobutyric-acid associated ion movements in rat sympathetic neurones, *J. Physiol. (London)* **365:**41–58.

Barker, J. L., and Ransom, B. R., 1978, Amino acid pharmacology of mammalian central neurones grown in tissue culture, *J. Physiol. (London)* **280:**331–354.

Boistel, J., and Fatt, P., 1958, Membrane permeability change during inhibitory transmitter action in crustacean muscle, *J. Physiol. (London)* **144:**176–191.

Bormann, J., Hamill, O. P., and Sakmann, B., 1987, Mechanism of anion permeation through channels gated by glycine and γ-aminobutyric acid in mouse cultured spinal neurones, *J. Physiol. (London)* **385:** 243–286.

Boron, W. F., and Russell, J. M., 1983, Stoichiometry and ion dependencies of the intracellular-pH-regulating mechanism in squid giant axons, *J. Gen. Physiol.* **81:**373–399.

Boron, W. F., Hogan, E., and Russell, J. M., 1988, pH-sensitive activation of the intracellular-pH regulation system in squid axons by ATP-γ-S, *Nature* **332:**262–265.

Brodwick, M. S., and Eaton, D. C., 1978, Sodium channel inactivation in squid axon is removed by high internal pH or tyrosine-specific reagents, *Science* **200:**1494–1496.

Busa, W. B., 1986, Mechanisms and consequences of pH-mediated cell regulation, *Annu. Rev. Physiol.* **48:** 389–402.

Byerly, L., Meech, R. W., and Moody, W. J., 1984, Rapidly activating hydrogen ion currents in perfused neurones of the snail *Lymnaea stagnalis, J. Physiol. (London)* **351:**199–216.

Carbone, E., Testa, P. L., and Wanke, E., 1981, Intracellular pH and ionic channels in the *Loligo vulgaris* giant axon, *Biophys. J.* **35:**393–413.

Chesler, M., and Chan, C. Y., 1988, Stimulus-induced extracellular pH transients in the *in vitro* turtle cerebellum, *Neuroscience* **37:**941–948.

Chesler, M., and Kraig, R. P., 1987, Intracellular pH of astrocytes increases rapidly with cortical stimulation, *Am. J. Physiol.* **253:**R666–R670.

Coombs, J. S., Eccles, J. C., and Fatt, P., 1955, The specific ionic conductances and the ionic movements across the motoneuronal membrane that produce the inhibitory post-synaptic potential, *J. Physiol. (London)* **130:**326–373.

Deisz, R. A., and Lux, H. D., 1982, The role of intracellular chloride in hyperpolarizing post-synaptic inhibition of crayfish stretch receptor neurones, *J. Physiol. (London)* **326:**123–138.

Dudel, J., 1977, Voltage dependence of amplitude and time course of inhibitory synaptic current in crayfish muscle, *Pfluegers Arch.* **371:**167–174.

Dudel, J., and Rüdel, R., 1969, Voltage controlled contractions and current voltage relations of crayfish muscle fibers in chloride-free solutions, *Pfluegers Arch.* **308:**291–314.

Dudel, J., Finger, W., and Stettmeier, H., 1980, Inhibitory synaptic channels activated by γ-aminobutyric acid (GABA) in crayfish muscle, *Pfluegers Arch.* **387:**143–151.

Dutar, P., and Nicoll, R. A., 1988, A physiological role for GABA_B receptors in the central nervous system, *Nature* **332:**156–158.

Eccles, J. C., 1964, *The Physiology of Synapses,* Springer, Berlin.

Edwards, C., 1982, The selectivity of ion channels in nerve and muscle, *Neuroscience* **6:**1335–1366.

Endres, W., Grafe, P., Bostock, H., and ten Bruggencate, G., 1986, Changes in extracellular pH during electrical stimulation of isolated vagus nerve, *Neurosci. Lett.* **64:**201–205.

Gallagher, J. P., Nakamura, J., and Shinnick-Gallagher, P., 1983, The effects of temperature, pH and Cl-pump inhibitors on GABA responses recorded from cat dorsal root ganglia, *Brain Res.* **267:**249–259.

Galler, S., and Moser, H., 1986, The ionic mechanism of intracellular pH regulation in crayfish muscle fibres, *J. Physiol. (London)* **374:**137–151.

Gerschenfeld, H. M., 1973, Chemical transmission in invertebrate central nervous systems and neuromuscular junctions, *Physiol. Rev.* **53:**1–119.

Goldman, D. E., 1943, Potential, impedance, and rectification in membranes, *J. Gen. Physiol.* **27:**37–60.

Gruol, D. L., Barker, J. L., Huang, L.-Y.M., MacDonald, J. F., and Smith, T. G., 1980, Hydrogen ions have multiple effects on the excitability of cultured mammalian neurons, *Brain Res.* **183:**247–252.

Gutknecht, J., and Tosteson, D. C., 1973, Diffusion of weak acids across lipid bilayer membranes: Effects of chemical reactions in the unstirred layers, *Science* **182:**1258–1261.

Hansen, A. J., and Zeuthen, T., 1981, Extracellular ion concentrations during spreading depression and ischemia in the rat brain cortex, *Acta Physiol. Scand.* **113:**437–445.

Hodgkin, A. L., and Katz, B., 1949, The effect of sodium ions on the electrical activity of the giant axon of the squid, *J. Physiol. (London)* **108:**37–77.

Iles, J. F., and Jack, J. J. B., 1980, Ammonia: Assessment of its action on postsynaptic inhibition as a cause of convulsions, *Brain* **103:**555–578.

Ito, M., Kostyuk, P. G., and Oshima, T., 1962, Further study on anion permeability of inhibitory postsynaptic membrane of cat motoneurones, *J. Physiol. (London)* **164:**150–156.

Iwasaki, S., and Florey, E., 1969, Inhibitory miniature potentials in the stretch receptor neurons of crayfish, *J. Gen. Physiol.* **53:**666–682.

Jack, J. J. B., Noble, D., and Tsien, R. W., 1975, *Electric Current Flow in Excitable Cells,* Oxford University Press (Clarendon), London.

Kaila, K., 1988, GABA-activated movements of formate and acetate: Influence on intracellular pH and surface pH in crayfish skeletal muscle fibres, *Ciba Found. Symp.* **139:**184–186.

Kaila, K., and Voipio, J., 1987, Postsynaptic fall in intracellular pH induced by GABA-activated bicarbonate conductance, *Nature* **330:**163–165.

Kaila, K., Mattsson, K., and Voipio, J., 1989a, Fall in intracellular pH and increase in resting tension induced by a mitochondrial uncoupling agent in crayfish muscle, *J. Physiol. (London)* **408:**271–293.

Kaila, K., Pasternack, M., Saarikoski, J., and Voipio, J., 1989b, Influence of GABA-gated bicarbonate conductance on potential, current and intracellular chloride in crayfish muscle fibers, *J. Physiol. (London)* **416:**161–181.

Kaila, K., Saarikoski, J., and Voipio, J., 1990, Mechanism of action of GABA on intracellular pH and on surface pH in crayfish muscle fibers, (submitted).

Katz, B., 1969, *The Release of Neural Transmitter Substances,* Liverpool University Press, Liverpool.

Kelly, J. S., Krnjević, K., Morris, M. E., and Yim, G. K. W., 1969, Anionic permeability of cortical neurones, *Exp. Brain Res.* **7:**11–31.

Kerkut, G. A., and Thomas, R. C., 1964, The effect of anion injection and changes in the external potassium and chloride concentration on the reversal potentials of the IPSP and acetylcholine, *Comp. Biochem. Physiol.* **11:**199–213.

Konnerth, A., Lux, H. D., and Morad, M., 1987, Proton-induced transformation of calcium channel in chick dorsal root ganglion cells, *J. Physiol. (London)* **386:**603–633.

Kraig, R. P., Ferreira-Filho, C. S., and Nicholson, C., 1983, Alkaline and acid transients in cerebellar microenvironment, *J. Neurophysiol.* **49:**831–850.

Krishtal, O. A., and Pidoplichko, V. I., 1980, A receptor for protons in the nerve cell membrane, *Neuroscience* **5:**2325–2327.

Krishtal, O. A., Osipchuk, Y. V., Shelest, T. N., and Smirnoff, S. V., 1987, Rapid extracellular pH transients related to synaptic transmission in rat hippocampal slices, *Brain Res.* **436:**352–356.

Krnjević, K., 1974, Chemical nature of synaptic transmission in vertebrates, *Physiol. Rev.* **54:**418–540.

Kuffler, S. W., and Eyzaguirre, C., 1955, Synaptic inhibition in an isolated nerve cell, *J. Gen. Physiol.* **39:** 155–184.

Llinás, R., Baker, R., and Precht, W., 1974, Blockage of inhibition by ammonium acetate action on chloride pump in cat trochlear motoneurons, *J. Neurophysiol.* **37:**522–532.

Lux, H. D., 1971, Ammonium and chloride extrusion: Hyperpolarizing synaptic inhibition in spinal motoneurons, *Science* **173:**555–557.

Lux, H. D., Loracher, C., and Neher, E., 1970, The action of ammonium on postsynaptic inhibition of cat spinal motoneurons, *Exp. Brain Res.* **11:**431–447.

McLaughlin, S. G. A., and Dilger, J. P., 1980, Transport of protons across membranes by weak acids, *Physiol. Rev.* **60:**825–863.

Mahnensmith, R. L., and Aronson, P. S., 1985, The plasma membrane sodium–hydrogen exchanger and its role in physiological and pathophysiological processes, *Circ. Res.* **56:**773–788.

Mason, M. J., Mattsson, K., Pasternack, M., Voipio, J., and Kaila, K., 1990, Postsynaptic fall in intracellular pH and increase in surface pH caused by efflux of formate and acetate through GABA-gated channels in crayfish muscle fibres, *Neuroscience* (in press).

Meech, R. W., 1979, Membrane potential oscillations in molluscan "burster" neurones, *J. Exp. Biol.* **81:** 93–112.

Meech, R. W., and Thomas, R. C., 1987, Voltage-dependent intracellular pH in *Helix aspersa* neurones, *J. Physiol. (London)* **390:**433–452.

Meier, P. C., Ammann, D., Morf, W. E., and Simon, W., 1980, Liquid-membrane ion-sensitive electrodes and their biomedical applications, in: *Medical and Biomedical Applications of Electrochemical Devices* (J. Koryta, ed.), Wiley, New York, pp. 13–91.

Melnik, V. I., Glebow, R. N., and Kryhanovski, G. N., 1985, ATP-dependent translocation of protons across the membrane of rat brain synaptic vesicles, *Bull. Exp. Biol. Med.* **99:**35–38.

Moody, W. J., 1980, Appearance of calcium action potentials in crayfish slow muscle fibres under conditions of low intracellular pH, *J. Physiol. (London)* **302:**335–346.

Moody, W. J., 1984, Effects of intracellular H^+ on the electrical properties of excitable cells, *Annu. Rev. Neurosci.* **7:**257–278.

Motokizawa, F., Reuben, J. P., and Grundfest, H., 1969, Ionic permeability of the inhibitory postsynaptic membrane of lobster muscle fibers, *J. Gen. Physiol.* **54:**437–461.

Mutch, W. A. C., and Hansen, A. J., 1984, Extracellular pH changes during spreading depression and cerebral ischemia: Mechanisms of brain pH regulation, *J. Cerebr. Blood Flow Metabol.* **4:**17–27.

Phillips, J. M., and Nicholson, C., 1979, Anion permeability in spreading depression investigated with ion-sensitive microelectrodes, *Brain Res.* **173**:567–571.

Phillis, J. W., and Ochs, S., 1971, Excitation and depression of cortical neurones during spreading depression, *Exp. Brain Res.* **12**:132–149.

Pontén, U., and Siesjö, B. K., 1966, Gradients of CO_2 tension in brain, *Acta Physiol. Scand.* **67**:129–140.

Raabe, W., and Gumnit, R. J., 1975, Disinhibition in cat motor cortex by ammonia, *J. Neurophysiol.* **38**:347–355.

Rice, M. E., and Nicholson, C., 1988, Behavior of extracellular K^+ and pH in skate (*Raja erinacea*) cerebellum, *Brain Res.* **461**:328–334.

Robinson, R. A., and Stokes, R. H., 1959, *Electrolyte Solutions,* Butterworths, London.

Roos, A., and Boron, W. F., 1981, Intracellular pH, *Physiol. Rev.* **61**:296–433.

Russell, J. M., 1983, Cation-coupled chloride influx in squid axon. Role of potassium and stoichiometry of the transport process, *J. Gen. Physiol.* **81**:909–925.

Sharp, A. P., and Thomas, R. C., 1981, The effects of chloride substitution on intracellular pH in crab muscle, *J. Physiol. (London)* **312**:71–80.

Siggaard-Andersen, O., 1974, *The Acid–Base Status of the Blood,* 4th ed., Munskgaard, Copenhagen.

Siggins, G. R., and Gruol, D. L., 1986, Mechanisms of transmitter action in the vertebrate central nervous system, in: *Handbook of Physiology,* Section I, *The Nervous System,* Volume IV (F. E. Bloom, ed.), American Physiological Society, Bethesda, pp. 1–114.

Spuler, A., Endres, W., and Grafe, P., 1987, Metabolic origin of activity-related pH-changes in mammalian peripheral and central unmyelinated fibre tracts, *Pfluegers Arch.* **408**:R69 (abstract).

Stadler, H., and Tsukita, S., 1984, Synaptic vesicles contain an ATP-dependent proton pump and show 'knob-like' protrusions on their surface, *EMBO J.* **3**:3333–3337.

Syková, E., 1988, Extracellular pH and stimulated neurons, *Ciba Found. Symp.* **139**:220–235.

Takeuchi, A., and Takeuchi, N., 1965, Localized action of gamma-amino butyric acid on the crayfish muscle, *J. Physiol. (London)* **177**:225–238.

Takeuchi, A., and Takeuchi, N., 1967, Anion permeability of the inhibitory post-synaptic membrane of the crayfish neuromuscular junction, *J. Physiol. (London)* **191**:575–590.

Takeuchi, A., and Takeuchi, N., 1969, A study of the action of picrotoxin on the inhibitory neuromuscular junction of the crayfish, *J. Physiol. (London)* **205**:377–391.

Thomas, R. C., 1976, The effect of carbon dioxide on the intracellular pH and buffering power of snail neurones, *J. Physiol.* **255**:715–735.

Thomas, R. C., 1984, Experimental displacement of intracellular pH and the mechanism of its subsequent recovery, *J. Physiol. (London)* **354**:3P–22P.

Thomas, R. C., 1988, Changes in the surface pH of voltage-clamped snail neurones apparently caused by H^+ fluxes through a channel, *J. Physiol. (London)* **398**:313–327.

Thomas, R. C., and Meech, R. W., 1982, Hydrogen ion currents and intracellular pH in depolarized voltage-clamped snail neurones, *Nature* **299**:826–828.

Urbanics, R., Leniger-Follert, E., and Lübbers, D. W., 1978, Time course of changes of extracellular H^+ and K^+ activities during and after direct electrical stimulation of the brain cortex, *Pfluegers Arch.* **378**:47–53.

Vanheel, B., De Hemptinne, A., and Leusen, I., 1986, Influence of surface pH on intracellular pH regulation in cardiac and skeletal muscle, *Am. J. Physiol.* **250**:C748–C760.

Voipio, J., Rydqvist, B., and Kaila, K., 1988, The reversal potential of GABA-activated current (E_{GABA}) may be sensitive to metabolic production of CO_2/HCO_3, *Eur. J. Neurosci.* ENA-Suppl. p. 61 (abstract).

Wanke, E., Carbone, E., and Testa, P. L., 1979, K^+ conductance modified by a titratable group accessible to protons from the intracellular side of the squid axon membrane, *Biophys. J.* **26**:319–324.

Wanke, E., Carbone, E., and Testa, P. L., 1980, The sodium channel and intracellular H^+ blockage in squid axons, *Nature* **287**:62–63.

Wichser, J., and Kazemi, H., 1975, CSF bicarbonate regulation in respiratory acidosis and alkalosis, *J. Appl. Physiol.* **38**:504–512.

B

Ca²⁺-Activated Cl⁻ Channels

Calcium-Dependent Chloride Currents in Vertebrate Central Neurons

Mark L. Mayer, David G. Owen, and Jeffrey L. Barker

1. INTRODUCTION

Anion-selective ion channels activated by inhibitory neurotransmitters such as γ-aminobutyric acid (GABA) and glycine occur in the soma-dendritic membrane of a majority of CNS neurons (see Chapters 7–9). Until recently, these were the only well-characterized anion-selective channels in vertebrate neurons and there was no evidence for any additional types of anion channel, although the existence of a leakage conductance permeable to Cl^- was considered to be probable. Some new work, on a variety of preparations ranging from amphibian egg cells to mammalian CNS neurons in tissue culture, has changed this picture with the discovery of a conductance mechanism that is selectively permeable to anions and activated by a rise in Ca_i^{2+} activity. The study of this conductance mechanism is in its infancy and, although some sophisticated biophysical techniques have been applied to the problem, the small conductances of these ion channels, our lack of selective pharmacological probes, and the added complexity of studying processes linked to changes in Ca_i^{2+} activity, have prevented any real understanding of the physiological functions of Ca^{2+}-dependent Cl^- channels in nerve cells. Similar studies in exocrine cells that secrete electrolyte solutions suggest that Ca^{2+}-activated Cl^- fluxes may participate in the production of tears in the lacrimal gland (Marty *et al.*, 1984).

Ca^{2+}-dependent Cl^- currents were first observed in experiments on rod inner segments isolated from salamander retina (Bader *et al.*, 1982), and showed several characteristic features that have been observed in all subsequent studies on vertebrate nerve cells: (1) slow activation evoked by depolarizing voltage steps to membrane potentials known to trigger Ca^{2+} entry, coupled with lack of pronounced inactivation during voltage steps that are hundreds of milliseconds in duration; (2) prolonged

Mark L. Mayer • Unit of Neurophysiology and Biophysics, Laboratory of Developmental Neurobiology, National Institute of Child Health and Human Development, National Institutes of Health, Bethesda, Maryland 20892. *David G. Owen and Jeffrey L. Barker* • Laboratory of Neurophysiology, National Institute of Neurological Disorders and Stroke, National Institutes of Health, Bethesda, Maryland 20892. *Present address of D.G.O.:* Electrophysiology Section, Wyeth Research UK Ltd, Taplow Near Slough, Bucks, United Kingdom.

inward tail currents on repolarization to -60 mV, when the intracellular anion concentration is raised above physiological values; (3) sensitivity to extracellular Ca^{2+} channel blockers.

Ca^{2+}-activated Cl^- currents were also discovered in experiments on *Xenopus* oocytes (Miledi, 1982; Barish, 1983) and in this preparation are expressed as a transient outward current activated by depolarization. Since intracellular injection of Ca^{2+} evokes a prolonged activation of the anion-selective conductance (Miledi and Parker, 1984), the transient response evoked by depolarization most likely reflects inactivation of Ca^{2+} entry. *Xenopus* oocyte membranes also contain muscarinic acetylcholine receptors that appear to trigger intracellular synthesis of polyphosphoinositides, with subsequent release of Ca^{2+} stores and activation of a Ca^{2+}-dependent Cl^- current (Kusano *et al.*, 1982; Oron *et al.*, 1985).

2. ISOLATION OF Ca^{2+}-DEPENDENT Cl^- CURRENT IN CNS NEURONS

The existence of several cation-selective conductance mechanisms activated by a rise in Ca_i^{2+} complicates the study of Ca^{2+}-dependent Cl^- currents, and this is why the latter currents were not discovered until recently in vertebrate neurons. Two procedures initially helped to offset this problem: (1) raising the $[Cl^-]_i$ above physiological values, such that at membrane potentials at which Ca^{2+}-activated K^+ conductances give rise to outward currents, Ca^{2+}-dependent Cl^- currents are inward; (2) block of outward K^+ currents with intracellular injection of Cs^+ and extracellular application of TEA^+. Using the first approach, slow depolarizing responses and prolonged inward tail currents due to activation by Ca^{2+}-dependent Cl^- currents were recorded in CNS neurons from the dorsal root ganglion and spinal cord (Mayer, 1985; Owen *et al.*, 1986). Typical examples are shown in Fig. 1. Of interest is the response triggered by a single action potential, suggesting that activation of the Ca^{2+}-dependent Cl^- current may occur during the normal electrical activity of CNS neurons.

If Ca^{2+}-activated Cl^- currents are activated as easily as the data in Fig. 1C suggest, some explanation is required for their relatively recent discovery. It seems likely that under physiological conditions in which the Cl^- equilibrium potential lies close to the resting potential, and in which rises in Ca_i^{2+} activate large outward K^+ currents, the simultaneous activation of a smaller Ca^{2+}-dependent Cl^- current would be hard to detect. Raising the a_{Cl}^i provides an experimental technique for recording Ca^{2+}-dependent Cl^- currents without the use of K^+ channel blockers, but is not an unambiguous means of identification, since activation of Ca^{2+}-dependent, nonselective cation-permeable channels could also generate an inward current (Kramer and Zucker, 1985). Under appropriate ionic conditions, it is possible to differentiate between inward currents carried by cation entry versus those carried by anion exit and thus to distinguish between the above conductance mechanisms. The appropriate experiments would be replacement of Na^+ and K^+ by salts of *N*-methyl-D-glucamine, a very large cation not believed to permeate Ca^{2+}-activated, nonselective cation-permeable channels, or replacement of extracellular NaCl with mannitol or sucrose. It

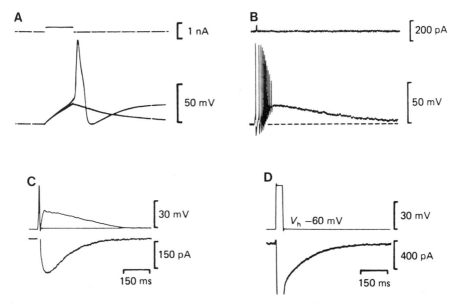

Figure 1. Slow excitatory responses mediated by $I_{Cl(Ca)}$. A, B and C show responses triggered by single action potentials in rat dorsal root ganglion neurons. In A and B a slowly rising depolarizing spike afterpotential was recorded using patch electrodes containing 140 mM KCl. This depolarizing response is excitatory, as shown by the burst of action potentials triggered in the trace shown in B: the first action potential was triggered by a brief current pulse injected through the recording electrode; the depolarizing afterpotentials triggered by this and subsequent action potentials summed to produce a sustained depolarization and high-frequency firing. Termination of the burst discharge probably reflects a combination of Na^+ current inactivation, and a rise in membrane conductance due to activation of I_K and $I_{Cl(Ca)}$. C and D show voltage-clamp recordings of slow inward tail currents triggered by a single action potential (C) and, in a Cs^+-loaded cell, a depolarizing voltage jump (D). The convenience of voltage-clamp recording, combined with pharmacological blockade of K^+ currents by intracellular perfusion with Cs^+ and extracellular application of K^+ channel blockers was used as an alternative to action potentials for activation of $I_{Cl(Ca)}$ in most experiments. (Modified from Mayer, 1985.)

seems sufficient to say that in intact cells the activation of calcium of several other conductance mechanisms greatly hinders the study of Ca^{2+}-activated Cl^- currents.

2.1. Ionic Mechanism

In sensory neurons from dorsal root ganglia, activation of the Ca^{2+}-dependent Cl^- current ($I_{Cl(Ca)}$) was studied using whole-cell recording. In this recording mode, the cell could be perfused in order to raise the a^i_{Cl}, and to block K^+ channels by perfusion with Cs^+. By employing a low concentration of EGTA in the intracellular solution, relatively small electrode tips, and a discontinuous voltage clamp, it was possible to record stable Ca^{2+}-dependent Cl^- currents for periods of at least 30 min. With this approach it was possible to show the slow activation and deactivation kinetics typical of Ca^{2+}-dependent anion conductances, and a U-shaped activation curve that peaked at $+10$ mV, and declined with further depolarization, suggesting dependence on prior Ca^{2+} entry (see Figs. 2 and 3B,C).

The sensitivity of $I_{Cl(Ca)}$ to the removal of Ca_o^{2+} or to block of Ca^{2+} entry with Cd^{2+} or Co^{2+} further suggests that activation depends on prior uptake of Ca^{2+} (Fig. 3), as does potentiation of $I_{Cl(Ca)}$ on raising the Ca_o^{2+}. In *Xenopus* oocytes, injection of Ca^{2+}, via an intracellular microelectrode, was used to demonstrate directly the Ca^{2+} dependence of $I_{Cl(Ca)}$ (Miledi and Parker, 1984). Together, such results leave little doubt concerning activation of $I_{Cl(Ca)}$ by a rise in $[Ca^{2+}]_i$.

The ion responsible for this slowly activating conductance mechanism in sensory ganglion and spinal cord neurons was initially identified by a process of eliminating a change in cation conductance, in conjunction with the expected behavior on reducing $[Cl^-]_o$ (Mayer, 1985; Owen *et al.*, 1986). In sensory ganglion and spinal cord neurons, with $[Cl^-]_i$ raised to around 140 mM, $I_{Cl(Ca)}$ is recorded as a slow, inward tail current, the reversal potential of which was estimated at approximately -10 mV in

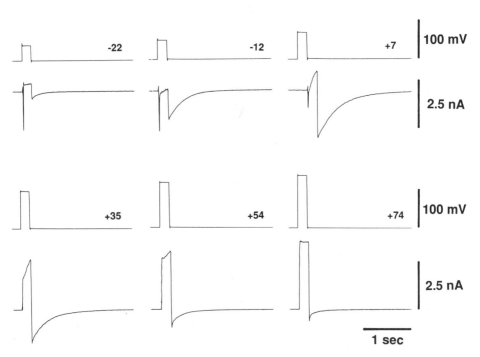

Figure 2. Activation of Ca^{2+}-dependent Cl^- currents in a sensory neuron perfused with CsCl under whole-cell voltage-clamp. Responses to depolarizing voltage steps 100 msec in duration were recorded from a holding potential of -60 mV. Activation of $I_{Cl(Ca)}$ occurs in a graded fashion (traces at -22, -12, and $+7$ mV), and is visible as a slowing rising outward relaxation, followed by a prolonged inward tail current on repolarization to -60 mV. With further depolarization the amplitude of the outward relaxation during the step decreases, and the tail current is of reduced amplitude; at $+74$ mV both responses are virtually abolished. Such results are reminiscent of the activation of Ca^{2+}-dependent K^+ current, and suggest activation of the inward tail current carried by efflux of Cl^- to be linked to prior influx of Ca^{2+} through voltage-activated Ca^{2+} channels.

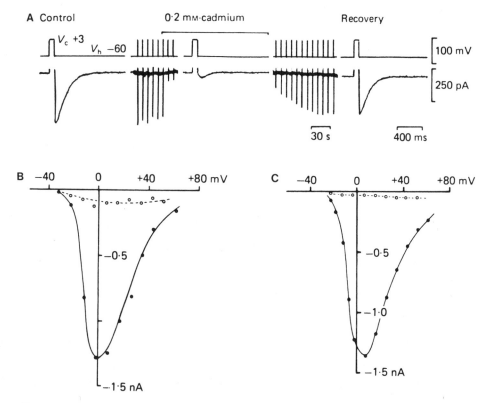

Figure 3. Calcium dependence of $I_{Cl(Ca)}$. A shows inward tail currents due to $I_{Cl(Ca)}$ recorded in a sensory neuron perfused with a KCl pipette solution, without the use of K⁺ channel blockers. Application of 200 μM cadmium reversibly blocks $I_{Cl(Ca)}$. B and C show activation curves for $I_{Cl(Ca)}$ measured from tail current data similar to those in Fig. 2; application of 2 mM cobalt (B), or replacement of external calcium by magnesium (C), also reversibly blocked $I_{Cl(Ca)}$. In both B and C, the filled circles represent control data and the open circles represent the test data. (Modified from Mayer, 1985.)

sensory neurons and −1 mV in spinal cord neurons (Fig. 4a). Replacement of $[Na^+]_o$ with tetraethylammonium or choline did not cause the hyperpolarizing shift in reversal potential expected for an inward current mediated by Na⁺ entry (Fig. 4C,D), while lowering $[Cl^-]_o$ by replacing Cl⁻ with isethionate or mannitol did cause the increase in tail current expected for an inward current mediated by Cl⁻ exit (Fig. 4B). Gradual collapse of the large transmembrane Cl⁻ gradient when $[Cl^-]_o$ was reduced complicated measurement of the change in reversal potential in these experiments. However, under better controlled experimental conditions in lacrimal gland cells, Evans and Marty (1986) were able to use reversal potentials to measure selectivity of the Ca²⁺-activated anion conductance for a series of seven anions, leaving little doubt that $I_{Cl(Ca)}$ is indeed a Cl⁻ current.

In the majority of nerve cells, E_{Cl} normally lies a few millivolts hyperpolarized to the resting potential, the primary exceptions being sensory and autonomic ganglion

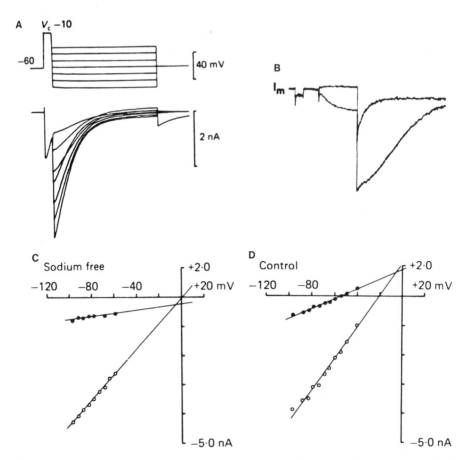

Figure 4. Ionic mechanism of $I_{Cl(Ca)}$. A shows the protocol used to estimate the reversal potential of I $_{Cl(Ca)}$: a brief depolarizing step from -60 to -10 mV was followed by repolarization to test potentials ranging from -30 to -90 mV; tail current was measured 20 and 800 msec after jumping to the test potential. Extrapolation of the resultant $I-V$ relationships gave the reversal potential for $I_{Cl(Ca)}$. B shows two superimposed traces recorded from a spinal cord neuron evoked using a similar protocol (holding potential -50 mV; activation at -20 mV; test potential -90 mV) recorded with 158 and 37 mM $[Cl^-]_o$, using isethionate as a Cl^- substitute. A depolarizing shift in the reversal potential for $I_{Cl(Ca)}$ is indicated by the inward relaxation at -20 mV, and the increase in tail current at -90 mV on lowering $[Cl^-]_o$. (C, D) $I-V$ plots from sensory ganglion neurons obtained by the protocol shown in A. Tail currents were measured 20 msec (○) and 800 msec (●) after jumping to the test potential. The reversal potential for $I_{Cl(Ca)}$ is represented by the intersection of the two lines. As can be seen, there was no difference ($E_{rev} = 5$ mV) when the cells were superfused with 145 mM TEA⁺ (C) or 145 mM Na⁺ (D). (Modified from Mayer, 1985.)

neurons in which active Cl^- uptake raises $[Cl^-]_i$ such that E_{Cl} is more positive than the resting potential (see Chapter 3). Although the physiological conditions that allow activation of $I_{Cl(Ca)}$ in nerve cells have yet to be determined, it seems probable that just as activation of an anion conductance by GABA and glycine causes inhibition in central neurons, so too will activation of $I_{Cl(Ca)}$. In sensory ganglion neurons, activation of $I_{Cl(Ca)}$ would be expected to be weakly depolarizing, and when $[Cl^-]_i$ is raised experimentally this is readily demonstrated (Fig. 1A–C).

2.2. Activation and Inactivation

In neurons from both spinal cord and sensory ganglia the activation of $I_{Cl(Ca)}$ follows relatively slow kinetics: during depolarizing voltage jumps to between -10 and $+10$ mV, activation can be described by a single-exponential process with a time constant of around 100 msec (Fig. 5A). Little inactivation occurs during depolarizations lasting several seconds, in contrast to the transient activation of $I_{Cl(Ca)}$ in *Xenopus* oocytes. Deactivation shows more complex kinetics. Following relatively brief periods of activation (10- to 60-msec steps to 0 mV), the decay of $I_{Cl(Ca)}$ follows single-exponential kinetics, with a time constant of around 200 msec at -60 mV (Fig. 5B). Membrane potential hyperpolarization increases the rate of deactivation of $I_{Cl(Ca)}$ such that, on average, the deactivation rate constant increases three-fold per 118 mV hyperpolarization. With longer periods of activation, the tail current decay shows complex kinetics. A progressive slowing of the decay kinetics with longer periods of prior activation, which is the general trend observed in cells from both spinal cord and sensory ganglion preparations, is complicated by the appearance of plateaus during which the decay is halted. It is likely that the cell's ability to regulate a^i_{Ca} following net uptake of large quantities of Ca^{2+} can greatly influence the decay kinetics of $I_{Cl(Ca)}$.

Such results raise some interesting questions. First, they emphasize that Ca^{2+} homeostasis has a strong influence on the kinetics of activation and deactivation of $I_{Cl(Ca)}$. This result suggests that Ca^{2+} homeostasis will play a pivotal role in determining when $I_{Cl(Ca)}$ becomes activated during physiological activity and, unfortunately, greatly complicates study of the conductance mechanism underlying $I_{Cl(Ca)}$ in intact

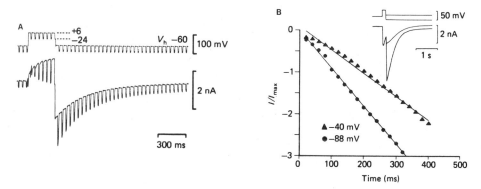

Figure 5. Activation of $I_{Cl(Ca)}$ is slow and voltage-dependent. A shows the increase in conductance accompanying activation of $I_{Cl(Ca)}$ during a step from -60 to $+6$ mV, 300 msec in duration, followed by slow deactivation on returning to -60 mV. B shows a semilog plot of the decay of $I_{Cl(Ca)}$ recorded at -40 and -88 mV (raw data shown in inset); the decay of $I_{Cl(Ca)}$ follows single-exponential kinetics and becomes faster with hyperpolarization, with time constants of 201 msec at -40 mV and 112 msec at -88 mV. (Modified from Mayer, 1985.)

cells. The slow decay kinetics of $I_{Cl(Ca)}$ following prolonged activation seems likely to reflect overloading of Ca^{2+} regulatory mechanisms, and perhaps even Ca^{2+}-triggered release of Ca^{2+} from intracellular stores. Under experimental conditions, the borderline between normal physiology and pathophysiological behavior is sometimes difficult to detect, but easy to cross. Second, these results suggest that the gating of

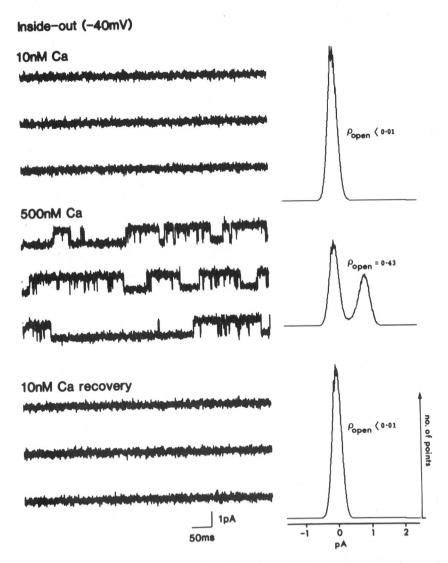

Figure 6. Single-channel recording of Ca^+-activated channels recorded in an inside-out patch from a spinal cord neuron bathed in symmetrical Tris-Cl solutions. When the cytoplasmic face of the patch was exposed to 10 nM Ca^{2+}, single-channel activity was absent (pipette potential −40 mV). Application of 500 nM Ca^{2+} activated a channel that showed bursts of openings of conductance 23 pS (current 0.9 pA).

$I_{Cl(Ca)}$ may be sensitive to the membrane potential, as is the gating of Ca²⁺-activated K⁺ currents; clearly further exploration of this requires control of $[Ca^{2+}]_i$.

Recently, Evans and Marty (1986) have studied the activation of $I_{Cl(Ca)}$ in lacrimal gland cells in which $[Ca^{2+}]_i$ was heavily buffered by perfusion with high concentrations of EDTA-Ca of known pCa. Their results confirm voltage-dependence of the gating of $I_{Cl(Ca)}$ which are suggested by the above results (see also Marty *et al.*, 1984); at 0.5 μM $[Ca^{2+}]_i$ the activation and deactivation time constants are single-exponential functions which change *e*-fold per 130 mV (see Fig. 5), and remain constant at any voltage even following long periods of activation, suggesting that the prolonged decay of $I_{Cl(Ca)}$ recorded in sensory and spinal cord neurons may have been due to poor control of $[Ca^{2+}]_i$.

2.3. Single-Channel Recording

At present, we do not have much information about the properties of the channels responsible for $I_{Cl(Ca)}$ in mammalian CNS neurons, but biophysical experiments on Ca²⁺-dependent Cl⁻ channels in *Xenopus* oocytes, using the patch-clamp technique, suggest that in this latter preparation these channels have a small conductance (3.6 pS in symmetrical 115 mM KCl). Fluctuation analysis of single-channel records gave prolonged open-time estimates, of 73 and 98 msec, in two patches exposed to 10 μM Ca²⁺ at the cytoplasmic face of the membrane (Takahashi *et al.*, 1987). Noise analysis of $I_{Cl(Ca)}$ currents obtained during whole-cell recording from lacrimal gland cells also gave estimates of 1–2 pS, and a lifetime of 230 msec during elevation of $[Ca^{2+}]_i$ to some unknown value in response to application of carbachol or the calcium ionophore A23187 (Marty *et al.*, 1984). In preliminary attempts to study the ion channels responsible for $I_{Cl(Ca)}$ in mammalian neurons, inside-out patches were isolated from spinal cord neurons, and bathed in Tris-Cl solutions on both faces of the membrane to eliminate current through cation-selective channels. Under these conditions, an ion channel of unitary conductance 23 pS, activated by pressure application of 0.5–1 μM Ca²⁺, was found in two patches clamped at +40 mV (Fig. 6). However, the increase in variance during $I_{Cl(Ca)}$ tail currents in mammalian neurons appears to be remarkably small, suggesting activation of another ion channel of lower conductance perhaps similar to that found in oocytes and lacrimal gland cells. Further work is required to resolve this discrepancy.

3. CONCLUSION

Biophysical experiments will be of great use in determining the Ca²⁺- and voltage-dependence of $I_{Cl(Ca)}$ in both intact cells and isolated membrane patches. For example, it would be helpful to know the peak $[Ca^{2+}]_i$ under the plasma membrane following a train of action potentials, and to know whether once activated by a brief $[Ca^{2+}]_i$ transient, $I_{Cl(Ca)}$ ion channels close with slow kinetics, as suggested by tail current recording following action potentials in sensory neurons. Clues as to the conditions required for activation of $I_{Cl(Ca)}$ in intact nerve cells should give some idea of the physiological processes $I_{Cl(Ca)}$ is likely to participate in. In this regard, it is of interest that data on lacrimal gland cells suggest that $I_{Cl(Ca)}$ channels require a higher $[Ca^{2+}]_i$ for activation than do larger-conductance Ca^{2+}-dependent channels (Marty *et al.*, 1984).

REFERENCES

Bader, C. R., Bertrand, D., and Schwartz, E. A., 1982, Voltage-activated and calcium-activated currents in solitary rod inner segments from the salamander retina, *J. Physiol. (London)* **331**:253–284.

Barish, M. E., 1983, A transient calcium-dependent chloride current in the immature *Xenopus* oocyte, *J. Physiol. (London)* **342**:309–325.

Evans, M. G., and Marty, A., 1986, Calcium-dependent chloride currents in isolated cells from rat lacrimal glands, *J. Physiol. (London)* **378**:437–460.

Kramer, R. H., and Zucker, R. S., 1985, Calcium-dependent inward current in Aplysia bursting pacemaker neurones, *J. Physiol. (London)* **362**:107–130.

Kusano, K., Miledi, R., and Stinnarke, J., 1982, Cholinergic and catecholaminergic receptors in the *Xenopus* oocyte membrane, *J. Physiol. (London)* **328**:143–170.

Marty, A., Tan, Y. P., and Trautmann, A., 1984, Three types of calcium-dependent channel in rat lacrimal glands, *J. Physiol. (London)* **357**:293–325.

Mayer, M. L., 1985, A calcium-activated chloride current generates the after-depolarization of rat sensory neurones in culture, *J. Physiol. (London)* **364**:217–239.

Miledi, R., 1982, A calcium-dependent transient outward current in *Xenopus laevis* oocytes, *Proc. R. Soc. London Ser. B* **215**:491–497.

Miledi, R., and Parker, I., 1984, Chloride current induced by injection of calcium into *Xenopus* oocytes, *J. Physiol. (London)* **357**:173–183.

Oron, Y., Dascal, N., Nadler, E., and Lupu, M., 1985, Inositol 1,4,5-trisphosphate mimics muscarinic responses in *Xenopus* oocytes, *Nature* **313**:141–143.

Owen, D. G., Segal, M., and Barker, J. L., 1986, Voltage-clamp analysis of a Ca^{2+}- and voltage-dependent chloride conductance in cultured mouse spinal neurons, *J. Neurophysiol.* **55**:1115–1135.

Takahashi, T., Neher, E., and Sakmann, B., 1987, Rat brain serotonin receptors in *Xenopus* oocytes are coupled by intracellular calcium to endogenous channels, *Proc. Natl. Acad. Sci. USA* **84**:611–615.

C

Voltage-Activated Cl⁻ Channels

Hyperpolarization-Activated Chloride Channels in Aplysia Neurons

Dominique Chesnoy-Marchais

1. INTRODUCTION

In Cl^--loaded *Aplysia* neurons, hyperpolarizing voltage jumps slowly activate an inward current that is carried by Cl^- ions (Chesnoy-Marchais, 1982, 1983). Recent reports suggest the existence of a similar Cl^- current in several other preparations. However, the more frequently studied inward currents activated by hyperpolarization are carried either by K^+ ions (Hagiwara *et al.*, 1976; Hestrin, 1981; Leech and Stanfield, 1981; Constanti and Galvan, 1983; Sakmann and Trube, 1984; Kurachi, 1985) or by both Na^+ and K^+ ions (Halliwell and Adams, 1982; Bader *et al.*, 1982; Mayer and Westbrook, 1983; Bader and Bertrand, 1984; DiFrancesco, 1985, 1986; Benham *et al.*, 1987).

The first part of this chapter summarizes the main results concerning the hyperpolarization-activated Cl^- current of *Aplysia* neurons. Most of the results were obtained with two-microelectrode voltage-clamp techniques (Chesnoy-Marchais, 1982, 1983); the channels responsible for this current were also identified in the outside-out configuration of the patch-clamp technique (Chesnoy-Marchais and Evans, 1986a).

The second part (Section 3) briefly reviews examples of hyperpolarization-activated Cl^- currents recently described in other preparations.

2. HYPERPOLARIZATION-ACTIVATED Cl⁻ CURRENTS IN APLYSIA NEURONS

As shown in Fig. 1, in *Aplysia* neurons, hyperpolarizing voltage jumps of a few seconds' duration can induce a slowly developing inward current. In the cells where most of the experiments have been performed, the A neurons of the cerebral ganglion, the amplitude of this current was generally very small when the intracellular microelectrode did not contain Cl^-. However, when the cell was impaled with a leaky KCl-filled microelectrode, the leak of Cl^- into the neuron progressively increased the

Dominique Chesnoy-Marchais • Laboratory of Neurobiology, École Normale Superieure, 75005 Paris, France.

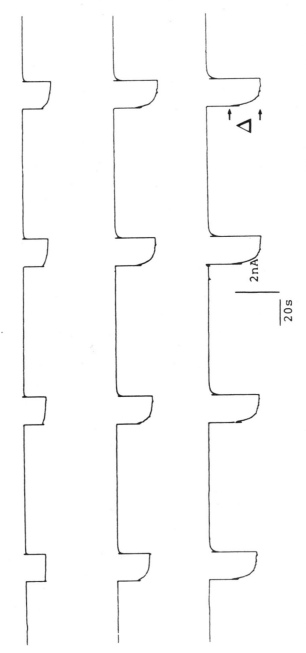

Figure 1. Progressive development of the hyperpolarization-activated inward current of *Aplysia* neurons (△) after penetration with a leaky voltage-recording microelectrode filled with 1 M KCl. Continuous current trace recorded while hyperpolarizing voltage jumps of 20 sec duration were regularly applied from −50 to −80 mV.

amplitude of this hyperpolarization-activated inward current (Fig. 1). This current usually became stable 20 to 30 min after penetration with the Cl^--filled microelectrode. The hyperpolarization-induced current could also be revealed by iontophoretic injections of Cl^- or Cl^--loading via an agonist-induced Cl^- conductance (see Fig. 4B in Chesnoy-Marchais, 1983).

This inward current, slowly activated by hyperpolarization, was clearly due to a progressive increase in conductance during the hyperpolarization. This was shown by the fact that the instantaneous current change induced at the end of a hyperpolarizing voltage jump (after development of the hyperpolarization-activated inward current) was larger than the instantaneous current change induced at the onset of this hyperpolarizing jump.

2.1. Cl⁻ Selectivity

The hyperpolarization-activated inward current of *Aplysia* neurons could be observed in the absence of external Na^+ and K^+, in the presence of external Cs^+, and after complete substitution of external and internal monovalent cations by Cs^+ using nystatin (see Fig. 16 in Chesnoy-Marchais, 1983). Thus, it seems quite different from the K^+ or nonselective cationic currents that are activated by hyperpolarization in other preparations.

The reversal potential, E_r, of the *Aplysia* hyperpolarization-activated inward current was measured by extrapolation. One of the methods that was used is illustrated in Fig. 2; the ratio of the amplitudes of the "on"- and "off"-relaxations (corresponding respectively to the activation and deactivation of the current) was measured for a series of hyperpolarizing jumps from a fixed membrane potential, V_H, to a variable membrane potential V, and this ratio was plotted as a function of V.

In the A cells (used in most experiments), the initial phase of the response to brief iontophoretic applications of cholinergic agonists was known to be a Cl^--selective conductance increase (see Ascher and Chesnoy-Marchais, 1982). Thus, in each experiment, the Cl^- equilibrium potential, E_{Cl}, could be determined by measuring the reversal potential of this cholinergic Cl^- response.

Independent measurements of E_r and E_{Cl} showed that E_r was always close to E_{Cl}. Variations of E_r with the intracellular and extracellular Cl^- concentrations are shown in Fig. 2B and C. Thus, the hyperpolarization-activated inward current of *Aplysia* neurons seems to be carried by Cl^-. Further evidence that the hyperpolarization-induced inward current is carried by Cl^- comes from the observation that a prolonged hyperpolarization, initially inducing a large inward current, could progressively lower the $[Cl^-]_i$, as expected from a prolonged net Cl^- exit. This property is illustrated in Fig. 3. In this experiment, the neuron was preloaded with Cl^- (by the leaky KCl-filled voltage-recording microelectrode), and the prolonged hyperpolarization induced a large hyperpolarization-activated inward current (note the slow time scale and the difference between the peak inward current and the instantaneous current level indicated by the arrow in Fig. 3B). The reversal potential of the Cl^--selective initial phase of the cholinergic response was close to -24 mV and became close to -38 mV after the prolonged hyperpolarization. Thus, the prolonged hyperpolarization had lowered

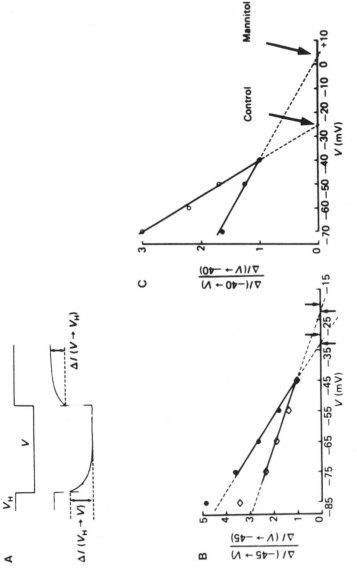

Figure 2. Measurement of the reversal potential (E_r) of the hyperpolarization-activated inward current. (A) The method illustrated consisted of measuring the ratio of the amplitudes of the "on"- and "off"-relaxations induced by hyperpolarizing jumps from V_H to a variable potential V: $\Delta I(V_H \rightarrow V)/\Delta I(V \rightarrow V_H)$. (B) E_r measurements from the same neuron for two different $[Cl^-]_i$ (obtained by successive iontophoretic injections). In each case, E_r (indicated by the lower arrow) was found to be very close to the corresponding E_{Cl} value (indicated by the upper arrow and measured as the reversal potential of the Cl^--selective cholinergic response of the neuron; see text and Fig. 3). (C) E_r measurements from another neuron for two different $[Cl^-]_o$ (610 mM in the control solution, 150 mM after substitution of the external NaCl by mannitol). The observed shift of E_r was close to that expected for E_{Cl} (E_{Cl} was not measured in this neuron). (Reproduced, with permission, from Chesnoy-Marchais, 1983.)

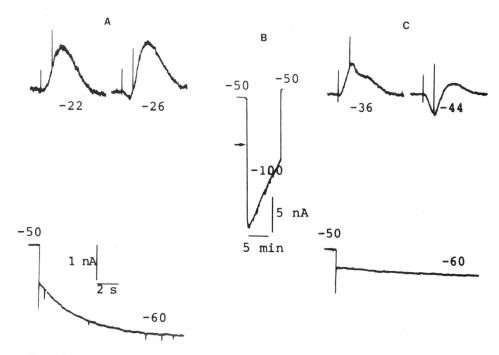

Figure 3. Effects of a prolonged hyperpolarization on the Cl⁻ equilibrium potential and on the hyperpolarization-activated inward current. This neuron was first Cl⁻-loaded by the leaky, KCl-filled voltage-recording microelectrode. (Upper records) Responses to iontophoretic applications of carbachol at the indicated membrane potentials, which allow E_{Cl} measurements before (A) and after (C) a hyperpolarization (B) to −100 mV lasting several minutes; the initial phase of the cholinergic response is a Cl⁻-selective conductance increase, whereas the second phase of the cholinergic response is a K⁺-selective conductance increase (Ascher and Chesnoy-Marchais, 1982). (Lower records) Inward current activated by hyperpolarization from −50 to −60 mV before (A) and after (C) the prolonged hyperpolarization to −100 mV. The current trace recorded during this long hyperpolarization is illustrated in B; the arrow indicates the current level observed at −100 mV at the time of hyperpolarization.

the $[Cl^-]_i$ from 235 mM to 135 mM. This reduction in $[Cl^-]_i$ was accompanied by a marked reduction of the hyperpolarization-activated inward current (see the slow decrease in inward current illustrated in B and the lower current records).

2.2. Voltage- and [Cl⁻]-Dependence

The voltage-sensitivity of the Cl⁻ conductance activated by hyperpolarization was studied by applying a series of hyperpolarizing voltage jumps from a holding potential at which this conductance was not activated. Thus, the steady-state conductance activated at the end of each hyperpolarization could be easily calculated by dividing the amplitude of each "on"-relaxation by the corresponding Cl⁻ driving force.

The time course of activation of this conductance was also voltage-sensitive, becoming more rapid with stronger hyperpolarizations. The slow hyperpolarization-induced current relaxations could usually be fitted by a single exponential. Furthermore, in the potential range where it was possible to record both an "on"-relaxation and an "off"-relaxation at the same membrane potential, the time constants of these two relaxations were identical. Thus, the kinetic properties of the slow activation–deactivation process of the hyperpolarization-activated Cl⁻ current could be characterized by a single time constant.

The steady-state value and activation time constant of the hyperpolarization-activated Cl⁻ conductance not only were voltage-sensitive but were also very sensitive to the $[Cl⁻]_i$.

Figure 4 summarizes the effects of both membrane potential and $[Cl⁻]_i$ on the steady-state value (B) and the activation time constant (C) of the hyperpolarization-

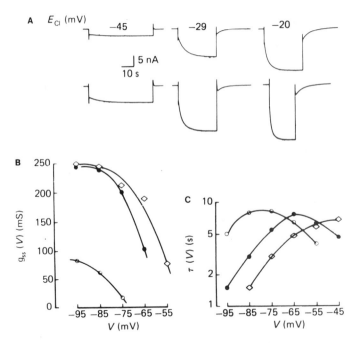

Figure 4. Effects of membrane potential and $[Cl⁻]_i$ on the hyperpolarization-activated Cl⁻ current. The neuron was first loaded with Cl⁻ leaking from the KCl-filled intracellular microelectrode, bringing E_{Cl} to a stable value of −45 mV. Then, after a series of hyperpolarizing pulses from −45 mV, Cl⁻ was injected by iontophoresis (through a double-barraled microelectrode) bringing E_{Cl} first to −29 mV and then to −20 mV. The same series of hyperpolarizing pulses was repeated for these two levels of $Cl_i⁻$. (A) Current records during voltage jumps from −45 mV to −75 mV (upper row) or −85 mV (lower row) for the three $[Cl⁻]_i$. (B) Voltage-dependence of the steady-state conductance $g_{ss}(V)$ for E_{Cl} = −45 mV (○), −29 mV (●), and −20 mV (◇). (C) Voltage-dependence of the time constant of activation for the three $[Cl⁻]_i$ (same symbols as in B). About 1 hr after the end of the second $Cl_i⁻$ injection, E_{Cl} was back to around −45 mV, and records very similar to the initial ones were obtained. (Reproduced, with permission, from Chesnoy-Marchais, 1983.)

activated Cl⁻ conductance. In this experiment, a series of hyperpolarizing voltage jumps was successively applied in the same cell for three different $[Cl^-]_i$ (obtained by successive iontophoretic Cl⁻ injections) estimated from the corresponding E_{Cl} values.

For each value of $[Cl^-]_i$, the steady-state conductance shows a threshold of activation, increases with hyperpolarization above this threshold, and saturates at large hyperpolarizations; the voltage-sensitivity of the activation time constant is bell-shaped, showing a maximum close to the threshold of activation.

When comparing the curves obtained for different $[Cl^-]_i$, it appears that not only the maximum steady-state conductance but also the threshold and time constant of activation are Cl⁻-sensitive. Thus, an increase in the $[Cl^-]_i$ facilitates the efflux of Cl⁻ through the hyperpolarization-activated Cl⁻ conductance in three ways:

- Lowering the threshold of activation
- Accelerating the activation kinetics
- Increasing the Cl⁻ driving force and the maximum steady-state conductance value

The high sensitivity to $[Cl^-]_i$ of the gating properties of this hyperpolarization-activated Cl⁻ current explains why it is much easier to study this current after Cl_i^--loading. However, there is also another way to reveal this current. In cells that are not Cl⁻-loaded, after a train of depolarizing jumps, a hyperpolarizing jump activates the Cl⁻ current even if it does not activate this current before the depolarizing train (Fig. 5A); in Cl⁻-loaded cells, a train of depolarizing jumps can reversibly increase the amplitude of the Cl⁻ current activated by a given hyperpolarizing jump (Fig. 5B). Although the precise origin of these effects is not clear [changes in the intracellular concentrations of Ca^{2+} and H^+ might be involved (see Chesnoy-Marchais, 1983)], these observations suggest that the hyperpolarization-activated Cl⁻ current can be activated under physiological conditions.

2.3. Other Properties of This Current

The hyperpolarization-activated Cl⁻ current is blocked by internal NO_3^- ions, which means in practice that internal NO_3^- ions are not permeant through these channels. However, it is not clear whether this $Cl^- - NO_3^-$ selectivity is a property of the open channel itself or whether it occurs at the level of the step responsible for the sensitivity of the gating of the channel to the $[Cl^-]_i$.

The hyperpolarization-activated Cl⁻ current is blocked by internal (but not external) applications of stilbenesulfonate derivatives such as DIDS and SITS.

The gating properties of this channel are sensitive not only to membrane potential and Cl_i^- but also to Cl_o^- and pH_o. Increasing the pH_o (from 7.6 to 9.0) made the activation of the current more difficult, by shifting toward more negative membrane potentials both the steady-state conductance–voltage curve and the activation time constant–voltage curve; reducing the pH_o (from 7.6 to 6.0) had reverse (but less pronounced) effects (see Figs. 10 and 11 in Chesnoy-Marchais, 1983). Reducing the pH_i strongly reduced the hyperpolarization-activated Cl⁻ current. The effects of decreasing the $[Cl^-]_o$ are similar to those found on increasing the $[Cl^-]_i$. This further

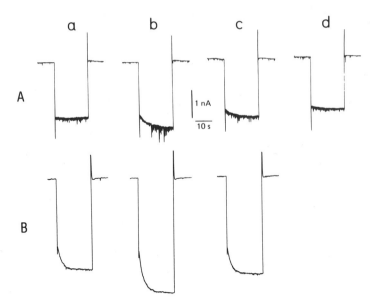

Figure 5. Induction or enhancement of the hyperpolarization-activated inward current by trains of depolarizing pulses. (A) Current traces recorded in a neuron that had not been loaded with Cl^-, during hyperpolarizing pulses from -60 to -100 mV before (a), 20 sec after (b), 6 min after (c), and 16 min after (d) a train of depolarizing pulses from -60 to $+5$ mV (pulses of 100 msec duration every 100 mses for 10 sec). (B) Current traces recorded in a Cl^--loaded neuron, during hyperpolarizing pulses from -45 to -75 mV before (a), 3 min after (b), and 13 min after (c) a train of depolarizing pulses from -45 to $+15$ mV (pulses of 20 msec duration every 20 msec during 10 sec). Calibration bars are for both A and B. (Reproduced, with permission, from Chesnoy-Marchais, 1983.)

supports the hypothesis that this hyperpolarization-activated Cl^- conductance could play a role in the regulation of the distribution of Cl^- ions (see Discussion).

2.4. Single-Channel Data

Using the outside-out configuration of the patch-clamp technique, it has been possible to identify the single channels responsible for the hyperpolarization-activated Cl^- current (Chesnoy-Marchais and Evans, 1986a). These channels could be observed under conditions where Cl^- ions were the only plausible current carrier (in Na^+-free and K^+-free external and internal solutions). They always appeared as long bursts of transitions between three current levels, the zero current level, a level of maximum conductance (γ) of about 10–15 pS, and a level corresponding to about half this conductance ($\gamma/2$) (Fig. 6B). Furthermore, these bursts were clearly activated by hyperpolarization, both their frequency and duration being voltage-sensitive (Fig. 6A).

Unfortunately, a detailed study of these channels has not been possible because they were observed only in a very small number of membrane patches. However, another type of Cl^--permeable channel was always observed in these membrane patches. This latter channel was not appreciably voltage-dependent and was nonselective, being permeable to Cl⁻, large anions like isethionate, methanesulfonate, and sulfate, and even small cations (see Chesnoy-Marchais and Evans, 1986b). The rarity of the activity of the hyperpolarization-activated Cl⁻ channel contrasts with the reproducible large amplitude of the hyperpolarization-activated Cl⁻ current recorded with the classical voltage-clamp technique. This contrast suggests that the activity of the hyperpolarization-activated Cl⁻ channels is regulated by some mechanism that must be strongly affected by disruption of the plasma membrane and/or modification of the intracellular compartment. However, as reviewed below, it has been possible to detect a similar kind of channel much more reproducibly in other cellular systems (in lipid bilayers, after incorporation of *Torpedo* electroplaque membrane fragments, and for example, in inside-out membrane patches from Cl⁻ secretory epithelial cells pretreated with adenosine, dibutyryl cAMP, and forskolin). The behavior of this type of channel in an isolated membrane patch might depend on its state before membrane excision. Note that in the case of the classical Ca^{2+} channel, it has recently been demonstrated that the persistence of activity in excised membrane patches is dependent on the phosphorylation state of the channel; in addition, if the original intact cell is pretreated with the Ca^{2+} channel agonist Bay K8644, the further activity of Ca^{2+}

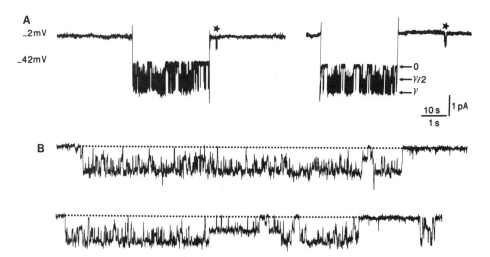

Figure 6. Characterization of the Cl⁻ channel activated by hyperpolarization in an outside-out membrane patch (CsCl, internal solution and mannitol, external solution; $E_{Cl} = +36$ mV). (A) Chart record of the current obtained during hyperpolarizing pulses applied every minute from −2 to −42 mV (note the very slow time scale). At −42 mV, long bursts of channel openings are frequently observed and the current levels corresponding to the conductance levels 0, γ/2, and γ are indicated by arrows. At −2 mV, the bursts of channel openings (indicated by stars) are much less frequent. (B) Two segments of the current trace recorded at −42 mV during the same experiment are shown at a faster time scale (300-Hz filter) in order to illustrate the rapid transitions between the conductance levels 0, γ/2, and γ which occur within a prolonged burst of openings. The resting level recorded between bursts is indicated by the dotted line. The lower trace has been selected in order to show the current level of the half conductance state, which is particularly apparent in the middle of this trace and is usually observed much more transiently. (Reproduced, with permission, from Chesnoy-Marchais and Evans, 1986a.)

channels in the excised membrane patch is retained much longer than without cell pretreatment (Armstrong and Eckert, 1987).

3. HYPERPOLARIZATION-ACTIVATED Cl⁻ CURRENTS IN OTHER CELLS

More than 20 years ago, it was reported that in crayfish muscle fibers, "the membrane develops a time-variant decrease in resistance when the fibers are hyperpolarized by some 15 mV or more," and that this is due to the activation by hyperpolarization of a Cl⁻ conductance (Reuben *et al.*, 1962; Ozeki *et al.*, 1966). Similarities between this crayfish muscle Cl⁻ conductance, the Cl⁻ conductance that seemed to be activated by hyperpolarization in frog muscle fibers at acidic pH$_o$ (Warner, 1972), and the hyperpolarization-activated Cl⁻ conductance of *Aplysia* neurons have already been discussed elsewhere (Chesnoy-Marchais, 1983). The Cl⁻ channels extracted from *Torpedo* electroplaque membranes (see Miller and Richard, this volume) were also known to give rise to slow current relaxations very similar to those found in *Aplysia* (with respect to time course, voltage-dependence, asymmetric sensitivity to Cl⁻ channel blockers, pH sensitivity, and Cl^-/NO_3^- discrimination). Single-channel recordings have revealed further similarities between the *Torpedo* and *Aplysia* voltage-activated Cl⁻ channels. The elementary conductance values derived for both channels are very similar, and in both cases the channels appear as long bursts of transitions between three current levels. This behavior has been studied in detail in the case of the *Torpedo* channel and was interpreted as showing that this channel has a dimeric structure (Miller, 1982; Miller and White, 1984).

More recently, the existence of hyperpolarization-activated Cl⁻ conductances has been shown or suggested in several other preparations.

3.1. Neurons

In inside-out patches of *Lymnaea* neurons, Geletyuk and Kazachenko (1985) have described Cl⁻ channels showing several conductance states. The elementary conductances of these states seem to be multiples of about 12 pS. Although it is not clear whether all the observed conductance states have the same properties, it has been reported that the gating of these channels is controlled in a complex way by Ca_i^{2+}, K_o^+, and membrane potential. According to the experimental conditions, these channels may be activated either by depolarization or by hyperpolarization.

In rat sympathetic neurons, hyperpolarizing voltage jumps can slowly activate an inward current which seems to be carried by Cl⁻, being detectable in the presence of Ba_o^{2+}, and augmented with Cl⁻ (rather than acetate)-containing electrodes. This current is blocked by 0.2 mM Cd^{2+} but persists (like the current found in *Aplysia*) in the presence of Co_o^{2+} or in Ca^{2+}-free EGTA external solutions (Selyanko, 1984).

A similar current, activated by hyperpolarization, Cd^{2+}-sensitive, persistent in the absence of Ca_o^{2+}, has been described in rat hippocampal pyramidal neurons, using a single-electrode voltage-clamp, while blocking most of the other voltage-gated conductances by a Ca^{2+}-free TTX-, Cs⁺-, and TEA-containing external solution

(Madison *et al.*, 1986). When the cell was penetrated with potassium methylsulfate electrodes, hyperpolarizing jumps activated an inward current only below -70 mV, whereas hyperpolarization-activated inward current was observed for all negative membrane potentials when the cell was penetrated with CsCl-filled electrodes (Fig. 7). Thus, this current seems to be carried by Cl^- and is reported to not require Cl^--loading for its detection. Note that the *Aplysia* hyperpolarization-activated Cl^- current can also be detected without Cl^--loading (after a train of depolarizing pulses).

In inside-out membrane patches from cultured *Drosophila* neurons, the presence of a channel of 8 pS elementary conductance, which is activated by hyperpolarization, selective for Cl^-, and blocked by internal SITS, has been mentioned (Yamamoto and Suzuki, 1987).

3.2. Cl⁻-Secretory Epithelial Cells

Recent single-channel studies have revealed the existence of a variety of Cl^- channels in Cl^--secretory epithelial cells. Some of these channels are activated by hyperpolarization. This is the case of the "small"-conductance Cl^- channel found in inside-out membrane patches from the luminal membrane of the dogfish rectal gland (Gögelein *et al.*, 1987). The elementary conductance of this channel is similar to that of the other hyperpolarization-activated Cl^- channels (8–11 pS). However, in this case, fluctuations between different conductance states have not been reported. This channel was observed after stimulating the cell with adenosine, dibutyryl cAMP, and

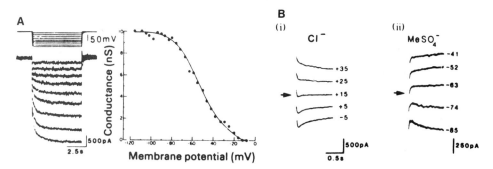

Figure 7. Cl^- current activated by hyperpolarization in hippocampal pyramidal neurons (external solution designed to block K^+, Ca^{2+}, and Na^+ currents, containing TTX, 0 Ca, TEA, and Cs). (A) Membrane currents resulting from a family of long-duration hyperpolarizing commands from a holding potential (V_H of -4 mV. A plot of the steady-state membrane conductance during the illustrated voltage command pulses (value plotted is the steady-state conductance during hyperpolarizing command pulses minus the steady-state conductance at V_H). CsCl-filled microelectrode. (B, i) Tracing of current relaxations produced by depolarizing voltage command steps from a V_H of -50 mV to the voltage indicated at the right of each trace (in mV) in a cell impaled with a KCl-filled microelectrode. (ii) Tracings of current relaxations produced by hyperpolarizing steps from a holding potential of -21 mV to the voltage (in mV) indicated at the right of each trace in a cell impaled with a KMeSo₄-filled microelectrode. Arrows indicate the approximate value of the reversal potential. (Reproduced, by permission, from *Nature*, Vol. 321, p. 695, copyright © 1986, Macmillan Magazines, Ltd.)

forskolin; its behavior has not been described in nonstimulated cells. Interestingly, this "small"-conductance channel is not blocked by a Cl⁻ channel blocker [5-nitro-2-(3-phenylpropylamino)-benzoate], whereas the "larger"-conductance Cl⁻ channel of the same preparation, which is activated by depolarization and is clearly stimulated by dibutyryl cAMP and forskolin, is blocked by this compound (Greger *et al.*, 1987).

One of the Cl⁻ channels found in inside-out membrane patches from canine tracheal epithelial cells was also reported to be slowly activated by hyperpolarization (Shoemaker *et al.*, 1986). The elementary conductance of this channel is reported to be 30–50 pS. However, this channel shows outward rectification, and if the slope conductance of its elementary current is measured in the range of negative membrane potentials, the elementary conductance value obtained seems close to 17 pS with 300 mM $[Cl^-]_i$. Some of the data illustrated by the authors (see the upper records of Fig. 3 in Shoemaker *et al.*, 1986) might suggest that these channels also show a dimeric behavior.

3.3. Amphibian Oocytes

In voltage-clamped amphibian oocytes, hyperpolarizing voltage jumps can induce two kinds of inward currents which are selective for Cl⁻. The hyperpolarization-activated Cl⁻ current described in *Xenopus* oocytes, "in the course of maturation" (Peres and Bernardini, 1983) clearly shows inactivation on a time scale of a few seconds and has been reported to be abolished by the absence of Ca_o^{2+}; thus, this current is different from the hyperpolarization-activated Cl⁻ currents found in *Torpedo* electroplaques and *Aplysia* or vertebrate neurons.

A transient hyperpolarization-activated inward current, similar to that found in *Xenopus* oocytes, was also found in *Rana* oocytes, but not in all animals and only during spring. However, the hyperpolarization-activated Cl⁻ current usually found in *Rana* oocytes, I_{Ir} (Taglietti *et al.*, 1984), looks strikingly similar to that found in *Aplysia*. This current does not inactivate and the threshold of activation of its conductance becomes less negative as the $[Cl^-]_i$ is increased (Fig. 8). In the presence of 1.8 mM $[Ca^{2+}]_o$, the addition of 1 mM Cd^{2+} reduces this current strongly (as it does in the case of the hyperpolarization-activated Cl⁻ current of vertebrate neurons); however, in the presence of 20 mM $[Ca^{2+}]_o$, this reduction no longer occurs. Such a competition between Ca^{2+} and Cd^{2+} could explain why the *Aplysia* hyperpolarization-activated Cl⁻ current was not abolished by the addition of 0.5 mM Cd^{2+} (unpublished observation; higher doses of Cd^{2+} are clearly toxic in *Aplysia* neurons). The hyperpolarization-activated Cl⁻ current of *Rana* oocytes, I_{Ir}, is suppressed during maturation (Taglietti *et al.*, 1984).

3.4. Mouse Astrocytes

In a few outside-out patches from mouse astrocytes, channels of small conductance (about 5 pS) were opened by hyperpolarization and frequently exhibited a half-conductance state (as the hyperpolarization-activated Cl⁻ channels of *Torpedo*

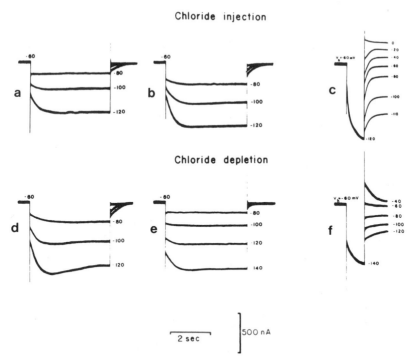

Figure 8. Cl⁻ current activated by hyperpolarization in frog oocyte (I_{Ir}). The cell was impaled with two microelectrodes filled with 2 M K-citrate plus one microelectrode filled with 3 M KCl. (a) The Cl⁻ current was elicited in control conditions while a braking current of 100 nA was applied through the K⁺ electrode to avoid leakage of Cl⁻ into the cell. (b) The same sequence of command pulses was applied following injection of Cl⁻ with a current of 800 nA for 15 min. (c) Example of tail current relaxations during determination of the reversal potential of the Cl⁻ current following iontophoretic injection of Cl⁻ (800 nA for 30 min) in an oocyte immersed in standard saline. (d, e) The decrease in Cl⁻ current amplitudes with decreasing $[Cl^-]_i$. In d, the Cl⁻ current was elicited in control conditions; in e, the same sequence of command pulses was applied in standard saline after 15 min during which all NaCl had been replaced by Na-isethionate in the external medium. The membrane potential was held at −60 mV throughout the experiment. Notice also the shift in activation threshold toward more negative membrane potentials following depletion of Cl_i^-. (f) Example of tail current relaxations during determination of the reversal potential of the Cl⁻ current in normal saline, in an oocyte depleted of Cl_i^-. (Reproduced, with permission, from Taglietti *et al.*, 1984.)

electroplaques or *Aplysia* neurons); these channels might also be the hyperpolarization-activated Cl⁻ channels (Nowak *et al.*, 1987).

3.5. Plant Cells

In a pond water green alga, *Chara inflata*, an inward current ("inward-2-current") is slowly activated by hyperpolarization on a time scale of a few seconds (Coleman and Findlay, 1985; Tyerman *et al.*, 1986). Local measurements of the $[Cl^-]_o$ adjacent to the cell, and vacuolar voltage-clamp experiments with variable $[Cl^-]_o$ have shown that this current is partly carried by Cl^- ions. The net Cl^- exit measured at a given membrane potential during activation of this current is sensitive to the pH_o, being smaller at basic pH than at acidic pH (as are the currents found in crayfish muscle, frog muscle, *Torpedo* electroplaques, and *Aplysia* neurons). Cl^--selective channels responsible for this hyperpolarization-activated Cl^- current have recently been identified in the cell-attached configuration of the patch-clamp technique (Coleman, 1986). Interestingly, several conductance levels were detected and the elementary conductance value of the lowest level was 7 pS. This elementary conductance is very close to the lowest elementary conductance of the voltage-gated Cl^- channels of *Torpedo* membranes or *Aplysia* neurons and to the elementary conductance values of the hyperpolarization-activated Cl^- channels of *Drosophila* neurons and dogfish epithelial cells.

In summary, hyperpolarization-activated Cl^- currents seem to exist not only in *Torpedo* electroplaque membranes, invertebrate and vertebrate neurons, but also in a wide variety of cell types, including muscle cells, epithelial cells, oocytes, and plant cells.

4. POSSIBLE ROLES OF HYPERPOLARIZATION-ACTIVATED Cl⁻ CURRENTS

One of the possible physiological roles of hyperpolarization-activated Cl^- currents could be to prevent excessive hyperpolarizations (Chesnoy-Marchais, 1982, 1983; Taglietti *et al.*, 1984). However, such a role could be served as well by any other hyperpolarization-activated inward current.

A more specific role for the hyperpolarization-activated *chloride* currents is suggested by the sensitivity of the gating of the *Aplysia* current to the $[Cl^-]_i$ [note that the same sensitivity has been observed in frog oocytes (see Section 3.3)]. Due to this sensitivity, this current facilitates the *net exit* of Cl_i^- when the $[Cl^-]_i$ increases. Such a role in the regulation of the $[Cl^-]_i$ might be particularly important in the case of the apical membrane of Cl^--secretory epithelial cells.

Another possible functional role, which might be particularly important in the case of excitable cells, can be envisioned from the data obtained from the *Torpedo* channel. In this case, it has been shown that the voltage sensitivity of the intraburst rapid fluctuations is *opposite* to the voltage sensitivity of the slow activation process; inside a long opening burst, the fraction of the time spent in the state of maximum conductance increases with depolarization. This property (which remains to be established in the case of other hyperpolarization-activated Cl^- channels) predicts that upon prolonged depolarization from a highly negative membrane potentials, the macroscopic Cl^- current would appear to be rapidly "activated" before decreasing more slowly (see Miller, 1982, p. 410). Thus, in cells where E_{Cl} is more positive than the resting potential E_m, there would be a voltage range between E_{Cl} and E_m where, upon depolarization, the inward current carried by these Cl^- channels would look for example like a Ca^{2+} current. In such cells, these Cl^- channels might play a role in amplifying small depolarizations.

REFERENCES

Armstrong, D., and Eckert, R., 1987, Voltage-activated calcium channels that must be phosphorylated to respond to membrane depolarization, *Proc. Natl. Acad. Sci. USA* **84:**2518–2522.

Ascher, P., and Chesnoy-Marchais, D., 1982, Interactions between three slow potassium responses controlled by three distinct receptors in Aplysia neurones, *J. Physiol. (London)* **324:**67–92.

Bader, C. R., and Bertrand, D., 1984, Effect of changes in intra- and extracellular sodium on the inward (anomalous) rectification in salamander photoreceptors, *J. Physiol. (London)* **347:**611–631.

Bader, C. R., Bertrand, D., and Schwartz, E. A., 1982, Voltage-activated and calcium-activated currents studied in solitary rod inner segments from the salamander retina, *J. Physiol. (London)* **331:**253–284.

Benham, C. D., Bolton, T. B., Denbigh, J. S., and Lang, R. J., 1987, Inward rectification in freshly isolated single smooth muscle cells of the rabbit jejunum, *J. Physiol. (London)* **383:**461–476.

Chesnoy-Marchais, D., 1982, A Cl^- conductance activated by hyperpolarization in Aplysia neurones, *Nature* **299:**359–361.

Chesnoy-Marchais, D., 1983, Characterization of a chloride conductance activated by hyperpolarization in Aplysia neurones, *J. Physiol. (London)* **342:**277–308.

Chesnoy-Marchais, D., and Evans, M. G., 1986a, Chloride channels activated by hyperpolarization in Aplysia neurones, *Pfluegers Arch.* **407:**694–696.

Chesnoy-Marchais, D., and Evans, M. G., 1986b, Non-selective ionic channels in Aplysia neurons, *J. Membr. Biol.* **93:**75–83.

Coleman, H. A., 1986, Chloride currents in Chara. A patch-clamp study, *J. Membr. Biol.* **93:**55–61.

Coleman, H. A., and Findlay, G. P., 1985, Ion channels in the membrane of Chara inflata, *J. Membr. Biol.* **83:**109–118.

Constanti, A., and Galvan, M., 1983, Fast inward-rectifying current accounts for anomalous rectification in olfactory cortex neurones, *J. Physiol. (London)* **385:**153–178.

DiFrancesco, D., 1985, The cardiac hyperpolarizing-activated current, i_f. Origins and developments, *Prog. Biophys. Mol. Biol.* **46:**163–183.

DiFrancesco, D., 1986, Characterization of single pacemaker channels in cardiac sino-atrial node cells, *Nature* **324:**470–473.

Geletyuk, V. I., and Kazachenko, V. N., 1985, Single Cl^- channels in molluscan neurons: Multiplicity of the conductance states, *J. Membr. Biol.* **86:**9–15.

Gögelein, H., Schlatter, E., and Greger, R., 1987, The "small" conductance chloride channel in the luminal membrane of the rectal gland of the dogfish (Squalus acanthias), *Pfluegers Arch.* **409:**122–125.

Greger, R., Schlatter, E., and Gögelein, H., 1987, Chloride channels in the luminal membrane of the rectal gland of the dogfish (Squalus acanthias). Properties of the "larger" conductance channel, *Pfluegers Arch.* **409:**114–121.

Hagiwara, S., Miyazaki, S., and Rosenthal, N. P., 1976, Potassium current and the effect of cesium on this current during anomalous rectification of the egg cell membrane of a starfish, *J. Gen. Physiol.* **67:**621–638.

Halliwell, J. V., and Adams, P. R., 1982, Voltage-clamp analysis of muscarinic excitation in hippocampal neurons, *Brain Res.* **250:**71–92.

Hestrin, S., 1981, The interaction of potassium with the activation of anomalous rectification in frog muscle membrane, *J. Physiol. (London)* **317:**497–508.

Kurachi, Y., 1985, Voltage-dependent activation of the inward-rectifier potassium channel in the ventricular cell membrane of guinea pig heart, *J. Physiol. (London)* **366:**365–385.

Leech, C. A., and Stanfield, P. R., 1981, Inward rectification in frog skeletal muscle fibres and its dependence on membrane potential and external potassium, *J. Physiol. (London)* **319:**295–309.

Madison, D. V., Malenka, R. C., and Nicoll, R. A., 1986, Phorbol esters block a voltage-sensitive chloride current in hippocampal pyramidal cells, *Nature* **321**:695–697.

Mayer, M. L., and Westbrook, G. L., 1983, A voltage clamp analysis of inward (anomalous) rectification in mouse spinal sensory ganglion neurones, *J. Physiol. (London)* **340**:19–43.

Miller, C., 1982, Open-state substructure of single chloride channels from Torpedo electroplax, *Philos. Trans. R. Soc. London Ser. B* **299**:401–411.

Miller, C., and White, M. M., 1984, Dimeric structure of single chloride channels from Torpedo electroplax, *Proc. Natl. Acad. Sci. USA* **81**:2772–2775.

Nowak, L., Ascher, P., and Berwald-Netter, Y., 1987, Ionic channels in mouse astrocytes in culture, *J. Neurosci.* **7**:101–109.

Ozeki, M., Freeman, A. R., and Grundfest, H., 1966, The membrane components of crustacean neuromuscular systems, *J. Gen. Physiol.* **49**:1335–1349.

Peres, A., and Bernardini, G., 1983, A hyperpolarization-activated chloride current in *Xenopus laevis* oocytes under voltage-clamp, *Pfluegers Arch.* **399**:157–159.

Reuben, J. P., Girardier, L., and Grundfest, H., 1962, The chloride permeability of crayfish muscle fibers, *Biol. Bull.* **123**:509–510.

Sakmann, B., and Trube, G., 1984, Conductance properties of single inwardly rectifying potassium channels in ventricular cells from guinea-pig heart, *J. Physiol. (London)* **347**:641–657.

Selyanko, A. A., 1984, Cd^{2+} suppresses a time-dependent Cl⁻ current in rat sympathetic neurone, *J. Physiol. (London)* **350**:49P.

Shoemaker, R. L., Frizzell, R. A., Dwyer, T. M., and Farley, J. M., 1986, Single chloride channel currents from canine tracheal epithelial cells, *Biochim. Biophys. Acta* **858**:235–242.

Taglietti, V., Tanzi, F., Romero, R., and Simoncini, L., 1984, Maturation involves suppression of voltage-gated currents in the frog oocyte, *J. Cell. Physiol.* **121**:576–588.

Tyerman, S. D., Findlay, G. P., and Paterson, G. J., 1986, Inward membrane current in Chara inflata. I. A voltage and time-dependent Cl⁻ component, *J. Membr. Biol.* **89**:139–152.

Warner, A. E., 1972, Kinetic properties of the chloride conductance of frog muscle, *J. Physiol. (London)* **227**:291–312.

Yamamoto, D., and Suzuki, N., 1987, Blockage of chloride channels by HEPES buffer, *Proc. R. Soc. London Ser. B* **230**:93–100.

The Voltage-Dependent Chloride Channel of Torpedo Electroplax

Intimations of Molecular Structure from Quirks of Single-Channel Function

Christopher Miller and Edwin A. Richard

1. INTRODUCTION

Our intention in this chapter is to describe the remarkable functional properties—some would say the eccentric behavior—of an anion transport protein found in the electric organ of electric rays, such as *Torpedo* and *Narke*. This is a Cl^--specific ion channel, commonly called the "*Torpedo* Cl^- channel." Like all ion channels, this protein catalyzes the electrodiffusive, thermodynamically downhill flow of ions across the membrane in which it resides, and thus should be classed as a membrane "leak." Lest this term be considered pejorative by those familiar with active transporters which work to build up transmembrane ion gradients, we should point out that cells have good reasons to maintain specific leaks across their membranes; for instance, a leak for Cl^-, in combination with a Cl^- concentration gradient, can give rise to a transmembrane voltage, as is, we imagine, the purpose of the *Torpedo* Cl^- channel.

We must state at the outset three sad facts about the study of this channel. First, we do not know with certainty the physiological function of the channel. Second, we have not even identified the membrane in which this channel is located. Third, we know nothing about the biochemical nature of the channel protein.

Given all of these deficiencies, what is this channel good for? Why do we study it? What, in fact, *do* we know about it? The *Torpedo* Cl^- channel exhibits a most unusual set of electrophysical properties, especially when examined at the single-channel level. This single-channel behavior may be studied in such detail that solid conclusions about the channel's molecular structure can be made, even in the total absence of biochemical information. In particular, we will review this channel's properties in order to deduce three structural features of the channel protein: first, that the channel's ion entryway is far removed from the lipid surface; second, that the channel

Christopher Miller and Edwin A. Richard • Howard Hughes Medical Institute, Graduate Department of Biochemistry, Brandeis University, Waltham, Massachusetts 02154.

complex is actually composed of two independently operating "protochannels"; and third, that each protochannel closes in two physically separated regions.

1.1. Macroscopic and Microscopic Measurements on Ion Channels

A unique and delightful characteristic of ion channel proteins is that their function may be observed at two fundamentally different levels: the macroscopic and the microscopic. A macroscopic measurement is made on a membrane containing many ion channels, preferably identical. What is observed in such a measurement is the average behavior of the whole ensemble of channels. However, for the past decade, techniques have been available that allow us to make microscopic measurements on ion channels. Here, ionic current across a membrane containing only a *single* channel molecule is observed. The ability to observe current through a single channel is a simple consequence of the very high turnover rates catalyzed by channel proteins, 10^6–10^9 ions/sec, many orders of magnitude higher than those catalyzed by other types of transport proteins.

With single-channel techniques, we get a close look at the behavior of a given channel, since the opening and closing of individual channels are observed directly. These events carry detailed information that is "averaged out" in a macroscopic measurement. While microscopic measurements yield more information about a channel's function, macroscopic measurements are often technically more convenient.

1.2. Methods for Studying the Torpedo Cl⁻ Channel

Nearly all studies of the *Torpedo* Cl⁻ channel to date have been carried out in model membranes, primarily in planar lipid bilayers. As illustrated in Fig. 1, this system consists of two aqueous chambers separated by a partition with a hole on which a lipid bilayer may be formed. Plasma membrane vesicles prepared from the *Torpedo* electric organ may then be fused into this bilayer, thus inserting the channels of interest. We call the side of the bilayer to which vesicles are added the "*cis*" side, and the opposite the "*trans*." Purely by convention, the *trans* side is defined as zero

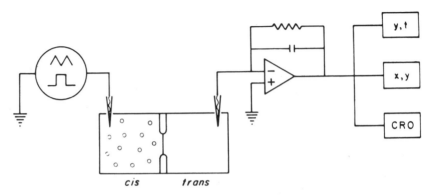

Figure 1. Planar lipid bilayer. Schematic diagram of a planar lipid bilayer system. Two aqueous chambers, labeled *cis* and *trans*, are separated by a partition with a hole on which is formed a lipid bilayer.

voltage. Although the *Torpedo* Cl⁻ channels insert into the bilayer with virtually 100% orientation, we still do not know the absolute sidedness of the channel, i.e., whether the *cis* side is equivalent to the cytoplasmic or external side of the channel in its physiological membrane.

The process by which membrane vesicles may be fused into planar bilayers has been extensively studied (Miller and Racker, 1976; Miller and White, 1980; Cohen *et al.*, 1980, 1983; Miller, 1982) and will not be reviewed here. Suffice it to say that a few simple conditions ensure reproducible fusion into planar bilayers: a high content of phosphatidylethanolamine and phosphatidylserine in the bilayer, 1–5 mM Ca²⁺ in the aqueous medium, and osmotic conditions leading to the swelling of the vesicles. When these conditions are met, the incorporation of *Torpedo* Cl⁻ channels in the bilayer occurs readily (Miller and White, 1980). Channel insertion is monitored by following the increase in electrical conductance of the membrane; our knowledge of this channel has emerged solely from studying the properties of this conductance.

The *Torpedo* Cl⁻ channel was discovered in a failed attempt to insert nicotinic acetylcholine receptor channels into lipid bilayers (White and Miller, 1979; Miller and White, 1980). The characteristic response originally observed upon adding electroplax membrane vesicles to a bilayer system under conditions promoting membrane fusion is shown in Fig. 2. At a constant applied voltage held across the bilayer, in the presence of Cl⁻, the current increases over several minutes in a series of abrupt jumps, each of which is followed by a slow relaxation to a nonzero level. Several lines of circumstan-

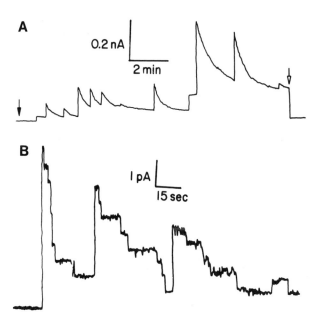

Figure 2. Fusion of *Torpedo* vesicles with planar bilayers. *Torpedo* vesicles were added to a planar lipid bilayer in the presence of 100 mM KCl, at a holding voltage of 30 mV, and current was monitored. (A) Individual fusion spikes observed under vigorous fusion conditions. (B) An example of the first three fusion spikes immediately after addition of vesicles to the bilayer.

tial evidence (reviewed in Miller, 1980) argue strongly that each of these conductance jumps represents the fusion of a *single* membrane vesicle carrying numerous ion channels. With *Torpedo* vesicles, the conductance jumps are not uniform, but follow a rather broad distribution, presumably following the wide distribution of vesicle size. The average conductance jump gives a conductance of about 6 nS (in 150 mM symmetrical [Cl⁻]), representing, as we will see, the insertion of a package of about 300 channels (White and Miller, 1979).

After 10–100 such fusion events have occurred, the fusion process can be suppressed by removing Ca^{2+} from the medium, or by changing the osmotic conditions (Miller and White, 1980). This macroscopic conductance then remains stable over tens of minutes, long enough to investigate the nature of the conductance induced by *Torpedo* vesicles.

The fact that the "typical" fusion event inserts several hundred channels would seem to preclude the possibility of observing single channels from this preparation. This is not the case, however, thanks to the very wide spread in the size of the fusion events. In particular, the first few fusion events observed after adding vesicles to a freshly formed lipid bilayer tend to be much smaller than an average event (perhaps reflecting the faster diffusion up to the bilayer through the unstirred layers of the smallest vesicles). Several examples of the *first* fusion events are shown in Fig. 2B; the conductances induced by these early events are roughly 100-fold smaller than the average case, and it can be seen that the relaxation phase of the fusion spike relaxes in discrete and uniform downward transitions. These are single Cl⁻ channels turning off.

It is therefore possible to observe single channels routinely by favoring the incorporation of small vesicles. After briefly sonicating the vesicle preparation to reduce the vesicle sizes, we often find that the first fusion event contains only a small number (1–5) of channels. After obtaining a membrane with a single channel, further fusion of vesicles is suppressed by changing the osmotic conditions and perfusing the vesicles out of the chamber. Single-channel characteristics can then be observed for up to an hour.

A second method for observing single Cl⁻ channels has recently been described exploiting the "bilayer-patch" method of Andersen (1983). *Torpedo* vesicles are fused into a planar bilayer, as with a macroscopic measurement. The chamber is then stirred vigorously for a few minutes, to encourage convective separation of the channels within the bilayer. A silanized glass pipette is then pressed onto the bilayer to isolate a single Cl⁻ channel. Besides providing a method to observe single Cl⁻ channels routinely, the glass pipette has superior noise characteristics to the planar bilayer system.

2. MACROSCOPIC CHARACTERIZATION OF THE TORPEDO Cl⁻ CHANNEL

All the initial studies of the *Torpedo* Cl⁻ channel were made at the macroscopic level, i.e., in planar lipid bilayers containing at least 1000 channels (White and Miller, 1979; Miller and White, 1980). It became evident very quickly that the conductance induced by *Torpedo* vesicles was especially appropriate to study, since only a single type of conductance pathway was inserted (or *survived* insertion) into the bilayers, and that these insertion events were highly *oriented*.

2.1. Slow Voltage-Dependent Gating

The macroscopic conductance is voltage-dependent, with the steady-state conductance low at positive potentials and high at negative potentials, with the transition between the high and low conductances occurring in a graded way over a range of 30–40 mV (Fig. 3). This behavior is reminiscent of many ion channels, which show voltage-dependent "gating" (conformational changes leading to channel opening and closing). In this case, however, the kinetics following a voltage change (say from 50 to −50 mV) are quite slow, on the time scale of seconds, not milliseconds.

2.2. Ion Selectivity and Conduction

The macroscopic conductance was found to be unusually selective for Cl⁻ ions (White and Miller, 1979). As long as Cl⁻ was present, fusion spikes were observed, regardless of the nature of the cation (Na⁺, K⁺, or Tris⁺). More importantly, the voltage-dependent conductance was found to be ideally Nernstian for Cl⁻ over cations when a salt gradient was present across the bilayer. Thus, cations do not permeate this

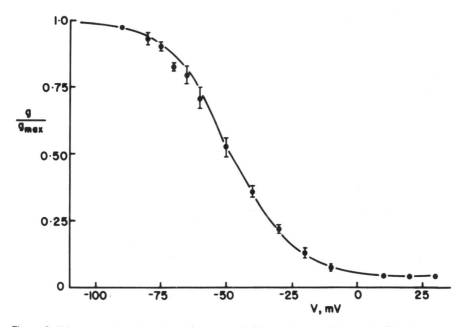

Figure 3. Voltage- and time-dependence of macroscopic Cl⁻ conductance. Macroscopic Cl conductance as a function of applied voltage is shown in membranes containing at least 500 channels.

conductance pathway. Given this fact, the interanionic selectivity was studied using bi-ionic conditions, i.e., with Cl^- on one side of the membrane and a test anion on the other. The result was clear: of all anions tested (F^-, Cl^-, Br^-, I^-, SCN^-, SO_4^-, OAc^-), only Cl^- and Br^- showed measurable permeability. Under comparable conditions, Cl^- is about three-fold more conductive than Br^- (Miller, 1982). This channel remains the most selective of Cl^- channels; no other anion-selective channel studied, either in planar bilayers by reconstitution methods or in the native membrane by electrophysiological techniques, shows more than a two-fold selectivity among the halides. Indeed, numerous "Cl^- channels" actually pass cations to a measurable extent (Colombini, 1980; Blatz and Magleby, 1985).

The conductance of the open channel was determined as a function of symmetrical Cl^- concentration (White and Miller, 1981). As with many ion-selective channels, the conductance–concentration relation follows a rectangular hyperbola, with a maximum conductance for the Cl^- channel of about 32 pS and a half-saturation constant of 75 mM Cl^-. It is notable that quantitatively identical results were obtained whether the channel was observed in a neutral membrane or one containing phosphatidylserine, a lipid with a negatively charged headgroup. This fact means that the channel's entryway for Cl^- is far removed (> 3 nM) from the lipid surface, so that the local negative surface potential, established by the charged lipid, does not influence the local Cl^- concentration near the channel's entryway. This is now hardly a surprising conclusion, as numerous membrane proteins are known to protrude far into the aqueous phase.

2.3. Reversible Blockers

While I^- and SCN^- do not carry current through the channel, they strongly affect permeation by Cl^-, acting as reversible competitive inhibitors, or "blockers" (Miller and White, 1980; White and Miller, 1981). The Cl^- current is reversibly reduced by SCN^- added in millimolar concentrations to the *cis* side of the channel, according to a simple reversible inhibition scheme. The channel acts as though the blocker binds in a rapid equilibrium and prevents Cl^- from entering. That SCN^- blocks the channel rapidly (much faster than the time response of the electronic detection system) was shown by the reduction in single-channel size by the blocker (White and Miller, 1981; Miller, 1982). At zero voltage, the K_i for SCN on the *cis* side is 3.3 mM; the blocker also works from the *trans* side, but with ten-fold lower affinity. I^- also blocks the channel from both sides, but about ten-fold less strongly than SCN^-.

Block by SCN^- and I^- is voltage-dependent. With SCN^- present on the *cis* side, the block becomes increasingly strong as voltage is made more negative, as though the blocking anion is being more strongly driven into the channel. With SCN^- on the opposite side of the membrane, precisely the reverse variation of blocking affinity occurs; block becomes stronger as voltage is made more positive. The effects are quantitatively explained by assuming that the channel contains two sites at which SCN^- can bind, each located about 35% of the way into the channel pore from its own side of the membrane.

The results are all consistent with the idea that this channel operates by a single-ion mechanism, in which at most one ion at a time can occupy the channel. While Cl^-

can traverse the entire conduction pathway, SCN^- can only enter from each side, but cannot get across the barrier somewhere in the center of the channel. We emphasize that this picture remains unproven; it is merely the most economical interpretation of the behavior of reversible blockers in this channel.

2.4. Irreversible Inhibition by DIDS

The compound DIDS has been known for some time to be a potent irreversible inhibitor of the Cl^-/HCO_3^- exchanger of the erythrocyte membrane, which is the best studied Cl^- transporter. DIDS also inhibits the macroscopic Cl^- conductance induced by *Torpedo* vesicles (White and Miller, 1979), as shown in Fig. 4. When DIDS is present on the *cis* side, the inhibition is irreversible, and occurs over a few minutes at a concentration of 1 μM. On the *trans* side, DIDS has no effect, even at 100-fold higher concentrations. This result (which clearly illustrates the nearly perfect orientation of the channels in the planar bilayer) shows that the channel exposes a specific DIDS-susceptible site to the *cis* solution. The reaction of DIDS with the Cl^- conductance shows simple bimolecular inhibition kinetics (Miller and White, 1980), with the rate of reaction linearly dependent on DIDS concentration.

We emphasize that DIDS is *not* a "specific Cl^- transport inhibitor," a misnomer given this molecule because of its great utility for studying the red cell Cl^-/HCO_3^- exchanger. DIDS is a very nonspecific bifunctional activated isothiocyanate compound, and as such is a good target for nucleophilic attack, especially (but not exclusively) by unprotonated amino groups. DIDS reacts readily with many lipids and proteins, including ion-pumping ATPases (Niggli *et al.*, 1982). The compound's negative charge may, in some cases, act to draw it near sites with an excess of positive charge, as may be expected to exist in anion-binding proteins, but in no way should this compound be mistaken for a general Cl^- channel inhibitor or label. Indeed, it is entirely possible that many sites on the channel protein are attacked by DIDS but that only one of these is crucial for the channel's activity. As we will see below, DIDS has been useful to us, but not because of any imagined specificity of the inhibition reaction;

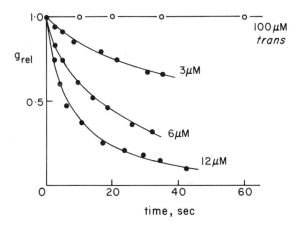

Figure 4. Inhibition of macroscopic Cl^- conductance by DIDS. Macroscopic conductance, as in Fig. 3, measured at -40 mV, after addition of DIDS to the concentration indicated. Except for the negative experiment labeled "*trans*," DIDS was added to the *cis* side. (Reproduced, with permission, from Miller and White, 1980.)

rather it is the irreversibility of the DIDS effect that has provided us with a tool to analyze the physical structure of this channel protein.

3. BINOMIAL GATING OF SINGLE CHANNELS: THE DOUBLE-BARRELED Cl⁻ CHANNEL COMPLEX

The conclusions described above were all reached on the basis of the macroscopic behavior of membranes containing many Cl⁻ channels. A picture emerged from these studies of a very conventional channel showing voltage-dependent gating, ion selectivity, blocking by analogues of permeating ions, and irreversible inhibition by nasty group-directed reagents. Initial work at the single-channel level (Miller and White, 1980) was in full harmony with these expectations. When observed at a depolarized holding potential, say +20 mV, conventional single-channel behavior is seen (Miller and White, 1980); channels fluctuate between two well-defined states of zero and 20 pS conductance (in 150 mM Cl⁻). These fluctuations are slow, on the order of a few seconds as are the macroscopic relaxation times, and so we were confident in describing this as a simple "open–closed" two-state channel (White and Miller, 1979) of unitary conductance of 20 pS. Under these conditions, we often saw multiples of this 20-pS channel, but never anything of smaller conductance.

However, at strongly negative voltages the *Torpedo* Cl⁻ channel exhibits a most astonishing characteristic (Fig. 5, top two traces) utterly unprecedented when first observed (Miller, 1982). Now, the channel fluctuations are much more complex. The channel switches between a zero-conductance state lasting several seconds and a "burst" of activity, also lasting seconds. During the burst, the channel rapidly fluctu-

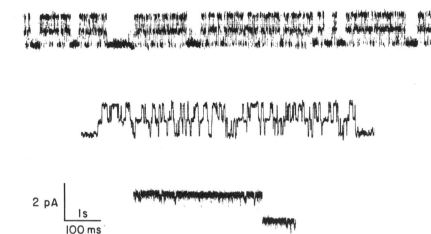

2 pA

1s

100 ms

Figure 5. Single Cl⁻ channels. A single Cl⁻ channel was observed in symmetrical 150 mM KCl solutions. Top and middle traces: −75 mV holding voltage. Middle trace is a ten-fold-expanded time scale trace, as indicated on the time bar. Lower trace: a channel closing event occurring 5 sec after switching the voltage to +70 mV.

ates among *three* conductance levels of 0, 10, and 20 pS. When the voltage is rapidly switched to a highly positive potential (Fig. 5, bottom trace), no such complex behavior is seen. An open channel conductance level of 20 pS is seen, until the channel turns off (never to reopen at this highly positive potential).

This puzzling behavior immediately raises several questions, which really reduce to the single question: what on earth is going on here? What is the meaning of the single-channel "substructure," i.e., of the three levels within a burst at −90 mV? Why do the conventional-looking channels seen at positive voltages so profoundly change their behavior at highly negative voltages? Is there any single way to view this bewildering single-channel behavior?

3.1. A Model: The Binomial Burst

From simple inspection of the single-channel record (Fig. 5), it is clear that we are dealing with at least four different states: the long-lived zero-conductance state, which we call "inactivated" (I), and the three short-lived substates of the burst, which we call "up" (U, 20 pS), "middle" (M, 10 pS), and "down" (D, 0 pS). Also by direct inspection of the channel records, we find that transitions occur according to the following scheme:

$$U \rightleftharpoons M \rightleftharpoons D$$
$$\diagdown \quad \diagup \quad \text{(1)}$$
$$I$$

Rapid transitions between U and D must proceed through M (Miller, 1982), and I can be reached from both U and M. Transitions between I and D may occur, but we have no way to detect them directly, since both states are nonconducting.

A simple physical model of the Cl⁻ channel helped to rationalize this unusual bursting behavior and immediately suggested numerous experimental tests (Miller, 1982). This "double-barreled shotgun" model of the channel is cartooned in Fig. 6. The channel is envisioned as a complex of two separate, distinct, and identical "protochannels." Each of these is a physical diffusion pore for Cl⁻ with its own "gate" operating on a millisecond time scale. A second gating process, operating on a time scale of seconds, acts to expose or occlude *both* protochannels simultaneously. We term the fast gating of the protochannel the "activation" process, and the slow gating of the double-protochannel complex the "inactivation" process.

It is this slower gating process that leads the double-barreled complex into and out of the I state in the scheme above. Once out of the I state, the channel enters a burst, whose three-level substructure is rationalized at once. The U, M, and D states correspond to configurations of the complex with both, one, or neither of the protochannels in the open state. The heart of this model is the postulate that the channel is a complex of two physically distinct Cl⁻ diffusion pathways, each with its own rapid gating process. Two auxiliary hypotheses were offered for the sake of simplicity (Miller, 1982): (1) that the protochannels are *physically identical,* and (2) that they operate *independently* of one another. We refer to this set of hypotheses as the "binomial burst" model of the Cl⁻ channel.

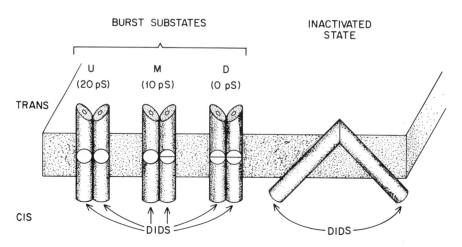

Figure 6. Cartoon of double-barreled Cl⁻ channel illustrating the four substates discussed in the text. The open circles represent the open state of the protochannel. The circles bisected with a horizontal line are intended to represent those protochannels as closed. U, M, D, and I are defined in the text.

3.2. Evidence Bearing upon the Binomial Burst Model

The above picture is appealing in its pithiness, but is it valid? Surely, there are many examples of channels with "substate" behavior in which the open channel fluctuates among several conductance levels (Colombini, 1980; Labarca and Miller, 1981; Hamill *et al.*, 1983; Bormann *et al.*, 1987; Fox, 1987). What is it about the Cl⁻ channel's substate behavior that emboldens us to propose a physical picture to account for it? The reason is that the binomial burst model makes several necessary, *a priori*, quantitative predictions about the characteristics of the three-level burst. None of these would rule out other substate models, but they are all absolutely *demanded* by the binomial burst model. Since they are all confirmed, we consider this a real triumph for a remarkably simple picture of an apparently complicated channel. We will now review these predictions.

3.2.1. Equal Spacing of Substate Conductances

The most obvious behavior expected from a pair of identical and independent protochannels is an equal spacing of the conductance levels observed. The U state should be exactly twice the M state, and the D state should have zero conductance. This prediction is satisfied to within 5% (Miller, 1982). Table 1 shows the conductances of the M and U states under a variety of experimental conditions. (D-state current was never any different than the I-state current, which was equal to the low background leak of the unmodified bilayer.) It is clear that regardless of the ion type, ion concentration, or the presence of a voltage-dependent blocker, the conductance levels are equally spaced. There is no particular reason why a "general" substate

Table 1. Conductance of Cl⁻ Channel Substates Measured
under a Variety of Ionic Conditions

Condition	γ_U (pS)	γ_M (pS)	γ_U/γ_M
150 mM Cl⁻ (low V)	18.5 ± 0.5	9.4 ± 0.1	1.97 ± 0.08
150 mM Cl⁻ (−135 mV)	15.9 ± 0.3	8.1 ± 0.2	1.97 ± 0.09
500 mM Cl⁻ (−90 mV)	21.9 ± 0.2	10.9 ± 0.2	2.01 ± 0.06
150 mM Cl⁻ + 1 mM SCN⁻	7.7 ± 0.1	3.7 ± 0.1	2.08 ± 0.09
150 mM Br⁻	6.1 ± 0.1	2.9 ± 0.1	2.07 ± 0.14

model, in which a *single* diffusion pore displays several conformations, each with its own conductance, would show equally spaced substates.

3.2.2. Binomially Distributed Substate Frequencies

Another strong prediction for a pair of independently opening and closing identical protochannels is that the probabilities of observing the three different levels be binomially distributed (Miller, 1982; Hanke and Miller, 1983). Suppose that under a given set of conditions, each protochannel has a "fundamental probability," p, of being in its open state. Then the probabilities of observing the three substates, f_D, f_M, f_U, are given by the frequencies of observing exactly 0, 1, or 2 channels open:

$$f_D = (1 - p)^2 \tag{1a}$$

$$f_M = 2p(1 - p) \tag{1b}$$

$$f_U = p^2 \tag{1c}$$

The substate frequencies, f_i, are measured directly from the single-channel record (within a burst). Furthermore, the fundamental probability p is directly determined as well:

$$p = \langle g \rangle / 2\gamma_o \tag{2}$$

where $\langle g \rangle$ and γ_o are the time-averaged conductance in the burst and the single-protochannel conductance, respectively. Therefore, a *single* measurement, that of time-averaged current during a burst, will demand an *a priori,* quantitative prediction of all three substate frequencies.

This prediction is fulfilled, as is shown in Fig. 7. Here, we display the predicted and measured substate frequencies at two different values of p, which we vary by changing voltage, as is discussed below. The data fit a binomial distribution well, and this has been our experience under all conditions (Miller, 1982; Hanke and Miller, 1983). Again, since a general substate model would make no particular demands for

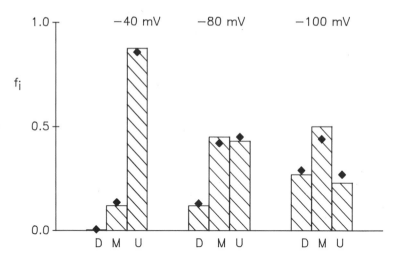

Figure 7. Binomial distribution of substates. Single Cl⁻ channel bursts were collected at the indicated voltages, and the observed frequencies, f_i, of the three substates were measured at each voltage (filled diamonds). The time-averaged current during the burst was also measured to obtain the value of protochannel open probability (Eq. 2), to calculate the values of f_i expected for a binomial distribution (hatched bars).

the substate frequencies, this result gives strong confirmation to the binomial burst model.

3.2.3. Kinetics of Transitions among Substates

A third set of predictions of the model concerns the rules for transitions among the three substates and the rates of these transitions. According to the model, the only way, while staying within a burst, of proceeding from the U state to the D state is through the M state. This is simply a statement of the fundamental assumption of Poisson processes, that two events cannot occur at *exactly* the same time; in order for two open protochannels to close, one of them must close first. However, we *do* sometimes observe transitions between the U and D states, and this observation would seem to rule out the model unequivocally.

But there is a practical problem to be considered before we pack up shop: the finite speed of the electronic measuring device. In order to achieve an acceptable signal-to-noise ratio, we must filter the signal, and so we cannot detect events shorter than about 1 msec. But this means that all U–M–D transitions with dwell times in the M state shorter than 1 msec will be detected as *direct* U–D transitions. Fortunately, given the characteristics of the amplifier and the observed M-state dwell times, we can calculate the expected probability of such false transitions between U and D, and this agrees well with what we actually measure (Miller, 1982). Therefore, we are still safe in assuming that the "binomial" transition rules apply.

The binomial burst model also makes a quantitative prediction about the rates of transitions among the three different levels. The dwell times in each level should be distributed exponentially, and this is the case (Hanke and Miller, 1983). Furthermore, if we define λ and μ as the fundamental rate constants of channel opening and closing, respectively, we demand that the observed mean dwell times in the three substates are given by

$$\tau_U = 1/2\,\mu \qquad (3a)$$

$$\tau_M = 1/(\lambda + \mu) \qquad (3b)$$

$$\tau_D = 1/2\,\lambda \qquad (3c)$$

Furthermore, the fundamental probability, measured directly from Eq. (2), is

$$p = \lambda/\mu \qquad (4)$$

Thus, the two unknown fundamental rate constants are given by *four* independently measured parameters, p and the three substate dwell times; the rate constants are overdetermined. We could, for example, calculate λ and μ using p and τ_D only (Eqs. 3c and 4), and then using τ_U and τ_M only (Eqs. 3a and 3b). If the binomial model is correct, these two different calculations of rate constants must agree. Such agreement is in fact observed, over a range of voltages where λ and μ vary considerably (Miller, 1982; Hanke and Miller, 1983).

3.2.4. DIDS Inhibition at the Microscopic Level

We have seen that DIDS irreversibly inhibits this channel. We do not have any idea of the mechanism of such inhibition. But it is tempting to guess that perhaps the anionic reagent interacts with the Cl⁻ conduction pore. If this is so, and if the channel complex is built as a pair of Cl⁻ conduction pores, then we *might* hope to observe single DIDS hits inhibiting the protochannels *one at a time*. This is in fact the case (Miller and White, 1984).

In Fig. 8 we show the action of DIDS at the single-channel level. Here, a single channel is observed for several seconds, and then a low concentration of DIDS is added to the *cis* side, with a few seconds of stirring. At about 5 sec after adding DIDS, a profound change in the burst occurs. Suddenly, the U state of the channel disappears! The channel still bursts, but the burst now shows only two conductance levels identical to the D and M states. After 20 sec more, the channel disappears, never to return.

We take this result as very strong evidence for the physical separation of the two protochannels, that the first DIDS hit inhibits one of the Cl⁻-conducting pores, leaving its twin unaffected, still opening and closing. Finally, the second DIDS hit destroys the remaining protochannel. This interpretation is strengthened by the fact that the probability of observing the isolated protochannel (between the first and second DIDS hits) is close to the fundamental opening probability calculated from the dimeric complex before the first DIDS hit (Miller and White, 1984).

Of course, this result was not required by the double-barreled channel picture;

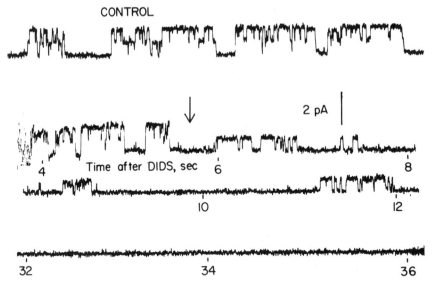

Figure 8. DIDS inhibition of single protochannels. A single Cl⁻ channel was observed at −80 mV (top trace), and then at "zero time" 10 μM DIDS was added to the *cis* side with vigorous stirring. At 4 sec, the stirrer was turned off, and the channel was observed for the next half minute. Arrow indicates the time at which the first protochannel disappeared, presumably as a result of a "hit" by DIDS. (Reproduced, with permission, from Miller and White, 1984.)

DIDS need not have exerted its effect at the single protochannel level. But it is a result very difficult to reconcile with any model postulating the substates as different conformations of a *single* pore.

3.3. Voltage-Dependence of Substates

We have left unresolved a basic puzzle: why does the single channel behave in a conventional way at positive voltages, with a single conducting state of 20 pS and slow gating, while showing this three-level "dimeric burst" at highly negative voltages? The reason becomes obvious in Fig. 9: the gating of the protochannels is voltage-dependent. At voltages more negative than −100 mV, the protochannels, when bursting, are most often closed; at more depolarized potentials, −70 mV to −40 mV, the protochannels begin to open more, and the three-level substate behavior becomes apparent. As voltage becomes more positive than −30 mV, the protochannel open probability approaches unity, and *both* protochannels remain open virtually all the time. At these voltages, then, only the U state is seen in the channel burst. The slow openings and closing of the inactivation gate, transitions between U and I, produce the appearance of a conventional single-channel record. But we must not be fooled (as we originally were): the simple 20-pS "open state" of this channel at +20 mV, say, represents the opening of a *pair* of protochannels, each with a conductance of 10 pS.

Once all this is understood, it is a simple matter to determine the probability of

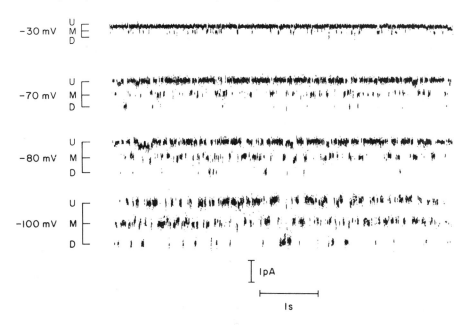

Figure 9. Voltage-dependent protochannel gating. Single-channel bursts examined at different holding voltages. The three substate levels, U, M, and D are indicated. (Reproduced, with permission, from Hanke and Miller, 1983.)

protochannel opening as a function of voltage ($p–V$ curve). This function follows a classical "open–closed" voltage-dependence (Labarca *et al.*, 1980), as shown in Fig. 10. Under standard conditions, at pH 7.3, symmetrical 150 mM Cl⁻, the "half-saturation voltage," V_o, is about -90 mV. This opening probability represents the protochannel's activation gate *within* a burst. The probability of *forming* a burst, determined by the inactivation gate, is also voltage-dependent, in the opposite direction, as shown as a dashed curve in Fig. 10. Thus, at very negative potentials, the channel is almost always in a burst, but the bursting protochannels are almost never open. In contrast, at positive potentials, the channel is almost always inactivated, but when a burst occurs, the protochannels are virtually always open. If activation and inactivation are independent, then the steady-state *macroscopic* conductance will be the product of the two curves of Fig. 10. Our early macroscopic measurements (White and Miller, 1979) are consistent with these expectations.

3.4. Activation by Protons

The activity of the *Torpedo* Cl⁻ channel is strongly affected by the pH of the medium in the *cis* aqueous solution (Miller and White, 1980). One dramatic effect of *cis* pH is on the protochannel $p–V$ curve (Hanke and Miller, 1983), illustrated in Fig. 11. Raising the pH closes the protochannel by shifting the $p–V$ curve to the right along the voltage axis, 50–60 mV per pH unit. The slope of the $p–V$ curve does not

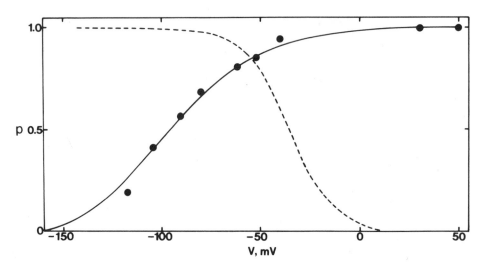

Figure 10. Activation and inactivation gates: voltage dependence. Solid curve with points: individual protochannel activation curve, i.e., probability of opening within the burst. Dashed curve: probability of burst formation. If activation and inactivation are independent processes, the macroscopic conductance-voltage curve will be the product of these two probabilities.

significantly change with pH. Thus, holding voltage fixed, an increase in the proton activity on the *cis* side of the membrane "activates" this channel.

A quantitative explanation for this effect was offered and tested by analysis of the protochannel kinetics (Hanke and Miller, 1983). The channel was considered to carry a dissociable group whose state of protonation controls the open probability according to the scheme below:

$$pK_c \quad \begin{array}{ccc} \text{CLOSED} & \rightleftharpoons & \text{OPEN} \\ \updownarrow & & \updownarrow \\ \text{CLOSED-H} & \rightleftharpoons & \text{OPEN-H} \end{array} \quad pK_o$$

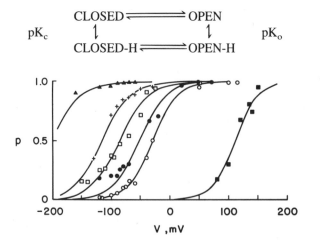

Figure 11. Effect of pH on protochannel gating. Protochannel activation curves are shown at varying pH values. Reading from the leftmost curve, pH was: 6.0, 7.0, 7.5, 8.0, 8.5, 10.4. (From Hanke and Miller, 1983.)

The idea is that the two pKs above are different, although the same titratable group is involved. To account for the results, we assigned a pK of 6 to the closed channel and a pK of 9 to the open channel. Thus, the open channel binds protons 1000-fold more strongly than the closed channel does, and so adding protons to the medium drives the channel toward the open state. A fully cyclic scheme was required in order to account for the kinetics of the single protochannels as a function of pH.

The reason this effect is interesting is that the data provide good evidence that the conformational change leading to protochannel opening alters the pK of a group on the channel protein by three units. The biochemical literature is replete with examples of large perturbations of pK upon conformational change, and in these cases the reason is often the movement of the dissociable group near to another charged residue. We speculate that such is the case for the Cl⁻ channel as well.

In summary, then, the assumption that this channel is built as a dimeric complex has allowed us to interpret naturally a diverse set of behaviors that would otherwise appear extremely complicated. The binomial burst model is a highly constrained version of the double-barreled picture of the channel complex. It is surprising that the assumptions of equivalence and independence of the two protochannels work so well; if the two "barrels" are close to each other in the same channel complex, we might expect the state of one to influence the behavior of the other. Therefore, we will not be surprised to find some violations to perfect binomial behavior in future studies on this channel.

4. APPARENT VIOLATION OF MICROSCOPIC REVERSIBILITY IN SINGLE-CHANNEL GATING

When examined at the single-channel level, the *Torpedo* Cl⁻ channel exhibits a most surprising phenomenon: a violation of detailed balance. Figure 12 illustrates the point. Here, a single channel is examined, in particular the pathway of entering and leaving the inactivated state. Notice that most channel bursts end from the M state, but begin in the U state. Time asymmetry is clearly present in the single-channel record. Described in another way by considering only those states observably communicating with the I state, we can say that there is a *net circulation* around the cycle (clockwise, as illustrated below):

This result means that the reactions involving the I state are *not at equilibrium*. Some form of free energy is necessarily being utilized to drive the time-asymmetric channel behavior. At first realization, this conclusion is a trifle upsetting, because the channel is being observed in a minimal reconstituted bilayer system, where there are no externally added energy sources like ATP to keep any reaction away from equilibrium. So how can this be? Where does the energy come from to keep the cycle above turning? We do not know the answer to this question definitively, but we do have a

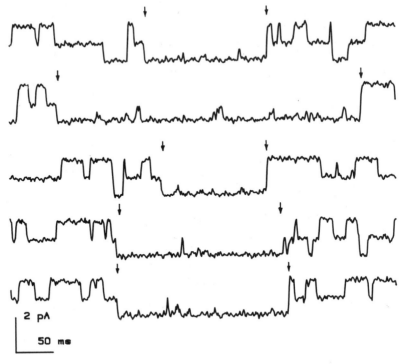

Figure 12. Time asymmetry of single-channel record. Single channels were recorded at −80 mV, and the states immediately preceding or following an inactivated state were examined. The figure displays five successive inactivated intervals (defined as nonconducting intervals longer than 100 msec). Arrows mark the states just before and just after each inactivated interval. The record shows a clear time asymmetry: the inactivated state is most often entered from the M state, and it is most often terminated by the U states.

plausible, and testable, explanation: that the energy comes from the Cl^- current through the channel, that the gating process is strongly coupled to Cl^- flowing through the channel down its thermodynamic gradient.

In fact, we consider that the view we have already developed of this channel leads naturally to models that almost *demand* a violation of detailed balance. To illustrate this point qualitatively (Fig. 13), let us consider a hypothetical single channel, of the conventional kind with only a single open state, which has the property we have postulated for the protochannel: that is contains *two* gating processes due to the operation of a slow inactivation gate and a fast activation gate. Let us further assume that the two gates are located on opposite sides of the channel, as pictured schematically in Fig. 13, which displays the four possible channel conformations, with activation and inactivation gates either open or closed.

According to the notation in Fig. 13, state 4 is the only conducting state, the only state with *both* activation and inactivation gates open. The key point to be considered here is that the three forms of the nonconducting channel can be either occupied or unoccupied by a Cl^- ion. Furthermore, the Cl^- ion occupying state 1 must come from

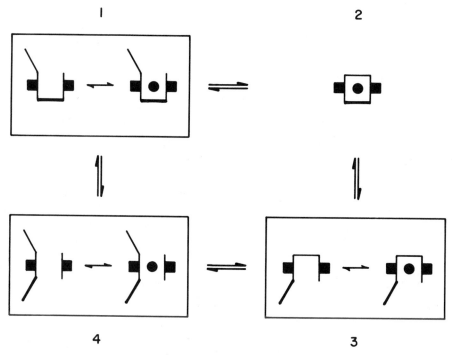

Figure 13. Model for coupling of gating to Cl⁻ gradient. A hypothetical channel spanning a membrane (solid bar) is considered to have two gates: a rapid "activation gate" (light line) and a slow "inactivation gate" (heavy line). Each gate can be either open or closed, as pictured. The various conformational states of the channel can be either unoccupied or occupied by a Cl⁻ ion, as shown by a filled circle. The ion occupancy reactions are assumed to be much faster than the gating reactions, and so the occupied and unoccupied states of a given conformation are "lumped" in the drawing. Note Cl⁻ can enter and leave state 1 only from one side of the membrane, and state 3 only from the other side of the membrane. State 4 is the only conducting state.

one side of the membrane (arbitrarily called the "outside"), while the ion occupying state 3 must come from the opposite side (the "inside"). The Cl⁻ ion in state 2 is occluded, and is not in communication with either side of the membrane. (For purposes of simplicity, we will not consider the additional state with both gates closed and the pore unoccupied.)

When there is no electrochemical gradient for Cl⁻ across the membrane (e.g., equal concentration, zero voltage), all reactions in Fig. 13 are at equilibrium; the number of transitions per unit time between contiguous states is the same in the forward and reverse directions. In other words, detailed balance holds everywhere. There is no net circulation around the cycle under these conditions.

Now consider the effect of removing Cl⁻ from the inside solution. Under these conditions, transitions can occur in both directions between states 1 and 2, states 1 and 4, and states 3 and 4. However, without Cl⁻ on the inside, transitions between states 2 and 3 are effectively one-way; state 2 can go to state 3, but the back reaction is never observed, since Cl⁻ dissociates rapidly as soon as the inside-facing gate opens. Once the system has traveled from state 1 to state 3 via state 2, it can return to state 1 *only* via state 4. Thus, a transmembrane Cl⁻ gradient produces a *net circulation* around the states of the system. For every turn of the cycle, a single Cl⁻ ion is transferred from a region of high electrochemical potential to one of low potential, through the *nonconducting* states of the channel; this is the energy source that keeps the cycle turning.

But what does this have to do with the Cl⁻ channel? If we were really observing a channel operating by the model described above, we would have no way of detecting the net circulation around the cycle; we would have no way of knowing which state the channel was in before it opened, state 1 or state 3. However, in a double-barreled channel, with both barrels sharing the same inactivation gate, information about the next-to-open state is available. The question is: just before the inactivation gate opens, is the protochannel's activation gate open or closed? The *first* state observed within a burst answers this question; if the U state is seen first, then *both* activation gates must have been open just before the inactivation gate opened. On the other hand, if the M state begins the burst, at least one of the activation gates was closed just before the burst.

In this way, by compiling the first and last states of a burst, we can identify the state of one protochannel's activation gate just before and just after the inactivated state. These are the observations that reveal the channel's violation of detailed balance, and demand that some sort of energy source is coupled to the gating process. The model above provides a natural explanation of this coupling: that the free energy of Cl⁻ ions falling through the *nonconducting* states of the protochannel drives the net circulation observed.

This model leads to several important predictions. First, the degree of asymmetry should decrease as the driving force for Cl⁻ (and hence the observed Cl⁻ current) decreases. (The asymmetry must disappear at the Cl⁻ reversal potential, where the channels cannot be observed, of course.) This prediction has been confirmed: in symmetrical Cl⁻, the degree of asymmetry increases with applied voltage. A second prediction is that the asymmetry must *reverse* when the current through the channel reverses. Cl⁻ ions flowing through the channel's nonconducting states in the opposite direction will necessarily turn the cycle in the opposite direction. We are presently attempting to test this crucial prediction.

The above arguments rationalize qualitatively the unusual time-asymmetric behavior observed for the single Cl⁻ channel. The ideas of Fig. 13 do not uniquely account for the channel's asymmetric gating. But we are impressed that the general picture of the channel containing two spatially separated gates does lead to an immediate prediction of this phenomenon. We therefore take the observed time asymmetry as evidence suggestive of the idea that the two gates are physically separated in space, and that a form of the channel exists in which a Cl⁻ ion is occluded between the two gates. In order for this model to account for the observed direction of asymmetry (Fig. 12), we would require the inactivation gate to be located toward the *cis* side of the membrane, and the activation gate toward the *trans*.

More complicated models predicting a violation of detailed balance can be constructed (Finkelstein and Peskin, 1984; Kirber *et al.*, 1985), but all of them require that ion flow through the channel is coupled to the opening and closing reactions. Indeed, Läuger (1983) has treated theoretically the general problem of nonequilibrium channel behavior arising from coupling of ion translocation to gating. The Cl^- channel provides an actual experimental example where a very simple mode of ion-gating coupling accounts for the observed violation of detailed balance at the single-channel level.

5. RELATION TO OTHER Cl⁻ CHANNELS

It is especially difficult to compare the *Torpedo* Cl^- channel with other anion-selective channels, simply because the characteristics of the *Torpedo* channel in its native membrane are unknown. Indeed, though we hypothesize on the basis of membrane purification and flux data (White, 1981; White and Miller, 1981) that this channel resides in the plasma membrane of the noninnervated face of the electrocyte, there is no solid evidence identifying the native location of the channel. Nevertheless, assuming that the channel as studied in planar bilayers faithfully reproduces its behavior *in situ,* it is worthwhile to compare this channel with others.

The most striking property of the *Torpedo* channel, when considered alongside other anion-selective channels, is its high selectivity for Cl^-. All other Cl^- channels studied in detail, including the GABA- and glycine-gated channels (Hamill *et al.,* 1983; Bormann *et al.,* 1987; see Bormann, this volume), the voltage-dependent large Cl^- channel of muscle (Blatz and Magleby, 1985; see Blatz, this volume) and epithelial (Nelson *et al.,* 1984) cells, and a Cl^- channel reconstituted from cardiac membranes (Coronado and Latorre, 1982), are really not Cl^- specific. Rather, they select for anions over cations, but allow many types of anions to pass readily. In contrast, the *Torpedo* channel allows only Cl^- and Br^- to permeate.

The multibarreled property of this channel has now been proposed for several other channels as well. A Cl^- channel from alveolar epithelial cells (Krouse *et al.,* 1986) shows complex multistate single-channel behavior that led to the proposal of a structure containing 6 protochannels in parallel. Likewise, a Cl^- channel from molluscan neurons (Geletyuk and Kazachenko, 1985) has been proposed to be made up of 16 protochannels. Recently, Hunter and Giebisch (1987) showed that a K^+ channel from kidney epithelia behaves quantitatively as expected for a four-barrel channel, with all four protochannels sharing a common gate. The *Torpedo* channel, however, remains the simplest and clearest example of a multibarreled structure. Identical single-channel behavior has been observed in electroplax membranes from all other *Torpedo* species tested (*californica, nobiliana, marmorata*). The significance of such a structure is a mystery to us, and we are unwilling to offer any speculation regarding its function, teleology, or evolutionary history.

6. FUTURE DIRECTIONS

Our detailed knowledge of the *Torpedo* Cl^- channel's single-channel behavior is emphasized all the more by our nearly total ignorance of the channel's physiology on the one hand and biochemistry on the other. The major challenge for future research on this channel is to purify the protein or to directly clone the genes involved. The lack of high-affinity channel-directed ligands or inhibitory antibodies has stalled our efforts in this direction. It is tempting to use DIDS as a ligand to try to get a biochemical handle on the problem, as have Taguchi and Kasai (1980), but this avenue of attack is a very tricky one, as DIDS labels many proteins in *Torpedo* membranes. Recently, Yench, Garcia, and Lodish (unpublished) have cloned a DIDS-binding protein from *Torpedo* and are presently using mRNA injection into oocytes to try to express the Cl⁻ channel. A successful result in this area would break the biochemical impasse, and would open up this channel to a full molecular analysis.

There are still numerous biophysical questions remaining, however, which can be attacked by single-channel analysis. In particular, we have focused most of our efforts to date on the protochannel's activation gate, its voltage- and pH-dependence, and kinetic behavior. Less attention has been given to the inactivation process, which operates on both protochannels simultaneously. We know that this process is also voltage and pH sensitive, and it would be interesting to know how tightly coupled the inactivation process is to protochannel activation. Are these two gating processes independent, or does the inactivation gate "know" about the state of the protochannel activation gates? These are the sorts of questions that can be made to yield to single-channel analysis.

Finally, the physiology of this channel, and even its location, remain obscure. It might be, as proposed (White, 1981), that this channel resides in the noninnervated-face plasma membrane of the electrocyte, and that it provides the cell with the very high-conductance resting potential required for its current-generating function (Bennett *et al.*, 1961). But as long as direct electrophysiological recordings from the electrocyte remain unachieved, this question will stay unsettled.

REFERENCES

Andersen, O. S., 1983, Ion movement through gramicidin A channels. Single-channel measurements at very high potentials, *Biophys. J.* **41**:119–133.

Bennett, M. V. L., Wurzel, M., and Grundfest, H., 1961, The electrophysiology of electric organs of marine fishes. I. Properties of electroplaques of *Torpedo nobiliana*, *J. Gen. Physiol.* **44**:757–804.

Blatz, A., and Magleby, K. L., 1985, Single chloride-selective channels active at resting membrane potentials in cultured rat skeletal muscle, *Biophys. J.* **47**:119–123.

Bormann, J., Hamill, O. P., and Sakmann, B., 1987, Mechanism of anion permeation through channels gated by glycine and γ-aminobutyric acid in mouse cultured spinal neurones, *J. Physiol. (London)* **385**:243–286.

Cohen, F. S., Zimmerberg, J., and Finkelstein, A., 1980, Fusion of phospholipid vesicles with planar lipid bilayer membranes. II. Incorporation of a vesicular membrane marker into the planar membrane, *J. Gen. Physiol.* **75**:251–270.

Cohen, F. S., Akabas, M. H., and Finkelstein, A., 1983, Osmotic swelling of phospholipid vesicles causes them to fuse with planar phospholipid bilayer membranes, *Science* **217**:458–460.

Colombini, M., 1980, Pore size and properties of channels from mitochondria isolated from *Neurospora crassa*, *J. Membr. Biol.* **53**:79–84.

Coronado, R., and Latorre, R., 1982, Detection of K⁺ and Cl⁻ channels from calf cardiac sarcolemma in planar bilayer membranes, *Nature* **298**:849–852.

Finkelstein, A., and Peskin, C. S., 1984, Some unexpected consequences of a simple physical mechanism for voltage-dependent gating in biological membranes, *Biophys. J.* **46**:549–558.

Fox, J., 1987, Ion channel subconductance states, *J. Membr. Biol.* **97**:1–8.

Geletyuk, V. I., and Kazachenko, V. N., 1985, Single Cl⁻ channels in molluscan neurones: Multiplicity of the conductance states, *J. Membr. Biol.* **86**:9–15.

Hamill, O. P., Bormann, J., and Sakmann, B., 1983, Activation of multiple-conductance state chloride channels in spinal neurones by glycine and GABA, *Nature* **305**:805–808.

Hanke, W., and Miller, C., 1983, Single chloride channels from Torpedo electroplax: Activation by protons, *J. Gen. Physiol.* **82**:25–45.

Hunter, M., and Giebisch, G., 1987, Multibarrelled K channels in renal tubules, *Nature* **327**:522–525.

Kirber, M. T., Singer, J. J., Walsh, J. V., Fuller, M. S., and Peura, R. A., 1985, Possible forms for dwell-time histograms from single-channel current records, *J. Theor. Biol.* **116**:1111–1126.

Krouse, M. E., Schneider, G. T., and Gage, P. W., 1986, A large anion-selective channel has seven conductance levels, *Nature* **319**:58–60.

Labarca, P., and Miller, C., 1981, A K⁺-selective three-state channel from fragmented sarcoplasmic reticulum of frog leg muscle, *J. Membr. Biol.* **61**:31–38.

Labarca, P., Coronado, R., and Miller, C., 1980, Thermodynamic and kinetic studies of the gating behavior of a K⁺-selective channel from the sarcoplasmic reticulum membrane, *J. Gen. Physiol.* **76**:397–424.

Lauger, P., 1983, Conformational transition of ionic channels, in: *Single Channel Recording* (B. Sakmann and E. Neher, eds.), Plenum Press, New York, pp. 177–189.

Miller, C., 1982, Open-state substructure of single chloride channels from Torpedo electroplax, *Philos. Trans. R. Soc. London Ser. B* **299**:401–411.

Miller, C., and Racker, E., 1976, Calcium-induced fusion of fragmented sarcoplasmic reticulum with artificial planar bilayers, *J. Membr. Biol.* **30**:283–300.

Miller, C., and White, M. M., 1980, A voltage-dependent chloride channel from Torpedo electroplax membrane, *Ann. N.Y. Acad. Sci.* **341**:534–551.

Miller, C., and White, M. M., 1984, Dimeric structure of Cl-channels from Torpedo electroplax, *Proc. Natl. Acad. Sci. USA* **81**:2772–2775.

Nelson, D. J., Tang, J. M., and Palmer, L. G., 1984, Single-channel recordings of apical membrane chloride conductance in A6 epithelia cells, *J. Membr. Biol.* **80**:81–89.

Niggli, V., Siegel, E., and Carafoli, E., 1982, Inhibition of the purified and reconstituted calcium pump of erythrocytes by DIDS and NAP-taurine, *FEBS Lett.* **138**:164–166.

Taguchi, T., and Kasai, M., 1980, Identification of an anion channel protein from electric organ of *Narke japonica*, *Biochem. Biophys. Res. Commun.* **96**:1088–1094.

White, M. M., 1981, Characterization of a Cl channel from *Torpedo* electroplax, Ph.D. thesis, Brandeis University.

White, M. M., and Miller, C., 1979, A voltage-gated anion channel from electric organ of *Torpedo californica*. *J. Biol. Chem.* **254**:10161–10166.

White, M. M., and Miller, C., 1981, Chloride permeability of membrane vesicles isolated from *Torpedo* electroplax, *Biophys. J.* **35**:455–462.

Chloride Channels in Skeletal Muscle

Andrew L. Blatz

The resting membrane conductance of amphibian and mammalian skeletal muscle is largely due to Cl^- ions (Hodgkin and Horowicz, 1959; Hutter and Noble, 1960; Palade and Barchi, 1977). This large anion conductance appears to stabilize muscle transmembrane voltage, as muscles with reduced Cl^- conductance exhibit myotonic-like trains of action potentials which lead to aberrant muscle function (Adrian and Bryant, 1974). A comprehensive review of Cl^- conductances in skeletal muscle and many other tissues has been provided by Bretag (1987). This chapter will focus only on mammalian and amphibian Cl^- channels at the single-channel level as observed with the patch voltage-clamp technique (Hamill *et al.*, 1981).

1. MAMMALIAN Cl^- CHANNELS

Single-channel patch-clamp experiments have revealed three types of Cl^--conducting channels in mammalian skeletal muscle (Blatz and Magleby, 1983, 1985). Two of these channel types exhibit some of the properties required of the channels underlying the large resting Cl^- conductance, but confirmation of whether or not they actually are responsible for this conductance is lacking. Mammalian Cl^- channels can be divided into two classes: (1) Cl^- channels with large single-channel conductances (over 200 pS in symmetric 140 mM Cl^-); and (2) Cl^- channels with much smaller single-channel conductances of less than 100 pS in symmetric 140 mM KCl. Both of these classes of Cl^- channels are voltage-dependent but their gating kinetics are entirely different.

1.1. Large-Conductance Cl^- Channels

Cl^- channels with the largest single-channel conductance of any ion channel studied with the patch-clamp technique were first described in tissue-cultured rat skeletal muscle (myotubes; Blatz and Magleby, 1983). These channels exhibited many fascinating properties and similar channels have now been reported to be present in a variety of tissues, including: rabbit urinary bladder (Hanrahan *et al.*, 1985), mouse

Andrew L. Blatz • Department of Physiology, University of Texas Southwestern Medical Center, Dallas, Texas 75235.

macrophages and chicken myotubes (Schwarze and Kolb, 1984), pulmonary epithelia (Schneider *et al.*, 1985; Krouse *et al.*, 1986), rat Schwann cells (Gray *et al.*, 1984), and rat mast cells (Lindau and Fernandez, 1986). Although there are some differences among the different preparations, these channels display many properties similar to those originally described in rat myotubes.

In myotubes, large-conductance Cl⁻ channels were observed in about 5–10% of inside-out membrane patches. In addition, large-conductance channels were usually not observed for several minutes following the formation of the inside-out patch configuration. Channel activity could often be elicited from silent patches by holding the membrane at large depolarized voltages for several seconds. These observations suggest either that a diffusible blocking agent must be washed away from the inside-out patch for channel activation, or that there is a direct effect of voltage on the onset of channel activity.

1.1.1. Gating Kinetics

Figure 1 demonstrates the typical response of a patch-clamped rat myotube membrane containing two large-conductance Cl⁻ channels to a voltage step (upper trace) from 0 mV to depolarizing and hyperpolarizing voltages. In this experiment, the inside-out membrane patch was bathed with a solution containing 140 mM KCl on both the formerly intracellular and extracellular surfaces. Under these conditions, no single-channel current events are observed at 0 mV, as there is no driving force on any of the ions present in the bathing solutions. When the membrane potential is stepped to a depolarized value, say +32 mV, as in the top current trace in Fig. 1, the membrane current jumps to a large outward value and then returns to the baseline in two discrete steps. In a similar manner, when the membrane is hyperpolarized, as in the bottom two current traces in Fig. 1, a large inward current occurs that returns back to the baseline in two steps. As will be shown below, these steps represent the opening and closing of two large-conductance Cl⁻-conducting channels. It is clear from the current records in Fig. 1 that the channels do not always remain closed during the voltage step, but can reopen, as is observed in the current trace in response to the −21 mV step. In this trace, the current relaxes toward the baseline as one of the channels closes, but then increases in an inward direction as the channel reopens. This channel then flickers between the open and closed states for the remainder of the voltage step. An important observation is that the current becomes much more noisy when the channels are in the open state than when the channels are closed. This indicates that the channels are actually opening and closing very rapidly during the voltage step and that the limited frequency of the patch clamp recording system is filtering these openings and closings, resulting in a noisy current when the channels appear to be open. Because these channels do not remain open during a sustained voltage step under these conditions, it was not possible to perform the detailed stationary single-channel kinetic analysis used on other channels such as the large-conductance Ca²⁺-activated K⁺ channel (Magleby and Pallotta, 1983a,b). However, a measure of the voltage-dependence of channel gating kinetics could be obtained by measuring the length of time a channel remained open following either depolarizing or hyperpolarizing voltage steps. It was found that these open time

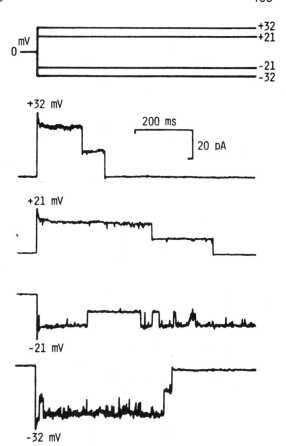

Figure 1. Single-channel currents recorded from an excised patch of plasma membrane (inside-out) from embryonic rat muscle cell (myotubes). Membrane potentials are expressed as the potential at the normal intracellular side minus the potential at the normal extracellular side. The membrane potential was held at 0 mV and then stepped as indicated in the schematic voltage traces (upper plot). The resulting current traces for each voltage step are plotted below. The small decline in current during the first 10 msec after each voltage step is a capacitive artifact. Upward current deflections correspond to outward currents. Room temperature (20–22°C). (Reproduced by permission from Rockefeller University Press; Blatz and Magleby, 1983.)

distributions following voltage steps could be described by single-exponential components. For a hyperpolarizing step from 0 mV to −30 mV the time constant of the distribution was 1.19 sec, while for a depolarizing step to +30 mV the currents declined with a much shorter time constant of 0.48 sec. Thus, these channels have a longer open lifetime during hyperpolarizing steps.

Although it appears from Fig. 1 that the channels are closed before the voltage step, open for a short time, and then close, another explanation is possible, and indeed seems to be the case. Since there is no driving force on the ions under these conditions (holding potential = 0 mV), it is impossible to know whether the channels are open or shut before the depolarizing or hyperpolarizing voltage step. To observe the possibility of channel activity at 0 mV, the [Cl⁻]ᵢ was raised to 400 mM. This places a net driving force on Cl⁻ even at 0 mV, and if the channels are active, then single-channel current events should be observed. It turned out that the channels were, in fact, open most of the time at 0 mV, and so the observed currents in response to voltage steps were probably due not to channel openings, but to increased current flow through already open channels when the driving force on the Cl⁻ was increased. This observation has no effect on the kinetic analysis presented above, as we were measuring channel lifetimes after the voltage step.

The response of the large-conductance Cl⁻ channels to a depolarization was not always as shown in Fig. 1, but sometimes occurred as shown in Fig. 2. With the membrane held at +40 mV, the patch current remains at the baseline level for several seconds while the current becomes noisier and noisier until the more discrete current steps, characteristic of single-channel openings and closings, occur. We call this phenomenon "megagenesis," and, as yet, have no real explanation for its occurrence. It is not noise due to membrane breakdown because it only occurs before channel activity becomes apparent. A very similar observation has been made for large-conductance Cl⁻ channels in frog skeletal muscle (Woll *et al.*, 1987). The overall conclusions about the typical gating kinetics of these channels are that the probability of a channel being in the open state is highest at zero membrane potential and that this probability is reduced for either positive or negative potentials·suggesting a bell-shaped distribution of open time probabilities.

1.1.2. Permeability Properties

The permeability properties of these large-conductance channels in rat myotubes were not studied in great detail. Current–voltage relationships were obtained in the presence of two different gradients of KCl across inside-out membrane patches. In both cases the formerly extracellular membrane surface was bathed with 140 mM KCl. When the solution bathing the intracellular surface contained 140 mM KCl (symmetrical conditions), the current–voltage relationship was linear and passed through zero current at 0 mV, as expected. When the $[Cl^-]_i$ was raised to 400 mM, the current now passed through zero at a positive potential of about +18 mV. This observation tells us two important things: the currents through these channels must be carried mostly by Cl⁻ because if K⁺ were carrying the current, raising the $[KCl]_i$ would have resulted in a negative reversal potential. The second conclusion from this observation is that these channels are not exclusively permeable to Cl⁻, but allow a significant amount of current to be carried by K⁺. This is demonstrated by the fact that if Cl⁻ were the sole current carrier, the current–voltage relationship should have passed through zero current at a more positive potential of around 25 mV instead of the observed 18 mV. A permeability ratio between Cl⁻ and K⁺ of about 4–6 can be

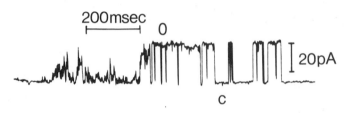

Figure 2. Onset of activity (megagenesis) of large conducting Cl⁻ channels in an excised, inside-out rat myotube membrane patch. Patch was held at +40 mV about 2–3 min after pulling the patch from the cell membrane. (Blatz and Magleby, unpublished observation.)

calculated from this datum, indicating that Cl⁻ is only about 4–6 times more permeant than K⁺ under these conditions. Similar results were obtained when the intracellular bathing solution was raised to 600 mM (Blatz and Magleby, 1983). It must be remembered that these experiments were performed under very unphysiological conditions, and so we do not know if a high cation permeability exists under more normal conditions.

1.1.3. Functions of Large-Conductance Cl⁻ Channels

Given the peculiar gating kinetics of these large-conductance Cl⁻ channels, which would be typically closed at normal resting membrane potentials, assigning a function to them is difficult. Blatz and Magleby (1983) presented a list of several possibilities including contributions to the resting Cl⁻ conductance, cell volume regulation, cell-to-cell communication channels, or even channels from mitochondria that have been somehow misplaced and inserted into the plasma membrane. None of these possibilities could be substantiated to any great degree and so the functions of these channels remain unknown.

1.2. Small-Conductance Cl⁻ Channels

At least two types of small-conductance Cl⁻ channels have been reported present in mammalian skeletal muscle (Blatz and Magleby, 1985). One type exhibits "fast" kinetics with open lifetimes on the order of 1 msec. This so-called fast Cl⁻ channel had a single-channel conductance of 45 pS in symmetric 100 mM KCl. The other channel had slower kinetics with open lifetimes of tens of milliseconds, and a somewhat larger single-channel conductance of 61 pS in symmetric 100 mM KCl. The channel with slower kinetics was observed only rarely, but the fast channel was present in the majority of membrane patches. These channels will be discussed together.

1.2.1. Gating Kinetics

Unlike the large-conductance Cl⁻ channels described earlier, the small-conductance channels are active throughout a given voltage step and do not show the pronounced inactivation of the large channels. Typical current records for the fast and slow Cl⁻ channels from an excised inside-out rat myotube membrane patch are shown in Fig. 3. Due to the small conductances of these channels, it was necessary to use very large concentrations of Cl⁻ in the intracellular bathing solutions. The large Cl⁻ concentrations increase both the channel conductance and the driving force on the permeant ions leading to a much better resolution of channel open and shut events. The difference in the kinetics of these two types of channels is clearly evident in Fig. 3. In Fig. 3A the fast Cl⁻ channel is presented at holding potentials of −30 and −80 mV, and the mean channel open time can be seen to be about 1 msec. A marked voltage-dependence of single-channel kinetics is also observed for the fast channel with the percent of time that the channel is open decreasing as the membrane potential is made more hyperpolarized. The slow Cl⁻ channel, presented in Fig. 3B at the same two

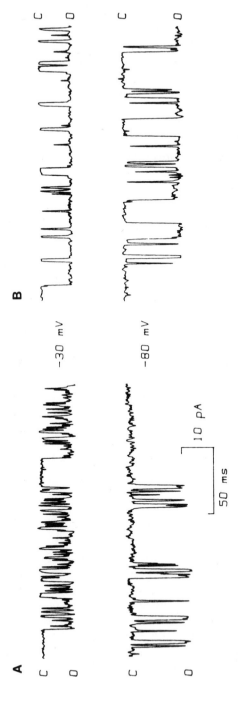

Figure 3. Single-channel currents recorded from excised patches of plasma membrane from cultured rat skeletal muscle. Currents from Cl⁻-selective channels with fast (A) and slow (B) kinetics are shown at the indicated membrane potentials. Closed (c) and open (o) channel current levels are indicated. Inward currents are plotted downward. (A) 1.6 M [KCl]$_o$. (B) 1.4 M [KCl]$_i$ and 100 mM [KCl]$_o$. Low-pass active filtering with cutoff of 1.6 kHz (24 dB/octave). Different membrane patch for A and B. (Reproduced by permission from Rockefeller University Press; Blatz and Magleby, 1983.)

holding potentials, on the other hand, has much less voltage-dependence, with the percent open time only slightly decreasing during the more negative voltage steps. The mean lifetime of the slow Cl^- channel is much longer than that of the fast channel, and in Fig. 3B can be measured to be about 10 msec. It should be noted that the current records shown in this figure were all filtered at the same cutoff frequency (1.6 kHz) and so the differences are not just artifacts of bandwidth limitations. Since it appeared so infrequently, the gating kinetics of the slow Cl^- channel could not be studied in detail. The fast Cl^- channel, however, occurred so often and remained active for long durations so that its kinetics could be thoroughly investigated.

One way of examining the kinetic behavior of ion channels is to collect large numbers of open and shut channel interval durations and then compare the distributions of these open and shut events with the predictions of kinetic reaction schemes. This was the method used by Blatz and Magleby (1986) to study the underlying kinetic reaction scheme of the fast Cl^- channel. If the underlying kinetic reaction schemes are Markovian (as they seem to be; McManus et al., 1985), then the distributions of open and shut intervals will be described by the sums of exponential components, with each component related to one of the open or shut kinetic states in the underlying kinetic scheme.

One requirement of such a kinetic analysis is, if there is more than one active channel in the membrane patch, that the individual channels do not interact with each other. In other words, one channel cannot know what the other channels are doing. This was demonstrated to be the case for the fast Cl^- channels by recording and digitizing extended segments of single-channel currents from patches that contained several active channels. When these data were plotted as an amplitude histogram (number of events of a given current magnitude versus current magnitude) the histogram could be accurately fitted with a binomial distribution which requires all channels to gate independently. Thus, unlike Cl^- channels from other tissues (Miller, 1982; Woll and Neumcke, 1987), fast Cl^- channels in mammalian muscle function independently.

Another requirement that must be met before a detailed kinetic analysis can be performed is that the channel under study be stable over the long periods of time required to collect large numbers of interval durations. An example of how channel stability can be measured is shown in Fig. 4. The top traces are plots of the moving mean duration for open and shut intervals. The means were calculated by adding the durations of either 50 open or 50 shut intervals and dividing by the number of intervals. These moving means are then plotted against the summed number of open or shut intervals. Since there is no way of predicting just how much open or shut variation to expect, one cannot tell whether the variation exhibited in Fig. 4 is excessive or merely represents the random fluctuations inherent in ion channel gating. As a measure of how much variation to expect, single-channel open and shut intervals were simulated by computer and treated to the stability analysis in the same way as the experimental intervals. The results of the analysis are shown in the lower two moving means in Fig. 4. Clearly, the experimentally observed channel activity is no more variable than would be expected. The moving mean plots in Fig. 4 represent only 4800 open and 4800 shut

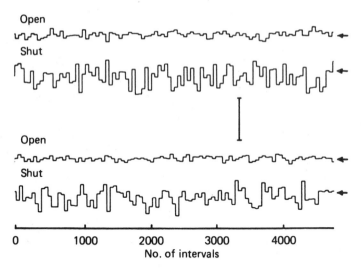

Figure 4. Testing for channel stability. Top two traces are moving means (see text) of open and shut fast Cl⁻ channel intervals plotted at 50-interval increments. Lower two traces are the moving means of simulated data treated exactly as the experimental data. Note that the vertical axis is logarithmic and represents one decade. Arrows indicate mean open and shut interval durations for both experimentally observed and simulated data sets. (Reproduced by permission from Cambridge University Press; Blatz and Magleby, 1986.)

intervals, but the entire 85,000 intervals used in the subsequent analysis were equally stable.

Although it is possible to analyze kinetic data from patches that contain more than one channel, the analysis is much simplified if the large numbers of open and shut events can be obtained from an individual channel in a patch that contains only that channel. In one experiment on the fast Cl⁻ channel in rat myotubes, Blatz and Magleby (1986) collected about 85,000 open and shut channel intervals at low temperatures. Figure 5 presents the distributions of the open and shut interval durations obtained from this experiment. The solid lines represent maximum likelihood fits of the sums of exponentials to the data. For the open intervals, the best fit was with the sum of two exponential components, a predominant component comprising 92% of the open intervals with a time constant of 1.5 msec, and a lesser component accounting for 8% of the intervals with a shorter time constant of 0.5 msec. Fitting these distributions with a single exponential gave significantly poorer fits, while fitting with three or more components did not improve the description of the data. For the shut distribution it was found that at least five exponential components were required for an adequate fit to the data, with time constants ranging from 0.06 to over 300 msec. Thus, based on these interval distributions, a minimal kinetic model for fast Cl⁻ channel activity must consist of at least two open states and at least five shut states. Using an iterative procedure, Blatz and Magleby (1986) found that, indeed, the experimentally observed

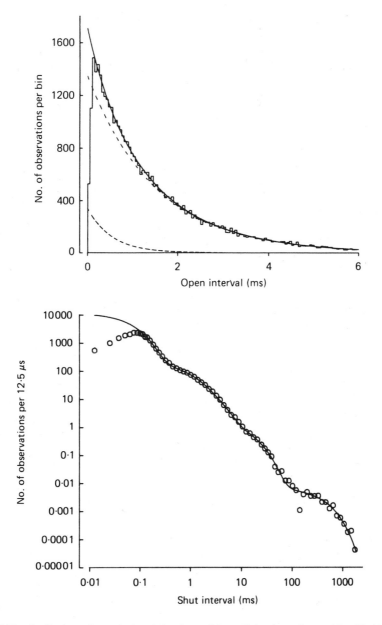

Figure 5. The distributions of open (top) and shut (bottom) interval durations of normal fast Cl^- channels are described by the sums of exponential components. The observed durations of 42,000 open and shut intervals are plotted as histograms. Note that the shut intervals are plotted as a log–log plot so that the wide range of durations and numbers of observations can be presented on a single graph. The solid lines in both plots are the best-fit sums of exponentials based on maximum likelihood fitting. Open distribution is fit with the sum of two exponential components, each of which are presented as dashed lines in the top graph. Shut distribution is fit with the sum of five components, the most prominent of which are responsible for the bumps in the fit (solid line). Membrane potential −40 mV. Temperature 7.6°C. (Reproduced with permission from Cambridge University Press; Blatz and Magleby, 1986.)

open and shut interval distributions could be fit very well by a kinetic reaction scheme with two open and five shut states:

$$C_7 \rightleftharpoons C_6 \rightleftharpoons C_5 \rightleftharpoons C_4 \rightleftharpoons C_3$$
$$\updownarrow \qquad \updownarrow$$
$$O_2 \qquad O_1$$

Although this scheme could provide an excellent fit to the open and shut interval distributions, Blatz and Magleby (1986) found that almost any kinetic reaction scheme with two open and shut states could adequately describe the data. In fact, these seven-state schemes were totally indistinguishable from one another. To distinguish among these models, another approach was required. This approach involved looking at the sequence of open and shut channel intervals rather than just how many of each occurred. McManus et al. (1985) reported that when the sequences of open and shut channel events were examined, proportionately more of the short open intervals occurred adjacent in time to long shut intervals, and, conversely, more long shut intervals occurred next to short open intervals. This phenomenon could not occur if the two open kinetic states were connected directly to a single shut state or if the two open states were connected to one another in series, thus ruling out a large number of kinetic reaction schemes that could fit the unconditional distributions equally as well. Further simulations of the predicted sequences of open and shut intervals ruled out many additional kinetic models (Blatz and Magleby, 1987, 1989).

An interesting observation about fast Cl⁻ channel kinetics occurred when analyzing the stability plots mentioned earlier. In about 1% of the current record the channel activity would change for 200–300 intervals in a row, with the mean open and shut interval durations falling to one-tenth of their normal values. We termed this behavior the "buzz" mode because of its characteristic appearance on chart records. Similar behavior has subsequently been observed in large-conductance Ca^{2+}-activated K^+ channels in myotubes (McManus and Magleby, 1988).

1.2.2. Permeability Properties

Similar to the large-conductance Cl⁻ channels in mammalian skeletal muscle, the small-conductance channels also seem to allow a measurable cation permeability. For both the fast and the slow Cl⁻ channels the Cl^-/K^+ permeability ratio was calculated to be about 5.

1.2.3. Functions of Small-Conductance Cl⁻ Channels in Mammalian Muscle

The fact that both the fast and slow small-conductance Cl⁻ channels were active at negative membrane potentials near the normal muscle resting potential led Blatz and Magleby (1985) to speculate that these channels could contribute to the large resting Cl⁻ conductance in these cells. However, these channels, as with the large-conductance channels, were not observed very often in cell-attached membrane patches. Also, the relatively high cation permeability is not entirely consistent with the channels underlying the resting Cl^- conductance. Whether these channels do contribute to the resting muscle conductance remains to be demonstrated.

2. *Cl⁻ CHANNELS IN AMPHIBIAN SKELETAL MUSCLE*

The only single-channel patch-clamp experiments on Cl^- channels in amphibian skeletal muscle describe a large-conductance, voltage-dependent, anion-selective channel (Woll *et al.*, 1987; Woll and Neumcke, 1987). In these experiments, the inside-out configuration of the patch-clamp technique was applied to enzymatically dissociated toe muscles of two species of frogs.

2.1. Gating Kinetics

Like the large-conductance Cl^- channels in rat skeletal muscle, the channels described in amphibian muscle were rarely if ever observed in cell-attached membrane patches, and usually several seconds or minutes were required following excision to the inside-out patch configuration before channel activity occurred. The activation of frog Cl^- channels could be induced by holding the inside-out patch at depolarized potentials for several seconds. Often, these channels became activated in a manner similar to the mammalian channels, with the current baseline becoming very noisy and flickery until complete single-channel open and shut events occur. In response to a large hyperpolarizing voltage step, the frog channels behaved much like the mammalian channels, as shown in Fig. 6. In this experiment, the membrane potential was held at -6 mV and stepped to -66 mV. In response to this hyperpolarization, the channels, which were quite active at -6 mV, began to close, and the current returned to near baseline level after several seconds. Also, as shown in Fig. 6, the frog Cl^- channels close more rapidly as the membrane potential is made more negative. Unlike the mammalian large-conductance channels, the frog channels do not have a bell-shaped voltage-dependence, but remain active in response even to large depolarizations. The presence of many subconductance states in the frog channels made a detailed kinetic analysis difficult, but Woll and Neumcke (1987) reported the interesting observation that the opening probabilities of the frog Cl^- channels are interdependent, that is, whether one channel in the patch opens depends on the open or shut state of the other channels in the patch.

2.2. Permeability Properties

Woll *et al.* (1987) found that frog large-conductance channels exhibit several conductance states with a predominant magnitude of about 260 pS, with other states of 77 and 328 pS when the patch was bathed in symmetric 110 mM NaCl. Due to bandwidth limitations, still other conductance states cannot be ruled out. Despite the difficulties imposed by the many conductance levels for these channels, Woll *et al.* (1987) were able to measure a Cl^- to Na^+ permeability ratio, based on reversal

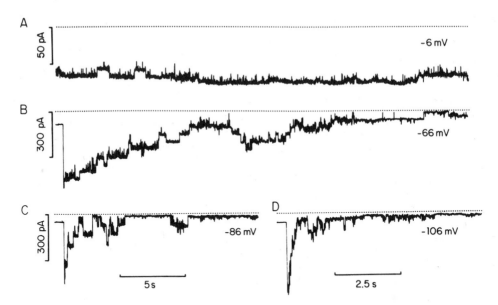

Figure 6. Single-channel currents recorded from excised inside-out patches from frog toe muscle surface membranes. At the beginning of each trace, the voltage was stepped from the holding potential of −6 mV to test potentials indicated next to the current traces. Patch pipet contained 110 mM NaCl and bath contained 500 mM NaCl. Temperature 10°C. (Reproduced with permission from Springer-Verlag; Woll *et al.*, 1987.)

potentials. As with the mammalian channels, the frog channels exhibit a marked permeability to cations with Cl^- being about three to five times more permeant than Na^+. Other anions were found to be permeant as well, including Br^-, aspartate, citrate, and glucuronate. The frog channels were also found to be blocked by zinc, tannic acid, and gallic acid. Interestingly, DIDS, a compound that blocks a variety of Cl^--conducting systems, did not block the frog Cl^- channels.

2.3. Functions

As with the mammalian Cl^- channels, it is difficult to ascribe a cellular function to the channels described in frog muscle. At resting membrane potentials, these channels would be largely inactive and so are unlikely to contribute much to the large resting Cl^- conductance, although, as Woll *et al.* (1987) point out, very few of the channels would be required to produce a measurable Cl^- conductance.

3. CONCLUSIONS

Much work remains to be done on this class of largely ignored ion channels. Perhaps the only generality that emerges from the studies quoted above is that all skeletal muscle Cl^- channels are only weakly selective, either for anions over cations or among anions. The separation used here of large and small Cl^- channels will probably not be possible with further experimentation, as Cl^- channels will undoubtedly be found with intermediate conductances. Although a great deal has been learned about single-channel kinetics from Cl^- channels, the exact functions of the channels described in this chapter remain to be discovered.

REFERENCES

Adrian, R. H., and Bryant, S. H., 1974, On the repetitive discharge in myotonic muscle fibres, *J. Physiol. (London)* **240:**505–515.

Blatz, A. L., and Magleby, K. L., 1983, Single voltage-dependent chloride-selective channels of large conductance in cultured rat muscle, *Biophys. J.* **43:**237–241.

Blatz, A. L., and Magleby, K. L., 1985, Single chloride-selective channels active at resting membrane potentials in cultured rat skeletal muscle, *Biophys. J.* **47:**119–123.

Blatz, A. L., and Magleby, K. L., 1986, Quantitative description of three modes of activity of fast chloride channels from rat skeletal muscle, *J. Physiol. (London)* **378:**141–174.

Blatz, A. L., and Magleby, K. L., 1987, Adjacent interval analysis of fast chloride channels in cultured rat skeletal muscle allows exclusion of several classes of kinetic schemes, *Biophys. J.* **51:**10a.

Blatz, A. L., and Magleby, K. L., 1989, Adjacent interval analysis distinguishes among gating mechanisms for the fast chloride channel from rat skeletal muscle, *J. Physiol. (London)* **410:**561–585.

Bretag, A. H., 1987, Muscle chloride channels, *Physiol. Rev.* **67:**618–724.

Gray, P. T. A., Bevan, S., and Ritchie, J. M., 1984, High conductance anion-selective channels in rat cultured Schwann cells, *Proc. R. Soc. London Ser. B* **221:**395–409.

Hamill, O. P., Marty, A., Neher, E., Sakmann, B., and Sigworth, F. J., 1981, Improved patch-clamp techniques for high-resolution current recording from cells and cell-free membrane patches, Pfluegers Arch. **391:**85–100.

Hanrahan, J. W., Alles, W. P., and Lewis, S. A., 1985, Single anion-selective channels in basolateral membrane of a mammalian tight epithelium, *Proc. Natl. Acad. Sci. USA* **82:**7791–7795.

Hodgkin, A. L., and Horowicz, P., 1959, The influence of potassium and chloride ions on the membrane potential of single muscle fibres, *J. Physiol. (London)* **148:**127–160.

Hutter, O. F., and Noble, D., 1960, The chloride conductance of frog skeletal muscle, *J. Physiol. (London)* **151:**89–102.

Krouse, M. E., Schneider, G. T., and Gage, P. W., 1986, A large anion-selective channel has seven conductance levels, *Nature* **319:**58–59.

Lindau, M., and Fernandez, J. M., 1986, A patch-clamp study of histamine-secreting cells, *J. Gen. Physiol.* **88:**349–368.

McManus, O. B., and Magleby, K. L., 1988, Identification and characterization of four modes of activity of the large conductance calcium-activated potassium channel in cultured skeletal muscle, *Biophys. J.* **53:** 145a.

McManus, O. B., Blatz, A. L., and Magleby, K. L., 1985, Inverse relationship of the durations of adjacent open and shut intervals for Cl and K channels, *Nature* **317:**625–627.

Magleby, K. L., and Pallotta, B. S., 1983a, Calcium dependence of open and shut interval distributions from calcium-activated potassium channels in cultured rat muscle, *J. Physiol. (London)* **344:**585–604.

Magleby, K. L., and Pallotta, B. S., 1983b, Burst kinetics of single calcium-activated potassium channels in cultured rat muscle, *J. Physiol. (London)* **344:**605–623.

Miller, C., 1982, Open-state substructure of single chloride channels from Torpedo electroplax, *Philos. Trans. R. Soc. London* **299:**401–411.

Palade, P. T., and Barchi, R. L., 1977, Characteristics of the chloride conductance in muscle fibers of the rat diaphragm, *J. Gen. Physiol.* **69:**325–342.

Schneider, G. T., Cook, D. I., Gage, P. W., and Young, J. A., 1985, Voltage sensitive, high-conductance chloride channels in the luminal membrane of cultured pulmonary alveolar (type II) cells, *Pfluegers Arch.* **404:**354–377.

Schwarze, W., and Kolb, H. A., 1984, Voltage-dependent kinetics of an anionic channel of large unit conductance in macrophages and myotube membranes, *Pfluegers Arch.* **402:**281–291.

Woll, K. H., and Neumcke, B., 1987, Conductance properties and voltage dependence of an anion channel in amphibian skeletal muscle, *Pfluegers Arch.* **410:**641–647.

Woll, K. H., Leibowitz, M. D., Neumcke, B., and Hille, B., 1987, A high-conductance anion channel in adult amphibian skeletal muscle, *Pfluegers Arch.* **410:**632–640.

Index

421